Technology for Inclusion and Participation for All: Recent Achievements and Future Directions

Katerina Mavrou · Pedro Encarnação
Editors

Technology for Inclusion and Participation for All: Recent Achievements and Future Directions

18th International Conference, AAATE 2025
Nicosia, Cyprus, September 10–12, 2025
Proceedings, Part I

Springer

Editors
Katerina Mavrou
European University Cyprus
Engomi, Cyprus

Pedro Encarnação
Universidade Católica Portuguesa
Lisbon, Portugal

ISBN 978-3-032-01627-0   ISBN 978-3-032-01628-7 (eBook)
https://doi.org/10.1007/978-3-032-01628-7

This Springer imprint is published by the registered company Springer Nature Switzerland AG
The registered company address is: Gewerbestrasse 11, 6330 Cham, Switzerland

Preface

The year 2025 heralds a pivotal moment for the assistive technology (AT) field. The United Nations Office of the High Commissioner for Human Rights issued the report "Rights of persons with disabilities and digital technologies and devices, including assistive technologies"[1], clearly affirming that "Access to assistive technologies is a human right and is essential for persons with disabilities, enabling independence, greater participation and improved quality of life." In parallel, 2025 marks the entry into force of the European Accessibility Act[2]. Assistive technology and accessibility are two faces of the same coin: while accessibility reduces environmental barriers, assistive technology empowers individuals to overcome remaining obstacles.

The year also marks the 30th anniversary of the Association for the Advancement of Assistive Technology in Europe (AAATE). Its foundation in 1995 was the outcome of a growing European network of experts who collaborated through the COST Actions 219 "Future Telecommunications and Teleinformatics Facilities for Disabled People" (1987–1996) and the "Technology initiative for disabled and elderly people" (TIDE) programme (1991–1994) and who recognized the need to establish a permanent forum for advancing the field. Since then, AAATE has pursued its mission of promoting assistive technology and inclusive design as enablers of human rights and full societal participation for all.

Within this context, the 18th AAATE biennial conference3 convenes under the theme "Technology for Inclusion and Participation for All: Recent Achievements and Future Directions." This reflects a focus on technology with a purpose—*supporting inclusion and participation*; on universal access as a human right—ensuring technology serves all; and on harnessing recent technological advances in the field of Assistive Technology. The conference was held in Nicosia, Cyprus, from September 10 to 12, 2025, and was co-organized by the European University Cyprus and the AAATE.

We received 224 contributions, out of which 107 were accepted to be included in this volume. Each submission was reviewed by at least three members of the international scientific committee. We extend our sincere thanks to all reviewers for their time, expertise, and dedication throughout the evaluation process. The accepted papers mirror the wide spectrum of the field, reflecting the richness and diversity of assistive technology research, development, policy, and practice. The volume comprises 20 chapters, which are briefly described below.

- *Advanced technologies for inclusion and participation in education and labour* present innovations using AI and digital tools to support neurodiverse individuals, autistic students, and people with learning disabilities across educational and vocational settings, while addressing employment challenges and media accessibility.
- *Accessibility in digital spaces* addresses the accessibility of digital content for blind and low-vision users, including adaptive eBook illustrations, accessible email applications, and guidelines for 360-degree media platforms.

- *Accessible indoor navigation systems* focus on indoor navigation and wayfinding technologies for visually impaired and mobility-impaired individuals, with contributions on smartphone-based systems, digital guidance design, and user-centered requirements analysis.
- *Advancing diagnosis and rehabilitation of neurodevelopmental disorders: The role of virtual reality, eye-tracking, and serious games* showcases VR-based interventions and smart wearables for supporting neurodiversity, including children with autism, neurological rehabilitation, and motor and cognitive development in older adults.
- *Artificial intelligence in care and support ecosystems* presents artificial intelligence solutions for residential care, caregiver support, and sensor-based systems promoting active aging, emphasizing personalization and real-time monitoring.
- *Artificial intelligence in higher education: Advancing inclusive environments, pedagogical approaches, and assistive technologies* examines the opportunities and risks of AI-driven educational technologies, accessibility solutions in publishing, and inclusive AI research, with particular attention to dyslexia, dyscalculia, and the role of disabled researchers.
- *Assistive technologies outcomes and impact measurement* addresses systemic challenges in assistive technology provision, access, and evaluation, presenting outcome frameworks, policy analyses, ethical discussions on AI, and studies on customized mobility devices and digital inclusion.
- *Assistive technology and accessibility for cognitive disabilities* investigates inclusive solutions for neurodiversity, including dyslexia, ADHD, cognitive accessibility, and simplified accessibility auditing, including AI-based Easy-to-Read adaptations and user-centered task management tools.
- *Assistive technology and accessibility training* focuses on inclusive education, curriculum development, and teacher training, featuring contributions on digital accessibility in higher education, disability simulation games, inclusive music education, and co-design approaches.
- *Assistive technology for sensory impairments* presents advances in tactile displays, multimodal interfaces, sign language translation, and XR-based subtitle systems for deaf, blind, and partially sighted users.
- *Assistive technology in the healthcare sector* discusses user-centered design for care robots, participatory digital health governance, safety innovations such as intelligent wheelchairs, and AI-assisted governance frameworks in healthcare delivery.
- *Assistive technology provision* highlights contextually appropriate assistive technology design and family-centered care frameworks, showcasing the AT4ALL program and the lived experiences of families caring for children with profound disabilities.
- *Augmentative and alternative communication* explores AAC strategies, AI-powered communication apps, teacher training platforms, professionals' awareness, and the evolving landscape of child-specific text-to-speech and neural speech synthesis systems.
- *Embracing change: Aging, digitalisation, and the future of care* addresses the intersection of aging, care innovation, and digitalization, presenting research on smart homes, digital home care, AI literacy, rights-based approaches, and care ethics.

- *Human-assistive technology interface* focuses on AI-powered solutions for speech recognition, gesture-based interfaces, and adaptive accessibility tools for individuals with complex mobility and communication impairments.
- *ICT to support inclusive education—Universal Design for Learning (UDL)*—examines inclusive educational technology for neurodiverse learners, digital accessibility for blind and visually impaired students, adaptive gaming, and teacher support tools for differentiated instruction.
- *Innovations in assistive technology to support mobility and participation* presents developments in wheelchair technologies, hand rehabilitation, transportation accessibility, mixed reality rehabilitation, paratransit, and urban mobility solutions for individuals with disabilities.
- *Museums for all: Co-design, inclusion and emerging technologies* explores how co-design, AI, and emerging technologies can promote inclusive access to museums and cultural heritage for individuals with intellectual disabilities.
- *Robotic assistive technology for children* examines the role of robots in inclusive education, play, therapy, and communication for children with disabilities, highlighting developments in educational robotics, inclusive play environments, and speech therapy interventions using robotics and virtual reality.
- *User-centered assistive technology co-design* explores participatory approaches in assistive technology development, highlighting co-design experiences with people with intellectual disabilities, AI-powered accessibility solutions in publishing, recommender systems for blind, partially sighted, and autistic users, tools for user-driven audio description, and clinical evaluation practices in industry.

Artificial intelligence is transforming many aspects of today's world, and the assistive technology field is no exception. Of the 107 accepted papers, approximately 60 explicitly address AI applications, ranging from machine learning and speech recognition to large language models, generative AI, and adaptive interaction systems. These contributions demonstrate AI's potential to develop assistive technologies capable of dynamically adapting to the person's varying needs and changing contexts, while also raising important questions related to reliability, transparency, bias, privacy, and equitable access. By actively involving people with disabilities as co-designers and decision-makers, AI-enabled assistive technologies can evolve into truly transformative tools that promote autonomy, inclusion, and participation for all.

We trust that the papers collected in this volume offer a comprehensive overview of the state of the art and valuable insights into how the field will continue to evolve in the years ahead.

Katerina Mavrou
Pedro Encarnação

Notes

1. United Nations Office of the High Commissioner for Human Rights (2024). Rights of persons with disabilities and digital technologies and devices, including assistive technologies. https://digitallibrary.un.org/record/4073939?ln=en&v=pdf

2. European Parliament & Council (2019). Directive (EU) 2019/882 on the accessibility requirements for products and services (European Accessibility Act). https://eur-lex.europa.eu/legal-content/EN/TXT/?uri=CELEX%3A32019L0882

3. This is the 18th edition because the AAATE conferences were preceded by three ECART (European Conference for the Advancement of Rehabilitation Technology) conferences.

About the Conference

Conference Chairs

Katerina Mavrou	European University Cyprus
Pedro Encarnação	Universidade Católica Portuguesa, AAATE

Programme Committee

Katerina Mavrou	European University Cyprus
Pedro Encarnação	Universidade Católica Portuguesa, AAATE
Eleni Theodorou	European University Cyprus
Dominique Archambault	Université Paris 8-Vincennes-Saint-Denis, AAATE
Tone Øderud	SINTEF Digital, AAATE
Georgios Kouroupetroglou	National and Kapodistrian University of Athens, AAATE
Emma Smith	Maynooth University, AAATE
Elena Stylianou	European University Cyprus, Nicosia Municipality Arts Centre (NiMAC) Marina GeorgiadouEuropean University Cyprus
Maria Meletiou-Mavrotheris	European University Cyprus

Organising Committee

Katerina Mavrou	European University Cyprus, AAATE
Eleni Theodorou	European University Cyprus
Evert-Jan Hoogerwerf	AIAS Bologna onlus, AAATE
Pedro Encarnação	Universidade Católica Portuguesa, AAATE
Priska Feichtenschlager	AAATE
Ralf Klug	Johannes Kepler University Linz, AAATE
Marina Georgiadou	European University Cyprus
Melina Christodoulou	European University Cyprus
Maria Mouka	European University Cyprus
Silvio Pagliara	University of Cagliari, Manè Social Enterprise, AAATE
Maria Meletiou-Mavrotheris	European University Cyprus
Elena Stylianou	European University Cyprus, Nicosia Municipality Arts Centre (NiMAC)
Marianna Efstathiadou	European University Cyprus
Maria Papazachariou	European University Cyprus
Spyros Spyrou	European University Cyprus
Ioulia Televantou	European University Cyprus
Ioanna Vekiri	European University Cyprus
Marios Vryonidis	European University Cyprus
Loizos Symeou	European University Cyprus
Marianna Christodoulou Devledian	European University Cyprus

International Scientific Committee

Adriana Rios Rincon	University of Alberta
Albert Moore Cook	Assistive Technology
Alex Kamadu	ISWP
Alexandros Pino	National and Kapodistrian University of Athens
Alexey Andrushevich	Lucerne University of Applied Sciences and Arts, iHomeLab, Lucerne University of Applied Science and Arts
Ana Correia de Barros	Fraunhofer Portugal AICOS
Ana Londral	NOVA School of Science and Technology, Nova University of Lisbon, Department of Physics
Andrea Bonarini	Politecnico di Milano
Andrea Petz	Johannes Kepler University, Institute Integriert Studieren

Andrew Paice	Lucerne University of Applied Sciences and Arts
Angela Riccio	Fondazione Santa Lucia, IRCCS, Assistive technology center, SARA-t; Neuroelectrical Imaging and BCI laboratory
Angelo Davalli	INAIL
Anis Rojbi	UFR eriTES, Université Paris8
Annalu Waller	University of Dundee
Arne H. Eide	SINTEF Digital Health
Branislav Gerazov	Faculty of Electrical Engineering and Information Technologies, Ss Cyril and Methodius University in Skopje
Carlos Duarte	Universidade de Lisboa
Cecilia Sik Lanyi	University of Pannonia, Department of Electrical Engineering and Information Systems
Célia de Sousa	Resource Center for Digital Inclusion-CRID
Celine Jost	Université Paris 8
Christian Galinski	Infoterm
Chryssoula Thodi	European University Cyprus
Clarisse Nunes	Instituto Politécnico de Lisboa, Escola Superior de Educação de Lisboa
David Banes	Access and Inclusion Services
Dikaios Sakelariou	University of Cardiff / European University Cyprus, School of Healthcare Sciences
Dominique Archambault	Université Paris 8
Donatella Mattia	Fondazione Santa Lucia
Elena Stylianou	European University Cyprus / Nicosia Municipality Arts Centre
Eleni Damianidou	European University Cyprus
Eleni Theodorou	European University Cyprus
Emma Smith	Maynooth University, Assisting Learning and Living institute
Evert-Jan Hoogerwerf	AAATE
Fabio Salice	Politecnico di Milano, DEIB – Department of Electronic, Information and Bioengineering
Francisco Florez-Revuelta	University of Alicante
Gabriel Pires	Instituto Politécnico de Tomar, Engineering Department
Georgios Kouroupetroglou	National and Kapodistrian University of Athens
Gottfried Zimmermann	Stuttgart Media University
Guillaume Thomann	Univesity Grenoble Alpes, Institute of Engineering and Management
Hanna Usatenko	DyvoGra Social Enterprise / UNICEF
Hille Maas	Arci Estonia OÜ

Hsin-Yi Kathy Cheng	Chang Gung University
Ioanna Vekiri	European University Cyprus
Iosif Klironomos	FORTH, Institute of Computer Science
Isis Truck	Université Paris 8
Ivan Traina	University of Verona, Department of Human Sciences
Jaime Moreira Ribeiro	Polytechnic University of Leiria, School of Health Sciences
Kalliopi (Kelly) Lappa	Hellenic Association of Assistive Technology
Katerina Mavrou	European University Cyprus
Katrin Nuppenau	Johannes Kepler University, Institute Integriert Studieren
Klaus Hoeckner	Access Austria
Klaus Miesenberger	Johannes Kepler University Linz
Krzysztof Dobosz	Silesian University of Technology, Department of Algorithmics and Software
Lieven De Maesschalck	Thomas More Kempen vzw
Lisbet Grut	SINTEF Digital, department of Health Research
Loizos Symeou	European University Cyprus
Lorenzo Chiari	Alma Mater Studiorum – Università di Bologna
Lorenzo Desideri	AIAS Bologna onlus
Luc de Witte	The Hague University of Applied Sciences, Centre of expertise Health Innovation
Luca Fanucci	Università di Pisa
Luís Azevedo	ANDITEC
Luís Filipe Garcia	Instituto Politécnico de Beja
Luis Montesinos	School of Engineering and Sciences, Tecnologico de Monterrey
Lukeshkumar Rajendra Bhuyar	All India Institute of Physical Medicine and Rehabilitation, Mumbai
Marcia Scherer	MPT Institute
Maria Kalli	European University Cyprus
Maria Meletiou-Mavrotheris	European University Cyprus
Maria Papazachariou-Christoforou	European University Cyprus
Marian Pădure Babeș-Bolyai	University in Cluj-Napoca, Special Education Department
Marianna Christodoulou Devledian	European University Cyprus
Marina Georgiadou	European University Cyprus
Mario Covarrubias Rodriguez	Politecnico di Milano
Marios Vryonides	European University Cyprus

Marta Figueiredo	RESNA – Rehabilitation Engineering and Assistive Technology Society of North America (USA); Escola Superior de Saúde do Alcoitão (PT)
Martin Cerny	VSB – Technical University of Ostrava
Maurice Grinberg	Assist Foundation
Melanie Schaur	Johannes Kepler University, Institute Integriert Studieren
Merita Hoxha	Aleksander Xhuvani University, Department of Foreign Languages
Nadine Vigouroux	University of Toulouse
Natasha Layton	Monash University, RAIL Research Centre
Nodoka Yamauchi	National Institute of Advanced Industrial Science and Technology (AIST)
Paolo Ciampolini	University of Parma
Pedro Encarnação	AAATE I Universidade Católica Portuguesa
Peter Burggraeve	VIVES University of Applied Science
Rafael Raya	Universidad San Pablo CEU
Reinhard Koutny	Johannes Kepler University Linz
Renata Pinjatela	University of Zagreb, Faculty of Education and Rehabilitation Sciences
Renzo Andrich	EASTIN (The Global Assistive Technology Information Network)
Riccardo Magni	GLIC Network
Roger Gassert	ETH Zurich, Department of Health Sciences and Technology
Romy Sels	Thomas More
Rosemary Joan Gowran	University of Limerick
Safaa Mahmoud Issa	Menoufia University
Sara Comai	Politecnico di Milano (POLIMI)
Sari Merilampi Satakunta	University of Applied Sciences, Faculty of Technology
Serenella Besio	Università di Bergamo
Silvana Contepomi	Argentine Assistive Technology Association
Silvia Mirri	Alma Mater Studiorum – Università di Bologna, Department of Computer Science and Engineering
Silvio Marcello Pagliara	University of Cagliari
Simon Judge Barnsley	Hospital NHS Foundation Trust, Assistive Technology Team
Spyros Spyrou	European University Cyprus
Stefano Federici	Università di Perugia
Susanne Dirks	TU Dortmund University

Takenobu Inoue	Nat. Rehabilitation Center for Persons with Disabilities & RESJA
Tânia Rocha	Universidade de Trás os Montes e Alto Douro
Tone Øderud	SINTEF Digital, Department of Health Research
Uta Roentgen	Zuyd University of Applied Science, Research Centre Technology in Care
Weiqin Chen	Oslo Metropolitan University
Yann Morère	Université de Lorraine, LCOMS
Zeynep Sahin Timar	Karadeniz Technical University

Contents

Advancing Diagnosis and Rehabilitation of Neurodevelopmental Disorders: The Role of Virtual Reality, Eye-tracking, and Serious Games

Artificial Intelligence in Care and Support Ecosystems

Artificial Intelligence in Higher Education: Advancing Inclusive Environments, Pedagogical Approaches, and Assistive Technologies

Assistive Technologies Outcomes and Impact Measurement

Assistive technology for sensory impairments

Advanced Technologies for Inclusion and Participation in Education and Labour

Evaluating the Role of an Assistive Online Toolkit to Support Autistic Students During the Transition into University

Marc Fabri[1(✉)], Priscila Ferreira Beni[2], Aimee Shepherd[1], and Dante Augusto Couto Barone[2]

[1] Leeds Beckett University, Leeds LS1 3HE, UK
m.fabri@leedsbeckett.ac.uk
[2] Universidade Federal do Rio Grande do Sul, Porto Alegre, RS 90010-150, Brazil

Abstract. Autistic students face significant challenges in higher education, particularly regarding navigation, sensory overload, and accessing appropriate support. This study explores the role of an Assistive Online Toolkit in addressing these barriers and promoting autonomy, well-being, and social adaptation. Through qualitative analysis of student experiences, findings reveal that navigation difficulties and sensory overload contribute to heightened stress and disorientation. The Toolkit's mapping and location features provided practical solutions, enhancing students' confidence and independence in navigating campus. Additionally, a lack of awareness about available support services created further obstacles, which the Toolkit helped mitigate by offering clear guidance on reasonable adjustments and self-advocacy strategies. The transition to university also brought challenges in academic adaptation and social integration, with many students experiencing isolation and anxiety. The Toolkit's resources helped normalize help-seeking behaviors and fostered a sense of belonging. These findings highlight the importance of digital assistive technologies tailored to autistic students' needs, reinforcing the necessity of framing autism within a disability paradigm to ensure access to legal protections, accommodations, and continued innovation in assistive solutions.

Keywords: Assistive Technology (AT) · eLearning and Education · Design for All and Universal Design

1 Introduction

Autistic students attend higher education in increasing numbers, due to advances in support during secondary education, greater awareness of autism and improved diagnosis rates. Although autism is not an indicator for academic ability, drop-out rates are high [1], especially during the first few months of study. Previous research [2, 3, 4] has shown that this transition period presents autistic students with a number of challenges, including academic barriers (executive functioning, rigid thinking patterns, difficulty seeking help), social barriers (difficulty making friends, not knowing how to interact

K. Mavrou and P. Encarnação (Eds.): AAATE 2025, *Technology for Inclusion and Participation for All: Recent Achievements and Future Directions*, pp. 3–10, 2025.
https://doi.org/10.1007/978-3-032-01628-7_1

with tutors, anxiety about being perceived as different), sensory challenges (noise and crowded places, unfamiliarity with surroundings, lack of quiet spaces) and general well-being issues (anxiety and stress, the impact of change, loneliness, academic pressure, effects of masking one's condition, burnout).

This study evaluates the effectiveness of an Assistive Online Toolkit designed to support autistic students during the transition into university. The Toolkit is in use at 19 European universities and provides students with information about what university is like, typical challenges for autistic students and how to pro-actively address them, personal stories written by autistic alumni, details of the university's physical environment including maps and directions, support staff and general offerings for autistic students. The Toolkit focuses on inclusiveness and positive messages, with frequent calls-to-action to develop the student's self-advocacy skills and autonomy. Its structure, visual layout and content were co-designed with autistic students [5].

2 Methods

During the summer of 2024, 13 students from five British and Irish universities that had adopted the Toolkit took part in a qualitative study around their experiences when starting university. Eleven participants completed an online survey, two took part in a semi-structured interview. Each participant was compensated for their time with a £20/€20 shopping voucher. Questions were designed by an autistic and a non-autistic researcher, revolving around the challenges and positive enablers students may encounter when starting their studies, the physical and sensory environment at university, planning and managing their studies, any support they received, and various aspects of health and well-being. We specifically asked how the Toolkit helped them during this time. Questions were piloted with an autistic student (who did not take part in the main study) and minor changes were made following the pilot. We followed Braun and Clarke's [6] method for reflexive thematic analysis of responses, using an inductive approach to identify patterned meanings within the dataset. Author 3 conducted interviews. Author 2 developed initial codes, then discussed and refined these with author 1. Authors 1 and 2 developed the themes and sub-themes.

3 Results

Analysis of the response data revealed the main difficulties and barriers faced by students and how these affected their experience at university, as summarised in Fig. 1.

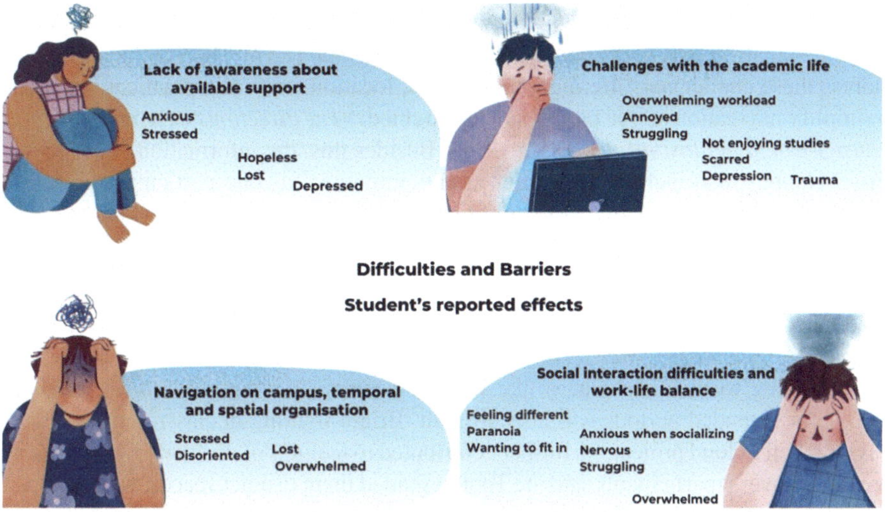

Fig. 1. Participant reported difficulties, barriers and effects

During the surveys/interviews, students shared information about their use of the Toolkit when transitioning into university. This allowed us to better understand how the Toolkit supported students with overcoming challenges and adapting to higher education. The analysis identified two major themes, further divided into five sub-themes, which illustrate the Toolkit's role in supporting students, as presented in Table 1.

Table 1. Themes and sub-themes illustrating the Toolkit's role in supporting students

Theme	Sub-Theme
1. Enabling navigation and managing sensory overload	1.1 Supporting navigation on campus 1.2 Addressing sensory needs
2. Awareness of support, adjustmentssss and well-being	1.3 Providing information about access to support, types of reasonable adjustments and useful strategies 1.4 Guidance on expectations and settling in 1.5 Guidance on well-being and sense of belonging

3.1 Theme 1: Enabling Navigation and Managing Sensory Overload

Sub-theme 1.1: Supporting Navigation on Campus. Navigating the university campus posed significant challenges for students, particularly in finding their way around new and unfamiliar lecture halls, buildings, and departments. These were described as *"a massive space"* with each department in its own building. One participant confided that *"finding my way to different buildings and classrooms was stressful. (P1)"*. Another mentioned that *"It was overwhelming... I felt disoriented (P5)"*. The Toolkit showed to address these challenges, offering map features, location tools, and practical resources to simplify navigation as one participant highlighted: *"The directions were really helpful 'cause I would get lost all the time... (P9)"*. Besides this, the information about routes across campus also increased confidence and autonomy, as one participant noted: *"I used the Toolkit's location and directions features a lot to help me get used to navigating campus alone (P7)."*

Sub-theme 1.2: Addressing Sensory Needs. Students reported significant sensory challenges in navigating university environments, particularly in crowded and noisy spaces like *"Queuing outside lecture rooms, where it was cramped and loud (P7)"* was overwhelming to them. In places like lecture theatres, *"Sitting very close to people (P10)"* for extended periods was also difficult. Bright lighting in classrooms and hallways, as well as loud projectors, further contributed to feelings of discomfort and sensory overload. Several participants said the Toolkit guided them to quiet spaces, which helped manage overstimulation. Similarly, it offered alternative directions that avoided noisy environments and reduced stimulation, as a student highlighted: *"Another thing that the Toolkit helped me to do is find directions to quieter routes. ... It helped avoid the overstimulation of moving around the campus at peak times. (P7)"*.

3.2 Theme 2: Awareness of Support, Adjustments and Well-Being

Sub-theme 2.1: Providing Information About Access to Support, Types of Reasonable Adjustments and Useful Strategies. Lack of awareness about available support and reasonable adjustments created significant obstacles for many students, particularly during their transition to university and the first year. Some stated they were unaware of the type of support and adjustments that were available to them, whilst others pointed out that insufficient or unclear information made them feel lost and anxious, leading to a stressful experience: *"The information on what they can do or support with and not having this information can be very difficult (P8)"*. In addition to these gaps, students also faced challenges with academic tasks and institutional procedures. Others described problems with time management.

For many students, the Toolkit was the starting point for navigating university services. As one participant pointed out: *"The Toolkit mentioned the student support, so I went there (P4)"*. It provided valuable information on how to find support, who to speak to, and the types of reasonable adjustments that are possible. This made students aware of the level of support they could expect, thereby raising expectations: *"It is really good, it helped me to understand the support that I can have at university... It also allows me to be able to get the support I needed in terms of advice and guidance (P8)"*.

Additionally, the Toolkit guided the students to get prepared for academic tasks more effectively, as one participant asserted when asked about how the Toolkit helped with planning: *"How to organize my time, like having my calendar in place now... How to manage my time better (P8)"*. The Toolkit also made the process of seeking help feel more accepted, and the right thing to do. This played a key role in reducing feelings of self-doubt or reluctance when requesting reasonable adjustments. A student highlighted this impact, stating, *"It just makes it feel as if you're not being a burden (P10)"*.

Sub-theme 2.2: Guidance on Expectations and Settling In. The transition to university was one of the most challenging periods for autistic students, particularly with regards to understanding "what to expect when joining university (P8)" and adjusting to a new environment. They reported feeling scared and uncertain about starting, and what awaited them. The shift to self-directed learning was particularly daunting and often led to difficulties in adjusting to increased academic demands. As one student stated, "The increased workload compared to what I was used to was also overwhelming (P1)". Additionally, "Managing to fit into such a big sort of student body (P1)" was described as frustrating and difficult. These challenges often resulted in a lack of engagement, emotional distress, and, in some cases, depression or even trauma.

The Toolkit was identified as a source of support in addressing these challenges by providing ample information about what to expect at university. It was described as "*very useful in being able to provide me information about different sections (P7)*", offering "*different categories of information that were very informative of the different factors and things to consider (P8)*". To assist with self-directed learning and academic demands, the Toolkit provided practical guidance on various aspects of university life, including exam tips and strategies for managing coursework. Many students found the alumni stories particularly helpful, as they offered insights from other autistic students who had faced similar experiences. One participant shared: "*Reading other autistic students' stories about starting university helped me understand a bit more about the process (P4)*". Another highlighted the importance of these narratives in shaping expectations, explaining that they helped *"what to expect and get the visions of what other people have [been] looking at going to university (P8)"*. For some, these resources were essential in navigating the overwhelming transition, as captured in one student's words: *"There's lots of resources which really helped in a difficult time (P8)"*.

Sub-theme 2.3: Guidance on Well-Being and Sense of Belonging. Starting university brought significant emotional and social challenges for autistic students. The overwhelming number of people, unfamiliar environments, and difficulty in *"socializing, making friends, feeling left out (P1)"* made it hard for them to form connections, leading to feelings of anxiety and isolation. The fear of being judged intensified these struggles further, as one student shared: *"I had paranoia that they would, like, say awful things about me (P9)"*. This sense of uncertainty about their place at university contributed to self-doubt, as another participant described: *"I was nervous, and I did feel like an impostor (P10)"*. Beyond social difficulties, anxiety about academic expectations and managing stress became a recurring issue, often leading to emotional distress. One student reflected on the impact, stating, *"If my mental health was down, so was my work, attendance, motivation, and overall health (P5)"*. Another described the emotional effect of these struggles, saying, *"I was very isolated, and it took a huge toll on my mental health (P2)"*. The challenge of maintaining a balance between study and personal needs also emerged as a major concern, as one participant noted: *"Balancing academic responsibilities with personal well-being was a struggle (P10)"*.

Regarding the emotional and social challenges of university life, the Toolkit assisted students with providing access to information about support services and social groups by and for autistic students, which some participants described as a consistent and reliable space for social interaction. One participant commented that *"the services section was also very helpful because that's where I found out about the autism social group, which became a constant in my year. It was a safe space I could go to hang out with similar yet very different people, most of whom I became friends with (P4)"*. For those struggling with isolation, that university's social group created valuable opportunities for connection.

Beyond facilitating social connections, the Toolkit also provided reassurance through first-hand experiences shared by autistic alumni. Reading about the challenges and coping strategies of others helped participants feel less isolated and more confident in their ability to adapt. One student expressed: *"I did enjoy reading the stories of others' experiences and feeling less alone. It gave me the confidence to not care so much about what others thought (P5)"*. The same student highlighted how these narratives helped with self-acceptance, stating, *"It was good to just support me in accepting who I am (P5)"*. Additionally, insights into alumni experiences of anxiety and their coping mechanisms were particularly valuable, as one participant noted that *"understanding about anxiety and exam stress was really good (P9)"*.

4 Discussion

The findings from this study highlight the significant challenges that autistic students face during their transition into higher education while also demonstrating the ways in which the Toolkit serves as an effective support mechanism. The data suggests that the Toolkit plays an essential role in helping students navigate university life. By offering structured guidance on both academic and social aspects, it not only supports self-directed learning but also facilitates social connections and emotional well-being.

A key takeaway from the study is that navigation difficulties and sensory overload remain significant barriers for autistic students, reinforcing prior research [2]. Many participants expressed difficulties in finding their way around large, unfamiliar university spaces, which resulted in heightened stress and disorientation. The Toolkit's mapping and location tools were particularly effective in mitigating these difficulties, not only providing students with a sense of control and familiarity in their new environments but also fostering greater autonomy. The guidance on quieter routes and sensory-friendly spaces helped students manage overstimulation, reducing anxiety and discomfort. These findings emphasize the importance of providing practical, easy-to-access resources that promote both confidence and autonomy, which in turn supports students' personal development towards independence and self-sufficiency [7].

Beyond spatial and sensory challenges, a lack of knowledge about available support services was identified as a significant obstacle for many autistic students. This gap in knowledge led to confusion, stress, and difficulty in arranging effective reasonable adjustments. The Toolkit helped bridge this gap by providing clear, accessible information about the types of support available, how to access them, what level of support a student can expect and request, and strategies for advocating for that support. Many participants reported that they would not have sought out university support services had they not first encountered the information in the Toolkit. This emphasizes the importance of proactive, student-centered resources that raise awareness of what "good support" looks like, facilitate access to necessary adjustments, and ensure that students receive the assistance they need to succeed. For the participants, the Toolkit became a catalyst for understanding and explaining their support needs to those who were able to put that support in place.

Autistic students also faced significant challenges in adjusting to the academic aspects of university life, particularly due to uncertainty about expectations, increased academic demands, and the difficulty of integrating into a large student body. These struggles often led to emotional distress and disengagement. The Toolkit eased this transition by providing clear information and practical academic guidance that helped students make sense of the new environment, manage their expectations, develop strategies, and feel more prepared. The inclusion of alumni experiences was particularly impactful, offering comfort, relatable advice, and a sense of shared experience that helped students feel less alone.

Finally, the Toolkit also played a role in fostering social adaptation and well-being. The transition to university is often associated with increased feelings of loneliness and anxiety for autistic students [4], and multiple participants in this study echoed this. The Toolkit's focus on guidance for social expectations, self-advocacy, and well-being was found to be particularly useful in alleviating these concerns. Participants noted that the Toolkit normalized the experience of seeking help and provided reassurance that they were not alone in facing these challenges, especially through the alumni stories. It also offered access to well-being support services and social groups, which were particularly welcome. This aligns with previous research emphasizing the importance of fostering a sense of belonging for autistic students to improve their overall university experience [8].

5 Limitations

The sample size of 13 participants was relatively small, which may limit generalizability of findings. Additionally, the study was conducted at a subset of English-speaking European universities, and the experiences of students at other institutions may vary depending on available support services, campus environments, and national higher education policies. Future research should explore the long-term impact of the Toolkit, as well as its effectiveness across more varied university settings.

6 Conclusions

Overall, this study contributes to the growing body of research on how best to support autistic students in higher education. It underscores the potential of an assistive technology like the Online Toolkit, in addressing the challenges autistic students face during key life transitions. The positive reception of the Online Toolkit suggests that similar digital resources should be implemented more widely, with an emphasis on developing self-advocacy, supporting self-directed learning and fostering autonomy. Additionally, its role in promoting social integration and well-being highlights the need for further innovation in digital support tools that move beyond the traditional understanding of the term "assistive technology" as something that focuses mainly on functional capabilities. Indeed, assistive technology can also play a transformative role in creating more inclusive educational environments. Future work should continue to refine and expand such initiatives to ensure that all autistic students have the necessary tools to thrive, whether at university, in the workplace or in their community.

This study supports the UN 2030 Sustainable Development Goals (SDGs), particularly SDG 4 – Quality Education and SDG 10 – Reduced Inequalities. By proposing an Online Toolkit for autistic students, it promoted educational equity, reduced barriers, and fostered inclusion in higher education.

Acknowledgements. The authors acknowledge the support of Leeds Beckett University EDI Research Development Funding (ref AM003779) which enabled study design and data collection; and CAPES, Brazil, through Priscila Ferreira Beni's doctoral scholarship (grant number 88887.719595/2022-00) which supported coding and thematic analysis.

References

1. Anderson, A.H., Stephenson, J., Carter, M.: A systematic literature review of the experiences and supports of students with autism spectrum disorder in post-secondary education. Res. Autism Spectrum Disorders **39**, 33–53 (2017)
2. Lambe, S., Russell, A., Butler, C., Fletcher, S., Ashwin, C., Brosnan, M.: Autism and the transition to university from the student perspective. Autism **23**(6), 1531–1541 (2019)
3. Cage, E., Howes, J.: Dropping out and moving on: a qualitative study of autistic people's experiences of university. Autism **24**(7), 1664–1675 (2020)
4. Davies, J., Bagnall, C.: 'A Good Transition Is Just Key': The Views and Experiences of Disability Support Staff in Supporting Autistic Students Transition to University. Autism in 5 Adulthood **7**(2), 155–170 (2024)
5. Fabri, M., Andrews, P.C.S., Pukki, H.K.: Using design thinking to engage autistic students in the creation of an online toolkit to help with transition into higher education. Assistive Technol. **10**(2), 102–114 (2016)
6. Braun, V., Clarke, V.: Reflecting on reflexive thematic analysis. Quali. Res. Sport, Exer. Heal. **11**(4), 589–597 (2019)
7. Lei, J., Russell, A.: Understanding the role of self-determination in shaping university experiences for autistic and typically developing students in the United Kingdom. Autism **25**(5), 1262–1278 (2021)
8. Pesonen, H.V., Nimienen, J.H., Vincent, J., Waltz, M., Lahdelma, M., Syurina, E.V., Fabri, M.: A socio-political approach on autistic students' sense of belonging in higher educ

Digital Skills, Reasonable Accommodation and Assistive Technologies for Inclusive Employment: The Case of Cyprus

Marina Georgiadou(✉)

European University Cyprus, 6 Diogenes 2404, P.O. Box: 22006, 1516 Engomi, Nicosia, Cyprus
m.georgiadou@euc.ac.cy

Abstract. Introduction: Employment is a key factor for social inclusion, yet persons with disabilities in Cyprus continue to experience exclusion from the labour market. Barriers such as limited digital skills, poor access to assistive technology, and weak enforcement of reasonable accommodation laws remain common.

Aim: This paper investigates the structural and practical barriers faced by persons with disabilities in accessing and sustaining employment in Cyprus, focusing on digital inclusion, assistive technologies and reasonable accommodation.

Methodology: A qualitative, exploratory approach was used, combining policy and document analysis, semi-structured interviews with representatives of organizations of persons with disabilities, and quantitative data from employers who completed the European Disability Forum's questionnaire.

Results: The study found low levels of employer awareness regarding disability rights and support frameworks. Only 10% of employers had hired a person with a disability, and most lacked internal policies on digital skills, assistive technology, or reasonable accommodation. These findings align with broader European Union (EU) trends but are intensified in Cyprus by a lack of national coordination.

Conclusions: Improving digital training, simplifying assistive technology funding, and strengthening employer engagement are essential for inclusive employment. Comparative practices from other EU countries highlight viable pathways for reform.

Keywords: digital skills · assistive technologies · disability employment

1 Introduction

The right to work is a fundamental human right recognized by the United Nations Convention on the Rights of Persons with Disabilities, which emphasizes access to work environments that are open, inclusive and accessible [1]. Employment is not only a source of income but also a means of social integration, personal development, and equal participation [2, 3]. Despite national legislation and inclusion strategies, persons with disabilities in Cyprus still face substantial barriers to securing and maintaining employment. In 2021, the employment gap between persons with and without disabilities

K. Mavrou and P. Encarnação (Eds.): AAATE 2025, *Technology for Inclusion and Participation for All: Recent Achievements and Future Directions*, pp. 11–17, 2025.
https://doi.org/10.1007/978-3-032-01628-7_2

in Cyprus was 27%, among the highest in the European Union (EU) [4]. Legal measures such as the Persons with Disabilities Law (N.127(I)/2000) and public sector hiring quotas (Law N.146(I)/2009) [5] have limited effect, often undermined by weak coordination and low employer engagement [6–8].

Digital exclusion remains a great barrier as more jobs demand digital competence, and persons with disabilities are left behind due to inaccessible training, high costs, and complex funding procedures [4, 6, 9–11]. Some EU countries, such as Austria and Spain, have adopted national strategies to support digital inclusion and inclusive hiring [7].

Reasonable accommodation is another area where Cyprus falls short. Although legally required under national and EU law [5, 13], many employers remain unaware of their obligations or lack guidance on implementation [7]. As a result, individuals are either excluded from hiring or work without adequate support [5].

These challenges are not unique to Cyprus. Across the EU, many employers lack formal inclusion policies, and public support is often poorly communicated [7, 14]. This paper explores the case of Cyprus using data from the European Disability Forum's (EDF) DATA project, combining organizations for persons with disabilities (OPD) interviews, employer surveys, and EU comparisons to offer realistic, rights-based recommendations for inclusive employment.

2 Methods

2.1 Study Design

A qualitative exploratory design [11] was used to examine barriers from both policy and lived experience perspectives. In addition to the qualitative data, this study also incorporates a structured questionnaire developed by the EDF to gather quantitative and qualitative insights from employers in Cyprus. This survey was part of the wider DATA project and was distributed nationally to employers across private and non-profit sectors. It aimed to capture organizational practices related to digital skills, access to AT and the provision of reasonable accommodations. This approach allowed the study to identify gaps between legislation and practice and to connect national trends with broader European data and insights from the 2024 European Disability Forum Summary Report [7].

2.2 Participant Sampling and Recruitment

Participants were selected using purposive sampling, a method commonly used in studies on disability and employment inclusion. All main Pancyprian OPDs were contacted and four representatives participated in semi-structured interviews. These OPDs included organizations representing a range of disabilities, selected for their active advocacy roles and national reach.

To capture employer's perspective, 60 private-sector businesses were contacted and 20 responded. The sample included employers from key sectors of the Cypriot labour market, particularly wholesale and retail (25%), accommodation and food services (20%), construction (10%), financial services (10%), and manufacturing (5%). While not representative of the entire market, the sample offers a diverse cross-section relevant to inclusive labour market dynamics.

Data Collection

Data was collected using three primary methods, document analysis, semi-structured interviews and employer questionnaire. All materials used were developed by the European Disability Forum and administered nationally [7].

Document analysis including national policies, legal frameworks, strategic action plans related to disability, employment and digital skills were reviewed. This included the Persons with Disabilities Law (N.127(I)/2000), the Recruitment Law (N.146(I)/2009) and the Annual Reports of the Department for Social Inclusion of Persons with Disabilities [5, 6]. Four semi-structured interviews with OPD representatives were conducted using a guide developed by EDF. Topics included national trends in disability employment, barriers to accessing assistive technologies (AT), and the practical realities of securing reasonable accommodations. The data collected was focused on hiring practices, awareness of AT, experience with accommodations and internal policies. Responses were submitted via email between May and September 2023.

2.3 Data Analysis

Data was analyzed using thematic content analysis [12]. Interviews transcripts were reviewed and coded inductively to identify recurring patterns and themes across participants' narratives. Employer survey responses were analyzed descriptively for frequency trends and open-text responses were grouped by theme. This approach allowed the study to identify convergence and divergence between OPD perspectives and employer practices. The national findings were further contextualized by comparing them with the EDF's cross-country analysis of 27 EU Member States [7]. This triangulation helped validate local insights while revealing how Cyprus aligns with or diverges from European norms in the areas of digital inclusion, AT access and accommodation policy.

3 Results

3.1 Key Findings

Digital skills

Digital exclusion emerged as a major employment barrier, especially as more jobs rely on digital tools. OPDs unanimously reported that their members often lack basic digital skills, which severely limits their job prospects, particularly as more work environments become technology driven. The situation reflects a broader European trend in which digital competence gaps disproportionately affect vulnerable groups [4]. No official statistics were available for persons with disabilities in Cyprus.

As shown in Table 1, only 10% offered digital training, despite 85% stating digital skills were not a hiring concern. OPDs also noted that available training programmes, such as those from the Cyprus Productivity Centre, often fail to address accessibility or disability-specific adaptations [9]. Both the Cyprus Productivity Centre representative and the Department of Social Inclusion of Persons with Disabilities confirmed via email, that no current training programme is available that could accommodate persons with disabilities.

Assistive Technologies

Access to AT remains limited due to bureaucratic barriers and cost. OPDs noted that public funding exists but is often difficult to navigate, leading many to cover expenses themselves. Furthermore, there is no public training offered for employees or employers on how to use assistive devices effectively [9].

Employer awareness of AT is similarly low. Only 5% ($n = 1$) of respondents were aware of any use of assistive technologies in their organization and 15% ($n = 3$) had any kind of policy on AT access, usually relying on public funding mechanisms. These numbers align with broader European patterns, where 81% of EU employers surveyed lack internal policies on AT [7]. OPDs also emphasized that Cyprus's small market size makes AT more expensive and harder to procure than in other EU countries.

Reasonable Accommodation

Despite legal obligation under the Persons with Disabilities Law (N.127(I)/2000), reasonable accommodations remain rare in practice [5]. OPDs reported that accommodations are often informal, or absent and individuals are frequently hired only if no adaptations are needed. Only 10% of employers reported having formal human resources (HR) policies on reasonable accommodation and just 5% were aware of public frameworks to support implementations. These figures point to insufficient communication around legal requirements and available resources. Cyprus reflects wider EU trends, where 76% of employers report lacking knowledge of national support mechanisms [4, 7].

Both OPDs and employers cited barriers such as unclear guidelines, limited financial support, and cost concerns. Additionally, the legal definition of "reasonable" remains vague, giving employers wide discretion to deny accommodations [5].

As shown in Table 1, employer responses revealed limited engagement with disability-inclusive practices. Only 10% had hired a person with a disability, and the same percentage had HR policies addressing inclusive hiring. None had participated in disability-related employment programmes. While 85% did not view digital skills as a hiring barrier, just 10% offered relevant training. Awareness and implementation of assistive technologies were also low: only 5% reported AT use among staff, and 15% had related policies. For reasonable accommodation, 10% had HR procedures in place, and only 5% were aware of public support frameworks. These findings indicate a widespread lack of employer engagement with inclusion policies, consistent with broader EU trends [7].

Table 1. Employer responses on disability-inclusive hiring, training and policy practices

Category	Indicator	Yes (%)	No(%)
Hiring & Inclusion	Has hired a person with a disability	10%	90%
	HR policy includes disability hiring guidelines	10%	90%
	Participated in a disability-related employment programme	0%	100%
Digital Skills	Offers digital skills training	10%	90%
	Digital skills not viewed as a hiring barriers	85%	15%
Assistive Technologies	Aware of AT usage by employees	5%	95%
	Has policy on AT access/support	15%	85%
Reasonable Accommodation	HR policy includes accommodation measures	10%	90%
	Aware of public support frameworks	5%	95%

4 Discussion

These findings mirror EU-wide patterns of fragmented implementation and weak coordination [7, 8, 14]. Issues like limited access to assistive technology, low digital literacy and the lack of proper workplace accommodations are not unique to Cyprus. These are problems that have been documented across Europe [7]. The 2024 EDF Summary Report revealed that 81% of employers in participating EU countries lacked formal policies on AT, while 76% were unaware of national support schemes for reasonable accommodation. These figures are reflected by the Cyprus sample.

However, the Cypriot context presents a pattern of fragmented implementation and weak institutional coordination, which together undermine the country's inclusion efforts. According to national statistics [7], only 50.2% of the general population in Cyprus have basic digital skills, and this rate is likely even lower among persons with disabilities [9]. Rather than the size of the labour market itself, the persistence of barriers appears linked to the absence of a national digital inclusion strategy that is responsive to the needs of persons with disabilities. This pattern is not unique to Cyprus; other EU countries with more extensive labour markets, such as Greece, Austria, and Spain, also report similar difficulties, especially when accessible training, employer engagement, and assistive technology provision are not addressed through coherent and enforceable policies [7, 8, 14]. The bureaucratic processes for accessing public subsidies, especially for AT, are lengthy and unclear, discouraging both employers and persons with disabilities from applying. Similar barriers have been noted in the implementation of the CRPD across smaller EU states, where the burden of administrative compliance falls heavily on individuals [15]. While legal frameworks such as the Persons with Disabilities Law exist [5], the frequent invocation of "disproportionate burden" clauses often allows employers to bypass their accommodation obligations [13].

Digital exclusion also stands out as a persistent challenge. Although digital skills are now essential for employability, most training programs in Cyprus fail to account for accessibility needs. This trend is consistent with research from Vicente and López [14], who found that persons with disabilities across Europe face compounded exclusion when digital literacy training is not adapted. The gap is even more critical in the context of a digitalizing economy, where lack of skills quickly translates into labour market marginalization [9].

Despite these limitations, examples from other EU countries offer practical models for reform. Austria has introduced mandatory inclusion audits for public employers and maintains a national funding scheme for workplace adjustments. Spain's Access to Employment (A2E) initiative provides targeted wage subsidies and support for on-the-job training [7]. These examples show that well-funded, clearly communicated, and accessible systems can significantly improve employment outcomes for persons with disabilities.

In contrast, Cyprus lacks a national framework that connects digital inclusion, employment accessibility, and employer incentives. This lack of coordination leads to scattered efforts—such as one-off training initiatives or subsidy schemes—that may sound helpful in theory but often fail in practice because they aren't aligned or properly followed up. For example, while some employers in this study said digital skills were "not a hiring concern," OPDs painted a different picture: many persons with disabilities aren't even getting the chance to apply for jobs because they lack access to digital tools and training in the first place. These gaps suggest accessibility should be built into all mainstream systems, not treated as an afterthought [6, 10].

Another issue is the role of employers, which often seems passive rather than actively resistant. Most aren't deliberately excluding persons with disabilities—they're just not aware of how to make inclusion happen. That's where stronger partnerships with OPDs and employment services could make a big difference. In countries like Finland and Denmark, such collaborations have helped shift workplace culture and build real, on-the-ground support for inclusive hiring [7].

To move forward, Cyprus needs a joined-up strategy that brings together better law enforcement, more accessible and inclusive education, and real employer engagement. And just as importantly, we need better data—data that tells us who is getting hired, who isn't, and why. Tracking outcomes by disability type and digital access would go a long way toward designing smarter, more targeted policies. These kinds of actions are already being called for by European experts [14] and international human rights bodies like the UN CRPD Committee [15].

5 Conclusions

This study confirms that persons with disabilities in Cyprus continue to face significant barriers when it comes to finding and maintaining meaningful work. Although there are legal frameworks intended to support inclusion, their practical application remains limited. Digital exclusion, limited access to AT, and the inconsistent provision of reasonable accommodations continue to keep many persons out of the labour market. To move forward, Cyprus needs to shift its focus toward implementation. This includes more access to digital skills training designed with accessibility in mind, simplifying and promoting public funding for assistive devices, and giving employers the tools and the incentives to create more inclusive workplaces.

Stronger political commitment, cross-sector coordination, and real OPD engagement are needed to make inclusion a reality. Further studies with broader samples and strengthened methodology would be valuable in validating these results and exploring long-term impacts of inclusion policies.

References

1. United Nations: Convention on the Rights of Persons with Disabilities (2006)
2. Banks, L.M., Polack, S.: The economic costs of exclusion and gains of inclusion of people with disabilities. CBM/ICED (2014)
3. Schur, L., Kruse, D., Blanck, P.: People with Disabilities: Sidelined or Mainstreamed? Cambridge University Press (2013)
4. European Commission: 2023 Country Report: Cyprus. Publications Office of the European Union (2023)
5. Republic of Cyprus: Persons with Disabilities Law (N.127(I)/2000) and Recruitment of Persons with Disabilities in the Wider Public Sector Law (N.146(I)/2009)
6. Department for Social Inclusion of Persons with Disabilities. Annual Report (2022)
7. Lulli, R., Sinicato, F., Couceiro Farjas, A.A.: Digital Skills, Accommodation and Technological Assistance for Employment: Supporting the inclusion of persons with disabilities in the open labour market. European Disability Forum (2024)
8. Mavrou, K., Liasidou, A., Tsakiri, M.: Striving for an Inclusive Labour Market in Cyprus. European Disability Expertise (2023)
9. European Commission. Digital Economy and Society Index (DESI) 2023 – Human Capital Report. Publications Office of the EU (2023)
10. Dobransky, K., Hargittai, E.: The disability divide in Internet access and use. Inf. Commun. Soc. **9**(3), 313–334 (2006)
11. Creswell, J.W., Poth, C.N.: Qualitative Inquiry and Research Design: Choosing Among Five Approaches (4th ed.). SAGE Publications (2018)
12. Braun, V., Clarke, V.: Using thematic analysis in psychology. Qual. Res. Psychol. **3**(2), 77–101 (2006). https://doi.org/10.1191/1478088706qp063oa
13. European Parliament: Report on the Implementation of Directive 2000/78/EC on Equal Treatment in Employment and Occupation (2022)
14. Vicente, M.R., López, A.J.: A multidimensional analysis of the disability digital divide. Inf. Soc. **26**(1), 48–64 (2010)
15. UN Committee on the Rights of Persons with Disabilities: Concluding Observations on the Initial Report of Cyprus. CRPD/C/CYP/CO/1 (2017)

Artificial Intelligence and Disability: Using ChatGPT to Improve the Well-Being and Understand the Needs of Students with Autism Spectrum Disorder

Eleni Damianidou(✉)

European University Cyprus, 6 Diogenous Str., 2404 Egkomi, Cyprus
e.damianidou@external.euc.ac.cy

Abstract. Artificial intelligence (AI) has the potential to transform education for students with Autism Spectrum Disorder (ASD) by providing tailored support that nurtures their potential and well-being. The aim of this study was to explore how ChatGPT can enhance the inclusion and well-being of students with ASD. To this end, a qualitative study was conducted, during which an intervention with ChatGPD was implemented in the Special Unit. Participants included six students, one teacher and three school assistants. Data were collected with participative observation and interview. It was found that there were recurring themes in the students' inquiries, preference for detailed and structured information, repeated focus on specific topics or areas of interest (e.g. cars), and questions related to social skills, emotional understanding, or coping strategies. ChatGPT served as a safe and non-judgmental medium for students to express emotional concerns or seek help. As a result, targeted interventions were designed and teaching strategies were tailored, so as to incorporate instructions, aids and guidance, based on the students' queries.

Keywords: Artificial Intelligence · ChatGPT · Autism Spectrum Disorder

1 Introduction/Background

Artificial intelligence (AI) is increasingly recognized as a useful tool for teachers, because of having the potential to transform education and improve the educational experience of students. AI-driven educational interventions may especially benefit diverse students because of having the potential to provide individualized support, facilitate the learning process and promote the students' autonomy and well-being [9]. Using AI may be highly beneficial for students with Autism Spectrum Disorder (ASD), who present a variability in learning styles, communication preferences, and sensory sensitivities. Since traditional educational approaches often fall short in fully meeting the above diverse needs, the use of AI technologies may provide teachers with a powerful educational tool that has the potential to bridge this gap by providing tailored support that adapts the learning process to each student's unique profile and thereby nurtures the students with ASD's learning potential and well-being [1].

K. Mavrou and P. Encarnação (Eds.): AAATE 2025, *Technology for Inclusion and Participation for All: Recent Achievements and Future Directions*, pp. 18–24, 2025.
https://doi.org/10.1007/978-3-032-01628-7_3

ChatGPT, as an advanced AI conversational model, can play a transformative role in educational settings. Being a language-based AI system, ChatGPT can serve as an interactive educational assistant that can facilitate a wide range of learning activities. This is possible, because ChatGPT can adjust its responses according to the user's comprehension level, pace, and interests through dynamic learning formats, which may result in creating a more inclusive, user-friendly and student-centered learning environment [9]. Thus, using ChatGPT may be of particular importance for students with ASD, because it can cater to the diverse student needs by supporting personalized learning, fostering creativity, and promoting engagement [2].

Hence, the benefits of ChatGPT may extend beyond using the tool for general purposes in the classroom and have particular significance for learners with ASD. Its capacity to provide consistent, judgment-free interaction makes it especially suited to support the learning preferences of students with ASD, who may thrive on repetition, predictability, and individualized attention [3]. Through role playing, simulations and open-ended discussions, as well as support for writing and idea generation, ChatGPT can offer -at least theoretically- unlimited opportunities to students with ASD for practice in areas where reinforcement is essential, such as language use, social communication, and executive functioning skills, without judgment [4]. It seems then that, by delivering a supportive, non-threatening environment for exploration and learning, ChatGPT may enhance educational equity and accessibility for students with ASD.

1.1 Aim and Research Questions

Based on the above, the aim of this study was to explore how ChatGPT can enhance the inclusion and well-being of students with ASD, aligning with international commitments to accessibility and inclusivity. It was assumed that ChatGPT could assist not only students with ASD, but also teachers in understanding the needs of students with ASD by analyzing the type and nature of questions students ask. Thus, it was hypothesized that this analysis could offer valuable insights into the students' cognitive preferences, challenges, and areas of interest.

Hence, the main research questions of this study were the following:

1. What are the recurring themes in the questions students with ASD ask ChatGPT?
2. Can students' interactions with ChatGPT provide insights into the emotional well-being and unmet needs of students with ASD?
3. What is the potential of ChatGPT to support the goals of inclusive education and digital transformation?

2 Methods

This was a qualitative study. The participants were a) six adolescent male students with ASD, enrolled in the Special Unit of a high school in Cyprus, b) one female teacher and c) three female school assistants. The participants were purposefully selected because of being the students, the teacher and the assistants in the same Special Unit. To safeguard anonymity of the participants, their demographic details are not presented in this study, because of the increased risk of their identification that is related to the existence of a limited number of Special Units in Cypriot high-schools.

The use of ChatGPT was part of the curriculum of the Special Unit, which was approved by the headteacher of the school, as indicated by the education law for students with special needs in Cyprus. Before using ChatGPT during the Social Studies lesson, the teacher and the school assistants were trained by the researcher, who has a specialty in teaching methods for students with disability. The aim of the training was to ensure that the teacher and the school assistants would be able to guide students through their inquiries.

The lessons for the students began with an introduction to ChatGPT and the way it could be used by the students. There were enough PCs for all the students in the classroom, who were then encouraged to use ChatGPT to ask questions to which they needed answers. The interaction of students with ChatGPT lasted for approximately half an hour. The teacher continued to use the tool for the purposes of Social Studies lessons for the rest of the school year.

The main tools to gather data were participative observation and interviews with the students, the teacher and the school assistants. Logs captured the questions asked, responses given, and patterns of interaction. Data was analyzed with thematic analysis. To reassure compliance with research ethics, the parents gave their informed consent, while the students agreed that the researcher could observe them and their discussions with ChatGPT. Moreover, the participants' anonymity was safeguarded.

3 Results

3.1 Students' Interactions with ChatGPT

The first theme was students' interactions with ChatGPT. It was found that there were recurring themes in their inquiries, preference for detailed and structured information, repeated focus on specific topics or areas of interest (e.g. cars and cooking), and questions related to social skills, emotional understanding, or coping strategies. Thus, students repeatedly asked ChatGPT: "what does it mean when someone says…". In addition, most students asked: "what should I do when…". Based on the students' discussions with ChatGPT, which reflected the students' cognitive preferences, learning challenges, and emotional needs, then the teacher planned social stories to enhance particular social skills and implemented interventions related to the students preferences.

For example, a student asked: "what does it mean when someone says, don't be a fool?" and ChatGPT answered "That phrase usually means someone is warning you not to make a bad decision or act without thinking. It's not a literal insult but a way to express concern". Then the student asked "one of the students in my school called me a fool during the school break. What do you think I should do?" and ChatGPT answered "That sounds hurtful. Sometimes people use the word 'fool' when they're frustrated or trying to tease someone. It's best to stay calm and talk to a teacher or trusted adult about how it made you feel". Based on the above inquiry regarding 'teasing' the teacher presented a related social story to the students and the appropriate solutions, so as to enable them to effectively manage such situations. The social story included cars and cooking, which were among the preferred themes of the particular students. At the same time, the school assistants started to be more alert during the breaks regarding likely events of teasing. In addition, the teacher talked with the headteacher about the likelihood of having teasing and bullying phenomena at school. As a result, the latter decided to act proactively and implement interventions with workshops and lectures for the students at school, so as to inform them about diversity and inclusion and prevent the use of offensive language and exclusionary behavior.

3.2 Insights into Emotional Well-Being and Unmet Needs of Students

Secondly, regarding emotional well-being and behavioral insights, ChatGPT served as a safe and non-judgmental medium for students to express emotional concerns or seek help. Observations highlighted changes in students' emotional behavior, such as increased confidence or reduced anxiety during and after their interactions with ChatGPT. In addition, there were minor communication improvements, evident in students articulating their thoughts and asking questions in a more organized manner.

For example, a student said to ChatGPT: "I feel like I don't fit in here" and ChatGPT answered: "That sounds really tough. A lot of people feel that way sometimes. Would you like to talk about what's been happening?". Then, the student turned to the school assistant and asked her to help him explain his feelings, who took the chance to talk with the student about emotions and how to express them. In the days that followed, according to the school assistant, "(Name of the student) seems more confident to talk about his worries after his talks with his friend ChatGPT".

Another student asked ChatGPT: "I get really angry when someone changes my plans. Can you help me?" and ChatGPT answered "It's normal to feel upset when plans change. You can try deep breathing or writing down how you feel. Would you like to practice a calming activity together?". Based on the above interaction, the teacher talked with the Physical Education teacher, who prepared some relaxing activities and taught relaxation techniques to the students of the Special Unit. In the days that followed, the teacher commented that "(Name of the student) seems to try to control his anger by using the deep breaths that (name of the Physical Education teacher) taught them. I also encourage all the students to practice deep breaths during my lesson, if something unexpected occurs and see them upset".

3.3 Potential of ChatGPT

The third theme referred to the insights gained, which could be useful for teachers and school assistants. It was found that the analysis of question logs and feedback from students helped them better understand the needs and preferences of students with ASD. As a result, targeted interventions were designed and teaching strategies were tailored, so as to incorporate instructions, aids and guidance, based on the students' queries.

For example, one student said to ChatGPT "I want to play with the other students, but I don't know what to tell them". Based on the above, the teacher understood that the student had difficulties initiating peer interactions, which is a common phenomenon among students with ASD. To support the students, the teacher designed and implemented role-play activities in the classroom. In this way, the student had the chance to practice possible conversations and learn how to respond to invitations from other students to play with them, in a structured, non-judgmental, safe and supportive environment.

Another interesting interaction of another student, who initially said to ChatGPT "No one wants to be my friend", was his inquiry the following days, after the social stories, the relaxing techniques and the role-play, when the student asked ChatGPT "How can I be a good friend to someone?" This was an important insight for the teacher and the school assistants, who noticed the change in question framing, i.e. from self-doubt to relational understanding, which suggested most likely an improvement in social maturity and increased confidence. As a result, the teacher started to recognize the student's development and began encouraging him to proceed to more independent peer interaction, while continuing to monitor and reinforce the student's skills in a less scaffolded setting. According to the teacher: "(Name of the student)'s queries about making friends opened for me a rich source of authentic data for shaping responsive, appropriate and meaningful educational practices and better understand his development". Thus, ChatGPT functioned not only as a conversational tool for the students, but also as a window for the teacher and the school assistants, who could see more clearly their internal world, and thereby proactively act to meet their social-emotional learning needs.

4 Discussion

The findings of this study highlighted the likely potential of AI-powered conversational tools, particularly ChatGPT, in supporting students with ASD in the educational setting of the Special Unit. Firstly, it was confirmed that students with ASD often experience challenges in interpreting social cues, initiating conversations, and navigating peer interactions [3]. Thus, the frequent engagement of students with ChatGPT to clarify ambiguous social phrases or seek advice on real-life social situations suggests a need for predictability, clarity, and structured guidance, which are hallmarks of ASD-related cognitive processing preferences [5].

On the other hand, the students' interactions with ChatGPT not only served as opportunities for learning for students but also were used as diagnostic tools for teachers and school assistants, who could identify more easily their need for key social skills, and thereby respond timely with tailored interventions, such as social stories and role-play. Importantly, by taking into consideration students' personal interests (e.g., cars and cooking) and including them in their stories and discussions with the students, the teacher and school assistants were able to increase the students' engagement and promote meaningful learning experiences, which is in agreement with the suggested teaching approaches for students with ASD [2].

Another important finding of this study was the potential of ChatGPT to provide a safe and non-judgmental environment, where students with ASD may feel more comfortable and thereby encouraged to express their feelings. Interestingly, some of the students with ASD were inspired to start articulating more clearly their emotions and adopt help-seeking behaviors, as evident when students approached assistants or teachers for further support following their AI-mediated conversations. Given that emotional regulation and expressive communication are often areas of difficulty for students with ASD [7], it seems that ChatGPT and its responses managed to open the path for developing social skills and learn practical strategies to improve their well-being and social understanding. Importantly, observing the students' queries sparked school-wide interventions, which benefited not only the students with ASD, but also the other students, because of fostering inclusive thinking.

One very important finding of this study was the potential of ChatGPT to be used as an indirect observational tool for teachers. Thus, the students' interactions with ChatGPT revealed cognitive patterns, emotional states, and developmental trajectories that may not have been immediately visible through traditional assessment methods. Thus, students with ASD often display a preference for precise, detailed information, suggesting a need for structured teaching methods [5]. In addition, questions that emphasize rules, logic, or predictable patterns highlight a preference for structured and logical subjects [2]. Repeated inquiries on a particular topic, as found in this study, indicate areas, where the students may struggle with comprehension or require reinforcement, and point to challenges in conceptual thinking, possibly suggesting a need for concrete examples or visual aids. In addition, students' inquiries related to special interests may be leveraged by teachers to engage students more effectively in other areas of learning [6], which was an important outcome of this study.

Questions about understanding emotions, social interactions, or navigating relationships may highlight areas where the student seeks support in developing social skills [7], as indicated in this study. Inquiries about managing stress, focus, or sensory overload may suggest specific challenges that require accommodation, such as quiet spaces or mindfulness techniques [8], which was an intervention that was implemented in this study.

In short, this study confirmed that AI can be effectively utilized to ensure that content is delivered in a way that resonates with the needs of students with ASD, to support communication and to enable understanding of social cues in a controlled, non-threatening setting [3]. More importantly, this study confirmed the importance of using technology not only as a teaching aid but as a tool for authentic assessment and responsive pedagogy [10]. When used ethically and thoughtfully, then, ChatGPT can complement teacher observation and help identify hidden needs, enabling timely and targeted interventions.

5 Implications for Practice

The use of AI in the education of students with ASD is not a replacement for interaction with teachers, but a powerful tool that may enhance teaching and learning. As found in this study, ChatGPT helped the teacher to support students with ASD and implement targeted interventions that were in line with their preferences and individual needs, because of opening a window from which the teacher and the school assistants could see into their internal worlds. This dual potential of ChatGPT as a tool for both supporting students and informing teachers puts AI in a central position regarding the implementation of inclusive education. On the other hand, it has to be noted that using AI as a teaching tool also raises important considerations related to data privacy, ethical use, and staff capability to make an effective use of AI. Hence, to include AI in teaching, it is essential to first train teachers to interpret the insights about students that may be gained from AI tools on the one hand, and on the other, to integrate them meaningfully, securely and responsibly into their everyday pedagogical practice.

6 Limitations

Although the findings of this study are interesting and promising, this study was based on a limited sample and was implemented in a particular context, which does not represent the general classroom. Hence, further research seems essential to explore the scalability, long-term impact and viability of such interventions across different educational settings and students with diverse needs.

7 Conclusion

Since this study found high levels of comfort and acceptance of ChatGPT by students with ASD, due to the tool's predictability, non-threatening nature, and instant feedback, it highlights the usability and usefulness of ChatGPT in classroom settings and indicates that it may be a helpful resource for supporting social-emotional learning and coping strategies for students with ASD. It seems then that this study supports inclusion by fostering accessibility, inclusivity, and digital transformation through the controlled use of AI. In addition, this study may have cross-cultural applicability, by providing a model for integrating AI into special education in other international contexts.

References

1. Lampos, V., Mintz, J., Qu, X.: An artificial intelligence approach for selecting effective teacher communication strategies in autism education. NPJ Sci. Learn. **6**(1), 25 (2021)
2. Mehta, P., Chillarge, G.R., Sapkal, S.D., Shinde, G.R., Kshirsagar, P.S.: Inclusion of children with special needs in the educational system, artificial intelligence (AI). In: AI-Assisted Special Education for Students With Exceptional Needs, pp. 156–185. IGI Global (2023)
3. Rakap, S., Balikci, S.: Enhancing IEP goal development for preschoolers with autism: a preliminary study on ChatGPT integration. J. Autism Dev. Disorders, 1–6 (2024)
4. Rakap, S.: Chatting with GPT: enhancing individualized education program goal development for novice special education teachers. J. Spec. Educ. Technol. **39**(3), 339–348 (2024)
5. Sideraki, A., Drigas, A.: Artificial Intelligence (AI) in Autism. Technium Sci. J. Soc. Sci. **26**, 262 (2021)
6. Li, G., Zarei, M.A., Alibakhshi, G., Labbafi, A.: Teachers and educators' experiences and perceptions of artificial-powered interventions for autism groups. BMC Psychol. **12**(1), 199 (2024)
7. Adako, O.P., Adeusi, O.C., Alaba, P.A.: Revolutionizing autism education: harnessing AI for tailored skill development in social, emotional, and independent learning domains. J. Comput. Cogn. Eng. **3**(4), 348–359 (2024)
8. Habibi, F., Sedaghatshoar, S., Attar, T., Shokoohi, M., Kiani, A., Malek, A.N.: Revolutionizing education and therapy for students with autism spectrum disorder: a scoping review of AI-driven tools, technologies, and ethical implications. AI and Ethics, 1–16 (2025)
9. Rahman, M.M., Watanobe, Y.: ChatGPT for education and research: opportunities, threats, and strategies. Appl. Sci. **13**(9), 5783 (2023)
10. Hoogerwerf, E.J., Mavrou, K., Traina, I.: The role of assistive technology in fostering inclusive education. Strategies and tools to support change. Routledge, London & New York (2021)

Participatory Development of an AI-Enabled Application for Translation into Plain Language. Insights from KARLA - A German Research Project with People with Intellectual Disabilities

Lukas Baumann(✉), Marie-Christin Lueg, Bastian Pelka, and Susanne Dirks

TU Dortmund University, 44227 Dortmund, Germany
lukas.baumann@tu-dortmund.de

Abstract. People with disabilities often encounter difficulties reading and understanding documents and correspondence from authorities, due to complex terminology. In this context, Artificial Intelligence (AI) technology is a promising way to facilitate independence and participation. This paper presents the KARLA project, a participatory research project from Germany (2024–2026) in which an AI-enabled application is being developed to support people with intellectual disabilities in understanding written text. Based on workshops and tasks in a messenger service, this paper reports the initial results of the participatory processes, including the design process of the app's icon and establishing a mutual understanding of the app's primary function. The results emphasise not only conducting joint research to achieve the project goal but also collectively managing all relevant project processes to frame a problem and facilitate appropriate team building.

Keywords: Participatory Research · Participatory Action Research · Artificial Intelligence · People with Intellectual Disabilities · Plain Language

1 Introduction

People with disabilities often encounter difficulties reading and understanding documents and correspondence from authorities, due to complex terminology. The use of easy-to-read (E2R) or plain language is a fundamental requirement for the inclusion of people with intellectual disabilities. In Germany, public sector bodies are legally obliged to explain the content of decisions, applications, and other documents in plain language (BGG § 11). Due to regulations and societal changes, the need for experts to adapt or create E2R documents will only grow in the future [1]. However, there is a discrepancy between users' needs and current practices. Currently, the adaptation of documents into E2R or plain language is mostly done manually and, therefore, time-consuming and oftentimes subjective [1]. Consequently, information and legal rights frequently remain inaccessible to individuals with disabilities, leading to a significant degree of dependency on others.

K. Mavrou and P. Encarnação (Eds.): AAATE 2025, *Technology for Inclusion and Participation for All: Recent Achievements and Future Directions*, pp. 25–32, 2025.
https://doi.org/10.1007/978-3-032-01628-7_4

2 Background

In this context, Artificial Intelligence (AI) technology is a promising way to facilitate independence and participation. For example, an evolving landscape of AI-driven language translation [2] is producing AI-driven solutions with the goal of offering the translation of texts into E2R or plain language. Especially in healthcare, research is done where large language models (LLM) like ChatGPT are utilised to translate information into plain or easy-to-read language [3–7].

In general, current trends indicate an increasing relevance of assistive technologies (AT) with AI-enabled functions in the forthcoming years [8]. Many authors claim that integrating AI into assistive technologies can improve support options [9–11]. According to Almufareh et al., "AI holds the potential to substantially enhance the quality of life for people with intellectual disabilities" [12]. However, the use of AI is also viewed critically, e.g., because it often reflects ableist perspectives [13]. Thus, involving and collaborating with user groups as active participants from the beginning of research and development processes is necessary in order to take their experiences, preferences and needs into account when designing AI-supported technologies [14]. This approach ensures that the developed solutions are more accessible and sustainable [15]. Selecting appropriate methods that foster effective collaboration is crucial for maximising the opportunities of participatory research and development. Currently, there is no participatory developed solution where an LLM is integrated into an application that is understandable and usable for people with intellectual disabilities.

3 The KARLA Project

This paper presents results from the KARLA project, a participatory research project from Germany (2024–2026) in which an AI-enabled application is being developed to support people with intellectual disabilities in understanding written text. The project has two main objectives. The first is to identify relevant application areas and necessary technical requirements. The second is to determine crucial aspects and methods for promoting participation in research. The project combines the technical possibilities of AI with the social and ethical aspects of participation and is in line with the characteristics of collaboration, iteration and openness to results as defined by Participatory Action Research (PAR) [16].

The objectives are developed by an inclusive project team. In order to assemble the peer-research team for the KARLA project, the project team recruited 13 people with intellectual disabilities who work or live in a social institution in Germany with the intention to be involved in *decisive power* regarding project development and implementation [17]. Recruiting took place between February and April 2024 via in-house information channels (mail, posters, video, etc.) with information on the project, its goals and the project staff in the collaborating social institution in Germany. No additional requirements for participation existed, allowing anyone interested to join the project.

Regular meetings and workshops are planned to ensure ongoing cooperation and communication between peer researchers and the other project staff (practitioners, software developers, designers and scientists). Additionally, communication between meetings and workshops is ensured via a Messenger service. Here, internal project information (e.g., workshop dates) and tasks between workshops and meetings are shared. These communication channels were agreed on during the first project meetings and workshops. Furthermore, peer researchers conducted a democratic election to select representatives responsible for attending meetings and ensuring continuous representation in project discussions and communication with individuals outside the project.

Eight inclusive workshops were held between June 2024 and April 2025, each with a different agenda, objectives and methodology. On average, around eight peer researchers attended the workshops. Four additional tasks were addressed within the Messenger service either as a preparation for an upcoming workshop or as a follow-up. After each workshop, a brief evaluation of all participatory methods was held, using an anonymous rating based on smileys. Additionally, the project staff asked the group for qualitative feedback. Table 1 presents a summary of three workshops and two Messenger service tasks conducted between August 2024 and February 2025 related to two crucial project results.

Table 1. Overview KARLA app icon & mutual understanding

Steps	Format & Date	Aims	Methodology
KARLA app icon	Messenger service task (August/September 2024)	Identifying relevant areas of application and functions, implications for corporate identity	Photovoice [18]
	Workshop (September 2024)	Implications for use, functions and corporate design of the KARLA app	Photovoice [18]
	Workshop (September 2024)	Introduction to AI and developing Corporate Design	AI-supported design challenge, Brainstorming [19]
Establishing a mutual understanding	Messenger service task (January 2025)	Finding a word to describe the primary function of KARLA	Think-Pair-Share [20]
	Workshop (February 2025)	Mutual understanding of the app's primary function	Adapted Speed-Dating [21], Think-Pair-Share [20]

4 Results

Based on Table 1, the results presented in this paper will focus on the participatory processes for two crucial non-technical aspects that significantly influenced the project: Designing the app icon and developing a mutual understanding of the app's primary function.

4.1 KARLA App Icon

At the beginning, one of the project's most important participatory tasks was establishing a participatory process for design-based decisions. In a first step, the project team decided to develop a corporate design. In order to achieve this, peer researchers were first introduced to a design proposal featuring a magnifying glass, which was used as a base for discussion. Additionally, the photovoice results presented in Lueg et al. [18] served as the foundation for the entire process (Messenger service task, August/September 2024). First, different ideas for icons were brainstormed and sketched. After discussing and evaluating all symbols, peer researchers agreed on a lightbulb with the letter K (for KARLA) and three stars as the key elements for designing a final icon (First workshop, September 2024). The peer researchers state that a magnifying glass symbolises searching for information, whereas a lightbulb represents understanding or receiving information. They also point out that the lightbulb is seldom used as an app icon, which is why it will make the project stand out.

In the next step, colours that peer researchers liked in other app icons were discussed. As a result, a fixed colour scheme of yellow and purple was implemented, and two versions with reverse colour options, each for the magnifying glass and the lightbulb, were voted on (Second workshop, September 2024). Every peer researcher, except one, chose the purple background version for both designs. In the final vote, the peer researchers unanimously selected the lightbulb design featuring a purple background and yellow foreground as the definitive design (see Fig. 1).

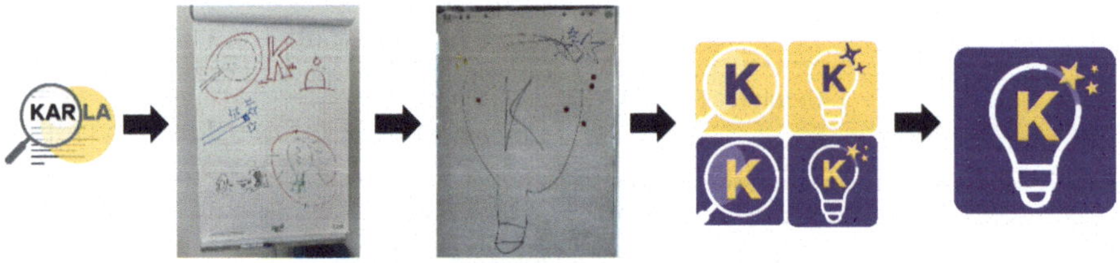

Fig. 1. Developing the KARLA App Icon

4.2 Establishing a Mutual Understanding

During the first workshops, it became evident that it was necessary to develop a mutual understanding of the app's primary function. Peer researchers disagreed with characterising KARLA solely as an app for translating written text into plain language. Moreover, the process of designing the app icon suggested that it is imperative for the peer researchers to understand the information.

Therefore, as a first step, the project staff asked the peer researchers to explain KARLA in their own words in a voice message (Messenger service task: *What does KARLA do?* January 2025*).* The project staff received eight voice messages from seven peer-researchers. Based on the voice messages, seven different keywords were extracted:

- Hilfestellung ("assistance"/"Support")
- Erklärhilfe ("explainer")
- Verstehen ("Understanding")
- (In leichte Sprache) umsetzen ("transform (into plain language)")
- entziffern ("decipher/decode")
- Erleichtern ("simplify")
- Übersetzen (translate")

During the workshop in February 2025, three groups engaged in discussions about these keywords using a modified speed-dating approach. Each group presented its results to the entire group and voted on a final word. The peer-researchers collectively decided on the word *Hilfe-Stellung* ("assistance"/"support"):

"(...) KARLA is there to assist someone who needs help with documents, such as forms and so on" (Peer-Researcher, voice message Messenger service task: *What does KARLA do?* January 2025; translated from German).

"(...) I would explain it by saying that the app (...) supports you in understanding difficult texts more easily. (...) That it makes it easier to cope with a leaflet or something (...)" (Peer-Researcher, voice message Messenger service task: *What does KARLA do?* January 2025; translated from German).

5 Discussion

The presented results demonstrate the positive impact of involving user groups in technology development processes. The adopted participatory approach not only led to better work results but also to more motivation and identification with the project among all project members and peer researchers. Even if participatory approaches often collide with organisational and structural conditions and the general feasibility of peer researchers' needs and wishes [15, 22], such methods of tailoring AI-enabled apps to the needs and capabilities of people with intellectual disabilities contribute to promoting inclusive artificial intelligence [23]. It becomes apparent that it is important to use not only methods suitable for research but also *process-organising* methods [24] to ensure a high level of participation. Additionally, the results highlight the importance of combining different methods and formats to achieve a successful outcome. In this context, a key aspect of the project is utilising the Messenger service. In line with Sattler et al. [25], using a Messenger service expands the range of methods, increases flexibility and saves resources.

However, the need for reminders indicates that additional efforts for clear instructions and support materials are needed. Furthermore, not all presented methods achieved a high level of participation. It needs to be discussed and analysed whether these methods are essential for the overall participation.

An important insight from the process of establishing a mutual understanding is that people with intellectual disabilities not only need simplified texts but also clarity about the necessary actions they need to take. This aspect also influences the development of a technical solution within the project, indicating that the KARLA app will benefit from offering more than just translating text into plain or E2R language. Moreover, as Freyer et al. [26] emphasise, the importance of additional research on implementation within current social and administrative systems needs to be discussed. This might include, for example, the importance of care workers in applying LLMS for plain language translation needs. In this context, the peer researchers highlighted the importance of involving relevant individuals, such as care workers or family members.

A thorough evaluation of the participatory approach will be essential for the further phases of the project. This will help to gain a more comprehensive understanding of the methods used and the emerging challenges. In addition, assessments from key stakeholders, such as family members or carers, will be collected to inform technology development. The systematic collection and analysis of participatory methods from other research projects with people with intellectual disabilities will help to answer the research questions effectively. Additionally, the project team agreed on finding a matching pictogram for the word *assistance/support*. Finally, the project will focus on the technical implementation and execution of participatory testing iterations.

6 Conclusion

This article provides an overview of the initial results of a participatory research and development project for the design of an AI-supported application for understanding written texts. In line with the transdisciplinary research process defined by Sattler et al. [25] & Lang et al. [27], the first results focus on non-technical aspects that helped to frame the problem and enable appropriate team building. The adaptation of procedures and methods to the needs of the peer researchers helps to ensure the accessibility of the project and to develop an application which can be used by the target group of people with intellectual disabilities. This includes, among other things, explaining all aspects of a research project (in this case, e.g., peer researchers, AI, etc.) and ensuring a common understanding of the concepts used. In addition, it was shown that identifying and motivating peer researchers is a key factor for the project outcomes. For a successful participatory project, it is essential not only to conduct joint research to achieve the project goal, but also to collectively manage all relevant project processes. Furthermore, it should be highlighted that, due to the PAR research characteristics of collaboration, iteration, and openness to results, an understanding of user needs and wishes was obtained. Based on the presented results and implications, the project team will proceed to implement a process for successfully developing the technical solution while also deepening the understanding of crucial aspects in participatory research.

References

1. Suárez-Figueroa, M.C., Diab, I., Ruckhaus, E., Cano, I.: First steps in the development of a support application for easy-to-read adaptation. Univ. Access Inf. Soc. **23**, 365–377 (2024). https://doi.org/10.1007/s10209-022-00946-z
2. Mohamed, Y.A., Khanan, A., Bashir, M., Mohamed, A.H.H.M., Adiel, M.A.E., Elsadig, M.A.: The impact of artificial intelligence on language translation: a review. IEEE Access. **12**, 25553–25579 (2024). https://doi.org/10.1109/ACCESS.2024.3366802
3. Lyu, Q., et al.: Translating radiology reports into plain language using ChatGPT and GPT-4 with prompt learning: results, limitations, and potential. Vis. Comput. Ind. Biomed. Art. **6**, 9 (2023). https://doi.org/10.1186/s42492-023-00136-5
4. Ayre, J., et al.: New frontiers in health literacy: using ChatGPT to simplify health information for people in the community. J GEN INTERN MED. **39**, 573–577 (2024). https://doi.org/10.1007/s11606-023-08469-w
5. Moons, P., Van Bulck, L.: Using ChatGPT and google bard to improve the readability of written patient information: a proof of concept. Eur. J. Cardiovasc. Nurs. **23**, 122–126 (2024). https://doi.org/10.1093/eurjcn/zvad087
6. Mondal, H., et al.: Assessing the capability of large language model chatbots in generating plain language summaries. Cureus. (2025). https://doi.org/10.7759/cureus.80976
7. Nasra, M., et al.: Can artificial intelligence improve patient educational material readability? a systematic review and narrative synthesis. Intern. Med. J. **55**, 20–34 (2025). https://doi.org/10.1111/imj.16607
8. De Freitas, M.P., Piai, V.A., Farias, R.H., Fernandes, A.M.R., De Moraes Rossetto, A.G., Leithardt, V.R.Q.: Artificial intelligence of things applied to assistive technology: a systematic literature review. Sensors. **22**, 8531 (2022). https://doi.org/10.3390/s22218531
9. Subashini, P., Krishnaveni, M.: Artificial intelligence-based assistive technology. In: Artificial Intelligence Theory, Models, and Applications. Auerbach Publications (2021)
10. Smith, E.M., Graham, D., Morgan, C., MacLachlan, M.: Artificial intelligence and assistive technology: risks, rewards, challenges, and opportunities. Assist. Technol. **35**, 375–377 (2023). https://doi.org/10.1080/10400435.2023.2259247
11. Bowker, L.: Plain language in the age of neural machine translation: an opportunity for translators. In: Peng, Y., Huang, H., Li, D. (eds.) New Advances in Translation Technology. pp. 163–180. Springer Nature Singapore, Singapore (2024)
12. Almufareh, M.F., Tehsin, S., Humayun, M., Kausar, S.: Intellectual disability and technology: an artificial intelligence perspective and framework. J. Disabil. Res. **2** (2023). https://doi.org/10.57197/JDR-2023-0055
13. El Morr, C., Kundi, B., Mobeen, F., Taleghani, S., El-Lahib, Y., Gorman, R.: AI and disability: a systematic scoping review. Heal. Inf. J. **30**, 14604582241285744 (2024). https://doi.org/10.1177/14604582241285743
14. Shuford, J.: Contribution of artificial intelligence in improving accessibility for individuals with disabilities. JKLST. **2**, 421–433 (2023). https://doi.org/10.60087/jklst.vol2.n2.p433
15. Dirks, S.: Empowering instead of hindering – challenges in participatory development of cognitively accessible software. In: Antona, M., Stephanidis, C. (eds.) Universal Access in Human-Computer Interaction. Theory, Methods and Tools. pp. 28–38. Springer International Publishing, Cham (2019)
16. Cornish, F.,et al.: Participatory action research. Nat. Rev. Methods Primers. **3**, 34 (2023). https://doi.org/10.1038/s43586-023-00214-1
17. Bethmann, A., Hilgenböcker, E., Wright, M.: Partizipative Qualitätsentwicklung in der Prävention und Gesundheitsförderung. In: Tiemann, M., Mohokum, M. (eds.) Prävention und Gesundheitsförderung, pp. 1–13. Springer, Berlin Heidelberg, Berlin, Heidelberg (2019)

18. Lueg, MC., Baumann, L., Dirks, S., Pelka, B., York, J.: Photovoice method for participatory software development – Insights from the KARLA project. In: Stephanidis, C., Antona, M., Ntoa, S., Salvendy, G. (eds) HCI International 2025 Posters. HCII 2025. Communications in Computer and Information Science, vol 2524 Springer, Cham (2025). https://doi.org/10.1007/978-3-031-94156-6_31
19. Wilson, C.: Brainstorming and beyond: a user-centered design method. Newnes (2013)
20. Lyman, F.T.: The responsive classroom discussion: the inclusion of all students. Mainstreaming Digest. **109**, 113 (1981)
21. Zimmerman, J., Forlizzi, J.: Speed dating: providing a menu of possible futures. She Ji. J. Design Econom. Innov. **3**, 30–50 (2017). https://doi.org/10.1016/j.sheji.2017.08.003
22. Balcazar, F.E., Keys, C.B., Kaplan, D.L., Suarez-Balcazar, Y.: Participatory action research and people with disabilities: principles and challenges. Canadian J. Rehabilitation. **12** (1998)
23. Tielman, M.L., Suárez-Figueroa, M.C., Jönsson, A., Neerincx, M.A., Cavalcante Siebert, L.: Explainable AI for all - A roadmap for inclusive XAI for people with cognitive disabilities. Technol. Soc. **79**, 102685 (2024). https://doi.org/10.1016/j.techsoc.2024.102685
24. Kasberg, A., Müller, P., Markert, C., Bär, G.: Systematisierung von Methoden Partizipativer Forschung. Bundesgesundheitsbl. **64**, 146–155 (2021). https://doi.org/10.1007/s00103-020-03267-9
25. Sattler, C., et al.: Participatory research in times of COVID-19 and beyond: adjusting your methodological toolkits. One Earth. **5**, 62–73 (2022). https://doi.org/10.1016/j.oneear.2021.12.006
26. Freyer, N., Kempt, H., Klöser, L.: Easy-read and large language models: on the ethical dimensions of LLM-based text simplification. Ethics Inf. Technol. **26**, 50 (2024). https://doi.org/10.1007/s10676-024-09792-4
27. Lang, D.J., et al.: Transdisciplinary research in sustainability science: practice, principles, and challenges. Sustain. Sci. **7**, 25–43 (2012). https://doi.org/10.1007/s11625-011-0149-x

Online Conversion Service of PDF/PowerPoint into Accessible Forms for Dyslexic Students

Katsuhito Yamaguchi[1](✉), Toshihiko Komada[1], and Masakazu Suzuki[2]

[1] Nihon University, Funabashi, Chiba 274-8501, Japan
eugene@sciaccess.net, komada.toshihiko@nihon-u.ac.jp
[2] NPO: Science Accessibility Net, Fukuoka 814-0001, Japan
msuzuki@sciaccess.net

Abstract. Incorporating new technologies into our OCR software for STEM, "InftyReader Online" (IRO) is developed, which is a web site to provide dyslexic students with an innovative online service to convert PDF/PowerPoint contents including STEM into various accessible forms. In IRO, Google Cloud Vision API is adopted as an OCR engine, which enables IRO to recognize multilingual STEM contents covering a wide range of languages. As a new type of accessible books for dyslexic students, "Fixed-layout DAISY" is also developed. Even if the page layout of the original document is very complicated, almost automatic conversion into Fixed-layout DAISY can be carried out. It should be a powerful/effective tool to realize reasonable accommodation in school examinations.

Keywords: Dyslexia · DAISY · PDF/PowerPoint · Online conversion

1 Introduction

Educational e-materials are usually provided in PDF or PowerPoint (PPT) format, but in many cases, it is hard for print-disabled students to access STEM (science, technology, engineering and mathematics) contents in these formats. Mathematical formulas, tables or other technical notations are usually inaccessible as they stand. Producing accessible alternative versions requires you a lot of time and efforts.

In Japan, "The Japanese Society for Rehabilitation of Persons with Disabilities (JSRPD)" [1] has been providing print-disabled students with e-textbooks in multimedia DAISY or accessible EPUB3 [2] (audio-embedded DAISY/EPUB3 as media overlay) since 2008. It now organizes 23 volunteer groups/organizations and produces around 300 titles of daisy textbooks for elementary and junior-high school (the ages of compulsory attendance at school in Japan). Now, around 30,000 print-disabled students (mostly ones with developmental reading disorder, namely, dyslexia) use those accessible textbooks. Unfortunately, however,

K. Mavrou and P. Encarnação (Eds.): AAATE 2025, *Technology for Inclusion and Participation for All: Recent Achievements and Future Directions*, pp. 33–40, 2025.
https://doi.org/10.1007/978-3-032-01628-7_5

it is still hard for many other dyslexic people in Japan to obtain accessible books. For instance, concerning senior-high school and higher education, no official organizations including JSRPD do not provide the dyslexic students with DAISY-version textbooks [3], and many students who do need accessible textbooks due to the severe dyslexia decide to give up the idea of going to senior-high school or college. Furthermore, even in elementary and junior-high school, it is hard for them to obtain accessible educational materials other than textbooks such as examination papers.

In terms of students in senior-high school and higher education, converting inaccessible contents into accessible ones for themselves (namely, self-supporting production of accessible contents) is essentially important, we believe. To give a practical/effective solution to that, we have developed "InftyReader Online" (IRO) under the Nippon Foundation's subsidy [4] that is contributed to welfare-aimed innovative projects. It is aimed at providing dyslexic people with an innovative service to convert their-uploaded PDF/PPT STEM document into various accessible forms including multimedia DAISY. In addition, to realize almost automatic conversion, we have also developed a new type of DAISY named "Fixed-layout DAISY".

Here, at first, we briefly review the current situation of online services to provide print-disabled people with accessible books and discuss some remaining problems. Next, our methodology to solve them practically is given. Finally, a preliminary result to evaluate our methodology is discussed.

2 State of the Art: Accessible Books for Dyslexic People

There are many online libraries for print-disabled people in all over the world. For instance, in USA, print-disabled students can read various types of accessible books, more-than-1,200,000 titles, in the possession of BookShare free of charge [5]. Similar libraries are also operated in many other countries. RoboBraille Projects offers an online service to convert inaccessible documents in several local languages into various accessible formats such as braille, DAISY, etc. [6]. Recently, reading environment for print-disabled people is definitely/remarkably improved.

As is well known, DAISY (Digital Accessible Information System) was originally designed for visually disabled people, and subsequently, dyslexic people have become to use [2]. As far as the dyslexic people are concerned, they can listen to contents with highlighting texts in multimedia DAISY, and it should be definitely the best solution as accessible books for them. However, DAISY is designed so that various types of print-disabled people can read the same contents, and some features of multimedia DAISY such as reflow, automatic aloud-reading are not absolutely necessary requirements for the dyslexic people. In addition, most popular DAISY/EPUB3 players in the world do not have a function to read out text in a picture. Instead, they would read out an alternative text for the picture if it were available. We also would like to point out that the picture representation would be rather better for dyslexic people. They can look, and the picture representation should be easier for them to understand if texts in the pictures were read out. They should prefer that to the alternative text.

3 Methodology (1): Fixed-Layout DAISY

To meet this demand, we have introduced a layer structure in DAISY. A picture on a page is represented in multiple layers, the second layer of which is an image that appears in the completely same layout as is on the original print document. On the transparent front layer, text-related information such as character code and coordinate on the page, font type, bounding box (rectangular area being circumscribed on a word), etc. is stored, and by making use of it, a DAISY (EPUB3) player can read out texts in the picture with a TTS (text-to-speech) voice together with highlighting them. This type of EPUB3 (DAISY) including multi-layer pictures is authenticated by an EPUB3 validator [7], and it should be regarded as a kind of accessible EPUB3. Actually, some EPUB3 (DAISY) players in Japan can read out texts in a multi-layer picture. JSRPD has already adopted this type of DAISY/EPUB3 textbooks and provided print-disabled students.

The layer structure allows us to develop a new type of accessible e-books named "Fixed-layout DAISY." In it, the whole page is treated as a multi-layer picture, the second layer of which has the same form as the original print document (the page image). A DAISY (EPUB3) player can read out any texts on the transparent front layer together with highlighting them as same as a multi-layer picture located locally on a page (See Fig. 1 and Fig. 2).

A main difference of Fixed-layout DAISY from ordinary DAISY is that it does not have the reflow function. The page layout is always kept as same as the original if it is magnified. It does not have information either in which order texts on the page should be read out. These features look disadvantage for print-disabled students to read the content. However, it should be pointed out that Fixed-layout DAISY has a certain merit. The textbook layout recently tends to be very complicated. Its page consists of not only Body texts but also many sub-texts such as foot notes or other marginal notes, balloons, figure captions, etc. Producing ordinary reflow-type DAISY requires you a lot of authoring process so that a DAISY player can read it out properly; that is, you have to arrange various text blocks on a page in a proper reading order. On the other hand, in Fixed-layout DAISY, such authoring process is not necessary, and it can be produced at extremely less cost than the reflow-type DAISY. Thus, as will be discussed later, Fixed-layout DAISY can be produced almost automatically from original PDF/PPT.

In Fixed-layout DAISY, its content cannot be read out automatically by a player. You need to click a position on a page where you want to start reading; then, the text is read out up to the end of the text block. Certainly, this mechanism may not work for visually disabled people; however, dyslexic people can look and click a place where they want to read. Fixed-layout DAISY should work for dyslexic people, we believe.

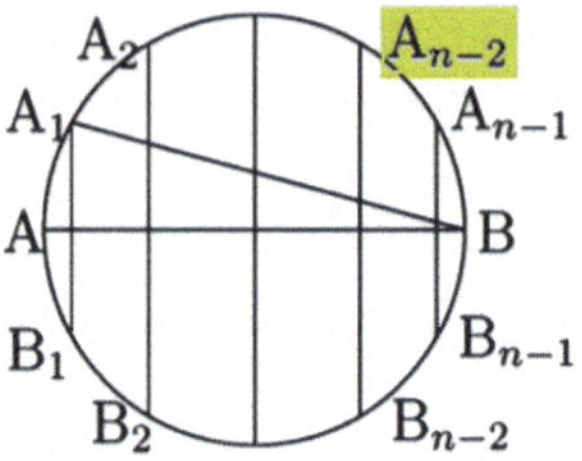

Fig. 1. Sample of Fixed-Layout DAISY (a)

Let us denote by S the surface area of a sphere of radius r, and by T the surface area of the right cylinder circumscribed on the sphere, that is, the base circle of the right cylinder is of radius r and its height is $2r$.

If we accept the result: "the surface area of the sphere is equal to the area of its great circle multiplied by four,"

$$S = 4\pi r^2.$$

On the other hand, since the upper-base area and the the lower-base area are both equal to πr^2, and the lateral area is equal to $2\pi r \times 2r$, we have

$$T = \pi r^2 + \pi r^2 + 2\pi r \times 2r = 6\pi r^2.$$

Fig. 2. Sample of Fixed-Layout DAISY (b)

In particular, Fixed-layout DAISY is extremely compatible with STEM contents since they include a lot of 2-dimentional technical notations such as pictures, figures, charts, tables, etc. In addition, Fixed-layout DAISY would rather be appropriate to produce the accessible version of a book other than textbooks such as comics, picture books, (Pictorial) maps, etc. An examination paper should also be compatible with Fixed-layout DAISY. In the examination paper, quick random access is more important than textbooks, and Fixed-layout DAISY enables that. It is quite obvious that a content in Fixed-layout DAISY is clearly identical with the original. It is also very important in examination papers.

4 Methodology (2): InftyReader Online

4.1 InftyReader

As is well known, ordinary OCR (optical character recognition) software cannot recognize properly technical notations such as mathematical formulas. To realize automatic conversion of STEM contents into accessible forms, we do need a special tool: "an OCR system for STEM" (or simply, mathematics OCR).

The end of 1990s, we started studying mathematics OCR and established its basic method in 2003 [8]. the protype version of our OCR software for STEM named "InftyReader" was released in 2004. We founded Non-Profit Organization: "Science Accessibility Net (sAccessNet)" and released the first commercial version [9] in 2007. For a long time, InftyReader was only one application in the world, which could recognize STEM documents properly. Recently, Mathpix is also available as mathematics OCR to convert printed STEM contents into editable forms [10], but it is not designed for accessibility purposes; its conversion result is not necessarily accessible for dyslexic people.

In its first version, InftyReader1, only clearly scanned page images were OCR targets. Only English and Japanese were accepted as a body text. The second version, InftyReader2 became able to recognize PDF. Some other languages such as Italian, Czech were added to an acceptable-language list. In the current version, InftyReader3, by introducing a PDF parser, recognition rate for "e-born PDF" (originally produced from digital contents such as Microsoft Word) has been remarkably improved. In e-born PDF, correct text-related information including character code is embedded, and many local languages represented in Unicode can be recognized [11].

However, the following points were still our remaining tasks that should be worked on.

- InftyReader is Windows software, but many dyslexic students do not own a Windows computer. Thus, there have been many requests to release other OS versions.
- How do we improve recognition accuracy of image (not e-born) PDF?
- How do we treat a local language represented in image or non-Unicode? Etc.

4.2 InftyReader Online

To carry out these tasks, we have developed "InftyReader Online" (IRO) that is a web site to provide you with an innovative online service to convert contents in PDF or PowerPoint including STEM into various accessible forms [12]. It works on a popular browser, and you can use it with any device other than a Windows computer.

In IRO, we newly adopt "Google Cloud Vision API" (GCVA) as an OCR engine [13]. Even if a body text were image or not represented in Unicode, or even if it were a unique local language, GCVA can it properly and give us character Unicode and correct coordinate on a page. It enables IRO to recognize multilingual contents covering extremely wide range of languages even if they were not represented in Unicode. In mathematics recognition, both the Google and our mathematics-OCR engines are used to improve accuracy.

GCVA charges money per processed pages, and we are forced to adopt a new system for charging in IRO. At first, when registering yourself, you can process 300 pages free of charge. We bear the cost for your trial use. The number equal to processed pages is consumed. When they are used up, and if you want to continue using the service, you need to purchase a license to increase convertible number of pages.

PDF can be converted into various different-type accessible forms: LaTeX source, Human-Readable (HR) TeX, Markdown text with AsciiMath, Microsoft Word with math object, ordinary (reflow-type) text-based DAISY/EPUB3 with MathML and Fixed-layout DAISY.

When a PDF file is uploaded to IRO, you are required to choose some settings: "Page Range", "Language", "Math Level", "Output File Formats", etc. Pressing "Start Conversion" button, the PDF file is recognized and converted automatically (See Fig. 3). PowerPoint (PPT) is once changed to PDF and converted into a specified accessible form(s).

Unfortunately, however, we cannot necessarily recommend you to select ordinary (reflow-type) DAISY as "Output File Formats". Automatic conversion should include some errors especially in arranging the order of text blocks. You would have to correct them manually. On the other hand, even if the page layout of the original document were very complicated, automatic conversion into Fixed-layout DAISY could be carried out without very limited recognition errors. Thus, Fixed-Layout DAISY is appropriate for online conversion service. Incidentally, another web site of ours named "Chatty Library" also provides print-disabled people with various other services [14,15] such as playing a DAISY book back with a popular browser, authoring DAISY contents, etc. You can correct conversion errors on it.

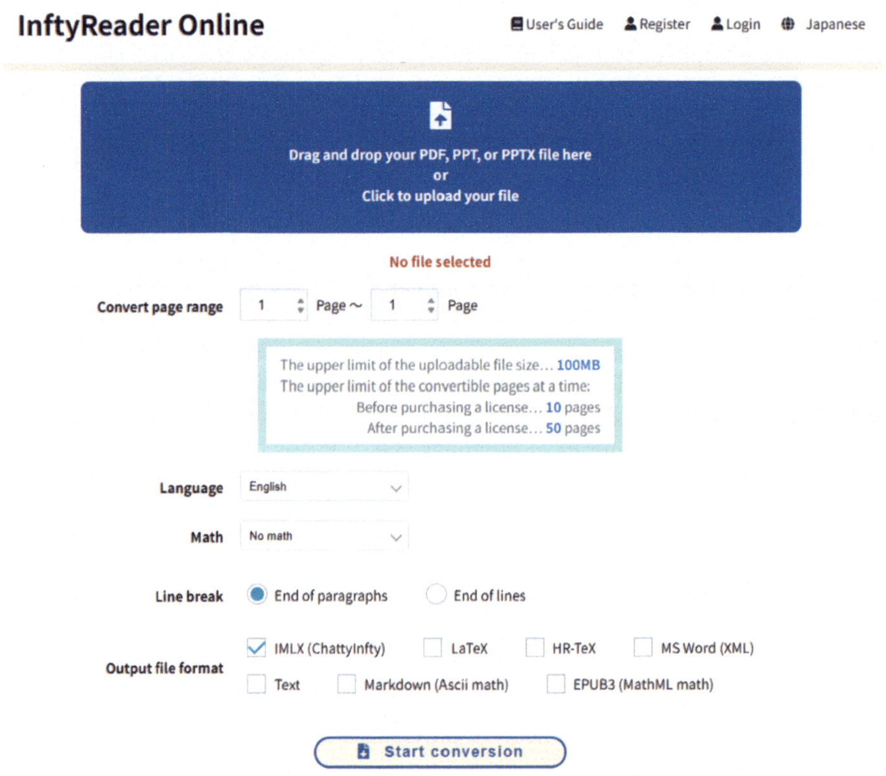

Fig. 3. InftyReader Online top page.

5 Result: Preliminary Evaluation

To see the effectiveness of Fixed-layout DAISY, we conducted a small evaluation test. In 2024, five teachers (one elementary and four junior-high schools in Tochigi Prefecture) took a short course on our web service and tried to produce the Fixed-layout-DAISY version of their examinations. Dyslexic students in their classes had those examinations, and the teachers evaluated whether or not the students did well on them.

We realized that dyslexic students in the elementary school achieve better scores than ordinary (paper-based) examinations, but as far as students in the junior-high schools were concerned, their performance were not improved significantly. In interviews with the teachers, they pointed out, to begin with, it seemed to be difficult for the students to understand what were asked in a problem even if its content were read out. The dyslexic student usually has a very poor experience of reading and may not be familiar with "written statements": wording, a vocabulary, etc. Thus, this year, we conduct another evaluation test; students are required to accumulate an experience of reading various contents in Fixed-Layout DAISY before having examinations. The result will be reported in the AAATE2025 conference.

However, we realized that majority of the teachers gave it strong support. Someone said that Fixed-layout DAISY should be very useful for students with slight dyslexia who were in a transition process to a reader of ordinary print books. In addition, they admitted that Fixed-layout DAISY should be easier to introduce as reasonable accommodation for examinations since Fixed-layout DAISY is clearly identical with the original PDF.

6 Conclusion and Future Tasks

InftyReader Online allows dyslexic people to convert inaccessible PDF/PowerPoint STEM contents into various accessible forms for themselves. They can use it on any device other than a Windows computer. Almost automatic conversion into Fixed-Layout DAISY is realized.

Our remaining important task is to develop a method to infer correct reading order. Using AI technologies, we have been repeating one trial and error after another to find a good mechanism to give a correct order to text blocks in a page image; however, unfortunately, not yet. If we found it, automatic conversion of PDF/PowerPoint into ordinary (reflow type) DAISY would be available on InftyReader Online.

Acknowledgments. We greatly appreciate support by the Nippon Foundation to achieve this project.

References

1. Japanese Society for Rehabilitation of Persons with Disabilities (Japanese only). http://www.dinf.ne.jp/doc/daisy/book/daisytext.html
2. DAISY Consortium. http://www.daisy.org/
3. Suzuki, M., Yamaguchi, K.: On support for dyslexic students in senior-high school and higher education. Stud. Health Technol. Inform. **306**, 137–143 (2023). https://doi.org/10.3233/SHTI230608
4. Nippon Foundation. https://www.nippon-foundation.or.jp/en/
5. BookShare. http://www.bookshare.org/
6. RoboBraille Projects. https://www.robobraille.org/
7. pagina EPUB-Checker. https://pagina.gmbh/startseite/leistungen/publishing-softwareloesungen/epub-checker/
8. Suzuki, M., Tamari, F., Fukuda, R., Uchida, S., Kanahori, T.: Infty – an integrated OCR system for mathematical documents –. In: Proceedings of the ACA Symposium on Document Engineering 2003, Grenoble, pp. 95–104 (2003)
9. Science Accessibility Net. http://www.sciaccess.net/en/
10. Mathpix. https://mathpix.com/
11. Suzuki, M., Yamaguchi, K.: Conversion of multi-lingual STEM documents in E-born PDF into various accessible E-formats. In: Miesenberger, K., Kouroupetroglou, G., Mavrou, K., Manduchi, R., Covarrubias Rodriguez, M., Penáz, P. (eds.) ICCHP-AAATE 2022. LNCS, vol. 13341, pp. 7–14. Springer, Cham (2022). https://doi.org/10.1007/978-3-031-08648-9_2
12. InftyReader Online. https://inftyreader.online/
13. Google Cloud Vision API. https://cloud.google.com/vision/
14. Chatty Library. http://www.chattylib.com/
15. Yamaguchi, K., Komada, T., Suzuki, M.: ChattyBox: online accessibility service using fixed-layout DAISY. In: Miesenberger, K., Peňáz, P., Kobayashi, M. (eds.) ICCHP 2024. LNCS, vol. 1475, pp. 167–174. Springer, Cham (2024). https://doi.org/10.1007/978-3-031-62846-7_20

UniNotesXR - Accessible Collaboration in Extended Reality a Spatial Whiteboard with 3D Audio User Guidance for the Visually Impaired

Elias Merzhäuser(✉), Gottfried Zimmermann, and Ansgar Gerlicher

Hochschule der Medie Stuttgart, Nobelstraße 10, 70569 Stuttgart, Germany
elias@merzhaeuser.com, gzimmermann@acm.org,
gerlicher@hdm-stuttgart.de

Abstract. The integration of 3D audio and visual accessibility features in Extended Reality (XR) whiteboards offers new possibilities for inclusive collaboration, especially for individuals with visual impairments. This research explores how these technologies can enhance user guidance and accessibility, utilizing a prototype developed through a participatory design process. Leveraging "UniNotesXR", a spatial XR whiteboard equipped with 3D audio navigation, the study evaluates usability and accessibility through empirical user testing with participants with visual impairments. The theoretical foundation includes insights into accessibility, XR technologies, and assistive tools such as earcons, screen readers, and visual enhancements like zoom and contrast settings. Methodologically an interactive prototype was employed to simulate collaboration scenarios. Supported by usability tests, interviews, and standardized questionnaires (UEQ and NASA-TLX). Results indicate that the combination of spatial audio cues and customizable visual aids significantly improves orientation and interaction on XR whiteboards, mitigating key barriers faced by users with visual impairments. However, the cognitive load imposed by excessive information and limitations of current hardware and software regarding feedback precision remain challenging. The findings provide actionable recommendations for optimizing accessibility in XR environments: the earcon design, 3D audio precision and the integration of visual feedback. Furthermore, results underscore the need for further research into multisensory feedback modalities, sonification mapping strategies for inclusive design and social aspects of accessible XR whiteboards.

Keywords: 3D audio · visual impairment · whiteboard · xr · accessibility

1 Introduction

Whiteboards are a popular platform for collaboration but continue to have accessibility deficits [1]. The 2.2 billion people with visual impairments, especially those 300 million with moderate to severe visual impairments, face significant barriers to using digital

K. Mavrou and P. Encarnação (Eds.): AAATE 2025, *Technology for Inclusion and Participation for All: Recent Achievements and Future Directions*, pp. 41–51, 2025.
https://doi.org/10.1007/978-3-032-01628-7_6

tools [2]. With an ageing global population, this proportion will continue to increase [3]. Digital accessibility is therefore a key factor for the equality and inclusion of people with disabilities in education and the workspace [4, 5].

Extended reality (XR) offers great potential for reducing such barriers and creating new opportunities for collaboration. However, the design of accessible XR solutions requires specific adaptations to meet the needs of users with visual impairments [6].

The aim of this thesis is to investigate how 3D audio in combination with visual accessibility features can improve user guidance for the visually impaired on XR whiteboards. While 3D audio as an assistive technology (AT) offers great potential for spatial orientation, there are only a few studies to date that systematically evaluate its use in combination with visual functions in XR environments related to whiteboards [7]. The focus of this work is the development and evaluation of a prototype that uses 3D auditory cues as a central navigation aid and uses visual functions as a supplement. The prototype was developed using a participatory design approach in collaboration with a visually impaired researcher.

The research questions aim to assess the effectiveness of 3D audio, evaluate its acceptance and usability, and investigate the synergies with visual features. The results of this research will help to improve the accessibility and user experience (UX) of XR whiteboards. This has the potential to significantly increase the participation of people with visual impairments in collaborative environments.

2 Methodology

The work follows the Research through Design (RtD) approach, in which design methods are used to explore research questions and develop solutions. RtD enables the systematic creation of artifacts that can be empirically evaluated [8–10]. In combination, a participatory design process was used to ensure that the needs of visually impaired users were directly incorporated into the development.

The research was conducted in two main phases. In the pre-study, scientific literature and technical possibilities were reviewed to define requirements for the prototype. Methods such as heuristic evaluation of existing XR technologies and analysis of field reports were used to formulate assumptions and derive basic design requirements from [7, 11, 12]. The developed prototype, UniNotesXR, is an XR whiteboard application that uses 3D audio as a central navigation aid and is complemented by visual accessibility features of the Apple Vision Pro (AVP) headset such as zoom, text size and contrast adjustment [13].

In the main study, the prototype was tested in an empirical evaluation with eight test subjects with moderate to severe visual impairments, but all with some degree of light perception. The focus was on investigating the user-friendliness and effectiveness of 3D audio in realistic application scenarios. The participants were familiarized in advance with the basic functions of the AVP to create an equivalent starting point. During the study, they completed various tasks in the whiteboard application focusing on the use of 3D audio and visual functions. The participants were given an introduction in the general interaction patterns the accessibility features before the performing the tasks. Apart from the visionOS system intrinsic lens calibration and the option for users to fine tune the accessibility features, a non-personalized setup was used for 3D audio (e.g. HRTF settings).

The data was collected using a multi-stage process. First, a usability test was carried out in which the test subjects completed tasks within the UniNotesXR environment. Participants ability to localize 3D audio sources was tested in the first task. In the second task the creation and repositioning of notes was evaluated. A collaboration scenario including comments was simulated in the third task (see Fig. 2). The usability metrics of success rate, error rate and user satisfaction were measured [14]. The User Experience Questionnaire (UEQ), a proven instrument for evaluating the pragmatic and hedonic quality of user experiences [15, 16], was used to record the subjective experience. To test the workload, the NASA-TLX (RAW) was also completed [17, 18]. In addition, structured interviews were conducted to record the participants' perspectives on the prototype in detail.

The quantitative analysis included the statistical evaluation of the success and error rates as well as the results of the UEQ and NASA-TLX. Qualitative content analysis according to Kuckartz [19] focused on identifying recurring themes and patterns in the interviews, taking into consideration the context of the responses to gain relevant insights. By combining these methods, a comprehensive understanding of the user experience with UniNotesXR was gained.

3 Pre-study Results

3.1 Acoustic Assistive Technology in XR

Screen readers such as VoiceOver are considered essential assistive technologies and offer comprehensive functions for exploring objects and environments, possible benefits in XR have been studied before [20–22]. Unity enables the integration of such technologies in XR applications, while headsets such as the Meta Quest currently do not offer integrated screen readers on default [23], the AVP being the exception. Audio descriptions and voice controls extend accessibility, although their availability and functionality vary greatly depending on the platform.

3.2 Visual Assistive Technology in XR

Visual adaptations such as enlarged text sizes and high-contrast displays are among the most important accessibility features [20, 24, 25]. Both the AVP and the Meta Quest support these options, although the configuration options for the AVP are more comprehensive, most importantly the zoom controller. Zoom and contrast adjustments facilitate the use of controls, while functions for stabilizing eye movements (e.g. for nystagmus) offer improved visual orientation.

3.3 Accessible 3D Audio Feedback

3D audio enables precise spatial orientation through technologies such as Head-Related Transfer Functions (HRTF) and dynamic sound simulations [26]. Studies show that 3D audio can help with navigation and localization in XR environments [27]. Challenges lie in the customization of HRTFs and calibration of sound signals for better vertical localization [10, 28]. Research confirms that 3D audio can enhance the usability of XR applications, especially by supporting orientation and navigation. Acoustic cues can improve both interaction with objects and social presence in collaborative environments [7, 29]. The preliminary study shows that 3D audio in combination with visual and other assistive technologies offers significant potential for the accessibility of XR applications. The challenges and opportunities identified will be specifically addressed in the further course of the work to develop a comprehensive and inclusive system.

4 Prototype

The UniNotesXR concept uses 3D auditory cues, so called "Earcons" as a central component to support user guidance for people with visual impairments in extended reality environments. These short, non-verbal audio signals are used specifically to convey information efficiently and facilitate orientation in the three-dimensional environment. Earcons have been proven in research to be an effective method for conveying information in 3D environments. Compared to verbal audio signals, they are less invasive and can reduce the cognitive load on users [21, 29, 30]. For this reason, implementation with VoiceOver was refrained from, as the function only reads out descriptions and does not allow interfaces for the integration of user-defined audio.

Earcons have already proven their efficiency, particularly in navigation and guidance systems for people with visual impairments [31]. Blattner et al. [32] define earcons as "auditory signals in human-computer interaction that convey information such as states or actions". In the context of UniNotesXR, they are used to indicate various objects and actions. Longer acoustic signals mark workspaces and task areas to emphasize their importance. Short signals mark new notes to make them easier to find. This differentiation enables a clear separation of object types and facilitates navigation in the workspace. The use of earcons enables users to quickly distinguish between objects and to detect their position acoustically. Studies show that short, concise audio signals are particularly effective, as they do not demand too much of the user's attention and at the same time enable precise orientation [33, 34]. Notes are the central elements of that users interact with and contribute to the board (see Fig. 1.). Work areas structure the space and serve as orientation points for tasks (see Fig. 1). They enable better allocation and navigation in three-dimensional environments. A chronologically sorted list of notes facilitates localization and interaction with elements (see Fig. 1). This concept, inspired by the whiteboard tools "Miro" and "Mural", reduces complexity for users. In team collaborations, personalized earcons can be used to identify team members. This supports the social aspect of collaboration and facilitates orientation within the group [7, 29]. Comments and notifications are facilitating the communication between board members (see Fig. 2).

The concept was first drafted in the prototyping tool Figma (see Fig. 1). Later the prototype was implemented with visionOS including the functionality required for fulfilling the tasks. Finally, the app was installed on the AVP for the user test (see Fig. 2).

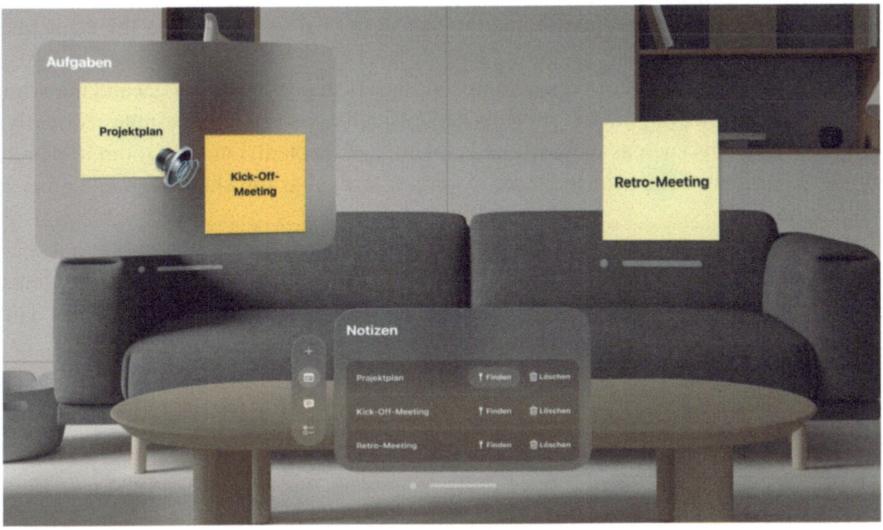

Fig. 1. Figma XR-Prototype

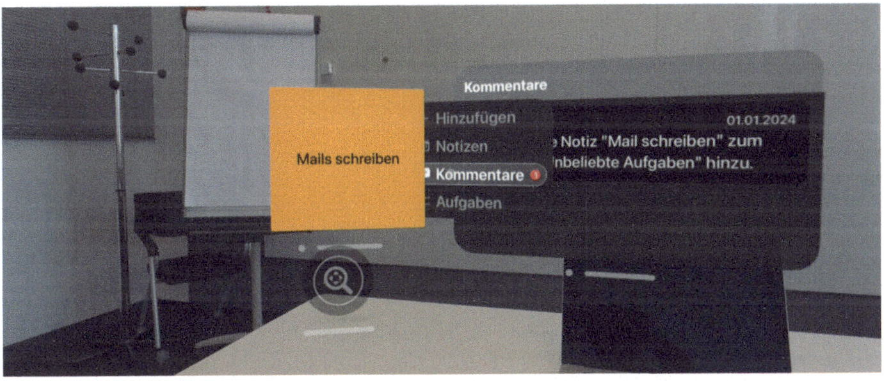

Fig. 2. Task 3 performed with the visionOS-Prototype on Apple Vision Pro

5 Main Study Results

5.1 Usability Test

Across all tasks, an average success rate of 0.7 and an error rate of 0.9 were measured (see Fig. 2). Recurring challenges included inadequate visual display, imprecise audio feedback and technical issues. The average user satisfaction was 3.5.

The first task investigated the perception of 3D audio feedback position and the visibility of the spherical visual indicator. The test analysis showed mixed results. The test subjects had difficulties with the vertical localization of the sound sources, while the horizontal localization was more precise. Visual limitations, such as the lack of visibility of the sphere on a bright background, made orientation more difficult. Most participants required several attempts or assistance. The success rate was 0.6, while the error rate was 1.0.

In the second task, performance improved: the success rate rose to 0.9 and the error rate fell to 0.7. However, the test subjects had difficulty positioning new notes correctly. Technical problems such as the keyboard closing unexpectedly and the zoom impaired the user experience. The acoustic signals were described as too short to enable precise localization.

In the third task, the test subjects showed similar challenges when adding notes as in the previous tasks. Vertical positioning remained a weakness, and technical problems such as the cursor disappearing or pixelated content due to zooming occurred again. The success rate was 0.8 and the error rate 1.2.

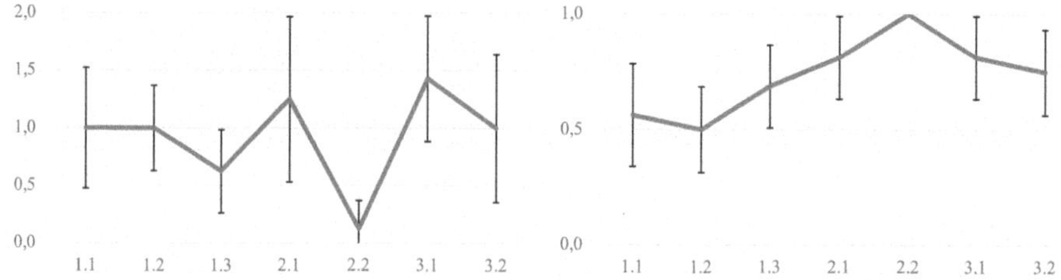

Fig. 3. Success rate (left) & error rate (right) - task 1–3 (confidence level 95%)

5.2 UEQ and NASA-TLX

The prototype generally received positive ratings with the overall attractiveness rated as excellent with an average score of 1.86 (see Fig. 3, left). The dimensions of transparency (1.25) and efficiency (1.07) were rated as above average. Weaknesses were seen in controllability (0.86), which was rated as below average.

The workload was rated as low with an average score of 25.83 out of 100 (see Fig. 3, right). Mental load was the highest at 43.13, followed by frustration (40.63). Physical load achieved the lowest value at 28.57 (Fig. 4).

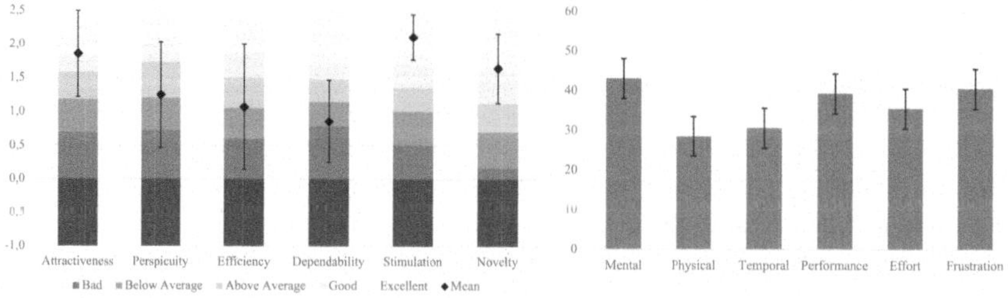

Fig. 4. UEQ (left) & NASA-TLX (RAW) (right)

5.3 Interview

The horizontal localization of the sound sources was rated positively, while the vertical localization was perceived as imprecise. The sound duration was described as too short, and parallel ambient noise made it difficult to perceive the signals. According to the participants, there is room for improvement in the coordination of the audio signals.

The text size and contrast were perceived as predominantly positive. Problems were encountered with the positioning of elements and the legibility of angled windows. The zoom function was rated as cumbersome to use and led to pixelated content.

6 Discussion

The results of the user test, the questionnaires and the interview were analyzed in the context of the research questions. The methods used provided credible insights, although the small sample size does limit generalizability.

The different tasks showed varying success rates and error rates, depending on complexity and visual support. Tasks without visual cues led to higher error rates, especially in vertical localization. While horizontal localization was rated as accurate with an error rate of only 0.1, the vertical axis showed clear weaknesses with an error rate of 0.7. This discrepancy underlines the need for improved audio and visual support. The learning curve played an important role, as subjects with more XR and whiteboard experience achieved better results.

The acoustic aids were generally rated positively (3.9/5), especially the horizontal localization. There were weaknesses in vertical perception, as the audio cues for this were insufficient. Quantitative data illustrates the deficits: the perception of height information remained problematic. The results confirm previous studies that found insufficient precision in generic HRTFs and vertical audio information [35]. The combination of audio and visual features was perceived as effective, which is also confirmed in other studies [29, 36, 37]. However, acoustic aids are only of limited use as a primary accessibility feature for people with severe visual impairment. Improvements in sonification strategies and customized audio output could address these limitations [38, 39]. Sound parameters like the pitch or temporal frequency could be used to indicate the elevation of audio sources.

The visual support was rated as good overall (4/5), similar to the earcons. Functions such as enlarged text sizes (4.6/5) and contrast adjustments (4.5/5) were positively emphasized. The two-handed zoom was praised as intuitive but had technical issues that led to mixed feedback. Fixed elements such as notes did not always meet contrast and font size requirements, which created significant barriers in XR environments. This also partially violates XR accessibility guidelines [40, 41]. Dynamic drop shadows or borders of free-standing 3D objects could provide better contrasts [42].

Although visual and acoustic signaling of comments was available, comments were often overlooked. Visual cues were too unobtrusive, and acoustic signals were drowned out by overlapping noise. The observations point to the need for clearer signaling.

The user interface was rated as user-friendly and easy to understand (4.3/5). The positioning of notes worked well, but the disappearance of the cursor and the obscuring of elements impaired the user experience. Technical problems such as unreliable keystrokes and inconsistent gesture recognition showed potential for improvement. The occlusion of windows remains a relevant problem in XR environments.

The high scores in the UEQ for attractiveness and stimulation reflect the novelty and technical innovation of the prototype. At the same time, low scores for controllability show the need for improved usability. The results of the NASA-TLX revealed a high mental load, which can be attributed to conceptual imperfections, technical deficits and the complexity of the tasks. This highlights the need for more intuitive operating concepts and optimization of accessibility functions.

7 Limitations

The results of the small sample of eight participants are not statistically representative and are only transferable to a limited extent. In addition to visual impairment, behaviors could also have been influenced by affinity for technology, cognitive abilities or the form of the day. The full potential of the 3D audio technology of the Apple Vision Pro could not be exploited without individual HRTFs. The tech-savvy sample further limits generalizability. Technical limitations, such as the lack of a screen reader with integrated earcons, impaired the realism of the test. The limited literature on XR whiteboards and 3D audio was challenging, which is why additional reports from the target group were analyzed. The prototype was only designed for basic navigation mechanisms, not for comprehensive productivity and collaboration scenarios.

8 Conclusion

Quantitative and qualitative data collected in this study show that 3D audio earcons are a valuable addition to user guidance in XR whiteboards by facilitating orientation, especially on the horizontal plane. However, precision in vertical localization remains a challenge, limiting the exclusive use of 3D audio as an accessibility feature. In combination with visual functions such as dynamic text size, contrast settings and zoom, 3D audio offers valuable support, especially for users with residual vision. The technology is perceived positively but needs improvements in the earcon design, 3D audio precision and integration with visual feedback. The test subjects generally rated the system positively and saw potential for use in team meetings. Criticism was voiced about the current technical maturity, the weight of the headset and the limited practical benefits.

Future research should focus on the further development of 3D audio technologies such as HRTF and AI-supported sonification to make the localization of sound sources more precise and improve user guidance. The integration of adaptive screen readers and standardized accessibility features based on familiar operating systems could increase usability. In addition, alternative zoom concepts and control mechanisms such as two-handed zoom offer potential for optimization. A context-based event chain for XR audio feedback can reduce the cognitive load, by only playing sound in an appropriate situation. Social and communicative aspects should be explored to a greater extent using avatars with spatial audio in virtual teams. Finally, participatory approaches with the visually impaired collaborators are essential in developing practical solutions with adequate usability and user experience.

References

1. Koutny, R., Günther, S., Dhingra, N., Kunz, A., Miesenberger, K., Mühlhäuser, M.: Accessible multimodal tool support for brainstorming meetings. In: Computers Helping People with Special Needs, vol. 12377, K. Miesenberger, R. Manduchi, M. Covarrubias Rodriguez, and P. Peňáz, Eds., in Lecture Notes in Computer Science, vol. 12377. , Cham: Springer International Publishing, pp. 11–20 (2020). https://doi.org/10.1007/978-3-030-58805-2_2
2. World Health Organization. Vision impairment and blindness. https://www.who.int/news-room/fact-sheets/detail/blindness-and-visual-impairment Accessed 10 Dec 2024
3. The International Agency for the Prevention of Blindness. Global Estimates of Vision Loss. The International Agency for the Prevention of Blindness. https://www.iapb.org/learn/vision-atlas/magnitude-and-projections/global/ Accessed 10 Dec 2024
4. Europäisches Parlament. Charta der Grundrechte Der Europäischen Union. https://www.europarl.europa.eu/charter/pdf/text_de.pdf Accessed 16 Jan 2025
5. B. für politische Bildung. Teilhabe und Inklusion. bpb.de. https://www.bpb.de/themen/inklusion-teilhabe/behinderungen/521497/teilhabe-und-inklusion/ Accessed 20 Oct 2024
6. Mucha, H., Hoppe, A., van de Camp, F., Anken, J., Schwarz, T., Müller, K.: InclusiveVR@Work: Inklusive Kooperationsumgebungen für Arbeitskräfte mit und ohne Sehbehinderung mittels Virtueller Realität. Presented at the Mensch und Computer 2023 - Workshopband, GI (2023). https://doi.org/10.18420/muc2023. https://dl.gi.de/handle/20.500.12116/42056 Accessed 24 Sep 2024
7. XR Access. The Value of Inclusively Designed XR Workplace Tools. XR Access. (2025). http://xraccess.org/bcxr-report/ Accessed 19 Jan 2025
8. Gaver, W.: What should we expect from research through design?. In: Proceedings of the SIGCHI Conference on Human Factors in Computing Systems, in CHI '12. New York, NY, USA: Association for Computing Machinery, pp. 937–946 (2012). https://doi.org/10.1145/2207676.2208538
9. Zimmerman, J., Forlizzi, J., Evenson, S.: Research through design as a method for interaction design research in HCI. In: Proceedings of the SIGCHI Conference on Human Factors in Computing Systems, in CHI '07. New York, NY, USA: Association for Computing Machinery, pp. 493–502 (2007). https://doi.org/10.1145/1240624.1240704
10. IxDF. What is Spatial Audio? — updated 2024. The Interaction Design Foundation.. https://www.interaction-design.org/literature/topics/spatial-audio Accessed 29 Dec 2024
11. Nielsen Norman Group. 10 Usability Heuristics for User Interface Design. Nielsen Norman Group. https://www.nngroup.com/articles/ten-usability-heuristics/ Accessed 19 Jan 2025
12. Nielsen Norman Group. 10 Usability Heuristics Applied to Virtual Reality. Nielsen Norman Group. https://www.nngroup.com/articles/usability-heuristics-virtual-reality/ Accessed 23 Jul 2024
13. Apple. Anpassen von Farbe und Kontrast auf der Apple Vision Pro. Apple Support. https://support.apple.com/de-de/guide/apple-vision-pro/tandd53c64c3/visionos Accessed 22 Jan 2025
14. Witte, F.: Metriken für usability-tests. in: metriken für das testreporting: analyse und reporting für wirkungsvolles testmanagement. In: Witte, F., ed., Wiesbaden: Springer Fachmedien, pp. 173–178 (2018). https://doi.org/10.1007/978-3-658-19845-9_25
15. Laugwitz, B., Held, T., Schrepp, M.: Construction and evaluation of a user experience questionnaire. In: HCI and Usability for Education and Work, A. Holzinger, Ed., Berlin, Heidelberg: Springer, pp. 63–76 (2008). https://doi.org/10.1007/978-3-540-89350-9_6
16. Schrepp, M., Hinderks, A., Thomaschewski, J.: Applying the user experience questionnaire (UEQ) in different evaluation scenarios. In: Design, User Experience, and Usability. Theories, Methods, and Tools for Designing the User Experience, Marcus, A., ed., Cham: Springer International Publishing, pp. 383–392 (2014). https://doi.org/10.1007/978-3-319-07668-3_37

17. German UPA, "NASA-TLX | German UPA. Accessed 15 Jan 2025. https://germanupa.de/wissen/fragebogenmatrix/nasa-tlx
18. NASA. TLX @ NASA Ames - Home. https://humansystems.arc.nasa.gov/groups/TLX/ Accessed 15 Jan 2025
19. Kuckartz, U.: Qualitative Inhaltsanalyse. Methoden, Praxis, Computerunterstützung. 4. Auflage. In: Grundlagentexte Methoden. Beltz Juventa, Weinheim; Basel (2018)
20. Hoppe, A.H., Anken, J.K., Schwarz, T., Stiefelhagen, R., van de Camp, F.: CLEVR: A Customizable interactive learning environment for users with low vision in virtual reality. In: Proceedings of the 22nd International ACM SIGACCESS Conference on Computers and Accessibility, in ASSETS '20. New York, NY, USA: Association for Computing Machinery, pp. 1–4 (2020). https://doi.org/10.1145/3373625.3418009
21. Das, M., Piper, A.M., Gergle, D.: Design and evaluation of accessible collaborative writing techniques for people with vision impairments. ACM Trans. Comput.-Hum. Interact. **29**(2), 9:1–9:42, (2022). https://doi.org/10.1145/3480169
22. Petrykowski, M., Berger, P., Hennig, P., Meinel, C.: Digital collaboration with a whiteboard in virtual reality. In: Proceedings of the Future Technologies Conference (FTC) 2018, Arai, K., Bhatia, R., Kapoor, S., eds., Cham: Springer International Publishing, pp. 962–981 (2019). https://doi.org/10.1007/978-3-030-02686-8_72
23. Meta. Learn about accessibility features for Meta Quest. https://www.meta.com/de-de/help/quest/articles/in-vr-experiences/oculus-features/accessibility-features/ Accessed 18 Jan 2025
24. Teófilo, M., Nascimento, J., Santos, J., Albuquerque, Y., Souza, A.L., Nogueira, D.: Bringing basic accessibility features to virtual reality context. In: 2016 IEEE Virtual Reality (VR), pp. 293–294 (2016). https://doi.org/10.1109/VR.2016.7504769
25. Kasowski, J., Johnson, B.A., Neydavood, R., Akkaraju, A., Beyeler, M.: A systematic review of extended reality (XR) for understanding and augmenting vision loss. J. Vis. **23**(5), 5 (2023). https://doi.org/10.1167/jov.23.5.5
26. Edwards, J.: Signal processing supports a new wave of audio research: spatial and immersive audio mimics real-world sound environments [special reports]. IEEE Sign. Process. Mag. **35**(2), 12–15 (2018). https://doi.org/10.1109/MSP.2017.2784881
27. Guezenoc, C., Seguier, R.: HRTF individualization: a survey. In: 145th Audio Engineering Society Convention (2018). https://doi.org/10.17743/aesconv.2018.978-1-942220-25-1
28. Xie, B.: Spatial sound: principles and applications. Boca Raton: CRC Press (2022). https://doi.org/10.1201/9781003081500
29. Ji, T.F., Cochran, B., Zhao, Y.: VRBubble: enhancing peripheral awareness of avatars for people with visual impairments in social virtual reality. In: Proceedings of the 24th International ACM SIGACCESS Conference on Computers and Accessibility, in ASSETS '22. New York, NY, USA: Association for Computing Machinery, pp. 1–17 (2022). https://doi.org/10.1145/3517428.3544821
30. Bazilinskyy, P., van Haarlem, W., Quraishi, H., Berssenbrugge, C., Binda, J., de Winter, J.: Sonifying the location of an object: a comparison of three methods. IFAC-PapersOnLine **49**(19), 531–536 (2016). https://doi.org/10.1016/j.ifacol.2016.10.614
31. Katz, B.F.G., et al.: NAVIG: augmented reality guidance system for the visually impaired. Virtual Reality **16**(4), 253–269 (2012). https://doi.org/10.1007/s10055-012-0213-6
32. Blattner, M.M., Sumikawa, D.A., Greenberg, R.M.: Earcons and icons: their structure and common design principles. Hum.-Comput. Interact. **4**(1), 11–44 (1989). https://doi.org/10.1207/s15327051hci0401_1
33. Betsworth, L., Rajput, N., Srivastava, S., Jones, M.: Audvert: Using spatial audio to gain a sense of place. in Human-Computer Interaction – INTERACT 2013, Kotzé, P., Marsden, G., Lindgaard, G., Wesson, J., Winckler, M., eds., Berlin, Heidelberg: Springer, pp. 455–462 (2013). https://doi.org/10.1007/978-3-642-40498-6_35

34. Torres-Gil, M.A., Casanova-Gonzalez, O., González-Mora, J.: Applications of virtual reality for visually impaired people. WSEAS Transactions on Computers archive (2010). https://www.semanticscholar.org/paper/Applications-of-virtual-reality-for-visually-people-Torres-Gil-Casanova-Gonzalez/f0e8f996e77a6ee10ff6326fc0caaa506c2fc70d Accessed 10 Sep 2024
35. Afonso-Jaco, A., Katz, B.F.G.: Spatial knowledge via auditory information for blind individuals: spatial cognition studies and the use of audio-VR. Sensors **22**(13), Art. no. 13 (2022). https://doi.org/10.3390/s22134794
36. Hu, X., Song, A., Wei, Z., Zeng, H.: StereoPilot: a wearable target location system for blind and visually impaired using spatial audio rendering. IEEE Trans. Neural Syst. Rehabil. Eng. **30**, 1621–1630 (2022). https://doi.org/10.1109/TNSRE.2022.3182661
37. Apavou, F., Bouchara, T., Bourdot, P.: Accessibility of shooting task for blind and visually impaired: a sonification method comparison. In: Proceedings of the 29th International Conference on Auditory Display (ICAD2024), Rensselaer Polytechnic Institute: International Community for Auditory Display, pp. 81–88 (2024). https://doi.org/10.21785/icad2024.012
38. Gao, Z., Wang, H., Feng, G., Lv, H.: Exploring sonification mapping strategies for spatial auditory guidance in immersive virtual environments. ACM Trans. Appl. Percept. **19**(3), 9:1–9:21 (2022). https://doi.org/10.1145/3528171
39. Bujacz, M., et al.: Sound of vision - spatial audio output and sonification approaches. In: Computers Helping People with Special Needs, Miesenberger, K., Bühler, C., Penaz, P., eds., Cham: Springer International Publishing, pp. 202–209 (2016). https://doi.org/10.1007/978-3-319-41267-2_28
40. Koch, S., Ableitner, T., Zimmermann, G.: Comparison of guidelines for the accessible design of augmented reality applications. In: Computers Helping People with Special Needs, Miesenberger, K., Kouroupetroglou, G., Mavrou, K., Manduchi, R., Covarrubias Rodriguez, M., Penáz, P., eds., in Lecture Notes in Computer Science. Cham: Springer International Publishing, pp. 89–98 (2022). https://doi.org/10.1007/978-3-031-08645-8_11
41. Joshue, O.C., Janina, S., Jason, W., Michael, C.: XR User Needs and Requirements. W3C (2021). https://www.w3.org/TR/xaur/ Accessed 26 Jan 2023
42. Valakou, A., Margetis, G., Ntoa, S., Stephanidis, C.: A framework for accessibility in XR environments. In: HCI International 2023 – Late Breaking Posters, vol. 1958, C. Stephanidis, M. Antona, S. Ntoa, and G. Salvendy, Eds., in Communications in Computer and Information Science, vol. 1958. , Cham: Springer Nature Switzerland, pp. 252–263 (2024). https://doi.org/10.1007/978-3-031-49215-0_31

Immersive Learning for Inclusive Workplaces: Evaluating Mixed Reality in Vocational Training for People with Neurodevelopmental Disorders

Laura Wuttke(✉), Sophia Goßheger, and Susanne Dirks

Department of Rehabilitation Sciences, Research Unit of Rehabilitation Technology, TU Dortmund University, Emil-Figge-Str. 50, 44227 Dortmund, Germany
laura.wuttke@tu-dortmund.de

Abstract. As digitization reshapes the labor market, integrating emerging technologies into vocational training for individuals with disabilities becomes crucial. This study explores the use of mixed reality (MR) to enhance learning experiences for neurodiverse trainees, specifically in hairdressing, kitchen and warehouse logistics training. Over a three-year initiative, interactive holograms were developed for HoloLens 2 in collaboration with IT specialists, trainers and rehabilitation researchers. The applications were tested with 16 trainees, employing usability assessments, participant feedback and video analysis. Results indicate a generally positive reception, with most participants finding the application easy to use and engaging. Notably, hairdressing trainees experienced smoother interactions, while some challenges arose in the kitchen and logistics scenarios. Participants reported increased motivation and knowledge acquisition, though a few experienced mild exhaustion. While MR technologies show promise in vocational training, institutional implementation requires addressing technical, financial, and personnel factors. Trainers play a key role in sustaining MR integration beyond research projects. The study underscores the potential of MR to make learning more engaging and accessible while highlighting the need for further research on optimizing usability and implementation in training institutions.

Keywords: Mixed Reality · Vocational Training · Neurodevelopmental Disorders · Usability

1 Introduction/Background

At a time when digitization is challenging employees with disabilities more than ever, international research as well as educational curricula imply the necessity of integrating digital media and current technologies in learning processes in order to equip learners with digital skills for the labor market 4.0 [1].

Mixed reality (MR) bears the potential of making complex and abstract learning contents accessible and provide the opportunity to practice skills in a safe environment and thus serves as valuable tool for neurodiverse learners or learners with intellectual

K. Mavrou and P. Encarnação (Eds.): AAATE 2025, *Technology for Inclusion and Participation for All: Recent Achievements and Future Directions*, pp. 52–58, 2025.
https://doi.org/10.1007/978-3-032-01628-7_7

disability [2]. Especially in vocational training MR technologies have proven to enhance work performance, increase motivation and make training more engaging due to their immersive nature [3]. Targeted use that is individually adapted to people's needs and the work process is an important prerequisite for this, as well as the close involvement of committed trainers. MR offers new ways to visualize learning contents: 3D holograms can represent complex theoretical content directly at the training location [4].

Therefore, in the project, MR technologies are being researched, developed and tested at three vocational training centers for people with disabilities. The three locations also offer three different fields of work in which the use of MR is being tested: hairdressing, kitchen and logistics. The project is a three-year initiative that began in February 2022 and specifically targets vocational fields that have seen limited digitization in the past but encompass a wide range of tasks and activities.

In the project, an IT service provider developed complex, interactive holograms that depict realistic learning content. The learning content was developed for HoloLens 2 and trainers, rehabilitation scientists, project coordinators and trainees (with disabilities) were involved in the development process. These holograms can visualize intricate concepts and enable learners to practice work processes or use the holograms as supplementary tools.

As the project partners offer a wide range of training programs, one subject-specific MR scenario was programmed for each of the three partner sites. These three learning scenarios differ greatly in terms of complexity and task format. In the kitchen learning scenario, different foods and kitchen utensils are familiarized with and put together for specific recipes. This prepares for cooking according to a recipe and prevents the waste of resources, as the process can be practiced with holographic foods before actually cooking. The scenario for hairdressers is also about avoiding wasting resources. There, cutting graphics are shown on a holographic head and can be practiced and created. The holographic head can also be used as a template for real haircuts. As mannequin heads are often used for practice in hairdressing, it is helpful to use holographic 3D practice objects first. For the warehouse/logistics trainees, holograms of size-adjustable storage systems with movement animations have been developed. In this way, even storage systems which are not physically available in a training camp can be depicted in real size or smaller size. There is also a fully holographic warehouse in which learners can practice storing diverse goods correctly by choosing the right storage system for each type of good.

This comprehensive approach enhances the learning experience but also highlights the importance of effective communication between trainers and trainees to ensure that the tools are used effectively and understood clearly. This paper deals with the development, testing and final evaluations of three different MR learning scenarios for neurodiverse learners in rehabilitation training for the general labor market.

2 Methods

After a first finished and usable version of an application was created in collaboration with trainers, trainees, researchers and IT specialists, it was tested in several stages. Once all bugs and difficulties with content and operation had been eliminated, testing with the trainees could take place. The participants were informed in advance about the procedure of the evaluation. It was ensured that the participants were aware of the scope of use of the generated data for the evaluation of the application. Further, the trainees were aware of the voluntary nature of their participation in the project. Data protection was ensured by anonymizing the answers.

First, the participants completed a questionnaire collecting sociodemographic information, technology affinity and details about their disabilities. Next, they tested the HoloLens application. During this phase, the researchers could view the participants' perspectives on a computer screen to provide guidance if needed. The sessions were also recorded via screencast, generating first-person videos for subsequent analysis. These videos underwent content analysis, focusing on themes such as usability (e.g., the number of attempts to click a button successfully) and immediate participant feedback. Following the testing phase, participants filled out a second questionnaire to share their opinions on the application. This questionnaire included modified items from the System Usability Scale, presented on a 5-point Likert scale ranging from "Strongly Disagree" to "Strongly Agree" as well as self-developed items. Additional comment fields allowed participants to provide further feedback and suggestions for improvement.

The findings from the first evaluation led to adjustments to the application in order to adapt it to the needs of the target group, which includes neurodiverse individuals (e.g., those with autism or ADHD). After these modifications had been completed, a second evaluation was conducted using the same methodology with minor adjustments. The results of this final evaluation were used to assess the application's effectiveness and to draw conclusions about the potential of mixed reality in vocational training for individuals with disabilities. The results shown in the next chapter are the results of the final evaluation.

3 Results

A total of 16 trainees participated in testing the three learning scenarios of the application. 12 male and 4 female participants aged 17 to 26 tested the application. The participants were trainees in the partnering institutions of the project. Both trainees who had already been included in the previous evaluation and those who had no experience with the application were involved. They were selected by the trainers to take part in the trial and did so during their working hours.

In order to handle the trainees' self-perception of their disability sensitively and avoid potential feelings of stigmatization, the questionnaire did not directly ask about their disability status. Instead, the trainees were asked in which area, if any, they had the most difficulties.

The most frequently mentioned categories were 'Difficulties with thinking, learning, understanding' (n = 8) and 'Difficulties with mind, emotions and moods' (n = 8). 4 participants stated that they had difficulties with their vision and 6 participants reported to wear glasses.

The results indicate a generally positive reception with trainees successfully navigating the application. Most participants agree that it is easy to open the app (with 13 participants choosing the best or second-best option on the 5-point-scale of agreement).

Also, the contrast of the MR elements and the font size of the instructions are rated positively, with the majority of participants choosing the best or second-best option for both topics and no participant choosing the worst or second worst option.

However, the question of whether the holograms could be tapped or moved well revealed a very heterogeneous range of opinions. All answer options on the 5-point Linkert scale were used, resulting in an average score of 3.3 out of a possible 5 points. The question arises as to whether this variance is due to the application, environmental factors such as WLAN or the battery status of the HoloLens, or to MR technology in general. The trainees' previous frequency of use of the HoloLens did not have a significant influence on how they rated hologram interaction. The trainees were located between first-time users and 4–6 previous uses of the HoloLens but no correlation could be identified between the number of uses and the evaluation of the interaction with the holograms.

Furthermore, the findings reveal differences between the three learning scenarios. While hairdressing trainees report smoother interactions with holographic elements, participants in the kitchen and warehouse logistics scenarios find certain interface elements more challenging to manipulate (see Fig. 1).

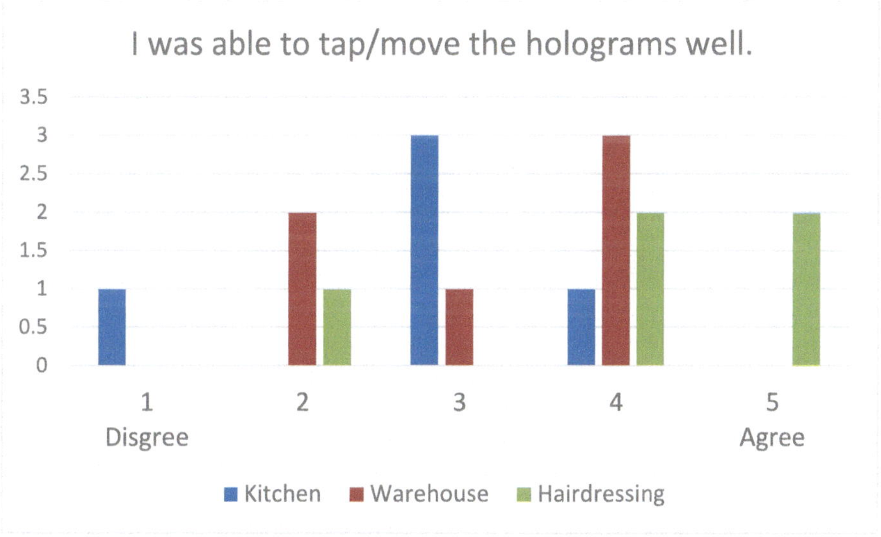

Fig. 1. Interaction with the Holograms divided for the different scenarios.

Most participants find the learning tasks understandable (with 15 participants choosing the best or second-best option). When the trainees were asked whether they knew more about the training-specific content taught in the scenario after using the app, the majority indicated increased knowledge (Fig. 2).

While the majority did not experience fatigue after using the HoloLens, two participants from the warehouse and kitchen groups report mild exhaustion. Regarding the

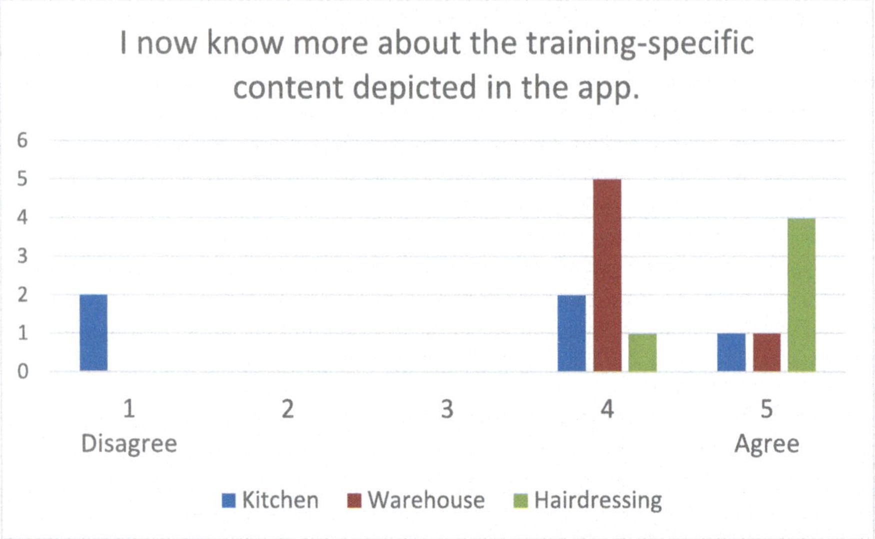

Fig. 2. Increase in knowledge reported by the participants divides for the different learning scenarios.

overall usability, ten out of sixteen trainees did not consider the application error-prone, though four participants perceived technical issues.

In the first evaluation round, the application achieved a System Usability Scale (SUS) score of 63.5. After targeted improvements to the application, a second assessment yielded a higher score of 69. The SUS is a standardized tool for measuring the usability of a system. Its results can be interpreted with the use of established benchmarks. According to Bangor et al. (2008) [5], scores below 50 are considered unacceptable while scores between 80 and 90 are rated as good and those above 90 as excellent. In contrast, Sauro (2011) [6] suggests that a score of 70 represents the typical average, with scores of 85 or above considered very good. Based on these classifications, the initial SUS score of 63.5 falls below average, while the improved score of 69 is just below the average threshold. Although it does not reach the level of high usability, the increase in SUS score indicates a modest yet measurable improvement in the application's usability for the targeted group of users between the prototype and the revised version.

Finally, results on the emotional perception of the app show that 15 of the 16 participants agree or partially agree with the statement that using the app is fun. The willingness to use the application more frequently is high, but opinions on wearing the HoloLens during work are mixed.

4 Discussion

While minor usability challenges and individual differences in perception are noted, the overall structure of the training scenarios proves effective. The evaluation suggests that the mixed-reality approach can be a valuable tool for vocational training.

However, it must be noted that due to the small sample size individual differences between the participants carry more weight. Therefore, the quantitative questionnaire results should not be considered entirely isolated. Analyzing the video material from the evaluation showed that factors such as environmental differences, trial difficulties, technical problems and personal factors had an impact on the evaluation and impact on the results can be predicted.

During testing with the HoloLens 2, several technical conditions affected the overall usability and reliability of the device in educational scenarios. A stable and high-speed internet connection is essential to ensure a continuous video stream from the HoloLens to a companion laptop, which is used to guide and support learners during their activities. If the connection is unstable or too slow, the video transmission becomes choppy or is interrupted entirely. Additionally, the video streaming process consumes a significant amount of battery power, leading to rapid battery depletion and limiting the duration for which the device can be used effectively. As the battery level drops, the quality of the video stream tends to deteriorate further. Moreover, prolonged use causes the headset to become noticeably warm, which is a disadvantage when the device is used consecutively with multiple learners. The extent of these issues varies depending on the learning scenario: they are less apparent in shorter or simpler sessions, such as the hairdressing scenario, but more problematic in longer and more complex tasks, such as those involving warehouse or kitchen activities.

However, when using the HoloLens in training, there is no need to stream the HoloLens view to a screen. Trainees can use the app on their own and practice scenarios as often as they like. This means that fewer errors should be expected when actually using the application in training.

This study presents promising insights into the use of Mixed Reality (MR) in vocational training; however, several limitations should be acknowledged. First, the sample size was limited to only 16 trainees. While this allowed for close observation and detailed data collection, the small cohort restricts the generalizability of the findings. A larger and more diverse group of participants would likely provide more robust and representative results, helping to reduce the weight of individual outlier performances.

Second, the current study did not include a direct comparison with more traditional vocational training methods. To more accurately evaluate the effectiveness of the MR application, future research should involve a control group using conventional approaches. Such comparative analysis is essential to determine whether MR-based training offers significant advantages or complements existing pedagogical strategies in a meaningful way.

5 Conclusion/Implications for the AT Field

The evaluation of the project results shows that immersive technologies such as MR have great potential in the training of people with disabilities. They can promote motivation and fun in the workplace and show trainees a new form of learning with improved accessibility of complex or abstract contents. Vivid MR visualizations can optimize traditional learning formats by saving resources, supplementing new content that could not be taught in the real world and enable practical, physical experiences. This leads to learners understanding the learning content better and provides the opportunity to practice and consolidate it again and again.

However, the question arises as to how the potential postulated by the project results and the current state of research on the subject can be effectively implemented in institutions for training people with disabilities. Many different factors need to be taken into account. These include the technical and infrastructural conditions of the facility, organizational and financial aspects as well as personnel aspects [7]. Trainers play a key role here, as only technologically skilled trainers who know how to integrate MR into training in a pedagogically meaningful way can ensure that MR technologies continue to be used effectively beyond research projects and develop their full potential. In the final months of the project, the focus will therefore be on equipping the trainers with skills that are needed for a sustainable use of the technology in the teaching context. Furthermore, recommendations for the use of the application and for future projects involving MR for people with disabilities in vocational training will be formulated.

References

1. Kultusministerkonferenz (KMK), Lehren und Lernen in der digitalen Welt. Die ergänzende Empfehlung zur Strategie "Bildung in der digitalen Welt". Sekretariat der Kultusministerkonferenz, Berlin (2021)
2. Köse, H., Güner-Yildiz, N.: Augmented reality (AR) as a learning material in special needs education. Educ. Inf. Technol. **26**, 1921–1936 (2023)
3. Wuttke, L., Bühler,C., Dirks, S. How can the new technologies of augmented reality, mixed reality and virtual reality support people with cognitive impairments and learning difficulties in the workplace? Submitted to Technology and Disability (2025)
4. Fehling, C.D.: Erweiterte lernwelten für die berufliche bildung. augmented reality als perspektive (2017). In: Thissen, F., (eds.). Lernen in virtuellen Räumen: Perspektiven des mobilen Lernens, pp. 125–142. Berlin, Boston: De Gruyter
5. Bangor, A., Kortum, P., Miller, J.A.: The system usability scale (SUS): an empirical evaluation. Int. J. Hum. Comput. Interaction **24**, 574–594 (2008)
6. Sauro, J., Lewis, J.R.: When designing usability questionnaires, does it hurt to be positive? In: CHI'11: Proceedings of the SIGCHI International Conference on Human Factors in Computing Systems 2011, pp. 2215–2224 (2016)
7. Dziarstek, L., Söffgen, Y., Wuttke, L.: Die rolle der ausbildenden bei der einführung digitaler technologien in der beruflichen bildung (Rahmenbedingungen, Akzeptanz, Kompetenzen). (2024). In: Heitplatz, V., Wilkens, L.: (eds.). Die Rehabilitationstechnologie im Wandel: Eine Mensch-Technik-Umwelt Betrachtung, pp. 459–473. Dortmund: Eldorado. https://doi.org/10.17877/DE290R-24357

Augmented Reality-Based Training for Gardening: Enhancing Skills for Individuals with Special Needs at Il Seme Social Cooperative

Mario Covarrubias[1](✉), Giulia Stevanoni[2], Samuele Sergio Fontana[2], and Giuseppe Andreoni[2]

[1] Mechanical Engineering Department, Politecnico di Milano, Milan, Italy
mario.covarrubias@polimi.it
[2] Desing Department, Politecnico di Milano, Milan, Italy

Abstract. Teaching gardening skills to individuals with special needs presents distinct challenges, largely due to diverse learning preferences and accessibility limitations. Augmented Reality (AR) offers a compelling solution by delivering interactive, adaptive, and immersive educational experiences. This study introduces the design, development, and evaluation of an AR-based training application, specifically tailored for gardening activities and delivered through the Microsoft HoloLens 2 device.

The application provides step-by-step visual and auditory instructions, dynamically adjusts content based on the user's abilities, and supports skill acquisition through engaging, simulated practice. To assess the usability and effectiveness of the tool, the System Usability Scale (SUS) protocol was administered to participants with varying cognitive and physical needs.

The paper details the conceptual framework, technical implementation, and insights gained from a pilot study involving users with diverse disabilities. Findings indicate that the use of AR—particularly via the HoloLens 2—significantly enhances accessibility, learning outcomes, and user engagement. Overall, the system demonstrates strong potential to empower individuals with special needs to build gardening skills and achieve greater independence.

Keywords: Augmented Reality · Training · Gardening Activities · Instructional Guides · Immersive Experience

1 Introduction

Individuals with mild cognitive impairments or minor psychological conditions often face exclusion from the workforce. While the economic impact is significant, the consequences on personal dignity and social participation —for both

K. Mavrou and P. Encarnação (Eds.): AAATE 2025, *Technology for Inclusion and Participation for All: Recent Achievements and Future Directions*, pp. 59–64, 2025.
https://doi.org/10.1007/978-3-032-01628-7_8

individuals and their families—are even more profound. A primary challenge encountered by these individuals is difficulty in remembering task sequences and executing steps correctly, which can lead to anxiety and psychological distress, even in seemingly simple work scenarios, especially when performed independently.

People with special needs frequently struggle to acquire the practical skills required for autonomy in daily work tasks. Conventional training methods often fall short of accommodating diverse learning preferences, accessibility demands, and specific cognitive or physical limitations. This project targets routine manual tasks—such as cleaning, harvesting, gardening, or assembling basic components—which are typically within the capability of individuals with special needs when proper support is provided.

Wearable and interactive technologies present a promising, cost-effective pathway to address these challenges. In particular, Augmented Reality (AR) has emerged as a powerful tool for creating personalized, immersive, and inclusive training experiences. AR technologies [1,2] enable context-aware learning environments that adapt to the user's pace and abilities [3,4], promoting both skill acquisition and user engagement [5].

This paper introduces the design and evaluation of an AR-based application developed to train individuals with special needs in cleaning-related tasks. The solution aims to increase their confidence, reduce dependency, and support long-term employability.

The project has been developed in collaboration with Il Seme, a social cooperative based in Como, Italy, committed to empowering individuals with special needs through practical, inclusive training programs in cleaning and gardening. Il Seme provides a nurturing environment where participants receive hands-on instruction and personalized support, enabling them to develop essential life and work skills. Through these initiatives, the cooperative fosters independence, enhances social inclusion, and encourages meaningful community engagement. Their innovative approach highlights the transformative potential of tailored, technology-enhanced training to promote dignity, autonomy, and societal contribution among people with special needs.

2 AR Application and Implementation

The Augmented Reality (AR) application has been specifically developed to support individuals with special needs in learning and performing gardening tasks. Emphasizing accessibility, adaptability, and engagement, the application is designed to address a variety of user requirements through the following key features:

- **Customized Instruction:** Training content is tailored to the user's abilities and preferences, offering personalized step-by-step guidance that ensures comprehension and comfort.

- **Virtual Task Simulation:** Users can practice gardening procedures—such as planting, watering, pruning, or tool handling—through interactive AR simulations in a risk-free, controlled environment.
- **Multisensory Feedback:** The application integrates visual prompts, audio instructions, and tactile cues to support diverse learning styles and sensory needs.
- **Inclusive Interaction Design:** The system features voice control, gesture-based commands, and image tracking technologies that activate context-specific video guidance, improving accessibility for users with varying physical or cognitive capabilities.

The AR application is built to be compatible with multiple platforms, including smartphones, tablets, and wearable devices like Microsoft HoloLens 2. Leveraging the Vuforia engine, the system ensures smooth and dynamic interactions triggered by 2D visual markers.

When the HoloLens recognizes a designated 2D marker—illustrated in Fig. 1—the AR engine launches an interactive sequence. This initiates a hands-on training experience aligned with the specific gardening task associated with that marker.

The development process began with a detailed task analysis to deconstruct typical gardening activities into their fundamental components. A professional gardener performed several operations—such as soil preparation, planting, watering, and cleanup—which were recorded and studied. This analysis allowed us to:

- Segment each gardening task into a sequence of basic, teachable actions.
- Identify the tools required for each step, along with their preparation and proper use.
- Consider contextual environmental elements essential for task setup and execution.

From this analysis, we produced:

1. Short instructional videos with audio narration for each basic gardening action.
2. A system of numbered, color-coded tags designed for placement in the training environment.

Each 2D image acts as an access point to an immersive micro-learning module, helping users progress through increasingly complex gardening tasks. As users interact with these markers, they are drawn into realistic practice scenarios, developing both procedural knowledge and problem-solving skills.

By combining interactive technology with carefully structured training content, the AR application offers a fresh and empowering approach to learning gardening tasks. It moves beyond traditional instruction methods, providing a meaningful and accessible learning journey tailored to the needs of individuals with special needs.

3 Pilot Study and Evaluation

A pilot study was carried out with a group of 10 volunteers (Fig. 2) to assess the usability and training effectiveness of the AR-based gardening application developed for individuals with special needs. The participants represented a diverse range of cognitive, sensory, and motor challenges. Each individual took part in guided training sessions using the AR system, followed by a structured feedback phase that included both quantitative and qualitative evaluation tools.

Fig. 1. 2D image markers triggering gardening task video guides.

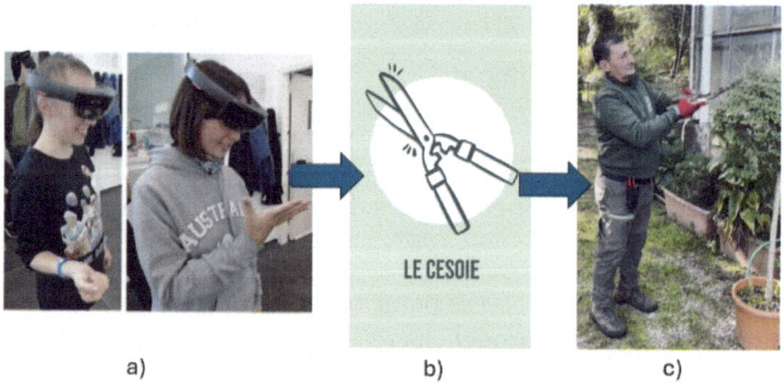

Fig. 2. Training session: HoloLens identifying a 2D marker (b) and activating a corresponding gardening task video (c).

Initial findings suggest that the AR application notably improved participants' ability to engage with and perform basic gardening tasks. Many users reported a heightened sense of confidence and independence when completing activities such as planting, watering, and tool handling. The step-by-step visual guidance and interactive simulations were highlighted as key elements that made the experience more intuitive and supportive of individual learning styles. While the overall response was positive, several participants noted difficulties related to the interface design, occasional technical delays, and the clarity of some instructions—providing valuable input for future iterations of the system.

To systematically evaluate user satisfaction and interface usability, we implemented the System Usability Scale (SUS)—a widely adopted and validated questionnaire designed to measure perceived usability across several domains including ease of learning, effectiveness, and overall satisfaction. Participants responded to a series of statements using a five-point Likert scale, where higher scores reflected a more favorable user experience.

In addition to the SUS data, qualitative insights were collected through open-ended questions and semi-structured interviews. This dual approach enabled a deeper understanding of the users' experience, highlighting both the strengths of the application—such as personalization and accessibility—and areas needing refinement, such as interaction design and technical reliability. These insights informed actionable improvements for enhancing the system's usability, functionality, and inclusivity.

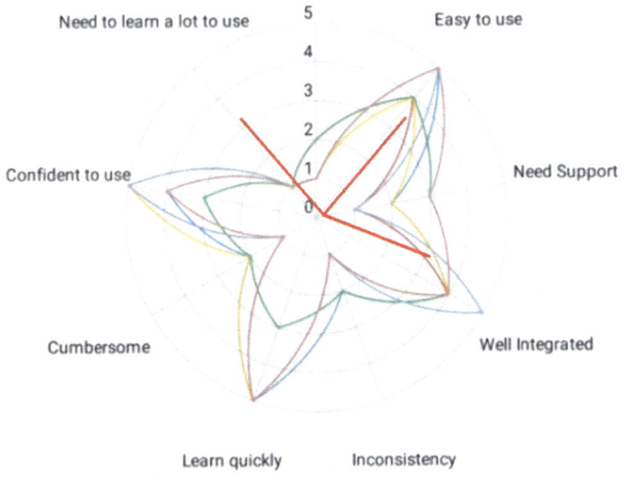

Fig. 3. System Usability Scale (SUS) evaluation results.

Figure 3 illustrates the results of the System Usability Scale (SUS) evaluation, represented as a radar chart. Each axis corresponds to a key usability dimension, including ease of use, need for support, system integration, learnability, and user confidence. The highlighted axes—Easy to Use, Well Integrated, and Need to Learn a Lot to Use—reflect particularly significant areas identified by participants. Higher scores on these axes indicate a positive user experience, while lower scores suggest potential areas for improvement. The chart visually summarizes user perceptions, supporting a comprehensive analysis of the AR application's strengths and usability challenges.

This pilot study confirmed the potential of AR as a practical and empowering tool for teaching gardening skills to individuals with special needs, paving the way for more inclusive and effective vocational training approaches.

4 Conclusions

In conclusion, this study introduces an Augmented Reality (AR) application specifically developed to support individuals with special needs in learning and performing gardening tasks. The tool aims to foster independence by helping users acquire practical skills through accessible, engaging, and personalized training experiences. The initial results highlight the application's effectiveness in addressing a wide range of cognitive, sensory, and physical needs, making gardening activities more approachable and inclusive. Moving forward, future developments will concentrate on refining the user interface, enhancing system stability, and extending functionality to cover a broader set of gardening scenarios. Ultimately, AR technology shows strong potential as a transformative tool for inclusive skill development, offering meaningful opportunities for empowerment and participation in everyday tasks.

Acknowledgements. Special thanks go to the Il SEME staff: Marta Fontanta, Marco Martinelli, Chiara Brambilla, Ilaria Liprino and Carlo Valli.

References

1. Geng, J., Rizzo, A., Seo, J., Jia, B., Lohr, C., Brooks, J.: Augmented reality-based job skills training for individuals with autism spectrum disorder (ASD): a preliminary investigation. J. Autism Dev. Disord. **50**(12), 4516–4525 (2020)
2. Lee, S., Han, J., Cho, H.: A study on the development of smart home cleaning system for people with disabilities using AR (augmented reality). In: 2019 IEEE 5th International Conference on Computer and Communications (ICCC), pp. 1375–1378. IEEE (2019)
3. Yu, X., Hao, Y., Yap, Y.L.: A study on the application of augmented reality in special education. In 2021 IEEE 12th International Conference on Software Engineering and Service Science (ICSESS), pp. 788–791. IEEE (2021)
4. Moore, D., Mihailidis, A., Kim, H.: The growing need for augmented reality for individuals with autism: a narrative review. J. Rehabili. Assist. Technol. Eng. **4**, 2055668317746179 (2017)
5. Amores, J.F., Borro, D.: AR4AD: augmented reality platform for teaching daily living skills to people with intellectual disabilities. In International Conference on Augmented Reality, Virtual Reality and Computer Graphics, pp. 308–319. Springer (2020)
6. https://www.ilseme.info/. Accessed 4 Feb 2024

Bridging the Digital Divide: Enhancing Internet Accessibility for Students with Special Needs via Application Easy Reading

Miriam Bursy(✉), Leevke Wilkens, Lukas Baumann, Luisa Kramschneider, and Susanne Dirks

Department of Rehabilitation Technology, TU Dortmund University, Dortmund, Germany
{miriam.bursy,leevke.wilkens,lukas.baumann,luisa.kramschneider, susanne.dirks}@tu-dortmund.de

Abstract. Media skills have become increasingly important in the education system. In Germany, the expansion of digital infrastructure and equipment in schools was prioritized until 2021, resulting in new challenges and a need for innovation. In this context, customizable applications and software offer great potential for pupils of different expertise levels. The Easy Reading app, originally developed as a browser plug-in, offers such customizability. Therefore, it was examined for its potential use in the classroom.

The presented study investigates the potential of media literacy education using individualized approaches through classroom observations and teacher interviews. Findings highlight promising opportunities but reveal that the prototype requires further refinement to unfold its full potential. The research underscores the importance of integrating technical and pedagogical considerations for effective innovation implementation.

Keywords: special education · students with special needs · Easy Reading

1 Introduction

The rapid development of information and communication technologies (ICT) over the last 10 years [1, 2] has increased the importance of media literacy in the educational system [3]. Due to the need to integrate digital media and media education into schools, German federal initiatives have promoted the integration of digital media into the education system. In these initiatives, the implementation of digital media is focused, aiming to offer all students the opportunity to use a digital learning environment and access the Internet by 2021 [4]. However, as Gräsel et al. [5] note, introducing new technologies in German schools has created new challenges that show that digitalization in this sector requires a high level of innovation.

Using the Internet requires a specific skill set, which includes reading and writing. Therefore, tools are needed to customize web pages. One such solution is the browser plug-in Easy Reading, which was developed in a previous participatory research and

K. Mavrou and P. Encarnação (Eds.): AAATE 2025, *Technology for Inclusion and Participation for All: Recent Achievements and Future Directions*, pp. 65–72, 2025.
https://doi.org/10.1007/978-3-032-01628-7_9

development project [6]. It makes digital content more accessible for people with intellectual disabilities (iD) through individual customization and has shown its potential as an innovative technology for improving media skills for students with special needs [7, 8]. Based on the Easy Reading Framework [6], a prototype app with six main functions for adapting websites was developed and tested in the EVE4all project. These six main functions are:

1. Maximizing and minimizing font
2. Maximizing and minimizing space between the lines
3. Contrast changes
4. Read aloud
5. Translating into different languages
6. Reading ruler (highlighting a single line in the text)

The following paper presents the results obtained from evaluating the mobile application of Easy Reading in special schools and assigns them to the resulting consequences for validation. In this context, applicability, usability, and acceptance are examined. Testing these constructs can provide information on whether the App can be sustainably launched and established in schools.

2 State of the Art

Inclusive media education aims to enable all pupils to use digital media equally to promote participation in, with, and through media [9]. Digital technologies introduce new learning opportunities but also make existing and new exclusion mechanisms visible in the school context [10]. Key challenges are the accessibility and usability of digital media [11], with barriers in technical equipment and didactic design. Even in industrialized countries, there are still restrictions on access to hardware and software [12].

Digital media can both enable and discourage participation and is influenced by a combination of individual impairments and structural barriers [9, 13]. They offer the potential for adapting learning processes and facilitating access to education if they are designed to be accessible and adapted to individual needs [13]. Basic skills, such as reading, perception, and fine motor skills, are also essential for digital participation. Students with iD especially need to be supported in their reading skills to develop digital skills [14, 15].

Students are dependent on the support of teachers and caregivers in order to take advantage of their digital participation opportunities [14, 16]. However, the role of teachers is also key to the successful implementation of digital media in schools. They must evaluate their didactic benefits, use them creatively, and teach students media skills [17]. Requirements are motivation, beliefs, self-regulatory skills, and technological knowledge [18].

3 Methods

The study aimed to identify which aspects of the Easy Reading app support students and teachers in special education schools and to determine whether and which technical and content-related further developments Easy Reading requires in this context. Three parameters were defined, according to which the investigation was structured: Usability [19, 20], applicability [21] and acceptance [22].

A qualitative research approach was used to investigate the potential of the developed app for supporting students with special needs (n = 64) and their teachers (n = 9) in digital learning scenarios. Data was collected through participant observation in teaching units and interviews with students and teachers.

Testing took place between September and November 2024 in seven classes at four special schools focusing on iD or physical and motor development (PMD) (see Table 1).

Table 1. Presentation of the surveyed sample

school's focus	Classes per school	Age of Students	Number of Students	Number of Teachers
iD	2	12–15 years	11	1
		10–11 years	9	1
iD	1	13–15 years	6	2
iD	2	15–16 years	12	1
		14–16 years	12	1
PMD	2	14–16 years	5	1
		14–16 years	9	2

For the participant observation, two classroom visits were conducted per class, during which the functionality and utilization of the software were evaluated and documented. Statements were also recorded if a declaration of consent was provided.

During the first visit, the Easy Reading App was introduced in regular lessons, and the students had the time to try its usage. In a second visit, the software was tested as part of a teaching scenario previously agreed upon with the teachers. The content of the lessons was based on the current teaching topic of the respective class.

During the second observation, 11 students shared their opinions about the Easy Reading App in class. Impulse questions were used, and the students were asked to comment on different aspects during their usage.

After the second observation, nine interviews with the respective teachers were conducted. They were asked for their opinions on Easy Reading, the special features they noticed during initial use and in class, and any suggestions for its development or improvement. The interviews and observation sheets were analyzed using qualitative content analysis by Mayring [23].

4 Results

The results from the observations and interviews are presented based on the parameters of usability, applicability, and acceptance.

4.1 Usability

During all observations and classes, the students showed little to no difficulty using the devices and were very interested in and enjoyed working with Easy Reading. In addition, the students understood the features of the individual functions but used them to different degrees. Students described the tools 'Reading ruler' and 'read aloud' as particularly helpful and fun, while the 'contrast' and 'translation' functions were rated as less helpful.

Although the teachers emphasized the potential of having the option to customize the app settings, students rarely used it. Overall, the students rated the app as helpful for their reading skills and for accessing websites.

If the students had difficulties with motivation and concentration, the teacher stepped in to accompany the task and encourage the use of Easy Reading.

In summary, teachers described the app as user-friendly and straightforward, especially due to features such as customizable font sizes and backgrounds that make reading easier. These functions are perceived as valuable support for students with different reading abilities. Zooming in and modifying the background was considered helpful for students with visual or cognitive impairment. Overall, teachers found the app's features helpful for their students, especially with longer texts.

However, the teachers also criticized long navigation paths in connection with the settings options and the app's instability. In another example, the app's frequent crashes were described as an obstacle to usability. They also emphasized that combining text reading and app operation could be a hurdle for inexperienced users.

As ideas for the further development of the app, the teachers suggested adding options for customizing reading speeds (e.g., different reading voices, and a repeat function). Another idea was to read sentences or words individually and add a function to simplify texts (e.g., using AI).

4.2 Applicability

Initially, some students encountered difficulties understanding the functions of the tools. Consequently, the assistance of the teacher and the observation team was necessary for all teaching units to use Easy Reading to complete the tasks. It was observed that students who were more familiar with mobile devices were more likely to use the gesture control for changing font size than the integrated functions in the Easy Reading app. The more intensive the functions were used, the easier it was for the students to remember the names and functions of the individual tools, even over a more extended time period.

It was also noticeable that a straightforward work assignment helped the students actively engage with the functions of Easy Reading and simultaneously work on the task. Overall, the students stated that Easy Reading helped them to complete the task.

During the interviews, the teachers identified different scenarios for using the app in the classroom. For example, the app could provide support when researching and presenting web content or offer the opportunity to digitally use recipes or other content.

Teachers suggested that the app could be used to support individual, low-achieving students in particular, who would experience more independence by using the app in lessons. In their opinion this could make teaching much easier in the long term as soon as the students are familiar with the functions and operation, as the app supports tasks such as researching one different topping. Initially, the teachers rated the distraction factor as high due to the app's unfamiliarity. However, in the long term, they highlight the potential to increase efficiency through familiarization and practice.

According to the teachers, once released, the Easy Reading app could also promote ease of use in different areas of students' lives, such as reading bus timetables.

4.3 Acceptance

Over the observation periods, it also became apparent that the students were becoming accustomed to working with Easy Reading. They also say they would like to use the app again and at home.

The teachers described how introducing the app into the classroom was perceived as motivating for the students but that the unfamiliar operation initially led to distractions: *"So it was certainly a great motivation to simply try something new. [...] So basically, it was actually a bit distracting because it was new" (RS-MIC-12-09L, pos.6).*

Unfortunately, errors repeatedly occurred within the application (e.g., not reading out a selected section), which sometimes led to frustration and negative reactions from the students, as they related errors to their abilities rather than to the devices or the software. The teachers also discussed the observed frustration of the students. In their opinion, it reduces the acceptance of the app. Nevertheless, the teachers would use the app in other contexts and recommend it to colleagues, as they rate the app useful for teaching. According to the teachers, using the app in the classroom context together with an exciting task increased the students' motivation to participate in lessons. According to their assessment, some students would probably have become demotivated more quickly without the app's support and would have abandoned the task.

Teachers also emphasize the added value of promoting students' independence. In the long term, the teachers believe that the Easy Reading app offers great added value, especially when used regularly and through individual adaptation to the students' needs. In their opinion, by constantly using and integrating the app into lessons, students could learn which functions help them and thus further consolidate acceptance of the app.

5 Discussion and Conclusion

The students' initial difficulties in using the Easy Reading app and its functions illustrate the importance of accessibility assistive technologies in the classroom. The observed difficulties challenge the usability and acceptance of the Easy Reading app. An introductory tutorial as part of the app could serve as a supportive mechanism to promote the independent use of digital media [24]. This could increase the learnability and effectiveness of an app [22], which suggests that a tutorial should be included as an essential part of the app. A tutorial could also help recognize individual functions and students' low use of individualization options. Due to the absence of such a tutorial in this prototype, students cannot see what functions the app has and what benefits they could gain from using it. As a result, the app could be perceived as complex by students and teachers, which increases the chance that the app will not be used in class [3, 20].

Furthermore, the observations showed that the Easy Reading app cannot compensate for students' limited reading and writing skills. However, these skills are crucial for digital participation and acquiring digital skills [13–15]. Additional features, such as a voice-activated search, could compensate for the lack of these skills and, at the same time, increase usability, which could improve classroom performance.

As the Easy Reading app is still in prototype status, incorrect output is not surprising but unfortunate. It became clear that the students frequently depended on support when problems occurred. It also became clear that they generally find it difficult to distinguish between errors caused by themselves and incorrect output from the app. This sometimes led to frustration and a feeling of inadequacy among the students. The role of teachers also became clear from these observations. As highlighted in other discussions on digital participation, the teachers' positive attitude and motivation also promoted the students' acceptance of using the app in this study [14, 16].

The use of new technologies requires a clear pedagogical structure and intensive support for both students and teachers. In addition, technical errors in the prototype require transparent communication with the students. The individual media skills, local digital infrastructure, and perceived usefulness of both teachers and students influence the successful integration of an app such as Easy Reading. Despite possible limitations, teachers and students rate the app as helpful overall.

For sustainable use, the absence of errors, a comprehensive tutorial and a place for support requests are essential. Whether the discussed measures and proposed changes or further developments will lead to long-term use of the app cannot be assessed at this time nor within this study's scope.

Acknowledgments. The EVE4all project is funded by the Federal Ministry of Education and Research as part of the funding program "VIP+" under grant agreement 03VP09660.

Disclosure of Interests. The authors have no competing interests to declare that are relevant to the content of this article.

References

1. Bundesministerium für Bildung und Forschung (BMBF): Verwaltungsvereinbarung Digital-Pakt Schule 2019 bis 2024 (2019)
2. Initiative D21 e.V.: D21-Digital-Index 2021/2022. Wie digital ist Deutschland? Jährliches Lagebild zur Digitalen Gesellschaft, 1st edn. D21-Digital-Index, vol. 9. Initiative D21, Berlin (2022)
3. Bosse, I.: Ethische aspekte inklusiver medienbildung. gleichberechtigter zugang zu information und kommunikation als voraussetzung. Commun. Soc. (2014). https://doi.org/10.5771/0010-3497-2014-1-6
4. Kultusministerkonferenz (Kmk): Bildung in der digitalen Welt. Strategie der Kultusministerkonferenz (2017). https://www.kmk.org/fileadmin/Dateien/veroeffentlichungen_beschluesse/2016/2016_12_08-Bildung-in-der-digitalen-Welt.pdf. Accessed 10 Jan 2025
5. Gräsel, C., Schledjewski, J., Hartmann, U.: Implementation digitaler Medien als Schulentwicklungsaufgabe. Zeitschrift Pädagogik (2020). https://doi.org/10.25656/01:23629

6. Feichtenschlager, P., Heumader, P., Miesenberger, K.: Final project progress report (easy reading) (2020)
7. Wüst, M., et al.: Easy reading – keeping the user at the digital original: assessment and potential use of easy reading for digital accessibility. In: Antona, M., Stephanidis, C. (eds.) Universal Access in Human-Computer Interaction. 17th International Conference, UAHCI 2023, pp. 469–487. Springer, Cham (2023)
8. Heumader, P., Miesenberger, K., Murillo-Morales, T.: Adaptive user interfaces for people with cognitive disabilities within the easy reading framework. In: Miesenberger, K., Manduchi, R., Covarrubias Rodriguez, M., Peňáz, P. (eds.) Computers Helping People with Special Needs. 17th International Conference, ICCHP 2020, Proceedings, pp. 53–60. Springer, Cham (2020)
9. Bosse, I., Haage, A., Kamin, A.-M., Schluchter, J.-R., GMK-Vorstand: medienbildung für alle: medienbildung inklusiv gestalten. Positionspapier der fachgruppe inklusive medienbildung der gesellschaft für medienpädagogik und kommunikationskultur e.V. (GMK). In: Brüggemann, M., Eder, S., Tillmann, A. (eds.) Medienbildung für alle. Digitalisierung, Teilhabe, Vielfalt. Schriften zur Medienpädagogik, vol. 55, pp. 207–219. Kopaed, München (2019)
10. Casale, G., Börnert-Ringleb, M., Hillenbrand, C.: Fördern auf Distanz? Sonderpädagogische Unterstützung im Lernen und in der emotional-sozialen Entwicklung während der COVID-19 bedingten Schulschließungen in den Regelungen der Bundesländer. Zeitschrift Heilpädagogik **71**, 254–267 (2020)
11. Ragnedda, M., Ruiu, M.L.: Digital Capital. A Bourdieusian Perspective on the Digital Divide. Emerald Publishing, Bingley (2020)
12. van Deursen, A.J., van Dijk, J.A.: The first-level digital divide shifts from inequalities in physical access to inequalities in material access. New Media Soc. (2019). https://doi.org/10.1177/1461444818797082
13. Heitplatz, V.: Einsatz digitaler Medien für Lernende im Bereich geistige Entwicklung. In: Betz, J., Schluchter, J.-R. (eds.) Schulische Medienbildung und Digitalisierung im Kontext von Behinderung und Benachteiligung, 2nd edn., pp. 65–80. Juventa Verlag, Weinheim (2022)
14. Alfredsson Ågren, K., Kjellberg, A., Hemmingsson, H.: Access to and use of the Internet among adolescents and young adults with intellectual disabilities in everyday settings. J. Intellect. Dev. Disabil. (2020). https://doi.org/10.3109/13668250.2018.1518898
15. Heitplatz, V., Bühler, C., Hastall, M.R.: I can't do it, they say! – perceived stigmatization experiences of people with intellectual disabilities when using and accessing the internet. In: Antona, M., Stephanidis, C. (eds.) Universal Access in Human-Computer Interaction. Applications and Practice, pp. 390–408. Springer, Cham (2020)
16. Ramsten, C., Blomberg, H.: Staff as advocates, moral guardians and enablers – using ICT for independence and participation in disability services. Scand. J. Disabil. Res. (2019). https://doi.org/10.16993/sjdr.608
17. Kultusministerkonferenz (Kmk): Strategie der Kultusministerkonferenz. Bildung in der digitalen Welt (2016). https://www.kmk.org/fileadmin/Dateien/veroeffentlichungen_beschluesse/2016/2016_12_08-Bildung-in-der-digitalen-Welt.pdf. Accessed 28 Feb 2025
18. Kiesler, N., Weers, C.: Medienbildung für Schule und Unterricht. MedienPädagogik: Zeitschrift für Theorie und Praxis der Medienbildung (2022). https://doi.org/10.21240/mpaed/00/2022.11.25.X
19. Nielsen, J.: Usability 101: introduction to usability (2012). https://www.nngroup.com/articles/usability-101-introduction-to-usability/. Accessed 8 Dec 2022
20. Brooke, J.: SUS: a 'quick and dirty' usability scale (1996). https://hell.meiert.org/core/pdf/sus.pdf. Accessed 2 Feb 2022
21. Davis, F.D., Bagozzi, R.P., Warshaw, P.R.: User acceptance of computer technology: a comparison of two theoretical models. Manage. Sci. **35**, 92–1003 (1989)
22. Sauer, A., Luz, F., Suda, M., Weiland, U.: Steigerung der Akzeptanz von FFH-Gebieten. Abschlussbericht. BMU-Druckerei, Bonn (2005)

23. Mayring, P.: Qualitative Inhaltsanalyse. Grundlagen und Techniken (Qualitative Content Analysis). Beltz, Weinheim (2015)
24. Lütje-Klose, B.: Schulische inklusion und sonderpädagogische professionalität. chancen und herausforderungen der digitalisierung. In: Ferencik-Lehmkuhl, D., et al. (eds.) Inklusion digital! Chancen und Herausforderungen inklusiver Bildung im Kontext von Digitalisierung, pp. 17–32. Verlag Julius Klinkhardt, Bad Heilbrunn (2023)

Impact of Interactive Sensory Ball on Children with Autism Spectrum Disorder: Enhancing Engagement, Self-regulation, and Teacher Support

Aleksandra Jankovic[1,2](✉), R. M. Chandima Ratnayake[2], and S. M. Samindi M. K. Samarakoon[2]

[1] Muzziball AS, Sverdrups gate 27, 4007 Stavanger, Norway
[2] Department of Mechanical and Structural Engineering and Materials Science, University of Stavanger, Stavanger, Norway
{aleksandra.jankovic,samindi.samarakoon}@uis.no, chandima.ratnayake@uis.com

Abstract. This study examines the impact of an Interactive Sensory Ball (ISB), incorporating light and music programs, on children with autism spectrum disorder (ASD). The minimum viable product (MVP) of the ISB prototype was individually tested with 12 children aged 7 to 10 years and their teachers across two primary schools. A case study-based action research methodology was used, incorporating the Single-Minute-Exchange of Die (SMED) approach with video footage to assess behavioral changes and investigate the impact of the ISB product features. Results indicate significant improvements in engagement, self-regulation, focus and social interaction. These findings highlight the potential of interactive sensory tools in supporting children with ASD and provide valuable insights for policymakers, educators, stakeholders in health and education sectors. The study emphasizes the need for innovations that boost engagement and support self-regulation in children with ASD, thereby improving teacher working conditions. This aligns with sustainability goals, enhancing conditions and learning potential in primary schools and ASD departments.

Keywords: Autism Spectrum Disorder (ASD) · Interactive Sensory Ball (ISB) · Self-Regulation · Inclusion · Sustainability

1 Introduction

The rapid advancement of innovative technologies presents new opportunities to address persistent challenges in education and healthcare, particularly for children with autism spectrum disorder (ASD) [1]. These developments highlight the need for multidisciplinary approaches that integrate technology, pedagogy, product design, and creative learning [2]. In schools, students with ASD often struggle with self-regulation, attention,

K. Mavrou and P. Encarnação (Eds.): AAATE 2025, *Technology for Inclusion and Participation for All: Recent Achievements and Future Directions*, pp. 73–81, 2025.
https://doi.org/10.1007/978-3-032-01628-7_10

and sensory processing, which affects both academic progress and social inclusion [3]. These difficulties also impact teachers, who must manage behaviour, foster engagement, and maintain inclusive classrooms, often under constrained resources [4].

Multisensory environments (MSEs) and technology-driven interventions have shown promise in supporting children with ASD, particularly by enhancing self-management and sensory responsiveness [5]. However, the effectiveness of MSEs depends on their capacity to meet individual needs [5, 6]. Advanced tools using VR, AR, XR, or AI offer personalized support but remain limited by high costs, mobility constraints, lack of teacher training, and incompatibility with everyday classroom routines [7–9]. Many also rely on screen-based interfaces, which may contribute to digital overexposure and affect attention and developmental health in young learners [10].

Teachers report high stress levels due to the challenges of managing complex student needs while ensuring inclusive, motivating learning environments [11]. This underscores the importance of assistive technologies (AT) that are not only responsive to learners' needs but also simple, flexible, and reliable for educators to use [12].

In parallel, sustainability and circular economy (CE) principles are increasingly recognized as essential to inclusive innovation [13]. Best practices now call for embedding social, economic, and environmental considerations early in the design process, ranging from material selection and durability to recyclability and responsible end-of-life management [13]. Recent guidelines for sensory tools and toys for children with ASD emphasize safety, adaptability, and environmental responsibility [14], aligning with global SDGs, particularly SDGs 3, 4, 10, and 12 [15].

This paper introduces the interactive sensory ball (ISB), a child-centred, educator-informed tool designed to support engagement, self-regulation, and inclusive classroom practice. Developed through an iterative, user-centred process, the ISB incorporates customizable sensory features and circular design principles. The study evaluates the ISB in real-world classrooms, contributing to research on adaptive sensory technologies and scalable, sustainable solutions in inclusive education.

2 Methodology

2.1 Research Design

This study employs a case study-based action research methodology to investigate the iterative design, testing, and refinement of an interactive sensory ball (ISB) for children with autism spectrum disorder (ASD). Conducted at Muzziball AS in Stavanger, Norway, the research was implemented in two specialized departments within public primary schools. Action research was selected for its ability to link theory and practice in real-world settings, enabling collaborative inquiry and iterative improvement based on user feedback [16]. This aligns with the need for participatory, inclusive approaches in educational technology, especially for neurodivergent learners.

The study also incorporates principles of embedded case study design [17], allowing in-depth analysis of individual sessions, teacher–child interactions, and feature testing outcomes within authentic classroom contexts. The research setting required adaptable methods to accommodate participant diversity, dynamic conditions, and ongoing product development.

The ISB development followed human-centered design principles, with teachers and students engaged as co-informants throughout. This participatory approach ensured that insights from daily pedagogical practices and observed behaviors directly informed ISB refinement. As Sanders and Stappers note, co-creation and contextual understanding are essential in designing inclusive, sustainable technologies [18].

2.2 Participants and Data Collection (Setting and Procedure)

The study involved 12 children aged 7–10 years, all formally diagnosed with ASD. Sessions took place in public-school classrooms adapted to reduce external stimuli while retaining typical educational features. Each child engaged with the interactive sensory ball (ISB) during a 15-min session alongside their teacher, with sensory programs randomly selected and periodically adjusted to provide varied audio-visual input. A trained facilitator managed the technology and program changes but refrained from direct instructional involvement to preserve natural teacher–student dynamics. Rather than imposing strict experimental controls, the study embraced the variability of real-world classrooms to enhance validity and inform the development of scalable, context-appropriate assistive technologies. Observations focused on child responses (physical, cognitive, emotional) and teacher interaction with the ISB, including ease of use, affective responses, and moments of social engagement.

Due to constraints on in-session evaluation by specialists, all sessions were video recorded. Four independent reviewers, including two child-behaviour specialists, assessed engagement, self-regulation, focus, and social interaction using predefined criteria. A multi-observer strategy supported reliability and reduced interpretive bias.

A structured risk evaluation matrix was used to identify participant-specific and contextual risks, consistent with AT safety requirements [19] and ISO standards. All participants were classified as "high risk" based on institutional criteria. The matrix informed session protocols, safety measures, and iterative ISB refinements (Table 1).

Table 1. Participant Risk-Evaluation Matrix

Risk Category	Low Risk	Moderate Risk	High Risk
Communication Skills	Child communicates effectively with some support	Child communicates but struggles in certain situations	Child has significant difficulty communicating effectively
Social Interaction	Child engages in social interactions with minimal support	Child shows limited social interaction and needs more encouragement	Child has difficulty or avoids social interaction, affecting relationships

(continued)

Table 1. *(continued)*

Risk Category	Low Risk	Moderate Risk	High Risk
Sensory Sensitivity	Child has mild sensitivity and manages well	Child has moderate sensitivity, affecting some activities	Child has severe sensitivity, significantly impacting daily activities
Behavioral Management	Minor behavioral issues, manageable in a regular setting	Moderate behavioral concerns needing structured support	Severe behavioral issues requiring specialized intervention
Emotional Regulation	Child generally self-regulates emotions	Child sometimes struggles with emotional control	Child frequently experiences emotional dysregulation affecting participation
Learning and Cognition	Child learns effectively with little support	Child has moderate learning challenges requiring tailored support	Child has significant learning challenges impacting overall progress
Adaptive Skills	Child performs daily tasks independently	Child requires reminders or some assistance	Child needs consistent support for daily tasks

2.3 SMED Approach and Video Analysis

An adapted version of the Single-Minute Exchange of Die (SMED) method, originally from manufacturing, was repurposed as a behavioral assessment tool to support rapid, iterative refinement of the interactive sensory ball (ISB) [20]. SMED enabled short-cycle evaluation of different ISB configurations, such as light, sound, and tactile feedback, based on real-time user responses. Each interaction acted as a "changeover point" to identify features linked to positive behavioral outcomes like engagement and self-regulation. All sessions were video recorded and reviewed by child behavior specialists, with multi-angle footage allowing detailed analysis of subtle cues, including facial expressions and social interactions. This combined approach provided a structured, responsive framework for assessing intervention impact and guiding user-informed design decisions.

2.4 Data Analysis

A mixed-method approach was employed to analyze the data collected from the ISB testing sessions. Quantitative analysis involved descriptive statistics and the use of the Wilcoxon Signed-Rank Test to compare pre- and post-intervention scores related to engagement, self-regulation, focus, and social interaction. This non-parametric test was selected due to the small sample size and the ordinal nature of the data obtained through 5-point Likert-scale observations.

Qualitative data, including observational notes and documented teacher behaviors, were subjected to thematic analysis. This method was used to identify recurring patterns and categories related to teacher engagement, perceived usability of the ISB, and the observed behavioral impact on students.

The integration of these quantitative trends and qualitative themes enabled a holistic understanding of the ISB's effectiveness and usability in classroom contexts. Furthermore, findings from both data streams informed iterative product improvements, particularly regarding the optimization of sensory outputs and content upgrades. Grounding development cycles in empirical evidence ensured that refinements remained directly responsive to user experience and contextual realities, thereby operationalizing the research process into a structured product feature refinement pathway (Fig. 1).

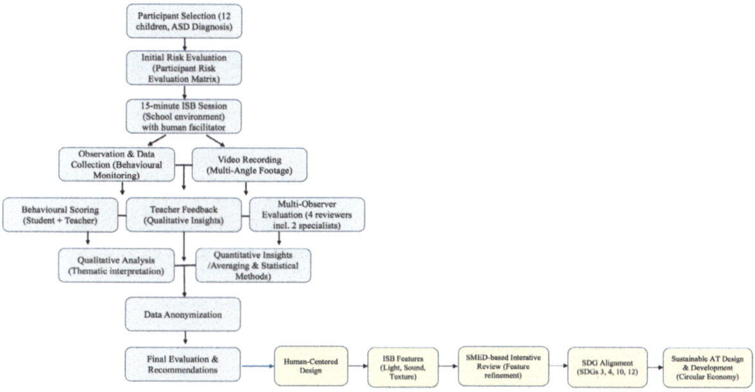

Fig. 1. Research and Product Feature Refinement Process

3 Results

3.1 Quantitative Results

Observed Student Behavior (Wilcoxon Signed-Rank Test). A Wilcoxon Signed-Rank Test was conducted on pre- and post-intervention scores for behavioral measures: Engagement, self-regulation, attention/focus, social interaction and overall behavior. The analysis revealed statistically significant improvements across all measures following the use of the interactive sensory ball (ISB). Participants' behaviors were scored using a 0–5 scale for each domain. Across the 12 participants, mean scores increased in all five behavioral areas during ISB use compared to baseline (Table 2, Fig. 2).

This uniform pattern of improvement across all 12 participants provides strong preliminary evidence of the tool's behavioral impact. To evaluate whether this observed improvement is statistically significant, we tested the following hypotheses:

- Null Hypothesis (H_0): There is no difference in scores with /without ISB
- Alternative Hypothesis (H_1): Scores increase with ISB

Table 2. Observation Score/Observational Domain Description

1. Engagement	*Actively interacts with the ISB (e.g., touching, manipulating, exploring)*
2. Self-Regulation	*Demonstrates calm body posture and reduced agitation during use*
3. Focus/Attention	*Sustains attention on ISB for an extended duration*
4. Social Interaction	*Makes eye contact with adult or peer during ISB use/ Shares or invites joint attention*
5. Overall Behavior	*Displays cooperative behavior*

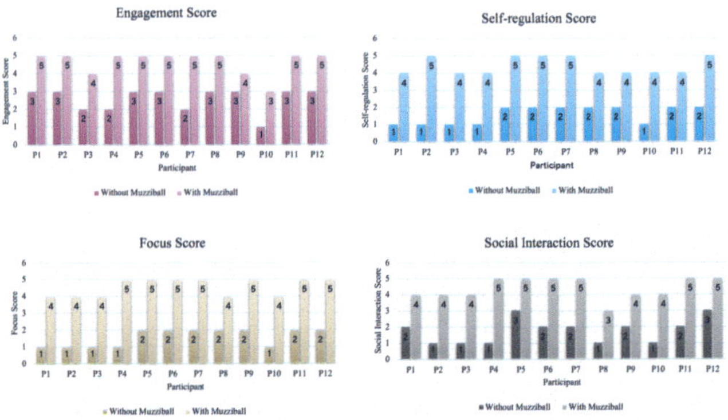

Fig. 2. Participant Behavior Scores

These results indicate a statistically significant improvement in all score categories following the introduction of ISB ($p < 0.05$). The consistent direction of change across all participants strengthens the case for a real effect, particularly in educational settings aiming to foster attention and participation in children with ASD (Table 3).

Table 3. Observed Student Behaviors

Behavioral Domain	*n*	*Mean without ISB*	*Std. Dev. Without ISB*	*Mean with ISB*	*Std. Dev. With ISB*	*Observed change*	*Wilcoxon Statistics*	*P-Value*
Engagement	12	2,58	0,64	4,66	0,62	2,08	0,00	0,001
Self-Regulation	12	1,58	0,49	4,41	0,49	2,83	0,00	0,001
Focus/Attention	12	1,58	0,49	4,58	0,49	3,00	0,00	0,001
Social Interaction	12	1,75	0,72	4,41	0,64	2,66	0,00	0,001
Overall Behavior	12	1,58	0,64	4,50	0,50	2,92	0,00	0,001

Observed Teacher Behavior (5-Point Likert Scale). To complement the analysis, researcher-led observations were also conducted using a structured 5-point Likert scale (0 = Not observed, 5 = Strongly and consistently observed) to assess teacher behaviors while facilitating ISB sessions. These observations focused on four domains relevant to classroom integration and educator experience (Table 4).

Table 4. Observed Teacher Behaviors

Observed Domain	Description	n	Mean Score (0–5)
Engagement	*Teacher actively participates in the session* *Teacher stays attentive and responsive throughout the session*	12	5,00
Device Ease of Use	*Teacher demonstrates smooth handling of the device* *Teacher requires minimal external support to operate ISB*	12	4,16
Positive Affective Reaction	*Teacher expresses enjoyment or satisfaction (e.g. smiles, laughter, positive comments)* *Teacher provides verbal/non-verbal cues that indicate approval of the device*	12	4,66
Positive Social Exchange	*Teacher builds connection or dialogue around the device use (e.g. joint attention moments)*	12	4,66

The high average scores across all four domains indicate that teachers were consistently engaged and demonstrated comfort using the ISB, while also exhibiting positive affect and social interaction with students. These findings support the ISB's potential for practical classroom implementation and underscore its value as an inclusive, educator-compatible tool.

3.2 Qualitative Results

To complement the quantitative findings, qualitative data from observer notes, teacher reflections, and field annotations were thematically analyzed, revealing consistent patterns across sessions. Students generally exhibited increased engagement, including prolonged attention and active participation, particularly among those typically less involved in classroom activities. The ISB appeared to support emotional regulation, with several students displaying calmer behavior and sustained focus, potentially due to the device's multisensory design and feature combination. Teachers noted that the ISB facilitated meaningful teacher–student interactions by serving as a shared point of focus that encouraged verbal and non-verbal communication. The device was reported as largely intuitive and feasible to integrate within standard classroom routines. However, some educators highlighted the need to explore adaptability possibilities, suggesting that customizable educational content could enhance the ISB's relevance across diverse learner profiles.

4 Discussion

This study highlights the potential of adaptive, human-centered assistive technologies to address key challenges in inclusive education. The ISB supported consistent improvements in engagement, focus, self-regulation, and social interaction among students with ASD. Teachers reported ease of use, minimal training needs, and enhanced interaction, underscoring dual benefits for student support and teacher well-being. An action research approach enabled iterative refinement, informed by classroom feedback, SMED methods, and co-design principles. Safety profiling and the evaluation matrix ensured risk awareness, while circularity was embedded through material durability, responsible design and feature customization possibilities. The project also aligns with Scherer's emphasis on stakeholder outcomes such as safety, usability, and satisfaction [21], reinforcing the ISB's relevance across user groups. These findings suggest the ISB can improve inclusion while also supporting system-level sustainability.

5 Conclusions/Implications for the AT Field

This pilot study provides strong evidence that the ISB enhances behavioral outcomes for children with ASD, with statistically significant gains and consistent positive feedback from educators. Its inclusive, low-barrier design benefits both students and teachers, aligning with SDGs 3, 4, 10 and 12. Beyond the classroom, the ISB contributes to circular economy thinking by addressing overlooked systemic factors, such as reducing teacher absenteeism linked to daily stress and minimizing indirect educational costs. These dimensions reinforce its value as a replicable, sustainable AT model. Future research should extend implementation, include academic and economic measures, scale impact across diverse student population, including both neurotypical and neurodivergent learners, and inform inclusive policy development.

References

1. Valencia, K., Rusu, C., Quiñones, D., Jamet, E.: The impact of technology on people with autism spectrum disorder: a systematic literature review. Sensors **19**, 4485 (2019). https://doi.org/10.3390/s19204485
2. Porayska-Pomsta, K., et al.: Developing technology for autism: an interdisciplinary approach. **16**, 117–127 (2012). https://doi.org/10.1007/S00779-011-0384-2
3. Ashburner, J., Ziviani, J., Rodger, S.: Sensory processing and classroom emotional, behavioral, and educational outcomes in children with autism spectrum disorder. American Occupational Therapy Association. https://doi.org/10.5014/ajot.62.5.564
4. Schultz, T.R., Able, H., Sreckovic, M.A., White, T.: Parent-teacher collaboration: teacher perceptions of what is needed to support students with ASD in the inclusive classroom. Educ. Train. Autism Dev. Disabil. **51**, 344–354 (2016)

5. De Domenico, C., et al.: Exploring the usefulness of a multi-sensory environment on sensory behaviors in children with autism spectrum disorder. J. Clin. Med. **13**, 4162–4162 (2024). https://doi.org/10.3390/jcm13144162
6. Unwin, K.L., Powell, G., Jones, C.R.: The use of multi-sensory environments with autistic children: exploring the effect of having control of sensory changes. Autism **26**, 136236132110501 (2021). https://doi.org/10.1177/13623613211050176
7. Dabat Pivotto, I., Matias, V., de Paula Ferreira, W.: CAVE automatic virtual environment technology to enhance social participation of autistic people: a classification and literature review. Res. Autism Spectrum Disord. **117**, 102453 (2024). https://doi.org/10.1016/j.rasd.2024.102453
8. Zhao, W., Xu, S., Zhang, Y., Li, D., Zhu, C., Wang, K.: The application of extended reality in treating children with autism spectrum disorder. Neurosci. Bull. (2024). https://doi.org/10.1007/s12264-024-01190-6
9. Habibi, F., Sedaghatshoar, S., Attar, T., Shokoohi, M., Kiani, A., Malek, A.N.: Revolutionizing education and therapy for students with autism spectrum disorder: a scoping review of AI-driven tools, technologies, and ethical implications. AI Ethics (2025). https://doi.org/10.1007/s43681-024-00608-1
10. Panjeti-Madan, V.N., Ranganathan, P.: Impact of screen time on children's development: cognitive, language, physical, and social and emotional domains. Multimodal Technol. Interact. **7**, 52 (2023). https://doi.org/10.3390/mti7050052
11. Eddy, C.L., Herman, K.C., Reinke, W.M.: Stress management programs for special education teachers. J. Emot. Behav. Disord. **32** (2024). https://doi.org/10.1177/10634266241234917
12. Fernández-Batanero, J.M., Montenegro-Rueda, M., Fernández-Cerero, J., García-Martínez, I.: Assistive technology for the inclusion of students with disabilities: a systematic review (2022)
13. Oldfrey, B., et al.: Could assistive technology provision models help pave the way for more environmentally sustainable models of product design, manufacture and service in a post-COVID world? Sustainability **13**, 10867 (2021). https://doi.org/10.3390/su131910867
14. Cañete, R., et al.: A new multi-criteria decision-making approach for the design and selection of materials and manufacturing processes of toys for children with autism. Mater. Today Commun. **40**, 109709 (2024). https://doi.org/10.1016/j.mtcomm.2024.109709
15. Allen, C., Metternicht, G., Wiedmann, T.: Initial progress in implementing the sustainable development goals (SDGs): a review of evidence from countries. Sustain. Sci. **13**, 1453–1467 (2018). https://doi.org/10.1007/s11625-018-0572-3
16. Reason, P., Bradbury, H.: The SAGE Handbook of Action Research. SAGE Publications Ltd., London (2008)
17. Yin, R.K.: Case Study Research and Applications: Design and Methods. SAGE Publications (2018)
18. Sanders, E.B.-N., Stappers, P.J.: Co-creation and the new landscapes of design. CoDesign **4**, 5–18 (2008). https://doi.org/10.1080/15710880701875068
19. Ryan, S.E.: Injury risk compensation in children with disabilities: could assistive technology devices have a dark side? Disabil. Rehabil. Assist. Technol. **5**, 199–208 (2010). https://doi.org/10.3109/17483101003602548
20. Shingo, S.: A Revolution in Manufacturing: the SMED System. Productivity Press, Portland (1985)
21. Scherer, M.J.: Technology adoption, acceptance, satisfaction and benefit: integrating various assistive technology outcomes. Disabil. Rehabil.: Assist. Technol. **12**, 1–2 (2016). https://doi.org/10.1080/17483107.2016.1253939

Accessibility in Digital Spaces

Flex Picture eBook: Software for the Creation of Accessible Interactive Images

Maximilian Punz(✉) and Danya Gharbieh

Institute Integriert Studieren, JKU Linz, Linz, Austria
{maximilian.punz,danya.gharbieh}@jku.at

Abstract. Flex Picture eBook utilizes interactive images, containing multiple levels of complexity which are adaptable to the demands of visually impaired children. However, creating these images requires a lot of time and technical knowledge [1]. To reduce this workload, a plugin suite for Adobe Illustrator [2] was developed, that allows artists to create these illustrations inside of the software they are used to. With these tools, they are able to imbue their work with multi-layer switching functionality, giving their users the ability to select one of 5 levels of accessibility, as well as allowing zooming in on elements they selected. This software also gives artists the ability of animating complex movements. All this can be done inside of Illustrator, without having to leave the software. This paper focuses on the guidelines developed for the creation of these images, as well as the software that was created to make the production of the illustrations possible.

Keywords: Accessible Images · Tools for the Creation of Accessible Media · Accessible Kids Books

1 Introduction/Background

The goal of the Flexibook Project is to present a novel genre of digital books that has accessibility built into its very essence. At the core of the idea are images that contain different layers of visual complexity, which can be seamlessly configured in real time. Each of these layers are based around accessibility guidelines aimed at creating media for people with low vision [3]. The first layer, (or level) contains the full complexity image, while the others get more and more simplified. The additional accessibility is achieved through various domains, for example by adding more contrast and reducing the number of colors, separating objects with intersecting silhouettes and gradually removing perspective from the illustrations, ultimately arriving at a version of the original that is the easiest to visually parse.

Different children with visual impairments may benefit from different forms of accessibility, and it is their place to choose which one suits them best [4]. This is the freedom we want to give them with this project.

K. Mavrou and P. Encarnação (Eds.): AAATE 2025, *Technology for Inclusion and Participation for All: Recent Achievements and Future Directions*, pp. 85–90, 2025.
https://doi.org/10.1007/978-3-032-01628-7_11

Another paper [1] deals with the creation process of the eBooks themselves, for which a separate software tool was created. This tool handles all of the backend code and dependencies for the book, packaging all of the files into a finished epub that can then be executed on any device. Great importance is placed on accessibility metadata and users cannot create an ebook without filling out everything that is necessary for an accessible experience.

2 State of the Art

The main way to make images accessible to people with visual impairments, is through using a combination of alternative descriptions (alt text, which is read through a screen reader) and by utilizing high contrast in the image itself [5].

Reducing image complexity also helps to make the image more user-friendly [6]. However, there has to be a balance found between turning an image into a nearly-abstract visualization that has optimal contrast and accessibility and the high visual fidelity that people expect from modern day web and eBook applications. Especially kids books which often contain hand drawn pictures, can be hard to visually parse [7]. The stroke of a brush or a fine pen can leave a mark that is often too thin to identify for everyone. This additional visual detail may look nice, but increases the complexity of a drawing, which therefore does not help in terms of how easy it is to extract the information from it. But, as stated above, reducing this complexity also removes details which are vital to display the artistic intent of a work. This paper will present a novel approach that allows for having both visual complexity as well as an abstraction of the image that only focuses on the most important parts of the image. This method also allows artists to create a discrete number of steps between the most complex and the simplest version, while giving the user the choice to switch to the level that fits their needs most.

Pixel-based Image creation software, such as Photoshop or MS Paint do not have features that facilitate the creation of accessible images. The former lets you add a description of the file as metadata, though this is embedded in such a way that alt text has to be added anyways, when importing the file into a website or software solution.

To create the images proposed for this project, the artists involved in the process used Adobe Illustrator, a tool that allows users to make vector graphics. It is important that they create them through a vector tool, because this allows us to zoom in on the details, without loosing the sharpness of the image. This is because these tools do not save the pixel data itself, but where the lines and shapes are located. The pixels are then calculated depending on the viewing angle. So when an image gets enlarged, the pixels get recalculated accordingly. In pixel-based software, the position of the pixels is set manually. So when an image is enlarged, the pixels become enlarged too.

The resulting image coming from Illustrator was then exported as an SVG file. In the old process, the image had to be drawn in 5 different files, which all needed to be exported separately. Then, a programmer manually copy and pasted the path information found in the SVG files, into a new xHTML file. Here, the layers have to be manually tagged- Meaning which details belong to the which level of complexity.

Sometimes these images where also drawn into one singular file, with paths of objects being used for multiple complexity levels. An example: A picture shows a ball. On the highest level of fidelity, the ball has a drawn outline, as well as a black and white pattern, that defines it as a soccer ball, a logo of the company that made the ball, as well as a drawn shadow, that shows the dimensionality of the object. In the more simplified version, the outline is kept (the circle that implies the shape of the object), with everything else being removed. Here everything would be drawn in one Illustrator file, and the information which objects belong where need to be communicated between artist and programmer.

This approach is complicated further, when looking at the way animations where done before the tool was developed: Artists drew the pages, then made visual representations of the animations with text describing the movement of the objects ("The red ball should move along the path for 5 s. It should rotate. Play a sound while it flies"), which were then passed off to the programmer again, who manually created the movements by adding SMIL animation to the raw path data. This was an incredibly difficult and time intensive process.

Research was done on software solutions that solve these problems, but none where found. Adobe After Effects has the ability to use SVGs and animate them, but it can only render them as video files (mp4 or avi files), which do not allow for the dynamic changing of visual complexity. In the Flex Picture format, objects can also be clicked on, which is not possible with mp4 files, as everything is in one frame and cannot be separated. These formats also make zooming harder, because they are pixel based: this means that they loose fidelity if the view gets enlarged, due to a limited number of pixels.

The tool developed for Adobe Illustrator allows the artist to assign which paths belong to which layers directly inside of the software, which reduces the need for an additional programmer. The animations can also be created inside of the suite, further reducing workload and complexity of the process.

3 Methods and User Involvement

To create the aforementioned images (see Fig. 1), a set of guidelines was developed which help illustrators figure out how to do these abstractions. These guidelines contain a theoretical framework, as well as step by step instructions. The former dive into the thought process behind the different layers, while the latter can serve as a detailed guide, explaining what details to extract and omit at specific levels.

This work was mainly based on the W3Cs "Accessibility Requirements for People with Low Vision" [3], the "Web Content Accessibility Guidelines" [9] as well as the "A to Z of assistive technology for low vision" by Perkins [10].

Because these images are SVGs that are controlled via JavaScript, they have to be bundled into a datatype that is able to contain all of the path information

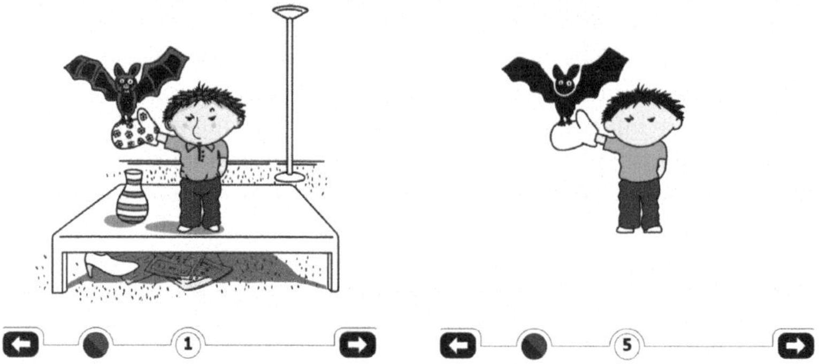

Fig. 1. Left: Version with the highest level of detail. Right: Version that is most accessible. Note that this version has no shading, texture or perspective, while in the other one these things do exist. The backgroud elements also got removed.

needed for the vector-graphics, as well as the code necessary to control them. For this, xHTML is used, because this is the file format compatible with EPUB, which allows us to bundle the images into books that can be viewed on any device. The second part of this project consisted of creating software that facilitates the creation of these interactive xHTMLs. While prototyping, the original images where created via programming. All 5 complexity layers had to be drawn in the same Illustrator file, ported to SVG and manually equipped with the interactive functionality. These new plugins should make the development of the graphics a lot easier. Drawing can still be done inside of the same program, but aspects like layer management, which previously had to be coded, are automated. The artist only has to determine which layer can be seen on which level in the final product. Zooming functionality can be added through a click of the button.

It is possible to add animation to one's creations- which also can be done directly inside of Adobe Illustrator, through the use of our plugins. Users only need to select objects and specify what should happen to them in simple-to-use menus. These instructions are then automatically translated to SMIL tags, a way of animating SVG-files, which are injected in the final images, all without the need to code.

The whole process was developed in collaboration with illustrators that work for a company specializing in accessible kids books (Les Doigts Qui Rêvent, located in Dijon, France). Most of their material is of a physical nature. Their books give kids a tactile experience by the addition of tangible textures, which are hand crafted in their workshop. The cooperation started through an interest of theirs to expand to digital media, and the idea for the plugins was created in a collaboration between illustrators and software developers. Their approach to creating such images was thoroughly analyzed and places in the process that took a lot of time, or had great technical difficulties, where to be improved with the software solutions presented here.

Early versions of the plugin where evaluated together with artists and illustrators and new decisions were made in collaboration, to ensure a prototype that is able to patch the technological holes in the creation pipeline and to be actually useful in their workflow.

Another publishing company who wanted to adapt a physical book to the format was brought on unto the project, which served as ample opportunity to check if the software actually solves the problems. In this context, people unfamiliar with the process got to try out the suite, and test how difficult it is to create something with it.

Plagued with some initial difficulties regarding the setup, the software then was able to be used to create such a book successfully.

4 Results

The outcome of this project is a software suite, as well as instruction materials on their use, including guidelines outlining the theoretical concepts.

Video tutorials explaining the functionality of the plugins were made, showing the development of an image, step by step, from the ideation phase, to the export of the finished xHTML file. The only prior knowledge needed to create such an image are visual arts skills, as well as a basic understanding of Adobe Illustrator, but no programming know-how is required.

If budgetary constraints allow it, thorough user-testing on the usability of the software will be done. Eventhough creators where involved in the process of its creation, the feedback was only anecdotal and based on their experience using it, rather than a more large scale user study.

5 Discussion

There are multiple points in the format that could be improved in further work. One problem is the fact that the plugin suite is built in Adobe Illustrator, an unwieldy tool that makes expanding it extremely difficult. It is also quite costly software, costing 26.16 Euros [11] (May 2025) in a monthly subscription model- so using an open source alternative would be optimal. This was not considered in the original project, because all of the partners where requesting Software for the Adobe Suite, as this was their preferred tool to work in. The code is publicly available and can be expanded to one's liking, however, Adobe makes this difficult.

The learning curve of their CEP Plugin programming environment (this is the environment in which extensions for their plugins are built) is extraordinarily steep, not least because most of the official documentation has been removed by the company in an effort to replace this system with a new one [12]. To this day, Adobe has not released the new version of this environment, placing programmers in a position which makes writing code for it incredibly difficult. Possible alternative open-source vector graphic tools could come in the form of the program Inkscape [13] or creating a from-scratch solution precisely aimed at creating accessible media, but this will be left for future work.

6 Conclusions/Implications for the AT Field

We strongly believe in the concept that the best way to spread accessible media is to make its creation as easy and painless as possible. If the right tools are available and the difficulty threshold is lowered far enough that a lot of creatives can produce these kinds of images, then their spread may come as a natural consequence.

Our hope is that the Flex Picture eBook format finds a broad use and many more children's books become available in it. We may not be able to significantly impact the spread of accessible media directly, but at least we were able to make the creation of this format a lot less difficult and time consuming.

References

1. Gharbieh, D., Punz, M.: Flex Picture eBook Builder—Simplifying the Creation of Accessible eBooks. In: Miesenberger, K., Bühler, C., Penaz, P. (eds.) Computers Helping People with Special Needs. LNCS, vol. 12345, pp. 1–13. Springer, Heidelberg (2024)
2. Adobe: Adobe Illustrator, Version 29.0. Adobe Systems Inc. https://www.adobe.com/products/illustrator.html. Accessed 12 May 2025
3. W3C: Accessibility Requirements for People with Low Vision. W3C Working Group Note (2016). https://www.w3.org/TR/low-vision-needs/. Accessed 12 May 2025
4. Gleason, C.: "Creatures of habit": influential factors to the adoption of computer personalization and accessibility settings. In: Miesenberger, K., Bühler, C., Penaz, P. (eds.) Computers Helping People with Special Needs. LNCS, vol. 12345, pp. 1–13. Springer, Heidelberg (2022). https://doi.org/10.1007/978-3-030-96648-5_11. Accessed 14 May 2025
5. Oh, U., Hwayeon J.: Image accessibility for screen reader users: a systematic review and a road map (2021). https://www.mdpi.com/2079-9292/10/8/953. Accessed 13 May 2025
6. Yale University: Usability & Web Accessibility- Images. https://usability.yale.edu/web-accessibility/articles/images. Accessed 14 May 2025
7. Lewis, M., Sturdee, M.: Accessibility of sketches. In: Sketching in Human Computer Interaction. Springer, Cham (2024). https://doi.org/10.1007/978-3-031-50136-4_8. Accessed 14 May 2025
8. Adobe: About file information (metadata). https://helpx.adobe.com/photoshop-elements/using/file-information.html. Accessed 14 May 2025
9. W3C: Web Content Accessibility Guidelines (WCAG) 2.1. W3C Recommendation (2024). https://www.w3.org/TR/WCAG21/. Accessed 12 May 2025
10. Lewis, V.: A to Z of assistive technology for low vision. Perkins (2024). https://www.perkins.org/resource/z-assistive-technology-low-vision/. Accessed 12 May 2025
11. Adobe: Plans and Pricing for Creative Cloud Apps and More. Adobe Systems Inc. (2025). https://www.adobe.com/at/creativecloud/plans.html?filter=graphic-design&plan=individual. Accessed 12 May 2025
12. erinferinferinf: The Future of CEP. Adobe Community Forum. https://community.adobe.com/t5/exchange-discussions/future-of-cep/td-p/11536877. Accessed 12 May 2025
13. Inkscape: Inkscape 1.4 (2024). https://inkscape.org/. Accessed 12 May 2025

Uses of Email Applications by People with Blindness and Low Vision

Mahmut Erdemli[1(✉)], Nika Muller[2], Murat Sahinbas[3], Tianyi Gu[4], Mark Cadigan[5], and Tanja Milosevic[5]

[1] DesignTest.net, Ottawa, ON, Canada
erdemli@gmail.com
[2] ESGInnovationLab.com, London, UK
[3] Google Cloud Partner, Istanbul, Turkey
[4] Algonquin College, Ottawa, ON, Canada
[5] GetBraille.com, Boston, MA, USA

Abstract. This study examines the accessibility of email applications for users with blindness or low vision, with a focus on non-visual features such as screen reader compatibility, alternative text descriptions, and voice-based AI interfaces. Through a survey and interviews, the research explored how persons with blindness and low vision use these features to access, compose, and manage emails, and identifies the barriers and challenges they encounter. The findings offer practical insights for improving email accessibility, emphasizing user needs and common barriers faced by users with blindness and low vision.

Keywords: accessibility · email · visual impairment · assistive technology

1 Introduction

The growing reliance on digital communication has introduced new challenges for individuals with blindness and low vision (BLV), particularly in the use of email applications. For users with BLV, composing messages, attaching files, organizing inboxes, searching for information, managing spam, and navigating complex conversation threads often require additional steps and workarounds compared to sighted users. Tasks such as formatting text, handling attachments, identifying important messages, and responding to meeting invites can be time-consuming and cognitively demanding due to inconsistent support for screen readers, lack of semantic structure, and insufficient integration of non-visual interaction methods. These barriers not only affect productivity and independence but also impact the overall user experience, highlighting the need for more inclusive and accessible email application design.

This study aims to investigate how effectively email applications integrate voice interfaces, screen reader compatibility, and keyboard navigation to enhance accessibility. The research will evaluate the usability of current email tools from the perspective of users with BLV, contributing their insights to the development of more equitable and accessible email interfaces.

K. Mavrou and P. Encarnação (Eds.): AAATE 2025, *Technology for Inclusion and Participation for All: Recent Achievements and Future Directions*, pp. 91–98, 2025.
https://doi.org/10.1007/978-3-032-01628-7_12

2 State of Art

Navigating email interfaces often presents challenges for individuals relying on screen readers or keyboard-only interaction. Tasks such as searching for messages, finding contacts, managing complex forms, and interpreting interface layouts can be particularly difficult [10]. Key factors contributing to perceivable navigation include text alternatives for images, logical tab order, adequate confirmation feedback, and clear headings with consistent user interface assets [11]. Common issues include non-semantic HTML, unlabeled buttons, and complex layouts that hinder efficient navigation for BLV users [10,11]. Screen readers process content sequentially, making navigation slow and inefficient in cluttered or complex layouts [8].

Organizing email effectively is challenging for all users, but new approaches like Google's "priority inbox" and the use of "labels" instead of traditional folders can be confusing for BLV users who are familiar with folder-based systems. As a result, there is a strong need for better search features and simpler navigation [10]. Many accessibility barriers could be addressed with straightforward design improvements [3]. Research on alternative interfaces, such as simplified HTML views and voice-based Gmail interfaces, suggests that reducing interface complexity can improve usability for BLV users [1,2]. Security is also a major concern, as BLV users are more likely to miss visual phishing warnings and other security cues [4,9]. Studies show that screen reader users often overlook traditional visual phishing alerts, increasing their vulnerability [4,9].

Voice-based email systems leverage speech-to-text and text-to-speech technologies to support users with BLV in managing email through voice commands, reducing barriers such as complex navigation, reliance on visual cues, and the need to memorize shortcuts [5–8]. These systems enable users to read, compose, and organize emails without depending on a keyboard or screen reader. While voice-based systems can enhance accessibility by allowing users to navigate messages, dictate content, and manage attachments using spoken instructions, their effectiveness is influenced by the accuracy of speech recognition and the clarity of system feedback. Common challenges include background noise, accent variability, occasional requirements for touch or keyboard input, and the inherently sequential nature of audio, which can complicate tasks such as identifying phishing attempts [4,9].

3 Methodology

This study aimed to identify the challenges and needs of users with BLV when interacting with email applications and their non-visual accessibility features. The objective was to determine how email applications can be improved to better support users in composing, managing, and navigating emails, with a focus on features such as voice interfaces, search functionality, spam handling, and attachment management.

3.1 Procedure

Participants were recruited through organizations serving individuals with BLV, as well as online forums and mailing lists focused on accessibility. After providing informed consent, participants completed an online survey designed to be accessible with screen readers and other assistive technologies. The survey took approximately fifteen minutes to complete. Following the survey, participants were invited to participate in a follow-up interview. Interviews were conducted remotely via phone or accessible online meeting platforms and lasted thirty to sixty minutes. Interviewers used a semi-structured guide to ensure consistency while allowing participants to elaborate on their experiences and suggestions.

3.2 Participants

Seventeen participants responded to the survey, ranging in age from twenty-one to eighty-two years. The mean age was forty point two four years, with a standard deviation of fourteen years. Participants were distributed across Europe (two), Africa (six), North America (five), Oceania (two), and Central America (two). Twelve participants reported blindness, and five reported low vision classified as "legally blind" (visual acuity less than twenty over one hundred). The onset of vision impairment among participants ranged from birth to fifty-four years of age, with a mean onset age of thirteen point seven years and a standard deviation of nineteen years.

3.3 Data Collection

A survey was designed to be accessible with screen readers and included single-select, multiple-select, multiple-choice, and open-ended questions. It gathered information on participants' demographics, types of email applications used, frequency and context of use, and specific assistive technologies employed (such as screen readers, voice commands, and magnifiers). Participants rated the accessibility of desktop, browser-based, and mobile email applications and provided suggestions for improvements. The survey also explored the use of voice AI assistants (e.g., Siri, Google Assistant) for email management and solicited feedback on desired features and challenges.

Follow-up interviews explored participants' experiences in greater depth, focusing on challenges related to composing emails, managing attachments, navigating inboxes, and using features such as spell check and meeting integration. Interviewers encouraged participants to elaborate on obstacles, strategies for overcoming barriers, and recommendations for improving non-visual email accessibility.

Participants were also asked to suggest conceptual designs or specific features they would like to see added or removed from existing email applications. Suggestions included introducing a simplified, clutter-free interface option, customizable keyboard shortcuts, improved voice command integration, and better-structured navigation with headings. Participants recommended removing unnecessary or redundant features that contribute to interface complexity, such as excessive folders or rarely used buttons, and emphasized the importance of clear, consistent labeling and feedback for all interactive elements.

All survey responses and interview transcripts were anonymized and securely stored for analysis.

3.4 Data Analysis

Quantitative survey data were analyzed using descriptive statistics to summarize usage patterns and accessibility ratings. Qualitative responses from both the survey and interviews were subjected to thematic analysis to identify recurring challenges, user needs, and opportunities for improvement in email accessibility for BLV users.

4 Results

4.1 Survey

The survey provided insights into the experiences of participants with vision impairments regarding their use of email applications, assistive technologies, and voice AI assistants.

Participants accessed email through various methods. Outlook was used by twelve participants, browser-based HTML interfaces by four, mobile apps like Gmail and Outlook by eleven, and voice assistants such as Siri and Google by one. Other methods included Apple's mail app. Assistive technologies that were used included screen readers (fifteen), voice commands (six), and magnifiers (two), with additional tools such as tactile keyboards and haptic feedback. Eleven participants used email daily, four used it a few times a week, and three used it less frequently. Common email applications included Gmail (ten), Outlook (twelve), and Apple Mail (five). Operating systems used were Windows (ten), Apple (seven), and Linux (one).

Accessibility ratings for desktop email applications were mixed. Two participants strongly agreed on accessibility, four agreed, and others expressed challenges with features like with complex and cluttered interfaces. For browser-based email applications, six participants agreed on accessibility, while others noted issues with navigating graphics and inconsistent layouts. Mobile email applications received feedback that included one participant strongly agreeing, four agreeing, and others highlighting difficulties with small screens and email summaries.

For desktop applications, participants recommended simplifying menus, improving descriptions, and adding more shortcut keys. For browser-based applications, they suggested ensuring proper semantic HTML and improving navigation for emails with graphics. For mobile applications, participants proposed adding keyboard shortcuts, simplifying navigation, and improving threading and folder management.

Voice AI assistants such as Siri (six) and Google (five) were used for tasks such as opening apps and creating reminders. Participants expressed a desire for AI to assist with reading, composing, and managing emails, as well as scheduling appointments. Other comments emphasized the need for user-centered design, simplified interfaces, and consistent updates to improve accessibility and efficiency.

4.2 Interviews

The interviews revealed eleven major themes regarding the experiences of BLV users with email applications, highlighting both challenges and opportunities for improvement.

- **Handling images:** Embedded images, especially those containing contact details in signatures, remain a significant obstacle. Screen readers frequently announce only generic labels like "graphic" or read out lengthy URLs, offering little useful information. Extracting details from such images often requires workarounds like OCR, which are not always reliable.
- **Tables and formatting:** Complex tables can confuse screen readers, making navigation within messages difficult. Formatting actions such as bold or italic text are slower to access via toolbars than through keyboard shortcuts. Participants expressed a preference for shortcuts, context menus, or gestures to streamline formatting tasks.
- **Managing attachments:** Locating attachments, particularly in long threads, is tedious and requires excessive navigation. Attachments in Gmail's web version, for example, appear at the end of emails, necessitating extra scrolling. On mobile apps, users often have to swipe repeatedly to reach attachments. Converting inaccessible formats like PDFs to accessible ones is a common pain point, leading some to request attachments in Word format. Workarounds, such as copying and pasting files in Outlook, are often used to save time.
- **Email threads and conversations:** Navigating lengthy conversation threads is challenging, especially when applications collapse messages or order them in confusing ways. Expanding each message individually is time-consuming, though some screen readers offer shortcuts to move through chat history more efficiently.
- **Interface changes and updates:** Frequent updates to email applications can disrupt established workflows. Unlabeled or relocated buttons and features force users to re-learn interfaces, increasing cognitive load. Updates that add steps to common tasks or remove familiar features, such as Outlook Classic or Gmail's simplified HTML view, are particularly disruptive.
- **Simplified interfaces:** The removal of Gmail's simplified HTML view was widely cited as a setback. Participants valued this option for its minimal styling and faster navigation, especially on slower connections. There is a strong desire for a return of such streamlined, HTML-based interfaces.

- **Managing email volume and spam:** High email volume and spam make it difficult to identify important messages quickly. Participants expressed a need for more effective filtering and organization tools to manage both personal and professional inboxes.
- **Navigation without structure:** Emails lacking clear heading structure force users to navigate line by line, making it hard to scan or jump between sections. Well-structured headings would improve navigation, especially in newsletters or itemized lists.
- **Excessive folders:** Too many folders, particularly default or auto-generated ones, increase organizational complexity. Participants prefer a basic set of folders and find similar names like "Junk" and "Trash" confusing.
- **Interface clutter:** Overly complex interfaces with numerous features and buttons make navigation cumbersome. Participants desire a simpler, less cluttered interface focused on essential functions.
- **Voice interfaces:** While some participants use voice commands for basic tasks, they often prefer keyboard shortcuts for efficiency. Key areas where AI and voice assistants could provide value include assisting with email composition and replies, suggesting context-aware responses, and improving the formatting and clarity of dictated text.

Participants discussed the potential for voice assistants and assistive technologies to address challenges in email use. AI-powered features such as summarizing long emails, organizing inboxes, and providing assistance for managing emails were frequently mentioned. There was interest in voice assistants helping with composing and replying to emails, suggesting responses, and improving the clarity of written text with AI assistance. Summarization features were valued for quickly understanding long emails or receiving daily overviews of meetings. Improved organization and search, such as filtering by category or keyword, were also highlighted. Some participants wanted voice assistants to support daily workflow with prompts, reminders, and recaps.

Despite this interest, concerns remain about the accuracy, reliability, and privacy of voice assistant solutions. Current use of voice assistants like Siri or dictation is limited to basic tasks, mainly due to concerns about errors. AI-based image recognition tools, such as Be My Eyes, are used to describe images in emails, but often require follow-up questions. Many users still prefer keyboard shortcuts for their efficiency, noting that voice commands can introduce errors that require correction.

Participants emphasized that any AI integration should prioritize user control, transparency, and compatibility with accessible, streamlined interfaces. Concerns were raised about the accuracy of AI-generated content, especially in summarizing messages or recognizing names, as well as the reliability of AI platforms—frequent updates can disrupt workflows or introduce unlabeled elements. Maintaining context throughout interactions and ensuring concise notifications were also important. Privacy and confidentiality were significant barriers, with some organizations restricting AI use due to data security concerns.

A direction discussed was integrating AI assistants with simplified, accessible user interfaces. This would allow users to interact via voice while retaining manual control through a clear UI. These insights highlight the need for email application design that goes beyond technical compliance, focusing on efficiency, clarity, and user control. The integration of AI-driven tools within simplified, accessible interfaces was seen as a promising direction for future development.

5 Conclusion

This study demonstrates that, while technical accessibility of email applications for people with BLV has improved, significant usability challenges remain. Participants consistently reported that while many email applications meet technical accessibility standards, they often fall short in terms of efficient usability for screen reader users. This gap results in tasks taking longer and requiring more effort compared to sighted users, impacting productivity and daily workflow. Participants reported that complex layouts, inefficient navigation, and inconsistent support for non-visual interaction continue to hinder independent and efficient email use. Key issues include difficulties with images, attachments, and security cues, as well as the impact of frequent interface changes and excessive interface clutter. The findings highlight the need for simplified, user-centered designs that prioritize robust keyboard and voice controls, clear structure, and reliable non-visual feedback. Ongoing collaboration with BLV users is essential to ensure that email platforms evolve to meet their needs, supporting both accessibility and productivity in digital communication.

References

1. Alkhattali, M.: Development of a computer application for handicapped people to use Gmail (2019)
2. Dasgupta, P., Magee, J.: HAIL Gmail: email with hierarchical adaptive interface layout. In: Antona, M., Stephanidis, C. (eds.) Universal Access in Human–Computer Interaction. Design and Development Approaches and Methods, pp. 324–332. Springer, Cham (2017). https://doi.org/10.1007/978-3-319-58706-6_26
3. Ilyas, M.: A study of web accessibility barriers for older adults, and heuristics evaluation of email websites based on web accessibility heuristics for older adults by AARP. **3**(5) (2012)
4. Khan, E.B., Atique, E., Javed, M.: Investigating phishing threats in the email browsing experience of visually impaired individuals. In: Extended Abstracts of the CHI Conference on Human Factors in Computing Systems, CHI EA2024, pp. 1–11. Association for Computing Machinery, New York (2024). https://doi.org/10.1145/3613905.3651076
5. Kumar, R., Singh, V., Akhtar, D.N., Verma, M., Srivastava, S.: Voice based email system for people with visual impairment. **4**(3) (2022)
6. Mujawar, T., Mule, P., Fartade, P., Bhoite, P., Ashtekar, N., Kurzekar, P.: Voice based Gmail service for visually challenged people. Int. J. Comput. Sci. Eng. **7**(3), 1084–1087 (2019). https://doi.org/10.26438/ijcse/v7i3.10841087

7. Nandgaonkar, V., Mane, S., Bhoite, S., Kakade, M., Bhosale, P.: A review on voice-based email system for visually impaired. **13**(1) (2024)
8. Sasikala, C., Rupalakshmi, M., Ahammad, S., Meghana, U.: Enhancing blind communication with voice-assisted email. In: 2024 International Conference on Expert Clouds and Applications (ICOECA), pp. 307–312 (2024). https://doi.org/10.1109/ICOECA62351.2024.00062
9. Sharevski, F., Zeidieh, A.: Assessing suspicious emails with banner warnings among blind and {low – vision} users in realistic settings. In: 33rd USENIX Security Symposium (USENIX Security 24), pp. 2083–2100 (2024)
10. Wentz, B., Hochheiser, H., Lazar, J.: A survey of blind users on the usability of email applications. Univ. Access Inf. Soc. **12**(3), 327–336 (2013). https://doi.org/10.1007/s10209-012-0285-9
11. Wentz, B., Lazar, J.: Usability evaluation of email applications by blind users. **6**(2) (2011)

Accessibility Guidelines for 360-Degree Platforms: Derivation and Implementation

Petya Ivanova(✉), Pirmin Gersbacher, and Gottfried Zimmermann

Stuttgart Media University, Stuttgart, Germany
p.petrova.ivanova@gmail.com

Abstract. 360-degree media is growing in prevalence in fields such as education, tourism, architecture, and leisure. However, in cases of negligence in accessibility, immersive experiences can impose significant challenges for persons with disabilities.

Currently, there are no official standards for 360-degree media. This aggravates the process of creating accessible immersive images and videos. The study aimed to review existing guidelines for immersive and non-immersive software and filter them by relevance to 360-degree media. As a result, we compiled a set of requirements that can serve as a tool for manual accessibility testing.

In order to detect common barriers among applications that implement 360-degree media, we reviewed the accessibility of three educational learning platforms - ThingLink, Uptale, and ImmerseMe. All three platforms fulfilled less than 50% of the requirements, which proves the relevancy of the topic. Based on the findings, we summarized recommendations for platform developers and authors of 360-degree experiences. Future research should focus on involving real users in order to expand the findings of this study.

Keywords: Guidelines · 360 · Immersive platforms · Evaluation

1 Introduction

360-degree images and videos capture real places and situations [1]. The camera operator films in an omnidirectional format, allowing the media consumers to explore different angles of the setting. This format supports 3 Degrees of Freedom (DoF), meaning movement in the space occurs solely through predefined teleportation points. Unlike the more immersive VR environments, users cannot influence their surroundings through their actions. To view the content, one can use a desktop or a mobile device. However, a VR headset is best suited for experiencing full immersion.

Because it allows users to simulate real-life scenarios, 360-degree media has become an educational tool that schools and universities use to enhance the traditional classroom environment [2].

K. Mavrou and P. Encarnação (Eds.): AAATE 2025, *Technology for Inclusion and Participation for All: Recent Achievements and Future Directions*, pp. 99–104, 2025.
https://doi.org/10.1007/978-3-032-01628-7_13

2 Related Work

Current accessibility research focuses on XR, or specializes in VR software. For instance, Meta [5] provides nine accessibility recommendations as a part of their Virtual Reality Checklist (VRC). However, fulfilling those guidelines is not mandatory for publishing an application. Regarding 360-degree platforms, there are no official overarching accessibility requirements that developers and content creators can follow. There is some initial research on the subject. However, it is limited to specific accessibility measures, namely captions [7] and audio descriptions [8]. Because we wanted to cover a variety of impairments in our research, we defined seven user groups – persons with visual, auditory, mobility, vestibular, speech, and cognitive impairments, and photosensitive epilepsy.

In a similar study, Zimmermann et al. [10] tested the platform ThingLink[1], using chapter 4 of EN 301 459 v.3.2.1. We expanded the findings of this paper by defining a more comprehensive set of guidelines. We included five more chapters of EN 301 459 v.3.2.1 and documents focusing on XR, VR and 360. To gain a better understanding of existing barriers in 360-degree applications, we tested two more platforms – ImmerseMe[2] and Uptale[3].

3 Methods

We chose to derive the guidelines deductively. We started with requirements applying to a broader spectrum of digital products and services, as they have been a subject of deeper research. We used EN 301 549 v3.2.1 [3] as the building block for the set of guidelines, because it provides a solid foundation for fulfilling legal accessibility requirements for applications in the EU [13]. We included chapters 4 (Functional performance), 5 (Generic requirements), 7 (ICT with video capabilities), 9 (Web), 11 (Software), and 12 (Documentation and support services), as they are the most relevant to 360-degree media. Consequently, we included sources focusing on XR - *XR Accessibility User Requirements* [4], VRC by Meta [5], and *Virtual Environments Accessibility Guidelines* [6]. Lastly, in order to specify the requirements, we incorporated two studies on 360-degree media - *Recommendations for accessible captions in 360 degree video* [7], and *Equal Entry Guidelines for Describing 360-Degree Video* [8]. To simplify the testing process and make the set of guidelines more easily perceivable, we assigned a category and a subcategory for all requirements. The category represents the impairment and the subcategory stands for the topic of the guideline.

By using the constructed set of guidelines, we manually tested three educational learning platforms – ThingLink, Uptale and ImmerseMe. For each application, we created a demo course containing as many accessibility barriers as possible. For testing, we used a desktop device with a Safari 18.1 browser, a mobile device with a Chrome 131 browser, Meta Quest 3 headset with and without controllers, and a Bluetooth 3.0 keyboard.

[1] https://www.thinglink.com/.
[2] https://immerseme.co/#home.
[3] https://www.uptale.io/en/.

After identifying the barriers across the three platforms, we proposed recommendations on fixing the specific issues within the platforms and general suggestions for ensuring the accessibility of similar applications. Furthermore, we provided workarounds for authors on avoiding some of the barriers and making the content more accessible.

4 Results

The final document contains 81 guidelines [11]. While testing the applications, the following barriers stand out:

- keyboard and screen reader incompatibility;
- missing text alternatives;
- lack of interaction methods for device-independent navigation;
- lack of audio descriptions;
- lack of captions for 360-degree media;
- lack of instructions;
- presence of unavoidable flashes or blinking;
- lack of text customization options;
- lack of magnification options;
- insufficient contrast;
- immutable sounds;
- unstoppable animations.

4.1 Platforms

Uptale's demo course presented 53 barriers [11]. The problems that require most attention are the incompatibility with screen readers and keyboards both in browser and in VR.

Compared to the other two platforms, Uptale offers a wider variety of functionalities. However, they tend to be less accessible. For example, the captions are not optimally executed, as there is no option for file upload. The author has to enter the lines of text for each timeframe manually. Furthermore, developers have given authors more freedom in designing the courses. For instance, it is possible to adjust the colors and sizes of the elements. Nonetheless, this leaves room for more barriers, especially if the author is not familiar with best practices for accessibility.

On the contrary, the ImmerseMe's team has prepared all the available lessons and the educators cannot change them. They can only choose the type of content to include or exclude from the program. For this reason, the authors have very little influence on the accessibility of the courses. However, they can still make the course inaccessible, if they choose to include only types of content that require speaking or hearing.

ImmerseMe has the least amount of barriers, namely 42. The biggest accessibility issue is a glitch that appears in the VR mode after exiting or finishing a course, which can pose a trigger for persons with photosensitive epilepsy or cognitive impairments.

ThingLink has 49 barriers. It is the only platform out of the three that provides an accessibility mode. This mode has multiple settings that accommodate different disabilities. The developers have focused predominantly on visual and cognitive impairments. The downside of this feature is that it is available only as a static web page, which is not comparable to the regular immersive experience.

4.2 User Groups

The group that would experience the most barriers are persons with visual impairments. There are 18 failed requirements affecting this user group found in all three applications (Fig. 1). However, proportionately to the total number of guidelines for visual impairments, 48,96% of them are unfulfilled. Another group takes the first place – persons with photosensitive epilepsy, with 66,67%.

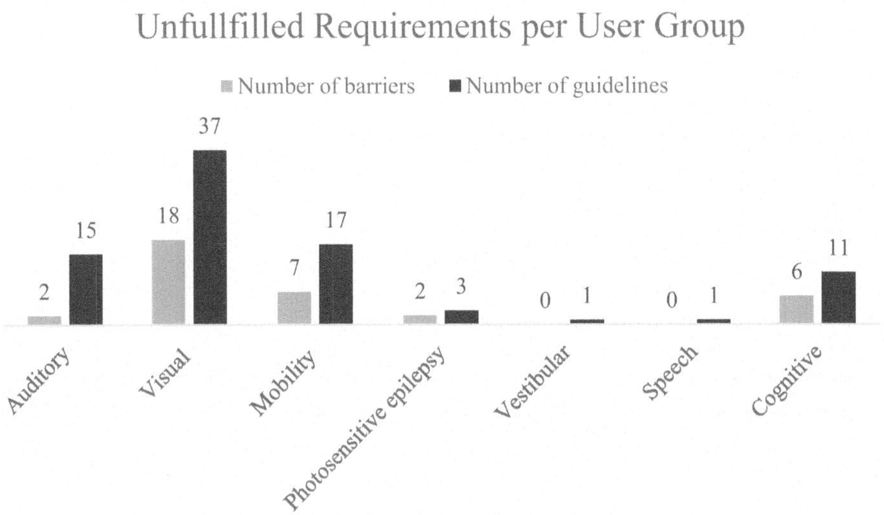

Fig. 1. Unfulfilled Requirements per User Group

5 Recommendations

5.1 Developers

After testing the three applications and identifying barriers, we provided general recommendations that developers of 360-degree applications can follow, as well as some platform-specific suggestions. The general recommendations can be summarized as follows:

- Allow authors to upload captions and synchronize them with the video [3, p. 35].
- Allow authors to upload a second audio file containing an audio description [3, p. 36].
- Provide a caption button and an audio description button on the same action level as other video and audio controls [3, p. 36].
- In cases of shaky footage, provide a feature that stabilizes it after upload. Alternatively, display a warning that the video involves excessive camera movements.
- Provide instructions on navigating the system [12].
- Ensure keyboard accessibility both in browser and in VR [12].

- When the user enables gaze-based interactions, display buttons for rotating the view on the screen.
- Allow users to rotate their view using the thumb sticks.
- Provide meaningful text alternatives [12].
- Ensure screen reader accessibility. If that is not possible, provide a custom screen reader.

5.2 Authors

Certain barriers are avoidable, if authors have the necessary knowledge in accessibility. Here is a summary of the workarounds and recommendations we formulated for the three applications.

- When adding tags, use an icon indicating the topic or purpose [3, p. 22].
- When placing a tag, check if the color contrast with the background is at least 3:1 [12].
- When using text tags or text boxes, ensure the contrast with the background stays above 4,5:1 [12].
- For tags that require the user's input, provide feedback in textual form.
- Avoid tags that require vision, hearing, speaking, or complex movements. If they are necessary for completing tasks, provide an alternative.
- Place tags in the vertical center of the environment, so that they do not require moving the head up and down.
- Avoid timers.
- Provide instructions for tasks [12], if they are not a part of the system.

6 Discussion

XR is still a new technology and is constantly evolving, which could explain the lack of clarity on the accessibility standards. The VR software itself misses some important accessibility features. For instance, many of the headsets on the market do not provide a system-wide screen reader, including Meta Quest 3 that we used for testing. Additionally, the Meta AI Assistant (Voice Command) feature on Meta Quest 3 is unavailable outside Canada and the US [9]. In-app magnification is also not possible. For those reasons, the browser modes of the applications tend to be more accessible.

During our research, we found that certain user groups are underrepresented. For instance, across all documents, we found only one requirement for persons with speech impairments and one for persons with vestibular impairments [11]. For this reason, additional studies are necessary for defining further guidelines for those user groups. Furthermore, testing with real users will highlight other issues we did not find during our theoretical testing process.

7 Conclusion

Using existing standards, requirements, and recommendations for web, software, VR and XR, we expanded initial research on the accessibility of 360-degree media. The document of guidelines covers accessibility measures for seven different types of impairments. To examine the current state of 360-degree platforms, we created demo courses in three educational learning platforms, implementing as many accessibility barriers as possible. Using the set of guidelines, we tested the demo courses and identified common accessibility problems. All three courses fulfilled less than 50% of the requirements. To address the barriers, we provided general and platform-specific suggestions for developers and workarounds for authors.

Looking forward, the development of accessibility standards for immersive environments should uphold the fast pace of their technological innovation.

References

1. Cameron, J., Gould, G., Ma, A.: 360 essentials: a beginner's guide to immersive video storytelling. Ryerson University Library, Toronto (2021). https://openlibraryrepo.ecampusontario.ca/jspui/bitstream/123456789/901/3/360-Essentials-A-Beginners-Guide-toImmersive-Video-Storytelling-1622558250_print.pdf
2. Wheeler, D.: Immersive Technologies: Moving Past the Hype. The Chronicle of Higher Education Inc. (2022). https://connect.chronicle.com/rs/931-EKA218/images/Immersive_Verizon_ResearchBrief.pdf
3. CEN, CENELEC, ETSI: EN 301 549 V3.2.1. Accessibility requirements for ICT products and services. (2021). https://www.etsi.org/deliver/etsi_en/301500_301599/301549/03.02.01_60/en_301549v030201p.pdf
4. Connor, J., Sajka, J., White, J., Hollier, S., Cooper, M.: XR accessiblity user requirements. W3C (2021). https://www.w3.org/TR/xaur/
5. Meta Horizon: Meta Quest Virtual Reality Check (VRC) guidelines (2024). https://developers.meta.com/horizon/resources/publish-quest-req
6. Center for Teaching Excellence: Virtual Environments Accessibility Guidelines. University of South Carolina (n.d.). https://sc.edu/about/offices_and_divisions/cte/teaching_resources/virtual_environments/ve_accessibility_guidelines/
7. Baum, F., Evans, M., Patnoe, C.: Recommendations for accessible captions in 360 degree video. W3C Final Community Group Report (2024). https://www.w3.org/community/reports/immersive-captions/CG-FINAL-360-captions-20240412/
8. Equal Entry: Equal Entry Guidelines for Describing 360-Degree Video (n.d.). https://equalentry.com/vr/equal-entry-guidelines-for-describing-360-degree-video/
9. Meta: Use and manage Meta AI assistant on Meta Quest (2024). https://www.meta.com/en-gb/help/quest/486179404363222/
10. Zimmermann, G., Derichs, D., Votkinb, T., Kauppinen, T.: Making 360-degree interactive content less inaccessible – authoring guidelines and lessons learned (2023). https://aaate2023.eu/wp-content/uploads/sites/26/2023/08/BookOfAbstracts-Prelim.pdf
11. Ivanova, P.: Accessibility of 360 educational learning platforms: guidelines and evaluation (2025). https://hdms.bsz-bw.de/frontdoor/deliver/index/docId/7231/file/BA.pdf
12. W3C: Web Content Accessibility Guidelines (WCAG) 2.2 (2023). https://www.w3.org/TR/WCAG22/
13. European Commission: Web accessibility directive—standards and harmonization (2023). https://digital-strategy.ec.europa.eu/en/policies/web-accessibility-directive-standards-and-harmonisation

Accessible Indoor Navigation Systems

A Systematic Review of Smartphone-Based Indoor Navigation Assistance for Blind and Visually Impaired People

Marzia Alizada(✉), Airi Lampinen, Peter Idestam-Almquist, and Thomas Westin

Department of Computer and Systems Sciences, Stockholm University, Stockholm, Sweden
{marzia.alizada,airi,pi,thomasw}@dsv.su.se

Abstract. Navigating complex indoor spaces is one of the challenges for individuals with blindness or visual impairments. Smartphone-based solutions offer a promising approach to addressing these challenges due to their affordability, availability, adaptability, and acceptability. This study systematically reviewed the literature on smartphone-based indoor navigation systems for blind and visually impaired individuals, focusing on technologies, techniques, and evaluation methods. Our review highlights that computer vision is an evolving technology in this domain, with Radio Frequency Identification and Internet of Things technologies serving as valuable complements. The findings underscore the potential of smartphones to enhance accessibility and inclusion, while also emphasizing the importance of a user-centered design approach in development efforts. By identifying trends and gaps, this study provides valuable insights for advancing smartphone-based assistive technologies for indoor navigation.

Keywords: Indoor Navigation · Indoor Localization · Blind Users · Visually Impaired · Smartphone

1 Introduction

Millions of people worldwide live with visual impairments and face daily challenges, particularly when navigating unfamiliar or dynamically changing indoor environments [1]. Navigation includes movement, orientation, and avoiding objects. Although GPS and location-based technologies have improved outdoor navigation for blind and visually impaired people (BVIP), indoor navigation remains a significant challenge due to the absence of satellite signals and complex indoor layouts. Thus, BVIP often depend on physical assistance from others, limiting their autonomy and complicating everyday tasks [2].

To address the challenges of indoor navigation for BVIP, there is growing interest in assistive solutions that are portable, affordable, and easy to use [3].

K. Mavrou and P. Encarnação (Eds.): AAATE 2025, *Technology for Inclusion and Participation for All: Recent Achievements and Future Directions*, pp. 107–114, 2025.
https://doi.org/10.1007/978-3-032-01628-7_14

Smartphones, in particular, present a promising platform due to their widespread accessibility and integration into everyday life [4]. Their relative affordability is especially important, as people with disabilities often face economic disadvantages. Compared to emerging technologies like Augmented Reality (AR) glasses—which may include built-in sensors but also introduce additional cost and unnecessary features, such as visual displays irrelevant for blind users—smartphones offer a more affordable and acceptable solution [5].

Despite ongoing innovation in assistive navigation technologies, there remains a lack of a comprehensive and up-to-date review specifically focused on smartphone-based indoor navigation tools for BVIP. Such a review would be a valuable resource for researchers, helping to consolidate existing knowledge, identify gaps, and guide future development efforts toward more effective and accessible solutions.

Several existing reviews have made valuable contributions to the field of navigation assistive technologies. For example, Zafari et al. [6] and Singh et al. [7] conducted comprehensive surveys on indoor localization and positioning systems, covering a broad range of technological advancements. While these studies provide a strong foundation, they do not focus specifically on BVIP, which may result in some important user-centered insights and technologies being overlooked. Similarly, reviews by Khan et al., Masal et al., Kuriakose et al., and Kim [8–11] offer helpful overviews of navigation solutions for BVIP. Yet, due to their wide scope, they provide limited depth on the most recent developments in *indoor* navigation. Other focused works, such as those by Façanha et al., Lian et al., Santos et al., and Xu et al. [12–15], have addressed critical domains—including virtual environments, outdoor navigation, and wearable devices—but do not fully explore real-world indoor navigation technologies that do not rely on specialized hardware.

More targeted reviews by Budrionis et al. [16] and Khan & Khusro [4] have highlighted the potential of smartphone-based solutions for BVIP, emphasizing their practicality and accessibility. However, Khan and Khusro's review, while informative, lacks a systematic approach, and Budrionis et al.'s study, though thorough, only covers literature up to 2018. Given the rapid pace of innovation in this field, including the presentation of toolkits such as Google ARCore and Apple ARKit in 2018, we see a need for an up-to-date, methodical review.

The present paper builds on the strengths of these earlier works by focusing specifically on smartphone-based indoor navigation solutions for BVIP. It provides a synthesis of recent developments, identifies challenges in conducting user studies, and highlights research gaps and limitations in the development of smartphone-based indoor navigation systems.

2 Method

The study adheres to the Preferred Reporting Items for Systematic Reviews and Meta-Analyses (PRISMA) guidelines, as outlined by Page et al. [17], which serve as a standard for conducting Systematic Literature Reviews (SLRs). PRISMA provides a structured approach to ensure that the review process is rigorous, reproducible, and minimizes the risk of bias. The SLR methodology under PRISMA involves several stages: formulating research questions, developing a robust search strategy, defining specific inclusion and exclusion criteria, and conducting a quality assessment of the selected studies. Each of these steps is carefully designed to ensure that only the most relevant and high-quality studies are included in the review, thereby improving the reliability and validity of the results.

The search strategy for this review was intentionally comprehensive, leveraging a diverse set of academic databases—including PubMed, ScienceDirect, Scopus, Taylor & Francis, ACM Digital Library, SpringerLink, and IEEE Xplore—to capture a broad range of peer-reviewed studies. To maintain the quality and academic rigor of the review, non-peer-reviewed content was excluded. A carefully constructed Boolean search string combined key terms and their synonyms to maximize relevance and coverage: ((blind OR "visually impaired" OR "visual impairment") AND (smartphone OR mobile OR sensors OR "computer vision" OR "computer-vision") AND ("indoor navigation" OR "indoor localization" OR "indoor positioning")). The search was restricted to publications from 2018 through August 30, 2024, to focus on recent advancements in this rapidly developing field. Beyond database queries, the strategy incorporated manual searching and snowballing techniques, including both backward and forward citation tracking, as outlined by Wohlin [18].

In a first phase, titles, keywords, and abstracts were screened; in a second phase, full articles were reviewed. Only English-language, primary research studies focused on indoor navigation proposing smartphone-based solutions for BVIP were included, i.e. non-smartphone solutions, such as wearables, were excluded.

To ensure the relevance and rigor of the selected studies, additional quality assessment criteria were applied. Studies were required to clearly describe their proposed solutions technically, include user evaluations, and—most importantly—involve blind or visually impaired participants. This last point was emphasized because using blindfolded sighted participants fails to capture the unique experiences and needs of BVIP [19], making their direct involvement essential for producing meaningful, applicable insights in research.

3 Result

After skimming and scanning an initial pool of 840 articles through multiple stages (title screening, abstract review, and full-text analysis) 16 articles [20–35] were included in this review. Based on the final selection, smartphone-based indoor navigation assistant systems for BVIP can be broadly categorized into three types: *sensor-based*, *vision-based*, and *hybrid* systems. Among the 16 articles, 9 employed sensor-based approaches [20,24,27–29,31,33–35], one article [21] proposed a vision-based system, and 6 studies [22,23,25,26,30,32] utilized hybrid approaches. While all three system types aim to acquire spatial data and environmental awareness, they differ significantly in terms of underlying technologies, sensor types, and processing techniques.

Sensor-based solutions are the most commonly employed method, typically leveraging a smartphone's built-in sensors. The specific advantages and limitations vary based on the types of sensors and technologies used. For instance, systems relying solely on inertial measurement units (IMUs) [28,33] often suffer from reduced wayfinding accuracy. In contrast, those incorporating LiDAR [27], Bluetooth [29,31], or WiFi [24] tend to offer better performance. However, the integration of multiple technologies introduces new challenges, such as the need for external installations (e.g., Bluetooth Low Energy [BLE] beacons or WiFi access points) and cloud-based data processing, which may raise concerns around scalability and privacy. Kuriakose et al. [21] presented a vision-based system utilizing a smartphone camera for data acquisition. This system, grounded in AR, offer unique benefits, such as multifunctional capabilities, including obstacle detection and object recognition in addition to navigation. However, this approach comes with trade-offs: being a relatively new and evolving technology, it can suffer from latency issues. Additionally, vision-based systems are generally resource-intensive, consuming significant battery power and computational resources. To overcome the individual shortcomings of sensor-based and vision-based approaches, several studies explored hybrid systems, which integrate the strengths of both. For example, smartphone cameras were combined with IMU sensors [22,25,30], Bluetooth [23,26], or LiDAR [32] to develop more robust systems capable of object recognition, obstacle detection, and improved localization accuracy.

A notable gap across the reviewed literature is the limited adoption of multimodal user interfaces. Although many blind users prefer interfaces that offer diverse feedback mechanisms [36], most systems still rely on sound-only feedback. However, a few studies [23,25,28,30–33] have experimented with multimodal feedback, that is, combinations of sound, haptic cues, visual displays, and gesture-based interaction, offering a more inclusive and user-friendly experience.

Among the 16 articles, 4 employed qualitative methods, 6 used quantitative approaches, and the remaining 6 adopted mixed-method designs. All studies included at least one blind or visually impaired participant. However, the user evaluation scenarios varied widely, typically tailored to highlight specific system features. Another major challenge is the lack of standardized evaluation metrics. The System Usability Scale (SUS), a widely recognized metric, was used in only a few studies [20,23,25,26]. Other metrics used include task completion time [20,29,31–34], localization accuracy [22,27,28], walking speed [22,23], statistical error [20,23,25,32], obstacle detection accuracy [28], and user satisfaction [31]. While some of these metrics are established standards in the literature, their inconsistent application makes it difficult to compare solutions across studies.

4 Discussion

The comparative analysis of indoor navigation methods reveals that each approach brings its own set of strengths and limitations. Vision-based systems, predominantly utilizing smartphone cameras, demonstrate strong potential for tasks such as object recognition and obstacle detection without the need for additional hardware. However, their dependence on computationally intensive machine learning models, coupled with the frequent lack of diverse and representative training datasets [21], presents significant challenges. Moreover, their high processing demands can lead to rapid battery depletion, raising concerns about their long-term practicality in everyday settings.

Sensor-based systems, on the other hand, tend to offer more reliable performance for landmark recognition and indoor positioning, particularly when complemented by external infrastructure such as BLE beacons or Wi-Fi access points [24,29]. These systems are generally more power-efficient and straightforward to deploy in prepared environments. However, their effectiveness is inherently tied to the presence of such infrastructure, which limits their adaptability across diverse locations. Hybrid approaches attempt to leverage the complementary benefits of both vision and sensor-based systems, aiming to enhance accuracy and broaden functionality. Nevertheless, these systems remain constrained by the availability of compatible sensors in the environment and the added complexity of integration.

User evaluation remains a persistent challenge in the development of indoor navigation systems for blind and visually impaired persons (BVIP) [37]. While involving real users is essential for gathering meaningful feedback and aligning systems with user needs, recruiting a sufficient and representative sample remains difficult. Yet many rely on blindfolded sighted individuals, who cannot accurately reflect the experiences of BVIP users. Additionally, the lack of standardized evaluation metrics and varied testing scenarios further complicates comparisons across studies [38]. Although metrics such as the SUS, task completion time, and localization accuracy are used, their inconsistent application hinders benchmarking, impedes commercialization, and limits broader adoption of these technologies.

5 Conclusion and Future Research

This review systematically examined 16 studies on smartphone-based indoor navigation systems for BVIP, offering a comprehensive overview of current technological approaches and user evaluation practices. We identified three main development strategies: vision-based, sensor-based, and hybrid systems, each with distinct strengths and limitations. By synthesizing existing work, this study not only reveals important gaps in the current literature but also offers directions for future research. One notable gap is the limited integration of multimodal interfaces and the lack of customization across modalities. Additionally, concerns about privacy for BVIP users are largely unaddressed in most of the articles. A critical finding is the persistent challenge of conducting robust user evaluations, primarily due to difficulties in recruiting BVIP participants and the absence of standardized assessment metrics.

We highlight two promising directions for future research: inclusivity and system robustness. *Inclusivity* can be advanced through the integration of artificial intelligence for multi-language support and the development of customizable, multimodal interfaces that accommodate diverse user needs. Future systems should move beyond sound-only modality by evaluating and integrating alternative feedback mechanisms, combining haptic, sound, and gesture-based cues, to reduce cognitive load and better suit individual preferences. One particularly promising avenue is the integration of these multimodal interfaces with emerging technologies. To enhance *robustness*, future efforts should focus on integrating machine learning for tasks, such as obstacle avoidance and Point of Interest (POI) detection, training models with more representative datasets, and enabling localized data processing. These strategies not only improve performance and efficiency but also address concerns related to privacy and scalability in real-world environments.

References

1. Engel, C., et al.: Travelling more independently: a requirements analysis for accessible journeys to unknown buildings for people with visual impairments. In: The 22nd International ACM SIGACCESS Conference on Computers and Accessibility, pp. 1–11, Virtual Event Greece. ACM (2020)
2. Jeamwatthanachai, W., Wald, M., Wills, G.: Indoor navigation by blind people: behaviors and challenges in unfamiliar spaces and buildings. Br. J. Vis. Impairment **37**(2), 140–153 (2019)
3. Abraham, C.H., Boadi-Kusi, B., Morny, E.K.A., Agyekum, P.: Smartphone usage among people living with severe visual impairment and blindness. Assistive Technol. **34**(5), 611–618 (2022)
4. Khan, A., Khusro, S.: An insight into smartphone-based assistive solutions for visually impaired and blind people: issues, challenges and opportunities. Univ. Access Inf. Soc. **20**(2), 265–298 (2021)
5. Westin, T., Romero, M., Palosaari-Eladhari, M., Bejnö, H., Rahmani, R.: Assistive augmented reality for adults on the autism spectrum with intellectual disability. In: Miesenberger, K., Penaz, P., Kobayashi, M. (eds.) Computers Helping People with Special Needs, pp. 257–266. Springer Nature Switzerland, Cham (2024)
6. Zafari, F., Gkelias, A., Leung, K.K.: A survey of indoor localization systems and technologies. IEEE Commun. Surv. Tutorials **21**(3), 2568–2599 (2019)
7. Singh, J., Tyagi, N., Singh, S., Ali, F., Kwak, D.: A systematic review of contemporary indoor positioning systems: taxonomy, techniques, and algorithms. IEEE Internet Things J. (2024)
8. Khan, S., Nazir, S., Khan, H.U.: Analysis of navigation assistants for blind and visually impaired people: a systematic review. IEEE Access **9**, 26712–26734 (2021)
9. Kuriakose, B., Shrestha, R., Sandnes, F.E.: Smartphone navigation support for blind and visually impaired people-a comprehensive analysis of potentials and opportunities. In: International Conference on Human-Computer Interaction, pp. 568–583. Springer (2020)
10. Masal, K. M., Bhatlawande, S., Shingade, S.D.: Development of a visual to audio and tactile substitution system for mobility and orientation of visually impaired people: a review. Multimedia Tools Appl. **83**(7), 20387–20427 (2024)

11. Kim, I.-J.: Recent advancements in indoor electronic travel aids for the blind or visually impaired: a comprehensive review of technologies and implementations. Univ. Access Inf. Soc., 1–21 (2024)
12. Facanha, A.R., Darin, T., Viana, W., Sanchez, J.: O&m indoor virtual environments for people who are blind: a systematic literature review. ACM Trans. Accessible Comput. (TACCESS) **13**(2), 1–42 (2020)
13. Lian, Y., Liu, D., Ji, W.: Survey and analysis of the current status of research in the field of outdoor navigation for the blind. Disabil. Rehabil. Assist. Technol. **19**(4), 1657–1675 (2024)
14. Dos Santos, A.D.P., Suzuki, A.H.G., Medola, F.O., Vaezipour, A.: A systematic review of wearable devices for orientation and mobility of adults with visual impairment and blindness. IEEE Access **9**, 162306–162324 (2021)
15. Xu, P., Kennedy, G.A., Zhao, F.Y., Zhang, W.J., Van Schyndel, R.: Wearable obstacle avoidance electronic travel aids for blind and visually impaired individuals: a systematic review. IEEE Access **11**, 66587–66613 (2023)
16. Budrionis, A., Plikynas, D., Daniušis, P., Indrulionis, A.: Smartphone-based computer vision travelling aids for blind and visually impaired individuals: a systematic review. Assist. Technol. **34**(2), 178–194 (2022)
17. Page, M.J., et al.: The Prisma 2020 statement: an updated guideline for reporting systematic reviews. BMJ **372** (2021)
18. Wohlin, C.: Guidelines for snowballing in systematic literature studies and a replication in software engineering. In: Proceedings of the 18th International Conference on Evaluation and Assessment in Software Engineering, EASE 14, pp. 1–10, New York. Association for Computing Machinery (2014)
19. Cattaneo, Z.: Imagery and spatial processes in blindness and visual impairment. Neurosci. Biobehav. Rev. **32**(8), 1346–1360 (2008)
20. Striegl, J., Felchow, J., Loitsch, C., Weber, G.: Accessible indoor orientation support by landmark-based navigation. In: Lecture Notes in Computer Science, vol. 14020, pp. 510–524. Springer Nature Switzerland, Cham (2023)
21. Kuriakose, B., Shrestha, R., Sandnes, F.E.: Exploring the user experience of an AI-based smartphone navigation assistant for people with visual impairments (2023)
22. Fusco, G., Coughlan, J.M.: Indoor localization for visually impaired travelers using computer vision on a smartphone. In: Proceedings of the 17th International Web for All Conference, vol. 2020, no. 8 (2020)
23. Nair, V., Olmschenk, G., Seiple, W. H., Zhu, Z.: ASSIST: evaluating the usability and performance of an indoor navigation assistant for blind and visually impaired people. Assistive Technol. **34**(3), 289–299 (2022)
24. Calle-Jimenez, T., Sanchez-Gordon, S., Lujan-Mora, S.: Indoor localization solution for users with visual disabilities. In: 2018 International Conference on Information Systems and Computer Science (INCISCOS), pp. 205–212 (2018)
25. Yoon, C., et al.: Leveraging augmented reality to create apps for people with visual disabilities: a case study in indoor navigation, pp. 210–221 (2019)
26. Costales, J.A., Vida, A.K.S., Albino, M.G.: Mobile based navigation system for visually impaired person using image detection with voice translation, pp. 32–36 (2023)
27. Sato, D., et al.: Navcog3 in the wild: large-scale blind indoor navigation assistant with semantic features. ACM Trans. Accessible Comput. **12**(3) (2019)
28. Ahmed, F., Mahmud, M.S., Yeasin, M.: RNN and CNN for way-finding and obstacle avoidance for visually impaired. In: 2019 2nd International Conference on Data Intelligence and Security (ICDIS), pp. 225–228 (2019)

29. Ahmetovic, D., et al.: Sonification of rotation instructions to support navigation of people with visual impairment. In: 2019 IEEE International Conference on Pervasive Computing and Communications (PerCom), pp. 1–10 (2019)
30. Jafri, R., Khan, M.M.: User-centered design of a depth data based obstacle detection and avoidance system for the visually impaired. HCIS **8**(1), 14 (2018)
31. Asakawa, S., et al.: An independent and interactive museum experience for blind people. In: Proceedings of the 16th International Web for All Conference, pp. 1–9, San Francisco. ACM (2019)
32. Kuribayashi, M., et al.: Corridor-walker: mobile indoor walking assistance for blind people to avoid obstacles and recognize intersections. In: Proceedings of the ACM on Human-Computer Interaction, vol. 6, no. MHCI, pp. 1–22 (2022)
33. Flores, G., Manduchi, R.: Easy return: an app for indoor backtracking assistance. In: Proceedings of the 2018 CHI Conference on Human Factors in Computing Systems, Montreal, QC, Canada, pp. 1–12. ACM (2018)
34. Giudice, N.A., Whalen, W.E., Riehle, T.H., Anderson, S.M., Doore, S.A.: Evaluation of an accessible, real-time, and infrastructure-free indoor navigation system by users who are blind in the mall of America. J. Vis. Impairment Blindness **113**(2), 140–155 (2019)
35. Luders, F., Striegl, J., Schmalfub-Schwarz, J., Loitsch, C., Weber, G.: Accessible adaptable indoor routing for people with disabilities. Volume 13341 of Lecture Notes in Computer Science, pp. 169–177. Springer International Publishing, Cham (2022)
36. Good, M.D., Whiteside, J.A., Wixon, D.R., Jones, S.J.: Building a user-derived interface. Commun. ACM **27**(10), 1032–1043 (1984)
37. Dey, A., Billinghurst, M., Lindeman, R.W., Swan, J.E.: A systematic review of 10 years of augmented reality usability studies: 2005 to 2014. Front. Robot. AI, 5 (2018)
38. Potortì, F., Crivello, A., Barsocchi, P., Palumbo, F.: Evaluation of indoor localisation systems: comments on the ISO/IEC 18305 standard. In: 2018 International Conference on Indoor Positioning and Indoor Navigation (IPIN), pp. 1–7 (2018)

Analysis of Indoor Wayfinding Accessible-Assistive Technologies for Mobility, and Blind and Visually Impaired Individuals

Siny Joseph[1](✉), Carmel Heydarian[2], Sarahelizabeth J. Baguhn[2], Arielle M. Silverman[2], and Chung Ho Philip So[3]

[1] Kansas State University, Manhattan, KS 66506, USA
Siny@k-state.edu
[2] American Foundation for the Blind, Arlington, VA 22206, USA
[3] Pace University, New York City, NY 10038, USA

Abstract. Navigating complex indoor environments remains a persistent challenge for mobility impaired (MI) and blind and visually impaired (BVI) individuals, despite advances in assistive technologies. This paper presents a large-scale, user-centered survey involving 109 BVI and 190 MI participants, providing a comprehensive examination of their indoor navigation experiences, preferences, and unmet needs. Results reveal that while BVI users often rely on general-purpose visual interpretation apps for indoor navigation, MI users are largely unfamiliar with existing solutions. Both groups articulated a strong desire for adaptable, real-time, and infrastructure-free navigation tools that align with their individual abilities and interaction preferences. Feedback on a conceptual system, MABLE, demonstrated high interest in its features particularly in unfamiliar environments and strong alignment with participants' "ideal" wayfinding solutions. These insights not only inform the development of MABLE but also underscore critical design considerations for inclusive, personalized, and context-aware assistive technologies for indoor wayfinding.

Keywords: Assistive technologies · Blind · Mobility Impairment · Visual Impairment · Wayfinding

1 Introduction

The global blind and visually impaired (BVI) and the mobility impaired (MI) populations face persistent challenges in navigation, particularly in indoor environments where orientation and mobility (O&M) barriers impact independence, accessibility, and social inclusion. According to the World Health Organization (WHO), approximately 253 million people worldwide are visually impaired, with 36 million classified as blind. Mobility impairments affect over 1 billion people globally, about 15% of the population, with many requiring mobility aids such as wheelchairs, walkers, or crutches to navigate their environment as reported by WHO. These individuals often experience significant

K. Mavrou and P. Encarnação (Eds.): AAATE 2025, *Technology for Inclusion and Participation for All: Recent Achievements and Future Directions*, pp. 115–124, 2025.
https://doi.org/10.1007/978-3-032-01628-7_15

difficulties in wayfinding, particularly in unfamiliar indoor spaces where traditional navigation tools like canes and guide dogs may be insufficient, or where stairs, narrow doorways, or uneven flooring pose risks for those using mobility aids. Research indicates that BVI individuals face significant accessibility barriers in public indoor spaces, with 65% reporting difficulties in navigating buildings such as malls, offices, and educational institutions [1]. Similarly, MI individuals frequently encounter environments that lack ramps, elevators, or accessible signage, making autonomous navigation impractical or unsafe [2].

The diverse needs of BVI and MI users also point to a pressing need for more inclusive and adaptable assistive technologies (AT). While some users benefit from haptic or auditory feedback, others may rely on visual cues, real-time localization, or voice guidance systems. Moreover, individual preferences, such as interface complexity, routing priorities, and mobility speed can vary widely across and within disability groups. As a result, designing a "one-size-fits-all" solution often leads to marginalization of users with intersecting or less commonly addressed needs [3]. Addressing these challenges requires technological solutions that not only enhance accessibility and safety but also respect individual autonomy, preferences, and usability expectations.

Over the past 25 years, there has been a significant increase in the development of ATs aimed at improving indoor wayfinding and navigation for people with disabilities (PWD). Some notable technology-driven solutions include technology-enhanced canes, AI-powered navigation apps [4, 5], LiDAR-based spatial mapping [6], and IoT-enabled smart environments that integrate artificial intelligence, computer vision, RFID, and haptic feedback to facilitate seamless and effective navigation, specifically in indoor environments [1, 7–10]. Additionally, robotic guides equipped with AI and sensors are being deployed in indoor environments such as airports and museums to assist PWD in real-time. These robots navigate alongside users, offering directions and helping them avoid obstacles. These technologies aim to facilitate access to information, enhance safety, and support social inclusion by addressing key challenges such as spatial awareness, orientation, and obstacle detection [11].

Despite continued advancements in ATs, research consistently shows that many PWDs are dissatisfied with existing indoor wayfinding solutions, which often fail to address their specific needs in real-world contexts [12]. While numerous AT solutions exist, their effectiveness is often limited by design assumptions that may not fully align with the lived experiences, practices, and needs of PWDs. Prior research studies for navigation ATs for PWDs have focused on small-scale evaluations of specific prototypes [7, 13], usability testing with limited participants [14], or app-level feature comparisons without substantial user input [15]. To help address this gap, a multidisciplinary team with lived experiences is developing a new conceptual system called MABLE (Mapping for Accessible Built Environments), which aims to address many of the challenges identified in existing solutions. To ensure this concept adheres to most of the pain points of the BVI and MI community, and their diverse needs, a large-scale user-centered survey was conducted and is presented in this study with reflections on existing tools, and envisioned solutions. Unlike works that focus solely on technical performance, this study triangulates feedback on current tools (e.g., GoodMaps, Aira), conceptual designs

(MABLE), and cross-cutting accessibility expectations such as infrastructure-free use, real-time environmental awareness, and hands-free interaction.

The objectives of this work are twofold, a) Identify user-desired features, interaction modalities, and environmental information needs that define ideal indoor navigation experiences; and b) Evaluate user perceptions of MABLE concept as a prospective indoor navigation solution, including its alignment with diverse user expectations and its potential impact on independent travel, especially for the BVI and MI individuals.

2 Methodology

2.1 Data Collection

This study is approved by the Kansas State University (KSU) IRB protocol# 10616 and The American Foundation for the Blind (AFB) IRB protocol# 00001619. Participants for the BVI survey were recruited from several blindness listservs as well as AFB's pool of participants from past studies interested in future research opportunities. Participants for the MI survey were recruited from listservs of organizations such as Cerebral Palsy Foundation, FSHD society, AIChE society, Easterseals and the Emerging Researchers Network. The survey was sent in October 2024 to the BVI community and from January-April 2025 to the MI community. A total of 109 responses were collected for the BVI survey and 190 responses for the MI survey. Participants in both surveys were given the option of being entered into a drawing to win one of 20 $50 gift cards.

2.2 Design of Survey

An accessible online questionnaire (created using the Qualtrics software) was designed and administered to the BVI community by AFB's research team and to the MI community by the MABLE team. The survey comprised of ten sections and approximately 40 questions. The survey flow incorporated skip logic to bypass questions irrelevant to certain respondents. Many of the questions in the survey were open-ended and a textbox was provided for responses. Additionally, participants had the option of typing or providing voice responses to open-ended questions. Other response options included selecting multiple options from a list and selections using a likert-scale. The survey was organized by questions about survey respondents' experiences with indoor navigation and currently available apps. Based on gaps they identified in the current landscape of solutions for indoor navigation, participants were then asked to assess and weigh in on the proposed MABLE concept (a video and description was provided) about its functionality and usefulness.

Video Conceptualization of MABLE. The "Welcome to MABLE!" video introduces MABLE, a conceptual indoor navigation system aimed at increasing independence and accessibility for PWDs, particularly those who are blind or visually impaired and mobility impaired, along with other disabilities. The video outlines MABLE's core purpose and vision, which is to provide real-time, infrastructure-free indoor navigation using personalized, hands-free guidance. It highlights several key features, including turn-by-turn navigation, integration with wearable technology and voice interfaces, and the ability to function without pre-installed beacons or signage. The technology stack behind MABLE is explained to leverage crowdsourcing, artificial intelligence, and robotics to automate indoor mapping and provide dynamic updates. The system is demonstrated through a prototype walkthrough and explained as being developed using a user-centered design approach informed by feedback from the BVI and MI community. Information is provided that it is currently being tested in various public and educational venues to support real-world use cases such as conferences, airports, and university campuses. The project's broader community impact is also emphasized, with support from the National Science Foundation's Convergence Accelerator Program and collaborations across research, accessibility, and disability advocacy groups.

2.3 Data Analysis

The study incorporated a mixed methods approach, with questions that were analyzed using quantitative and qualitative methods. No statistical analyses were used for the quantitative natured questions due to a lack of significant differences when examining different groups. For example, the researchers expected a comparison between white cane users and guide dog users would reveal a difference in indoor wayfinding methods and preferences, but no significant difference was noted. Qualitative data was analyzed using the constant comparison method [16]. The responses were coded individually to identify recurring themes and sub-themes. After individual researchers could not classify responses with new codes, they came together to reach consensus on each response and finalized common themes and sub-themes.

2.4 Participant Profile

Key characteristics of the survey participants, which offer valuable insights into their responses to the survey questions, include factors such as age, education level, phone preferences, assistive aids used, and the age of onset of disability as shown in Table 1.

3 Results

3.1 Current Indoor Wayfinding Apps Feedback

Use of Apps for Wayfinding in Indoor Environments. About 61% of BVI survey participants responded "Yes" when asked if they had ever used an app for indoor wayfinding (whether that app was intended for it or not). Among the users with MI, 34% reported using apps both in indoor and outdoor environments, even though their responses suggest familiarity with outdoor apps only

The top 3 apps that are currently used by BVI survey participants for indoor wayfinding are Be My Eyes (42%), Seeing AI (39%), and AIRA (35%). Notably, all of these apps are built to provide visual interpretation for a wide range of tasks, not just indoor wayfinding, but indoor wayfinding was a common use case for them.

MI survey participants were unfamiliar with indoor wayfinding apps and were surprised to learn of some exisiting apps. They mentioned Goodmaps (32%) and Google

Table 1. Characteristics of BVI and MI survey respondents

Characteristics	BVI	MI
Age	64% between the ages of 25 and 54, approximately 26% were 55 years or older	33% between the ages of 25 and 54, approximately 65% were 55 years or older
Education	62% of participants hold a Bachelor's, Master's, or Doctoral degree	72% of participants hold a Bachelor's, Master's, or Doctoral degree
Phone Preferences	84% of respondents used iPhones, with some reporting use of both iPhone and Android devices, while 9% exclusively used Android phones	63% of respondents used iPhones, while 37% used Android phones
Aids and Assistive Devices	Most respondents used a white cane (86%), followed by human guides (69%) and guide dogs (23%)	71% used ambulatory assistance devices such as mobility cane, walker, crutches or custom orthosis, prosthetics, exoskeleton and 56% used a wheelchair (manual or power), and human aid constituted 18%
Onset of Disability	82% of respondents were blind or had low vision before the age of 30. Additionally, 56% were either born with impairment or developed it before the age of 2	5% of respondents were born with mobility impairment, while 51% developed after the age of 40

maps (36%) which are used primarily in outdoor environments. Around 21% of the respondents mentioned familiarity with other outdoor apps such as Wheelmap, Access Now, iAccess Life, and AroundMe.

3.2 Description of Dream Solutions

Ideal Solution for Wayfinding in Indoor Environments. Analysis of open-ended responses from participants with MI and BVI revealed several overlapping themes in conceptualizing an ideal wayfinding system. While each group emphasized unique functional needs based on their specific access barriers, several shared priorities emerged, particularly related to application design, physical accessibility, and navigation customization.

Both groups overwhelmingly emphasized the need for robust, feature-rich digital wayfinding tools. Among BVI participants, application design was the most frequently mentioned theme (93 indications), with participants envisioning tools that offer GPS-like functionality tailored to indoor and complex environments. Similarly, MI participants expressed a desire for indoor navigation apps that support real-time location tracking, barrier-avoidant routing, and multimodal output (e.g., visual and auditory guidance).

Both groups stressed the importance of precision in routing instructions, with BVI participants highlighting the need for orientation-friendly descriptions (e.g., clock face directions), and MI users requesting route filters to avoid stairs, steep inclines, or long distances.

Participants across both groups expressed the need for systems that adapt to their individual abilities and preferences. BVI respondents suggested incorporating adjustable output formats and input preferences, while MI participants proposed customizable profiles that account for fluctuations in physical ability. This need for adaptability highlights a shared concern with reducing cognitive and physical burden while navigating unfamiliar environments.

Another shared theme involved leveraging and improving upon existing tools. Several BVI participants referenced a desire for wayfinding systems modeled after mainstream tools such as Google or Apple Maps. Likewise, MI participants emphasized the need for pre-visit tools and building previews like 3D scans or walkthroughs that mirror current consumer navigation platforms.

While BVI participants focused primarily on digital application design, a subset (fewer in number) expressed a desire for improved built environments as a complementary or alternative solution. Physical accessibility was a dominant theme among MI respondents. They strongly emphasized the importance of infrastructure improvements such as automatic doors, wide passageways, accessible restrooms, elevators, and rest seating. Both groups shared the perspective that digital wayfinding solutions alone are insufficient without accessible physical infrastructure.

A smaller number of BVI respondents (12 indications) expressed interest in alternative hardware-based solutions, such as wearable devices that integrate AI independently of smartphones. In contrast, MI participants more frequently requested compatibility with common mobile devices and interfaces that support both hands-free and accessible interaction methods (e.g., voice control, large buttons, or ground-level sensors).

Both groups expressed a broader desire for systems that enable independent and dignified navigation. Several BVI and MI participants discussed avoiding unfamiliar environments due to anxiety about accessibility or past negative experiences.

3.3 Evaluation of MABLE Concept

Use of MABLE in Familiar and Unfamiliar Environments. Overall, most BVI and MI participants said they were likely to use this application, with many more people interested in using it in unfamiliar environments. There were not significant differences in this pattern based on age, employment status, onset of disability. For BVI individuals, 83% of people were at least somewhat likely to use MABLE in familiar environments, and 93% in unfamiliar environments. For MI individuals, 69% of people were at least somewhat likely to use MABLE in familiar environments, and 86% in unfamiliar environments. Table 2 provides a breakdown of the use of MABLE in familiar and unfamiliar environments across BVI and MI individuals.

Types of Indoor Environment for MABLE Use. BVI and MI participants indicated their selections based on a list provided to them with the question. The top five environments identified across both populations were hotels, conference centers, airports,

Table 2. Use of MABLE in familiar and unfamiliar environments

	Familiar %		Unfamiliar %	
	BVI	MI	BVI	MI
Very Likely	54% (n = 52)	38% (n = 75)	76% (n = 77)	65% (n = 128)
Somewhat likely	29% (n = 28)	31% (n = 81)	17% (n = 17)	21% (n = 42)
Somewhat unlikely	12% (n = 12)	15% (n = 29)	4% (n = 4)	2% (n = 3)
Very Unlikely	5% (n = 5)	7% (n = 13)	3% (n = 3)	5% (n = 9)

shopping malls and office buildings. BVI participants scored atleast 75% (n = 80) and more for each of these environments while MI particpants scored on average of 55% (n = 95). BVI particpants additionally indicated a prospective or future workplace at 56% (59 responses) and 20% (20 responses) even considered it likely to use in their current workplace.

Additionally, many participants selected the "other" option and provided a list of environments they would use the app including: hospitals/medical facilities, stadiums/concert venues, bus/metro/train stations, museums, college campuses, grocery stores and supermarkets, theaters, community centers, restaurants, gymnasiums, and municipal buildings like city hall.

Landmark Identification with MABLE. Figure 1 breaks down the percentages of respondents interested in finding different landmarks. This figure splits the respondent groups as BVI users of guide dogs and non-guide dog users, and MI users due to the slightly different travel needs of each. Guide dog BVI users for the most part unanimously found value in the app locating all of the listed landmarks at equal importance. Non-guide dog BVI users and MI users varied in the level of importance to finding the landmarks. Figure 1 is a heatmap showing the importance of each landmark across the three user groups. The darker shades indicate higher importance. Building entrances/exits, restrooms, restaurants/cafes, elevators are important landmarks across all user groups. BVI users valued information desks, rooms, store entrances and stairs, while MI users valued information about ramps, resting areas and accessible entrances/exits.

Apart from the listed landmarks as ranked by the users, some additional landmarks were suggested of value. They are information about the location of first-aid kits, water fountains/water bottle refill station, trash cans, train lines/security lines, terminal gates in airports/terminals at train stations, ticket machines at train stations, commuter rail platforms, service dog relief areas, baby changing areas or family restrooms, vending machines, and contents of grocery store aisles.

Phone Use Preference with the App. The phone orientation preferences of participants consisted of two options; hands-free or holding the phone. Within the hands-free option, the choices were to wear it or keep it away in a purse or pocket. The hands-free option implied the use of voice to interact with the app while holding the phone aligned with typing preferences.

BVI group: The majority of the respondents 74% (n = 78) chose the hands-free option to keep it away, while 45% were open to wearing it and another 41% didn't mind

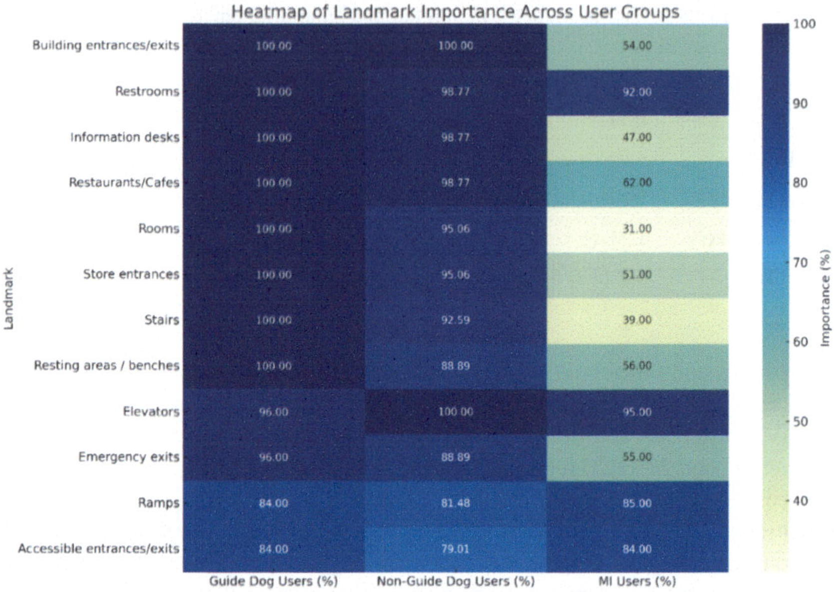

Fig. 1. Landmark importance across user groups

holding the phone. MI group: 58% (n = 118) prefer to type on the phone while 42% liked to use voice to interact with the phone.

MABLE Alignment with Ideal Dream Indoor Wayfinding Solution. Of the 105 BVI and 177 MI survey respondents on MABLE's alignment with their ideal wayfinding solution, 'mostly aligned' was chosen by 40% BVI and 29% MI respondents respectively, while 31.43% BVI and 48% MI participants reported that MABLE is 'completely aligned' with their dream solution.

MABLE Feedback. Four codes were developed from the BVI and MI participants who responded, and 73 coded responses were identified for BVI and 80 for MI users. The highest indicated category was that the participants wanted additional features with MABLE (44 indicators for BVI and 40 for MI users), including audio descriptions and the ability to limit veering while traveling and specificity such as stalls that can fit a wheelchair. This was followed by participants finding the question difficult or they were unable to answer the question (16 indicators for BVI and 34 for MI participants) due to not being able to try MABLE or not having enough information to provide an informed decision. An additional 11 BVI and 7 MI participants preferred MABLE as-is, with one participant stating, "I love everything about it," and another stating, "It's perfect for me." Finally, 3 BVI and 2 MI participants felt like they did not require assistance from using MABLE, where two of the three BVI participants wanted accessible spaces/buildings to begin with so that there was no need for a product like MABLE.

4 Discussion

This study provides valuable insight into the experiences, expectations, and unmet needs of BVI and MI individuals regarding indoor navigation technologies. Through a mixed-methods survey approach, the findings affirm the urgent need for customizable, infrastructure-independent, and multimodal ATs that support independent wayfinding in indoor environments.

While numerous navigation tools exist, many BVI and MI users remain underserved. Participants identified current solutions such as Seeing AI, AIRA, and GoodMaps, were limited in design specifically for indoor use or did not sufficiently account for diverse physical access needs [5, 8, 9]. Particularly among MI participants, there was a notable lack of awareness or usability of these tools. The strong interest expressed in MABLE, especially for use in unfamiliar environments (93% BVI, 86% MI), underscores a critical gap in the current AT landscape. Participants found MABLE's proposed features, such as real-time localization, barrier-aware routing, and multimodal interaction to be aligned with their expectations for an ideal indoor wayfinding system [1, 11].

A recurring theme across both user groups was the importance of adaptable interaction. BVI participants overwhelmingly preferred hands-free operation through voice commands, aligning with prior work emphasizing the value of voice interfaces in assistive contexts [3, 7]. MI participants exhibited more varied preferences, with many favoring manual interaction modes such as touch and text input. This variation highlights the need for systems like MABLE to offer highly flexible and accessible user interfaces that accommodate diverse motor and sensory abilities, echoing the principle of universal design [10].

While BVI participants focused heavily on software-driven solutions, MI participants repeatedly emphasized the interplay between physical infrastructure and digital navigation tools. This distinction suggests that MABLE's effectiveness may depend not only on its technological capabilities but also on its ability to interface with, or augment, physical accessibility cues, such as elevators, ramps, and accessible entrances. Additionally, the call for MABLE to identify nuanced spatial features like service dog relief areas or rest zones reinforces the importance of contextually rich, landmark-based navigation.

5 Conclusion

The findings from this study underscore a shared and urgent demand among BVI and MI populations for next-generation indoor wayfinding systems that emphasize personalization, accessibility, and real-time adaptability. MABLE addresses many of these needs, as evidenced by strong user alignment and enthusiasm expressed.

However, successful implementation of MABLE or any similar system will require moving beyond generic design assumptions to thoughtful integration of user feedback, especially around interaction preferences, landmark salience, and contextual use cases. Future development should include participatory design approaches with diverse disability communities to ensure the technology not only meets functional needs but also reflects the lived realities and autonomy goals of PWDs.

In advancing assistive navigation tools like MABLE, designers, developers, and policymakers must recognize that indoor mobility is not merely a technical challenge, but a deeply human one, intertwined with rights to independence, inclusion, and equal access to space.

Acknowledgements. The authors would like to acknowledge the U.S. National Science Foundation (award number #2345057) for support. The authors also appreciate the assistance of Drs. Nils Hakansson and Ted Conway in contacting organizations that serve people with mobility impairments for survey participant recruitment.

References

1. Ran, L., et al.: RFID-based indoor localization systems for visually impaired users. Sensors **17**(8), 1872 (2017)
2. Imrie, R.: Universalism, universal design and equitable access to the built environment. Disabil. Rehabil. (2012)
3. Hersh, M.A.: The design and evaluation of assistive technology products and devices. Appl. Ergon. (2015)
4. Azenkot, S., Fortuna, E.: AI navigation apps for visually impaired users: a usability study. In: Proceedings of the ACM CHI Conference (2019)
5. Bai, X., et al.: Wearable AI vision systems for navigation assistance. IEEE Trans. Neural Syst. Rehabil. Eng. **28**(4), 1025–1037 (2020)
6. Zhu, J., et al.: AI-based depth-aware navigation for visually impaired users. ACM Trans. Interact. Intell. Syst. **11**(3), 1–20 (2021)
7. Ahmetovic, D., et al.: NavCog: AR-based indoor navigation for blind users. ACM Trans. Accessible Comput. **9**(2), 1–25 (2016)
8. Kammoun, S., et al.: IoT-based smart indoor navigation for blind users. Sensors **19**(3), 514 (2019)
9. Yang, C., et al.: Haptic feedback navigation systems for visually impaired pedestrians. IEEE Trans. Haptics **11**(2), 293–304 (2018)
10. Dakopoulos, D., Bourbakis, N.G.: Wearable obstacle avoidance systems for visually impaired individuals. IEEE Trans. Syst. Man Cybern. **40**(1), 25–35 (2010)
11. Cazini, J., Frasson, A.C.: Voices project: technological innovations in social inclusion of people with visual impairment. J. Technol. Manag. Innov. **8**, 13 (2013)
12. Bhowmick, A., Hazarika, S.M.: An insight into assistive technology for the visually impaired and blind people: state-of-the-art and future trends. J. Multimodal User Interfaces **11**(2), 149–172 (2017)
13. Azenkot, S., Rector, K., Ladner, R., Wobbrock, J.: PassChords: secure multi-touch authentication for blind people. In: Proceedings of the 14th International ACM SIGACCESS Conference on Computers and Accessibility (ASSETS 2012), pp. 159–166. Association for Computing Machinery, New York (2012)
14. Kalia, A.A., Legge, G.E., Roy, R., Ogale, A.: Assessment of indoor route-finding technology for people who are visually impaired. J. Vis. Impairment Blindness **104**(3), 135–147 (2019)
15. Zheng, X., Maredia, A., Zahabi, M. A scoping literature review and content analysis of navigation apps for blind and visually impaired users. In: Proceedings of the Human Factors and Ergonomics Society Annual Meeting, vol. 66, no. 1, pp. 2295–2299 (2022)
16. Silva, P.R., et al.: Constant comparison. In: Given, L.M. (ed.). The SAGE Encyclopedia of Qualitative Research Methods, vol. 1, pp. 114–115. SAGE Publications (2008)

Creating User-Centered Guidelines for Digital Guidance Systems

Kevin Pakula(✉), Klaus Honold, and Gottfried Zimmermann

Hochschule der Medien, 70569 Stuttgart, Germany
pakula@hdm-stuttgart.de

Abstract. Digital guidance systems hold great potential for supporting visually impaired individuals in their orientation and mobility (O&M). Despite advances in assistive technologies, many systems fail to adequately reflect possible user requirements. This study aims to synthesize and assess requirements for the design and implementation of digital guidance systems that align with user needs. A systematic literature review identified 25 potential requirements for such systems. These requirements were then validated through expert interviews employing a Likert-style scale.

The findings highlight the importance of user independence and the role of secondary devices in complementing primary mobility aids like the white cane. The results emphasize adaptability, simplicity, and the inclusion of personalized information delivery. By integrating these guidelines, future systems can better address user needs and enhance accessibility in diverse environments. This work contributes to the field of assistive technology by providing insights regarding functional requirements for the development of effective digital guidance systems.

Keywords: Guidance systems · orientation and mobility · assistive technologies · user-centered design · accessibility

1 Introduction

Assistive technologies for orientation and mobility (O&M) are used to improve the navigational capabilities of visually impaired and blind people. To be adopted into everyday life such devices should address the needs and restrictions of the users. Otherwise, the device might be considered an obstacle due to lacking mode of operation and informative capabilities [1]. The use of such systems distinguishes between "primary" and "secondary" devices. Primary devices can be used independently and replace established systems like the white cane. Secondary devices, on the other hand, serve only as a supplement to other systems. Assistive technologies in O&M are categorized into Electronic Travel Aids (ETA) and Electronic Orientation Aids (EOA) among other possible categories. ETAs are typically used to complement other systems, such as the white cane, by providing information collected through sensors. These systems capture obstacles and environmental characteristics, offering the advantage of access to additional environmental information. EOAs, on the other hand, aim to improve orientation

K. Mavrou and P. Encarnação (Eds.): AAATE 2025, *Technology for Inclusion and Participation for All: Recent Achievements and Future Directions*, pp. 125–132, 2025.
https://doi.org/10.1007/978-3-032-01628-7_16

and navigation, either before or during travel. Instead of capturing direct environmental information through sensors, technologies such as GPS are used to determine the user's position and orientation, thus supporting navigation [2].

Kim (2024) employed a systematic literature review, analyzing modern ETAs. While various challenges exist that ETAs address, there is no comprehensive system that fully covers the O&M needs of individuals with visual impairments. [3] According to Bhowmick and Hazarika (2017), although there has been growth in O&M research, the number of assistive devices improving the users' quality of life and having a societal impact remains limited [4]. Hoogsteen et al. (2022) observed that the development of digital guidance systems is largely determined by the possibilities of contemporary technologies, while diverting from the users' needs. They further studied the information needs of blind users [5]. Khan and Khusro (2021) also concluded, that understanding the needs of possible users is a remaining challenge while discussing smartphone-based assistive solutions [6].

To design future digital guidance systems, it is crucial not to merely follow technological trends. Without actively engaging with users and incorporating their specific needs, there is a risk that these systems will fail to gain widespread acceptance. This issue can be addressed by thoroughly understanding users and their individual demands. Research into specific guidelines regarding O&M already exists; for instance, previous studies have explored information requirements and potential implementations [7, 8]. Based on this background, this work focuses on discussing additional guidelines on functional and non-functional properties of digital guidance systems.

2 Methods

To gather requirements, a systematic literature review was conducted using the keywords 'wayfinding', 'navigation', 'parameters' and 'accessibility'. The ACM Digital Library and SpringerLink databases were utilized for the collection of literature. The search was limited to publications between January 2019, and October 2024, to reduce the influence of outdated technologies. After screening, n = 28 suitable results remained, from which requirements for digital guidance systems were synthesized. Conference proceedings, books, book reviews, magazines, and newsletters were excluded. Results were collected automatically using the Zotero Connector. Subsequently, the automatically filtered results were manually reviewed for titles and duplicates, and unsuitable titles were removed. Thematic filtering ensured that the contents pertained to guidance systems. This approach ensures the inclusion of both technical requirements for guidance systems and usability requirements for the target audience (Fig. 1).

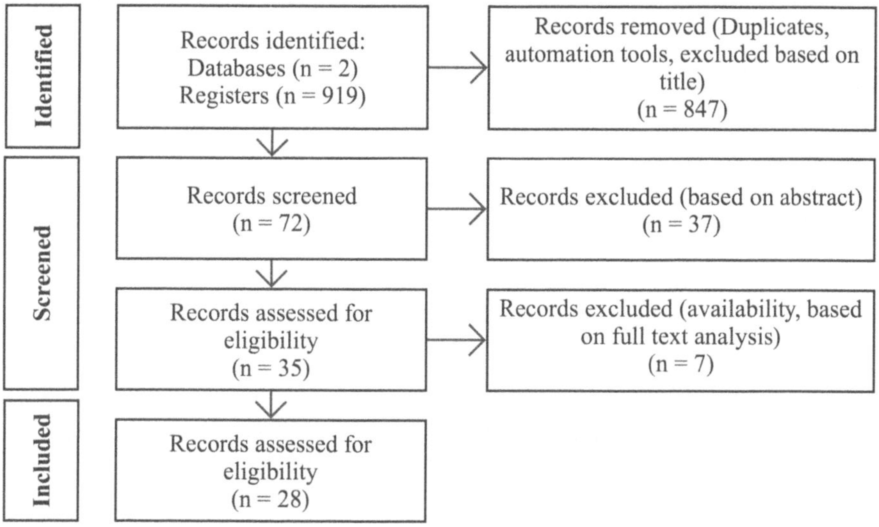

Fig. 1. Flowchart showing the filtering process of the systematic literature review

The synthesized requirements were discussed in interviews with O&M trainers, users of O&M assistive technologies, and advisors on accessibility to incorporate possible user needs. A Likert-style scale was included in the interviews for evaluation purposes. The rating scale was adapted to the following levels: strongly relevant (1), relevant (2), neutral (3), irrelevant (4), and strongly irrelevant (5). Parametric statistics were applied, allowing for the analysis of means and standard deviations [9]. This approach identifies expert disagreements or requirements rated as neutral to irrelevant. Such ratings were further evaluated based on the interviewees' comments and feedback. As the Likert-style scale supplements the expert interviews, it provides context for the responses and supports interpretations.

3 Results

Table 1 lists the 25 synthesized requirements resulting from the systematic literature review, with Likert-style scale ratings providing numerically ratings ranging from 1 (strongly relevant) to 5 (strongly irrelevant). The individual interview partners are listed as I1-I6. A total of six interviews were conducted. The median across all ratings is ($\tilde{x} = 1$), the mean is ($\varnothing = 1.33$), and the average standard deviation across all requirements is ($\sigma = 0.93$).

Table 1. Synthesized requirements including results from the Likert-style scale

Requirement	I1	I2	I3	I4	I5	I6
Adaptable output	1	2	1	1	1	2
Adaptable input	1	1	1	1	1	4
Multimodal in- and output	3	1	1	1	2	4
Minimalist system design	1	1	2	2	2	2
Filtered information output	2	1	1	1	1	1
Retention of habits	3	3	2	3	5	4
Accessible Design	1	2	1	1	1	1
Transparent system behavior	1	1	1	1	1	1
Ensuring physical safety	1	2	1	1	3	1
Ensuring privacy and data security	1	3	3	4	3	1
System behavior reproducibility	2	2	2	1	2	1
Error compensation	1	2	1	1	2	1
Timely information	1	2	1	1	1	1
Inclusion of O&M parameters	2	2	1	1	1	1
Mitigation of system limitations	2	1	1	1	2	1
Transparent navigation	1	3	1	1	1	1
Low pricing	1	3	2	3	1	3
Simple and reliable usage	2	2	2	1	1	1
Social compatibility	1*	3	2	5	4	4
Integrability	4	2	1	1	1	1
Market orientation and external consistency	3	2	1	2	5	1
Scalability	1	1	1	1	1	1
Promotion of situational awareness	5	3	2	1	2	1*
Avoidance of sensory overload	1	1	1	1	1	1
Contextual information	2	2	3	5	5	1

Twice an interview partner argued that a requirement could be rated either 'strongly relevant' or 'strongly irrelevant' based on situation and personal preference. To represent possible minority needs and preferences, those requirements will be regarded as 'strongly relevant', as after further confirmation this rating was applied. Those ratings are represented in chart Table 1 as 1*. In total, (n = 13) requirements received an average rating of 'relevant' or more relevant with ($\sigma < 0.8$). These requirements were also noted as highly relevant during interviews, often with additional comments. (n = 5) requirements had a standard deviation ($\sigma \geq 0.8$) with a median ($\tilde{x} \leq 2$) and a mean ($\varnothing \leq 2.0$), indicating that they were generally rated as more relevant, albeit with higher variability. (n = 7) requirements with a standard deviation ($\sigma \geq 0.8$), a median ($\tilde{x} > 2$), and a mean ($\varnothing > 2.0$) were rated by experts as less relevant. The high standard deviation suggests disagreements among experts, with occasional ratings as more relevant. Another scenario to examine would be requirements with ($\sigma < 0.8$), ($\tilde{x} > 2$), and ($\varnothing > 2.0$). In this case, the requirement would be deemed less relevant, and the low variance would indicate consensus among experts. However, this scenario did not occur among the collected requirements.

4 Discussion

Evaluations that were fundamentally rated as relevant and exhibited low standard deviation were not subject to adjustments by the interview participants. Requirements with a median of ($\tilde{x} < 2$) and higher standard deviation were nonetheless justified as relevant by the experts during the interviews. In some cases, targeted adjustments and additional specification of these requirements were suggested. Fundamentally, these requirements can also be considered suitable following the interviews.

Requirements with ($\sigma \geq 0.8$), a median ($\tilde{x} > 2$), and a mean ($\varnothing > 2.0$) should be discussed regarding their suitability to be included in possible guidelines.

For the requirement 'Retention of habits' it was noted that habits can be learned, and it is not always possible to account for everything (I2). Another participant warned that habits could be considered negative (I3). One respondent countered that a system should not handle such aspects (I4), or that it should not be concerned with them (I5). Comments revealed a generally negative attitude among interview participants toward this requirement. Instead, the system should compensate for user behavior and deviations. This is already addressed by requirements such as "Error compensation". Users' habits can also be accommodated through system personalization and filtered information.

'Ensuring Privacy and Data Security' was frequently argued against. Users should be competent in handling their data responsibly (I1, I2, I3, I5). This competence should be developed independently of the use of assistive technology (I2). Additionally, it was noted that strict data protection restrictions could make certain functions impossible (I4). Since a digital guidance system does not access critical data, such as banking information, this requirement is less relevant (I5). Two participants rated this aspect as very relevant; however, one argued against strict data protection measures in favor of greater user responsibility (I1). There was a call for greater independent use of new assistive technology as frustrations regarding overbearing data protection requirements were mentioned. While the use of such systems requires a multitude of personal data, data protection should be considered, yet current implementations in digital systems were seen as frustrating. Due to such responses, the requirement is not considered further, while ethical use of personal data in digital guidance systems should be discussed in future work.

For the requirement 'Low pricing' it was observed that users are willing to pay more for good solutions (I2). Expensive but high-quality solutions are usually more usable than free but subpar ones (I4). One participant advised against subscription models (I2), while another pointed out that not all users have the same financial means, making costs an important factor (I4). Based on comments and ratings, low costs are not critical. Instead, appropriate financing models and tailored costs should be considered. To address users with limited financial resources, the requirement should be reframed from to 'Appropriate pricing'.

'Social compatibility' was considered a very individual requirement multiple times (I1, I3, I5). It was noted that it can either be strongly relevant or not relevant at all (I1). 'Design for all' was cited as a rationale for this requirement by one participant (I3), while another stated that specialized solutions are not always problematic, as the function of assistive technology is more significant (I2). Since the functionality of assistive tools takes precedence to individual preference, the relevance of social compatibility is low as a requirement. While it may be of interest to certain user preferences, based on the interviews it cannot be considered a relevant requirement.

For 'Market Orientation and external Consistency' it was argued that alongside functional aspects, system design should align with market trends. Physical systems should have a modern design comparable to other contemporary technologies and should not appear outdated or cheap (I2). One participant countered that market orientation could lead to a loss of originality and disregard for cultural differences, such as varying cane techniques (I5). While external consistency was not criticized, market orientation was discouraged. Instead, better design was advocated. The requirement could therefore be reframed as 'appealing design and external consistency' to reflect the mentioned feedback.

'Promotion of situational awareness' was noted as highly context- and user-dependent (I2). Several participants argued that users should rely on their own senses (I1, I5, I6). One respondent countered that this requirement could inhibit sensory reliance, as users might wait for system-provided information instead (I1). Additionally, good sensory perception can compensate for many limitations. Therefore, the relevance of this requirement depends heavily on the context and user (I6). The risk that users might not rely on their own senses was a key reason for neutral or less relevant ratings. The requirement is suitable if this risk can be mitigated. Additionally, users' demand for high independence should be considered. This can be achieved through requirements like 'Adaptable output' and the 'Avoidance of sensory overload'. As in some interviews, users requested features improving situational awareness, it should still be considered a requirement.

For 'Contextual information' two participants argued that information should not be delivered contextually but should instead be shown explicitly (I4, I5). This issue was exemplified by audio guides in museums (I4). Often, additional information is required for interpretation (I5). Another participant noted that this depends on the type of information, as some require more context (I3). Two other participants expected that interpretation should be left to users (I2, I6), supported by their sensory perception (I6). Personalized filtering of information is already included as a requirement, but examples like 'Promotion of situational awareness' and 'Contextual information' represent potential functional requirements for information output. This requirement should promote user independence by incorporating their sensory perception. Since such information is deemed highly relevant in specific circumstances, the requirement remains appropriate under the condition of adaptable information.

There were also comments indicating additional needs. Regarding compensation for system limitations, it was noted that users should not assume reliable functionality (I5).

One participant mentioned that safety functions are desirable but that users should never rely solely on such systems (I3). Across the discussed requirements, the independence of users from digital systems was frequently emphasized. Based on these comments, user independence should be further reflected. Thus, a key requirement for digital guidance systems is to be designed to not restrict the use of primary mobility aids to ensure user independence.

A preliminary assessment was carried out using the gathered requirements. The apps "Here WeGo", "WeWALK", and "Lazarillo" were tested in real-world conditions on an iPhone XR (iOS 18.3). Those Apps were mentioned by the interview partners explaining aspects that were later summarized into given requirements. The test was conducted by a visually impaired individual, using VoiceOver as a screen reader and Zoom for magnification.

All three apps offered scalable and customizable input and output options, allowing flexible use. Personal settings could be adjusted via user accounts, and both virtual keyboards and voice input were available for interaction. However, some multimedia overlaps, such as VoiceOver and internal voice outputs, were noted. WeWALK addressed cognitive load by utilizing a minimalist design, while all three apps allowed filtering of information output, with a reduced exploration mode recommended for densely built areas. Aspects regarding adaptability and personalization were evident in all apps. The apps were generally intuitive in handling and responsiveness. However, their navigation systems varied based on location and map data, with occasional issues in timing and accuracy. The requirements were only partially met, and improvements in navigation are needed to address contextual challenges. In WeWALK and Lazarillo, the additional clock system for directional orientation proved helpful. Both apps sometimes struggled with correctly announcing points and orientations on the map, emphasizing the need for personalization and improvements in reliable, timely information.

Notably, all three apps required some degree of residual vision to interpret map displays. HereWeGo showed accessibility issues, as some menu displays were not recognizable or navigable with VoiceOver.

5 Conclusion

The requirements were assessed by a diverse group of experts, which ensured a broader representation of different experiences. However, bias cannot be ruled out.

While rating the requirements, the interviewees did not regard synergetic effects and possible interrelationships of shown requirements. Holistically rating interrelated requirements would mitigate possible bias. Thus, future work should explore emerging interrelations.

Aspects regarding individual preferences have shown the need for adaptable design. Requirements aimed at improving the information delivered to the users were either rejected or deemed to be not suitable for every user, based on individual preferences. As they serve functional purposes, they should still be seen as a requirement, to ensure functionalities for all possible users, future work should further discuss possible system designs to improve user independence and preference.

The findings of the study indicated that requirements like "Retention of habits", "Privacy and data security", and "Social compatibility" were less important, with experts stressing the significance of user independence and the distinction between functional needs and personal preferences. Adaptability emerged as a key factor, as well as the importance of designing systems that support user autonomy, ensuring that information output facilitates independence and reliance on sensory abilities. The preliminary assessment showed that certain requirements were already met in current systems. While all apps offered valuable assistance regarding safety and error handling, they did not fully compensate for the need for personal spatial awareness. The ability to assess and verify environmental information remains crucial. Moving forward, the integration of adaptable system designs will be vital in promoting user independence and meeting individual needs.

Acknowledgements. The primary author wants to extend sincere thanks to Dr. Thorsten Schwarz for his extended support and time supervising the work leading to this research.

Disclosure. There were no competing interests regarding this work.

References

1. Fernandes, H., et al.: A review of assistive spatial orientation and navigation technologies for the visually impaired. Univ. Access Inf. Soc. **18**(1), 155–168 (2019). https://doi.org/10.1007/s10209-017-0570-8
2. Wiener, W.R., Welsh, R.L., Blasch, B.B. Foundations of Orientation and Mobility. American Foundation for the Blind, Arlington (2010)
3. Kim, I.-J.: Recent advancements in indoor electronic travel aids for the blind or visually impaired: a comprehensive review of technologies and implementations. Univ. Access Inf. Soc. (2024). https://doi.org/10.1007/s10209-023-01086-8
4. Bhowmick, A., Hazarika, S.M.: An insight into assistive technology for the visually impaired and blind people: State-of-the-art and future trends. J. Multimodal User Interfaces **11**(2), 149–172 (2017). https://doi.org/10.1007/s12193-016-0235-6
5. Hoogsteen, K.M.P., Szpiro, S., Kreiman, G., Peli, E.: Beyond the cane: describing urban scenes to blind people for mobility tasks. ACM Trans. Access. Comput. **15**(3) (2022). https://doi.org/10.1145/3522757
6. Khan, A., Khusro, S.: An insight into smartphone-based assistive solutions for visually impaired and blind people: issues, challenges and opportunities. Univ. Access Inf. Soc. **20**(2), 265–298 (2021). https://doi.org/10.1007/s10209-020-00733-8
7. Ahmetovic, D., et al.: Sonification of navigation instructions for people with visual impairment. Int. J. Hum.-Comput. Stud. **177**(C) (2023). https://doi.org/10.1016/j.ijhcs.2023.103057
8. Chelladurai, P.K., et al. An exploratory study on the usability and features of indoor navigation apps for the blind and visually impaired. In: Proceedings of the 2nd International Conference of the ACM Greek SIGCHI Chapter. ACM, Athens (2023). https://doi.org/10.1145/3609987.3609998
9. Norman, G.: Likert scales, levels of measurement and the "laws" of statistics. Adv. Health Sci. Educ. **15**(5), 625–632 (2010). https://doi.org/10.1007/s10459-010-9222-y

Advancing Diagnosis and Rehabilitation of Neurodevelopmental Disorders: The Role of Virtual Reality, Eye-tracking, and Serious Games

Sensors in Inclusive Virtual Reality Games for People with Intellectual Disability

Daniel Tapia Huechao, Linus Ekberg, and Thomas Westin(✉)

Department of Computer and Systems Sciences, Stockholm University, Stockholm, Sweden
data6267@student.su.se, thomasw@dsv.su.se

Abstract. People with intellectual disability are often neglected in the design of emerging technologies, such as virtual reality (VR). This qualitative study, in the form of an exploratory experiment, investigated how VR gaming experiences can be designed using various sensors to be more inclusive for people with intellectual disabilities (ID). Data was collected through observation and semi-structured interviews with a VR game prototype. Six participants (n = 6), all of whom were staff working with individuals with disabilities, took part in the study. The findings suggest that VR controllers, when designed to replicate simple, concrete, and repeatable patterns, were well-received. Joystick had mixed results and were considered less intuitive due to design choices, but could still contribute to inclusive VR environments. Voice (not speech) control was the most promising option, with room for improvement in terms of accuracy and response time. An important conclusion is that the design of VR headsets should be improved to reduce the feeling of confinement and enhance accessibility. Finally, the study emphasizes the importance of integrating input methods early in the design process to create more inclusive experiences.

Keywords: virtual reality · intellectual disability · game development · gaming experience · accessibility

1 Introduction

Virtual reality (VR) has evolved from its early applications in healthcare and education during the 1990s to become a mainstream medium for entertainment, training and social interaction. However, despite advances in hardware and software, VR game experiences often overlook the needs of adults with intellectual disabilities (ID), who may struggle with complex sensor-based interactions. The problem this study addresses is the lack of explicit knowledge on how VR games can be adapted to improve inclusion and accessibility for users with ID; in particular, we ask: how can different sensor input methods in VR be designed to create more inclusive gaming experiences for this population?

K. Mavrou and P. Encarnação (Eds.): AAATE 2025, *Technology for Inclusion and Participation for All: Recent Achievements and Future Directions*, pp. 135–143, 2025.
https://doi.org/10.1007/978-3-032-01628-7_17

Existing accessibility guidelines rarely address VR's unique combination of motion tracking, gaze detection, and controller or voice input. Miranda-Duro et al. [1] found that fewer than 10% of commercial VR titles meet basic accessibility criteria; Heilemann et al. [2] highlight fragmented recommendations that fail to cover sensor-based controls. Additionally, Reski and Alissandrakis [9] demonstrated that input modalities, ranging from standard hand controllers to gaze interfaces, can significantly affect usability for individuals with cognitive or motor limitations. Broader critiques emphasize that assistive technology is often an afterthought in design rather than being integral to the process [3,4].

In VR, "sensors" refer to any hardware or software that detects user actions, whether physical movements, button presses, or voice commands, and translates them into in-game responses. These sensors can either create barriers or open up new avenues for interaction. For example, finely tuned motion controllers may overwhelm users with limited dexterity; well-designed voice or gaze controls could provide more intuitive alternatives. Early discussions with disability support staff confirmed a stark disparity: although video game titles exist for children with ID, few are accessible or specifically tailored to adults in this demographic.

People with ID often face challenges in abstract reasoning, memory, and task sequencing [5], underscoring the need for interfaces that emphasize simplicity, repetition, and clear feedback. Prior work has examined sensor accuracy, such as gaze-tracking latency, and its impact on usability [6,7]; yet little research explores how these technical parameters intersect with the lived experiences of adults with ID in VR contexts.

2 Methods

This study employed an exploratory experimental design, combining observation with semi-structured interviews. Participants were purposefully selected, focusing on professional caregivers and support staff, which minimized ethical risks. Thus, individuals with intellectual disabilities (ID) were not included in this first study, as we assumed a tangible prototype where responses can be observed was needed, rather than abstract concepts communicated by speech. Observation is an inclusive method [8] that could also be used to involve people with ID in later studies.

A VR prototype[1] (Fig. 1) was developed in two iterations to assess user interaction and apply theoretical insights. Data collection included concurrent think-aloud observation of day activity center staff (DAC, a facility governed by Sweden's LSS Act) to capture real-time user thoughts. Semi-structured interviews complemented this approach, providing deeper insights. This was a pragmatic approach where validity was considered as far as possible from staff close to, rather than directly from, people with ID. Furthermore, to enhance reliability, structured protocols and interview guides were used. Recordings were transcribed and coded, and inductive thematic analysis was applied to identify codes, categories and themes.

[1] Prototype demo video: https://streamable.com/a62nx3.

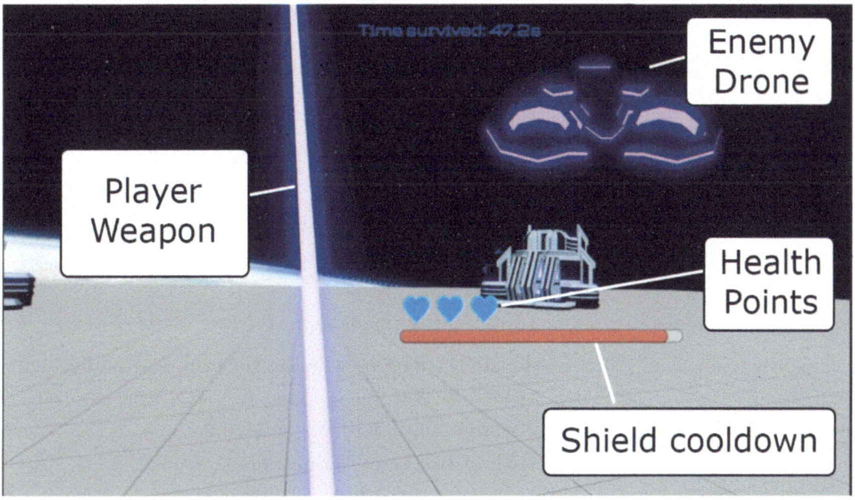

Fig. 1. Screenshot of the VR prototype showing the player's beam saber and an enemy drone. The interface displays the health points and the shield cooldown. Graphics used: TRON-ish low-poly drone, Bachklamk, CC-Attribution (CC BY 4.0). Energy Cell Factory, MRowa, CC-Attribution (CC BY 4.0). Mace Windu's Lightsaber, NahuelB, CC-Attribution (CC BY 4.0)

The study examined the usability and accessibility of different input methods for VR gaming, focusing on their suitability for individuals with intellectual disabilities. Six staff members ($n = 6$) from DACs participated in testing three input methods: VR controllers, a joystick with an additional button, and voice control. DACs are providing work-like activities for adults with ID (and often other disabilities, such as autism) to participate voluntarily, who are far from the job market and are not engaged in education. For analysis and reporting, staff members were anonymized and referred to using the codes P1 through P6.

3 Results

The results from this study were summarized in a set of themes (and categories), as follows: 1) Inclusive game experience for adults with ID (VR controller immersion, Joystick immersion, Voice immersion, Audiovisual immersion); 2) Accessibility and challenges for adults with ID (VR controller ergonomics, Joystick ergonomics, Voice ergonomics, Game mechanics, and Learning); 3) Suggestions for improvements (Language and instructions, More functionality, Dynamic game design) (Table 1).

Table 1. Thematic analysis

Theme: Inclusive game experience for adults with ID	
Category	Code
VR controller immersion	Sense of immersion with VR controllers
	Felt easier to handle
	Felt sensitive and responsive
	Felt intuitive
	Fun to use
Joystick immersion	Bulky and difficult to use
	Joystick response was less natural, especially for sword movement
	Joystick mode felt less engaging
	Risk of becoming monotonous
Voice immersion	Easy to use but not as enjoyable
	Voice command delay broke immersion
	Voice input felt repetitive and monotonous
Audiovisual immersion	Sci-fi theme was exciting and fitting
	Dark theme helped highlight the sword
	Sounds and vibrations enhanced immersion
Theme: Accessibility and challenges for adults with ID	
Category	Code
VR controller ergonomics	Easier to understand and use for people with ID
	Appreciated for mimicking real-life movements
	Requires less physical effort
Joystick ergonomics	Requires less physical movement
	Hard to use both hands simultaneously
	Challenging to manage multiple functions
	May be easier for some, but less stimulating
	Takes time to get used to
Voice ergonomics	Can be frustrating if not responsive
	Requires precise language, which may be hard for users with ID
	Lack of feedback increases difficulty
Game mechanics and learning	Challenging but not overwhelming
	Tutorial was clear and helpful
	Important to offer options for diverse needs
	Reasonably difficult for most users
Headset ergonomics	Headsets may feel confining or uncomfortable
	Users with ID may need more time to learn the controls

(*continued*)

Table 1. (*continued*)

Theme: Suggestions for improvements	
Category	Code
Language and instructions	Important to have the game in Swedish, especially for users with ID
	Need for simple and clear instructions
	Staff should receive setup guidance
More functionality	Ability to pause the game is needed
	Voice commands could activate based on sound level
	Ability to skip parts of the tutorial
	Customizable sound effects
	More visual aids to support understanding
	Clearer visual effects like making the shield more visible
Dynamic game design	Simpler versions with fewer enemies or more lives could increase accessibility
	More variety in environments would improve engagement

From this thematic analysis, the most important results were that VR controllers need to be designed to replicate simple, concrete, and repeatable patterns, to be accessible for the target group with ID. It was also found that joysticks had mixed results, such as not being as intuitive and easy to use in VR. Voice control (pitch and volume but not words or speech commands) was the most promising option, with room for improvement in terms of accuracy and response time.

3.1 Controller Immersion

All participants described the VR controllers as intuitive and well-suited for users with ID, primarily due to their alignment with natural hand movements and the minimal physical effort required to interact with the game. The controls were experienced as simple yet engaging, with several participants (P1, P2, P4, P6) emphasizing that the controllers made the experience fun and accessible. For these users, the immersive quality of swinging a virtual object and receiving immediate feedback through vibrations and in-game responses helped reinforce the feeling of control and enjoyment. One participant noted that VR controls would likely be easier for someone with ID to manage compared to other input methods, while another highlighted their intuitive nature as a key strength. Others (P3, P5) acknowledged the simplicity of the interaction, though with slightly less enthusiasm and cautioned that adding more buttons or functions could risk reducing accessibility for the target group.

Participants generally found the joystick less intuitive and harder to use compared to the VR controller. Five out of six (P1, P2, P3, P4, P6) described it as clunky or awkward, especially when simultaneous input (button + joystick) was required. They noted that this complexity could be particularly difficult for users with ID, as it demands fine motor coordination and the ability to multitask. The overall impression was that joystick-based input may create confusion and frustration, especially in fast-paced or precision-based tasks.

Participants had varied perceptions of the immersive potential of voice input. For some (P2, P6), the voice control system provided a novel and engaging experience, particularly due to its hands-free nature, which contrasts with conventional physical input methods. The method was described as accessible and intuitive, especially for users who may have difficulties handling controllers. However, other participants (P1, P4, P6) noted that the repetitive nature of using voice, particularly when the system lacked consistent responsiveness, could become tedious and potentially reduce user engagement over time. These insights indicate that while voice input can enhance immersion through minimal effort and novelty, its repetitive structure and dependence on technical reliability may limit its long-term appeal.

3.2 Controller Ergonomics

Participants generally perceived the VR controllers as ergonomically accessible, with a gameplay difficulty that was considered appropriate and well-balanced. The controllers were described as easy to operate, offering a level of challenge that was stimulating yet not excessive. Both P1 and P2 highlighted this equilibrium, noting that the interaction was neither too complex nor too simplistic. However, P2 also suggested that while the controls were manageable for them personally, individuals with ID might require additional support to navigate certain tasks. P3 noted that although some aspects of the interaction were straightforward, situations involving multiple simultaneous actions could increase cognitive load. To enhance accessibility for users with ID, P3 recommended clearer instructions and the incorporation of pauses between interactive elements, allowing for reflection and comprehension.

Participants generally perceived the joystick as more complex to use, particularly due to the need to coordinate both hands simultaneously. This complexity raised concerns about its suitability for individuals with ID. Several participants (P1, P3, P4, P5) noted that using a joystick in combination with a button press could be cognitively or motorically demanding, especially when actions had to be synchronized. While the joystick was described as manageable when limited to a single large button and a simple motion, additional complexity could quickly become overwhelming. One participant (P6) pointed out that despite these challenges, some individuals with ID might still find the joystick suitable, depending on personal abilities and preferences.

Voice control was widely regarded as a promising alternative for users with ID, primarily because it removes the need for fine motor skills or coordinated physical actions. Participants (P2, P5) observed that pitch- and volume-based voice input could be more intuitive than standard button-based systems, particularly for users who struggle with traditional interfaces. However, the effectiveness of this method was strongly linked to its responsiveness. Several participants (P1, P3, P6) described how minor delays or inconsistent feedback led to confusion and disengagement, especially when users were unsure whether their input had been registered. These findings underline that while voice input holds ergonomic advantages, its accessibility depends on real-time feedback and system precision to prevent user frustration and maintain clarity.

In addition to input methods, some participants raised concerns regarding the usability and comfort of the VR headset itself, which also includes displays for visual output. P3 specifically highlighted that the physical experience of wearing the headset could be a barrier for individuals with ID. The feeling of being enclosed or the unfamiliarity of the device might cause discomfort or hesitation. This discomfort, in turn, could discourage users from engaging with the technology or make it more difficult for them to focus on the experience. P3 emphasized the importance of preparatory routines and gradual exposure to help users become familiar with the headset before gameplay.

4 Discussion

The staff expressed a positive experience of the VR controllers, as being natural, responsive, precise and intuitive, analogue to actual body motion and voice, with a direct link between cause and effect, which is a typical good design for the target group. This can be contrasted with [9] who could not find significant differences between different input methods; however, their study did not include disabled people. Similarly, [10] found that immersion was not necessarily better in VR compared to PC desktop environments, also without considering people with ID for whom immersion may be even more crucial. The staff also expressed that VR controllers required guidance and time to be useful, as it is a new type of interaction for this group, supported also by [11]. Further issues were also discussed regarding joysticks and voice input.

5 Conclusion for the AT Field

Based upon the discussion of the results and related research, it can be concluded that the design of VR headsets may need to be improved to reduce the feeling of confinement, enhance accessibility and inclusion for the target group. Furthermore, a tentative VR design guideline for people with ID can be defined as: Replicate actual motion or voice patterns that the person with ID can perform. This requires the patterns to be concrete, easy to understand and repeat, and provide direct cause and effect between the actual and virtual worlds. Finally, the study emphasizes the importance of integrating input methods early in the design process to create more inclusive experiences.

While this study provides important information based on the delimitation of views from staff with experience working with the target group, there is a need for more participatory research directly with the target group. However, such research still needs the participation of staff to ease communication with researchers and support persons in the target group based on their individual needs.

Future research suggestions include a more direct user-centric study with participatory design, preferably involving a diversity of users with different needs, and examining more variables to provide a clearer picture of whether different input methods can be useful. Further research should also take into account the ethical and practical challenges experienced by this target group by developing different methods that allow participation on their own terms. Small-scale design workshops with visualizations and qualitative experiments with observation [8] are probably possible, as such research can be done individually and without requiring participant abilities such as speech and imagining abstract concepts of sensors and virtual reality. Conducting large, quantitative studies to enable generalizable results is a methodological challenge with this target group, as they have challenges participating in larger groups, in unfamiliar environments such as labs, or answering to surveys with texts and abstract concepts.

References

1. Miranda-Duro, M.C., Concheiro-Moscoso, P., Viqueira, J.L., Nieto-Riveiro, L., Domínguez, N.C., García, T.P.: Virtual reality game analysis for people with functional diversity: an inclusive perspective. Proc. **54**, 20 (2020). https://doi.org/10.3390/proceedings2020054020
2. Heilemann, F., Zimmermann, G., Münster, P.: Accessibility guidelines for VR games-a comparison and synthesis of a comprehensive set. Front. Virtual Reality **2**, 697504 (2021). https://doi.org/10.3389/frvir.2021.697504
3. Wobbrock, J.O., Gajos, K.Z., Kane, S.K., Vanderheiden, G.C.: Ability-based design. Commun. ACM **61**(6), 62–71 (2018). https://doi.org/10.1145/3148051
4. Mott, M., et al.: Accessible by design: an opportunity for virtual reality. In: 2019 IEEE International Symposium on Mixed and Augmented Reality Adjunct (ISMAR-Adjunct), pp. 451–454. IEEE, Beijing (2019). https://doi.org/10.1109/ISMAR-Adjunct.2019.00122
5. Luna, R.E.: Intellectual disability. American Psychiatric Association. https://www.psychiatry.org/patients-families/intellectual-disability/what-is-intellectual-disability
6. Teófilo, M., Lucena, V.F., Nascimento, J., Miyagawa, T., Maciel, F.: Evaluating accessibility features designed for virtual reality context. In: 2018 IEEE International Conference on Consumer Electronics (ICCE), pp. 1–6. IEEE (2018). https://doi.org/10.1109/ICCE.2018.8326167
7. Rojo, A., Cortina, J., Sánchez, C., Urendes, E., García-Carmona, R., Raya, R.: Accuracy study of the oculus touch v2 versus inertial sensor for a single-axis rotation simulating the elbow's range of motion. Virtual Reality **26**, 1651–1662 (2022). https://doi.org/10.1007/s10055-022-00660-4
8. Stephanidis, C.: Design for all. In: The Encyclopedia of Human-Computer Interaction. Interaction Design Foundation, Aarhus (2014). https://www.interaction-design.org/literature/book/the-encyclopedia-of-human-computer-interaction-2nd-ed/design-4-all

9. Reski, N., Alissandrakis, A.: Open data exploration in virtual reality: a comparative study of input technology. Virtual Reality **24**, 1–22 (2020). https://doi.org/10.1007/s10055-019-00378-w
10. Yildirim, C., Carroll, M., Hufnal, D., Johnson, T., Pericles, S.: Video game user experience: to VR, or not to VR?. In: 2018 IEEE Games, Entertainment, Media Conference (GEM), pp. 1–9. IEEE, Galway (2018). https://doi.org/10.1109/GEM.2018.8516542
11. Gluck, A., Solini, H., Maiti, K., Brinkley, J.: Evaluating 3D printed VR controller prototypes to increase VR accessibility for older adults. In: 2022 IEEE 3rd International Conference on Human-Machine Systems (ICHMS), pp. 1–6. IEEE, Orlando (2022). https://doi.org/10.1109/ICHMS56717.2022.9980720

Usability of Smart Wearable for Adult and Paediatric Neurological Upper Limb Re/Habilitation: Perspectives of Users and Allied Health Professionals

Sharon Borg Schembri(✉), Buhagiar Nathalie, Bonello Matthew, and Farrugia Philip

University of Malta, Msida MSD 2080, Malta
sborg04@um.edu.mt

Abstract. DigiClap device, is a 3D-printed smart wearable designed to enhance upper-limb function for individuals with motor disabilities, particularly those with cerebral palsy and similar conditions. DigiClap integrates motion sensors with gamified augmented reality exercises to improve user engagement and rehabilitation outcomes. A mixed-methods study, conducted at a Rehabilitation Centre in Cyprus, involved 20 participants and 2 allied health professionals. The findings highlighted high user motivation and engagement, with 55% of participants finding DigiClap easy to use and 90% appreciating its design. However, challenges were noted in its usability for individuals with severe impairments, indicating the need for personalized adjustments to the device. Allied health professionals found DigiClap efficient and beneficial but suggested improvements in haptics and game responsiveness.

Keywords: Neurological Re/Habilitation · Smart Wearables · Serious Games

1 Introduction

1.1 Upper Limb Motor Disabilities

Motor disabilities caused by conditions such as cerebral palsy (CP), stroke, and Parkinson's disease significantly impair upper limb function, affecting activities like eating, dressing, and writing. A range of medical conditions can lead to upper limb motor disabilities. For instance, cerebral palsy (CP), a non-progressive neurological disorder resulting from abnormal brain development, or damage to the developing brain, resulting in poor coordination, stiff or weak muscles, and tremors. Symptoms may also include sensation, vision, hearing, swallowing, and speaking problems, and can range from mild to severe [1].

Similarly, stroke and Parkinson's cause upper limb movement difficulties due to damage to the central nervous system. Stroke is a serious life-threatening medical condition that happens when the blood supply to part of the brain is cut off. Symptoms

K. Mavrou and P. Encarnação (Eds.): AAATE 2025, *Technology for Inclusion and Participation for All: Recent Achievements and Future Directions*, pp. 144–151, 2025.
https://doi.org/10.1007/978-3-032-01628-7_18

include weakness, numbness or paralysis in face, leg and arm, loss of speech, impaired vision and problems with gait [2]. Parkinson's is a progressive disorder that affects the nervous system and the parts of the body controlled by the nerves. Symptoms may start with a slight tremor in one hand. Other symptoms include stiffness and slow movement. As the condition progresses the symptoms increase in their severity [2].

1.2 Re/Habilitation Intervention

Occupational therapy and physiotherapy are widely recognised as core rehabilitative and habilitative interventions aimed at improving hand function, motor control, and strength to support individuals in performing activities of daily living. Habilitation focuses on helping individuals acquire or enhance new skills necessary for daily functioning, particularly when these skills have not yet been developed, while rehabilitation is oriented toward restoring lost or impaired functions following injury or illness [3]. Central to these interventions is the use of meaningful, goal-directed activities that are motivating and relevant to the individual. Research emphasises the importance of repeated opportunities to practice motor movements in realistic and functional contexts to support neuroplasticity and motor learning [4]. Play is often used as a therapeutic medium in pediatric populations due to its natural engagement and adaptability. Since play is also an activity that continues into adulthood it is also often used with adults as a therapeutic medium in adult rehabilitation. Engagement in motivating activities facilitates the integration of multisensory feedback—visual, auditory, and somatosensory—which supports motor planning and execution. To promote generalization of skills, therapists frequently prescribe home-based programs to increase practice opportunities [5].

In addition to rehabilitative approaches, assistive technology (AT) is often used as an intervention strategy when full motor recovery is not possible or to enhance independence during the rehabilitation process. AT serve as compensatory tools that enable individuals to perform tasks that might otherwise be challenging or impossible. According to the World Health Organization [6], assistive technology encompasses any item, equipment, or product system that increases, maintains, or improves the functional capabilities of individuals with disabilities. This includes a spectrum of solutions from low-tech adaptations (e.g., built-up utensils, splints) to high-tech innovations (e.g., robotic exoskeletons, powered orthoses, eye-gaze control systems). Rather than replacing therapeutic intervention, AT often works in synergy with rehabilitation by supporting participation, fostering autonomy, and enabling access to meaningful occupations when remediation alone does not meet functional needs.

1.3 Technology as a Re/Habilitative Tool

Advances in technology, particularly serious games and wearable devices, have introduced innovative rehabilitation methods that offers a fun and motivating way to learn movement [7]. Additionally, these games offer multi-sensory feedback through visual, auditory, and sometimes haptic stimulation. Commercial devices such as Microsoft Kinect, Nintendo Wii, or Leap Motion are designed to capture the individual's natural movements without the need to attach sensors to the body [8]. However, these devices may not be suitable for re/habilitation purposes or be accessible for individuals with upper limb disabilities since they are not adaptable or customisable to meet the particularities and clinical needs of these users. Additionally, these devices are not designed to measure changes in movement control and range of motion over a period of time [8].

Other devices have been designed for re/habilitation purposes aimed to facilitate hand therapy in children and adults with CP and other neurological disorders. These devices are classified in the literature as supportive wearable devices or smart wearables [9]. Examples of supportive wearable devices include passive and active exoskeletons. Passive exoskeletons generally make use of elastic bands to provide a fixed amount of movement support at any given moment that can be controlled to provide variable levels of movement support. Active orthoses or exoskeletons utilise motors or actuators to provide variable levels of movement support. Table 1 gives examples of passive and active exoskeleton.

Table 1. Examples of Supportive Wearable Devices

Supportive Wearable Device	**Type**	**Description**
Wilmington Robotic Exoskeleton(WREX) [9]	Passive	
AMADEO©	Active	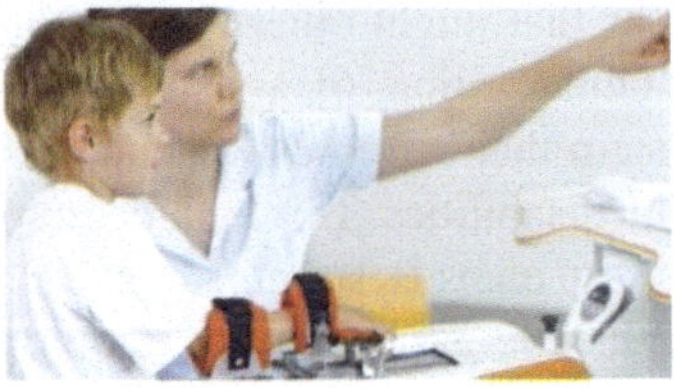

Smart wearable devices (SWD) incorporate sensors that collect continuous data across time, tasks, and environments [9]. These wearable devices are becoming very popular in the medical and re/habilitation field as they help to collect data over a period of time and throughout the individual's activity. These devices differ from the commercial devices mentioned above in the aspect of personalisation of the training program, data analysis, feedback, and adaptability to the home environment in their design. Table 2 illustrates these examples. These involve the utilisation of a glove that has motion sensors and is worn on the hand and connected to a serious game. The design of these devices is also aimed to be user-friendly to be used at home, thus complementing the therapy sessions and increasing the practice opportunities. On the other hand, these gloves are quite bulky and are in contact with almost the full surface of the hand. The latter can cause tactile sensitivities in some individuals. Additionally, it can be difficult to adapt to hands with severe movement difficulties as it lacks customisability as additive manufacturing is not considered for the glove design [10–12]. Finally, the cost of these devices is quite expensive making it difficult for them to be procured by rehabilitation centers with restricted financial resources or for home use [12].

Table 2. Examples of Smart Wearable Devices

Smart Wearable Device	Picture
FlintRahab's Music Device [10]	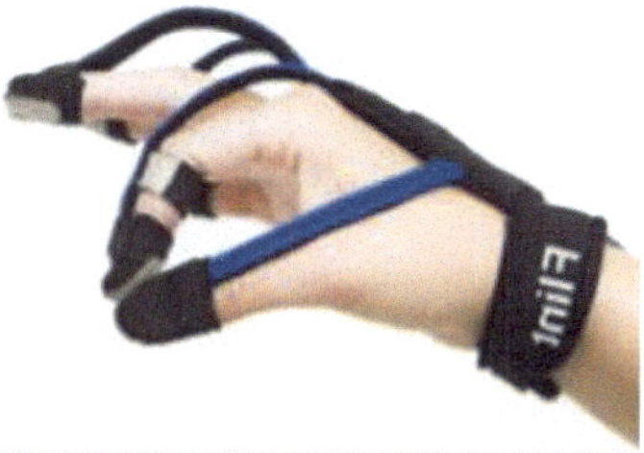
Raphael Smart Glove (adults) [11]	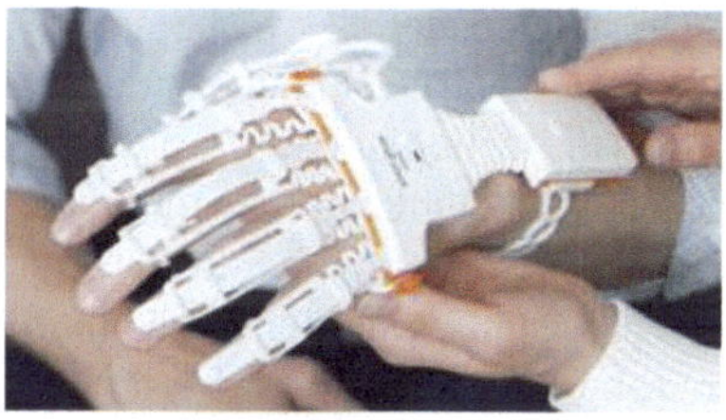
Raphael Smart Glove (kids) [11]	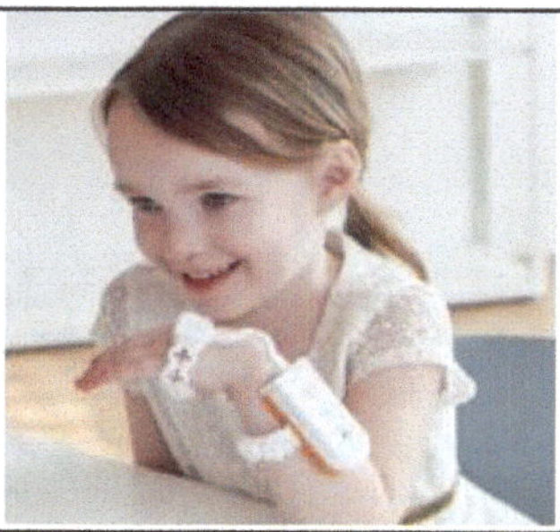

1.4 DigiClap

DigiClap, developed within the SMARTCLAP and CURACLAP projects through a multidisciplinary collaboration between occupational therapy and engineering experts, is a wearable assistive device designed to promote active upper limb engagement in children with cerebral palsy and other motor disabilities through interactive game play. The device comprises 3D-printed rings with embedded motion sensors linked to a control unit, which communicates wirelessly with a mobile device running a serious game enhanced with augmented reality (Fig. 1).

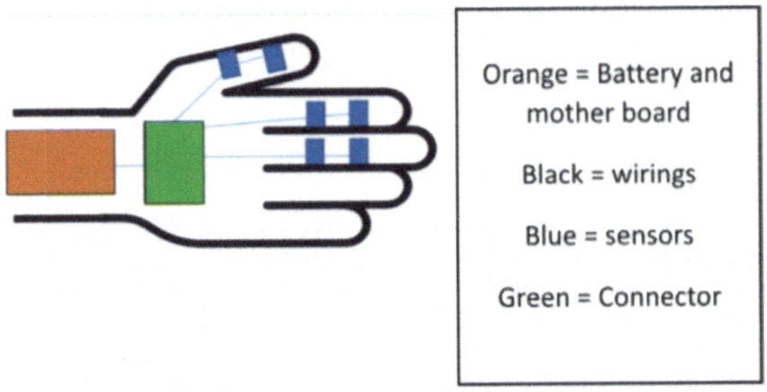

Fig. 1. Schematic representation of DigiClap

DigiClap, when compared to other devices is developed to present a less intrusive and child-friendly wearable device that can capture with high accuracy the movement of individual phalanges while the individual is interacting with a serious game in an Augmented Reality (AR) environment. It in fact features modularity and personalisation allowing finger modules to be added or removed according to the goals of the user and therapist. Initial studies conducted as part of the development of the device which involved the testing of DigiClap with children with CP and their occupational therapists indicated that DigiClap has good potential for re/habilitation purposes. Additionally, it was viewed as attractive to the user since it is lightweight, colourful and portable [13]. These studies highlighted a number of improvements including making the device more robust, to develop more interactive and motivating games with auditory feedback (Fig. 2).

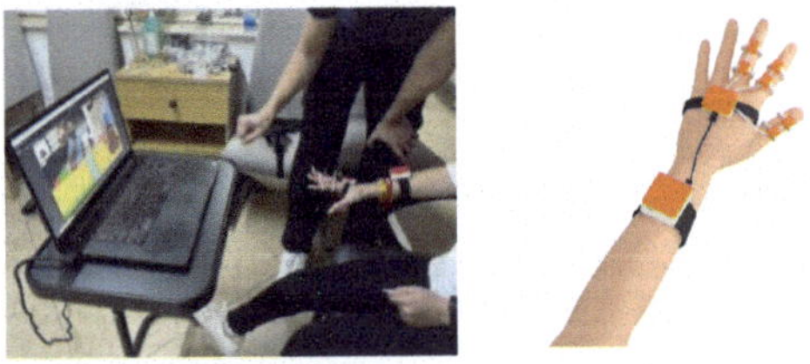

Fig. 2. 3-D Model diagram and photo of DigiClap

As a low-risk, Class I active medical device under the EU Medical Device Regulation (MDR 2017/745) and Council Directive 93/42/EEC, DigiClap adheres to relevant safety and classification standards.

In conclusion, this study evaluates DigiClap's usability, user experience, and rehabilitation potential, offering insights into its role in addressing the diverse needs of individuals with neurological conditions.

2 Methodology

This study took place at a NicoMed Rehabilitation Center (NRC), Cyprus to obtain a larger sample of participants with Cerebral Palsy and other neurological conditions. The study employed a convergent mixed-methods design to evaluate the DigiClap device, focusing on usability and user experience. Quantitative data were collected through structured observations and user questionnaires, while qualitative data were gathered from open-ended questions and thematic analysis following Braun & Clarke six-step strategy. Twenty individuals (4 children; 16 adults) with neurological conditions affecting the upper limb (mean age: 68.53 for adults; 11.5 for children) were recruited via snowball sampling [14]. The range of neurological conditions included cerebral palsy, right and left hemiplegia, ataxia and Parkinsons. Two physiotherapists trained in using DigiClap participated in the study.

Participants played three serious games, a burger, cake and pizza making game while wearing the DigiClap. Sessions were based on a standardised protocol to minimise variability. Feedback was obtained using the DigiClap Observation Form, to provide a structured framework to record specific behaviors, actions, and prompting that happened during the sessions. A semi-structured user questionnaire adapted for adults and children [15] were used to obtain perceptions on user experience. The modular UEQ+ questionnaire was utilised to obtain the allied health professionals experience based on several scales, including efficiency, novelty, visual aesthetics, usefuleness, stimulation, intuitive use and haptics [16]. Quantitative data were analysed descriptively, while qualitative insights were coded thematically to cross-verify findings. Ethical approval was obtained, and informed consent was secured from all participants in accordance with international guidelines.

3 Results and Discussion

3.1 User Experience and Perceptions

Analysis of the observations and open-ended questions yielded 4 themes:

Motivation and Enjoyment: The gamified approach enhanced engagement, with 60% of participants rating the device as highly enjoyable. The is an important factor as it can lead to higher adherence to therapy sessions. Requests for more challenging games and diverse themes were noted.

Usability and Engagement: While 55% found the device easy to use, challenges arose for participants with very limited hand movements or associated visual difficulties. In fact these participants rated the device as difficult to use. This was also reported in the physiotherapists observations.

Personalisation: Most participants (80%) found DigiClap comfortable, although customisation for smaller hands was suggested. Additionally, they found the serious games fun, but requested more features to increase or decrease the game challenge to meet the needs and preferences of the user.

Rehabilitation Implications: DigiClap device was perceived as beneficial by 60% of the participants, particularly in terms of improving hand strength and motor coordination. Additionally, it was perceived that DigiClap increased attention and focus and tactile awareness of the hand, which helps increase movement through mindfulness. The device's effectiveness was less clear, for those presenting with very limited hand movements and visual impairment suggesting that additional features or different games may be necessary to meet their needs.

3.2 Allied Health Professionals Perceptions and User Experience

The allied health professionals appreciated DigiClap's efficiency and motivational aspects but highlighted issues with durability, aesthetic appeal, and the need for better haptic feedback and hidden wiring. Suggestions included incorporating customizable difficulty levels and additional games.

4 Conclusion/Implications for the AT Field

DigiClap demonstrates potential as an innovative tool for upper-limb re/habilitation, particularly for enhancing motivation, adherence to therapy and to increase mindful movement. Feedback from participants in this study, highlighted the need for:

- Increased accessible features in the serious games to meet the needs of users with limited hand movements and additionally disability like visual impairment.
- More diverse and challenging game options.
- Enhanced durability and design aesthetics.

DigiClap can also be further develop as an assistive device to access computer or other technology. The personalised design and engaging serious games contributed to positive user experiences.

Acknowledgment. This work was supported by the CuraClap project, funded by Xjenza Malta (project reference R&I-2019-003-A). We also extend our gratitude to NicoMed Rehabilitation Center (NRC), Cyprus, and all the participants who contributed to this study.

References

1. Sadowska, M., Sarecka-Hujar, B., Kopyta, I.: Cerebral palsy: current opinions on definition, epidemiology, risk factors, classification and treatment options. Neuropsychiatr. Dis. Treat. **16**, 1505–1518 (2020). https://doi.org/10.2147/NDT.S235165
2. World Health Organization. ICD-10: international statistical classification of diseases and related health problems: tenth revision, 2nd edn. World Health Organization (2004). https://iris.who.int/handle/10665/42980
3. American Associaltion Occuptional Therapy (AOTA). Tool from AOTA, APTA, and ASHA helps evaluate habilitative and rehabilitative insurance benefits (2024). https://www.aota.org/advocacy/issues/health-care-reform/health-care-reform---eval-tool-habilitative-rehabilitative#:~:text=Habilitative%20services%20help%20a%20person,sick%2C%20hurt%2C%20or%20disabled. Accessed 9 May 2025

4. Missiuna, C., Mandich, A.D., Polatajko, H.J., Malloy-Miller, T.: Cognitive orientation to daily occupational performance (CO-OP). Phys. Occup. Ther. Pediatr. **20**(2–3), 69–81 (2001). https://doi.org/10.1080/J006v20n02_05
5. Verhaegh, A.P., et al.: Parents' experiences with a home-based upper limb training program using a video coaching approach for infants and toddlers with unilateral cerebral palsy: a qualitative interview study. BMC Pediatr. **22**(1), 380 (2022)
6. World Health Organization (WHO) & United Nations Children's Fund (UNICEF). Global report on assistive technology. World Health Organization and the United Nations Children's Fund, Geneva (2022). https://www.who.int/publications/i/item/9789240062887
7. Cisneros, T.V.P., Brons, A., Kröse, B., Schouten, B., Ludden, G.: Playfulness and new technologies in hand therapy for children with cerebral palsy: scoping review. JMIR Serious Games **11**(1), e44904 (2023)
8. Gomez-Portes, C., Lacave, C., Molina, A.I., Vallejo, D.: Home rehabilitation based on gamification and serious games for young people: a systematic mapping study. Appl. Sci. **10**(24), 8849 (2020)
9. Lobo, M., Hall, M., Greenspan, B., Rohloff, P., Prosser, L., Smith, B.: Wearables for pediatric rehabilitation: how to optimally design and use products to meet the needs of users. Phys. Ther. **99**(6), 647–657 (2019). https://www.ncbi.nlm.nih.gov/pmc/articles/PMC6545272/. Accessed 9 Sept 2024
10. Flint Rehab. MusicGlove Clinic Suite: Hand Therapy with a Beat - Flint Rehab. https://www.flintrehab.com/product/musicglove-clinic/. Accessed 09 Sept 2024
11. RAPAEL Smart Kids - Hand rehabilitation system by Neofect MedicalExpo, Medicalexpo.com. https://www.medicalexpo.com/prod/neofect/product-123611-940538.html. Accessed 9 Sept 2024
12. SMARTCLAP. Deliverable 2.1 - literature review on existing related product service system. Project funded by the Fusion R&I Technology Development Programme 2020 (R&I-2019–003-T), by XJENZA Malta (2021)
13. Bonello, M., Buhagiar, N., Farrugia, P., Mercieca, J.: Unveiling the impact of the SMARTCLAP project on habilitation. Comput. Struct. Biotechnol. J. (2024). https://doi.org/10.1016/j.csbj.2024.06.001
14. Creswell, J.W., Creswell, J.D.: Research Design, Qualitative, Quantitative, and Mixed Methods Approaches, 6th edn. Sage, USA (2023)
15. Read, J.C., MacFarlane, S.J.: Using the fun toolkit and other survey methods to gather opinions in child computer interaction. In: Proceedings of the 2006 Conference on Interaction Design and Children, pp. 81–88. ACM (2006)
16. Schrepp, M., Thomaschewski, J.: UEQ + handbook for the modular extension of the user experience questionnaire. V.6. (2023)

Creating a Virtual Environment for the Development of Social and Communication Skills of Children with Autism

Cecilia Sik-Lanyi[1,2](✉) and Barna Fogl[1]

[1] University of Pannonia, Veszprem 8200, Hungary
lanyi.cecilia@mik.uni-pannon.hu
[2] Hungarian Research Network (HUN_REN), Budapest 1052, Hungary

Abstract. This project aims to create a virtual environment that supports the development of children with autism through creative and skill-building activities. The software runs on the Meta (Oculus) Quest 2 VR headset. The development utilized the Godot Engine, primarily leveraging GDScript and C# programming languages. The application is based on the concept of "serious games", designed to make learning more engaging and enjoyable for children. Autism is a sensitive topic affecting a growing number of families worldwide. The use of virtual environments in the development of children with autism not only opens up new possibilities but also helps children and their families in their daily lives. The primary target audience includes children with autism and their families, as well as the professionals supporting them, such as special education teachers, psychologists, and therapists. The application aims to teach various skills and abilities to children in an enjoyable and engaging environment, enhancing the overall experience. The application developed within the scope of our project is intended to improve the quality of life and support the development of affected children. The virtual environment was tested with a psychologist and other testers, and the results and feedback were positive.

Keywords: Virtual Environment · Serious Games · Ocolus Quest 2 · Children with Autism

1 Introduction

In recent years, the use of virtual reality for therapeutic purposes has grown tremendously. Over the past 20 years, a large number of research has been conducted demonstrating that the use of technology can aid in neurocognitive rehabilitation for people with autism spectrum disorders. VR reduces the social pressures that affect these people and provides a realistic environment for more effective treatment, which may reduce the duration of rehabilitation. Current studies include large-scale training interventions, including training in social adaptation and communication skills, emotional skills, daily living skills, and cognitive functions [1].

K. Mavrou and P. Encarnação (Eds.): AAATE 2025, *Technology for Inclusion and Participation for All: Recent Achievements and Future Directions*, pp. 152–157, 2025.
https://doi.org/10.1007/978-3-032-01628-7_19

The school system can be difficult for children with autism, as it is not designed within their developmental framework. It is important to create an environment specifically designed for these children during the learning process because they have difficulty understanding abstract ideas or representations. Emotional factors can also have a greater impact on the learning outcomes of children with special needs, as they often suffer from emotional regulation disorders and other accompanying emotional problems. Therefore, their learning environment should be experience-oriented, and a strong emphasis should be placed on emotional consequences and emotional reactions. VR-based pedagogical platforms provide an opportunity to meet these special needs. In a virtual environment, children are exposed to different, concrete experiences in a realistic context. This allows them to learn from their mistakes without suffering real consequences. VR technology with these user-centric features and capabilities holds promise to become a uniquely empowering technique in the field of special education [2].

2 Methods

The method consisted of three steps. First, based on literature research, the discovery of games with similar goals was carried out. The second step was the development of the game. It was followed by the testing process.

Development was informed by a comprehensive review of related serious games, including *New Horizon* and *Autism & Beyond*. However, our application differs in its fully immersive VR format, multi-level emotional design, and gradual integration of decision-making and social-emotional reasoning in a first-person narrative setting.

User testing was conducted with a sample of 12 individuals: 2 children diagnosed with ASD, 4 neurotypical children, 5 adults (including caregivers), and 1 child psychologist. The testing process followed an informal, observational study design. Participants interacted with the full game on Meta Quest 2 and were assessed using three criteria:

- **Usability:** Ease of navigation and interface interaction.
- **Engagement:** Sustained interest and enjoyment.
- **Skill Development Potential:** Observable instances of emotion recognition, empathy, and conflict resolution behaviors.

Post-session interviews and structured feedback forms were used to gather qualitative data. Positive responses were especially noted regarding the use of storytelling to teach emotional regulation. Notably, testers reported that children often prioritized empathy over logic in conflict scenarios, suggesting successful internalization of the game's developmental goals.

2.1 Literature Review/State of the Art

During the literature review, several serious games that focus on improving the social and communication skills of children with autism were found. These serious games help to slow down or prevent the decline of old age in various ways. These games, as reference points, inspired and helped us develop our project.

The "New Horizon" serious game is a two-dimensional mobile application that contains puzzle-like games. The main game area is an infinite universe with an infinite number of planets to explore. The main character, Jimmy, travels through the universe in his spaceship. As he lands on the planets, he must complete mini-games to progress. This game helps children with autism develop cognitive skills, including logic. Of course, the further the player approaches the level, the more difficult the tasks will be, but since the game gradually becomes more difficult, this not only does not discourage children but also gives them a sense of achievement [3].

An example of this type of serious game is the app "Autism & Beyond", created by researchers at the Duke Institute for Health Innovation [4]. The main goal of the "Autism & Beyond" app is to promote early detection of autism and to educate parents on autism-related issues. The game involves children participating in interactive activities, such as recognizing faces and interpreting emotions. The app records the children's responses and behavior, which can help experts and doctors better understand the characteristics and developmental patterns of autism spectrum disorder.

2.2 Developing the Game (the Technology Used)

Choosing the right development environment was essential during the development of the project. For this project, we used the advantages of using VR glasses when creating the application. The Godot Engine was chosen for the development of the game, as it has several advantages in terms of integration with the Meta (Oculus) Quest 2 VR glasses. The Godot Engine was chosen specifically for its lightweight performance, open-source nature, and modular integration with the Meta Quest 2, which requires no external sensors or base stations (in contrast with HTC Vive or Valve Index). This hardware choice supports easier deployment in homes and schools and keeps costs significantly lower, improving accessibility for families and institutions alike [5].

3 Results

3.1 The Game

The application consists of several levels, during which the child must first help objects/toys, then animals, and finally people to complete various tasks. Each level gradually becomes "more difficult", so that the children's abilities and self-confidence increase with each level. By completing various small tasks, the child can develop their emotional intelligence, empathy, and skills for participating in the community. During the tasks, the children interact with characters in the virtual environment and learn to interpret different emotions and situations, which contribute to the development of skills that can be applied in real life.

The game begins in a calm and familiar environment, in a children's room, so the user is not exposed to too many stimuli. After pressing the "Start Game" button on the initial menu, the game starts. In the first level, they learn how to handle and how to react to a new environment. Starting from the children's room, during the game, the child can explore increasingly larger areas and encounter new challenges, so that in addition to emotional development, their spatial orientation skills can also improve. There are two interactable characters in this space: the teddy bear and "Robi the robot" (Fig. 1). In addition, there are interactable windows and the "battery". In this level, only Robi the robot moves after the player hands him the battery.

The central element of the second world is an interactive story based on the recognition and understanding of feelings, during which the player takes part in a journey full of emotions. The story focuses on four kittens who play with a ball of yarn - according to their current mood and emotions. The player's task is to observe these situations, recognize the emotions, and draw parallels with feelings experienced in real life. The player is not alone in this challenge: they are helped by the kittens' wise and experienced grandmother, who supports the player with her story in mapping the world of feelings. The background of the story is a magical, fairytale forest, full of life and detail. In the forest, there are not only the kittens and Grandma, but also other exciting characters, such as a friendly and curious pig, and some other special figures, who all contribute to the charm of the world. The atmosphere of the forest reflects a summer day (Fig. 2). This world is not only fun for the player but also educational, as it provides an opportunity to learn about emotions and interpret them more consciously.

The third world is the most difficult and complex part of the game, in which the user finds himself in a more complicated decision-making situation. The story focuses on a pair of siblings, a little boy and a little girl, between whom a conflict develops. The mother of the siblings, desperately looking for a solution, but not knowing how to handle the situation, asks the user for help to create a balance between the children's different desires. The setting is a lively and colorful playground (Fig. 3). The conflict is based on the fact that both siblings want to play together, but they would choose different games. The task is not for the user to take sides, but to find a more logical, balanced solution that takes into account the wishes of both children. The goal is for the siblings to cooperate and find common ground together, thereby learning deeper thinking and the ability to compromise.

Fig. 1. The robot in the 1st world

Fig. 2. Kitten in the 2nd world

Fig. 3. The playground in the 3rd world.

3.2 Testing

Comparative Analysis: While games like *New Horizon* focus on cognitive puzzles and *Autism & Beyond* emphasizes early detection, our application is distinguished by its emotionally driven narrative structure and direct engagement with empathy and social reasoning tasks in real time. Unlike 2D mobile apps, the VR environment allows embodied interaction and deeper immersion, which literature suggests is crucial for children with ASD in overcoming sensory and attentional challenges. The game consists of three narrative worlds of increasing complexity:

1. Children's Room – introductory level to teach interaction, reduce sensory overload, and build confidence through simple object manipulation.
2. Fairytale Forest – intermediate level focusing on emotion recognition using animated animal characters in narrative scenes.
3. Playground Conflict – advanced level designed around complex human social dilemmas, requiring negotiation and compromise.

Each level builds upon the previous, reinforcing learned emotional and social concepts through repetition and variation. Feedback indicated that children with ASD responded positively to the familiar, low-stimulus design and appreciated the gentle emotional tone.

Scalability and Accessibility: The game is designed to be deployed in multiple environments: therapy sessions, classrooms, and homes. Because it uses affordable and wireless VR hardware and open-source software, it is scalable and financially accessible. Future development may include adjustable difficulty levels for various autism severity profiles, multilingual support, and compatibility with additional VR platforms. Feedback from educators and therapists suggests strong potential for integration into existing special education curricula.

User Tesing: The user testing process involved a total of 12 individuals, including 2 children with ASD, 4 neurotypical children, 5 adult participants (including caregivers), and 1 psychologist. The evaluation focused on three main criteria: usability, engagement, and skill development. Observations showed that the children—especially those with ASD—could navigate the environment with minimal guidance and maintained consistent interest across levels. One particularly notable outcome was that children often prioritized emotional reasoning over logical outcomes, for example in conflict-resolution tasks, which reflects successful engagement with the game's developmental goals.

Although the current version does not employ standardized psychometric scales, structured feedback and behavioral observation forms were used to assess participant responses. Future versions aim to integrate more formal evaluative frameworks in collaboration with clinical partners.

In comparison to existing serious games such as New Horizon and Autism & Beyond, this project offers a more immersive and interactive format by utilizing full 3D virtual reality, narrative progression, and embodied first-person interaction. While New Horizon focuses on cognitive mini-games and Autism & Beyond targets early detection, our game focuses on emotional intelligence, empathy, and compromise—skills crucial for everyday social integration.

4 Conclusion

The developed virtual environment aims to promote the emotional development and social skills of children with autism. The game was tested by 12 observers (including 1 psychologist too). For example, during the testing phase, we observed that children often prioritized empathy over logic in resolving conflicts. We were particularly pleased with this feedback, as practicing empathy helps develop important emotional skills, which demonstrates the higher purpose of the game. In this way, the game not only develops problem-solving skills but also children's emotional intelligence. The weight of the details and decisions makes this world a particularly valuable element of the game, as it encourages both thinking and emotional immersion.

Overall, the game's goal is not only to provide entertainment, but also to support the development of emotional intelligence, social skills, and decision-making abilities. The different levels of the three worlds build on each other and help children gradually and playfully encounter the everyday challenges of life. The lively, imaginative environment and lovable characters provide a stimulating backdrop for children to become more open to learning and communication, while they can safely try out different play situations.

The application is scalable and adaptable: its modular design allows for the addition of new worlds, integration of multiple languages, and tailoring to different ASD severity levels. Because it relies on affordable consumer-grade VR equipment and free software, it has the potential for widespread implementation in therapeutic settings, special education programs, and even home use. Considerations around financial accessibility, hardware availability, and inclusive design were central to development.

While this is a prototype and not yet a certified therapeutic tool, it represents a promising step toward developing user-friendly, narrative-driven interventions that combine technology and pedagogy in meaningful ways. Further iterations will involve co-design processes with therapists, teachers, and caregivers to ensure maximum practical impact.

References

1. Karami, B., Koushki, R., Arabgol, F., Rahmani, M., Vahabie, A.H.: Effectiveness of virtual/augmented reality-based therapeutic interventions on individuals with autism spectrum disorder: a comprehensive meta-analysis. Front Psychiatry **12**, 665326 (2021). https://doi.org/10.3389/fpsyt.2021.665326
2. Zhang, M., Ding, H., Naumceska, M., Zhang, Y.: Virtual reality technology as an educational and intervention tool for children with autism spectrum disorder: current perspectives and future directions. Behavioral Sciences. Behav. Sci. **12**(5), 138 (2022). https://doi.org/10.3390/bs12050138
3. Carlier, S., Paelt, S.V., Ongenae, F., Backere, F.D., Turck, F.D.: Empowering children with ASD and their parents: design of a serious game for anxiety and stress reduction. Sensors **20**(4), 966 (2020). https://doi.org/10.3390/s20040966
4. Autism & Beyond Combining mobile technology with ML. https://dihi.org/project/autism-beyond/
5. Fogl, B.: Creating a virtual environment for the development of autistic children. (in Hungarian: Virtuális környezet létrehozása autista gyermekek fejlesztésére.) MSc thesis work, University of Pannaonia, Veszprém, Hungary (2024)

A New Path to Recovery: Virtual Reality-Driven Treadmill Exercises

Patrícia Szabó[1(✉)], Kinga Pete[2], and Cecilia Sik-Lanyi[1]

[1] University of Pannonia, Egyetem str. 10, Veszprem 8200, Hungary
szabo.patricia@mik.uni-pannon.hu

[2] Eötvös Loránd University, Egyetem square 1-3, Budapest 1053, Hungary

Abstract. This project presents the development of a virtual reality (VR) application designed to enhance treadmill-based walking by immersing users in dynamically changing natural environments. Traditional treadmill workouts are often repetitive and uninspiring, leading to decreased motivation and adherence. Our solution addresses this by synchronizing a VR experience with the user's walking pace on the treadmill, offering a continuously evolving landscape that replicates outdoor settings such as forests, valleys, and fantasy-themed terrains. The primary goal of this system is to support physical rehabilitation, particularly for individuals recovering from injuries, strokes, or neurological conditions that affect mobility. By providing an engaging and stimulating environment, the application can aid in therapy adherence and improve overall patient outcomes. Beyond rehabilitation, the system serves as an innovative alternative to conventional exercise, offering users a more enjoyable and immersive way to stay active. Key features include real-time terrain generation, personalized user profiles, adaptive difficulty levels, and performance tracking. A local database stores user progress and session statistics without requiring internet connectivity. The application is currently undergoing testing to evaluate its impact on user motivation, perceived exertion, and rehabilitation effectiveness. Preliminary results suggest promising benefits in both clinical and recreational contexts.

Keywords: Virtual Reality · Treadmill Rehabilitation · Immersive Exercise

1 First Section

Modern rehabilitation technologies are playing an increasingly important role in healthcare, particularly in improving motor coordination, physical condition, and mental well-being. New technologies, such as virtual reality (VR), open up new possibilities for the development of medical processes. VR-based rehabilitation allows patients to perform exercises in a safe, simulated environment. This not only makes therapy more effective but also more motivating, as the virtual world provides interesting and enjoyable experiences. A similar VR walking simulator has been used for rehabilitation purposes in strokes and multiple sclerosis patients [1]. As a result of using the VR application, patients' motivation increased, as well as the outcomes of the therapy applied. Another

K. Mavrou and P. Encarnação (Eds.): AAATE 2025, *Technology for Inclusion and Participation for All: Recent Achievements and Future Directions*, pp. 158–165, 2025.
https://doi.org/10.1007/978-3-032-01628-7_20

study focused on comparing the differences between real-world and virtual hiking [2]. The measurements were carried out with 10 healthy participants (2 males and 8 females, aged 24–53). The participants' step length and walking speed were measured using the Gait Rite mat. The results showed that the average walking speed and step length were greater during real-world hiking. The virtual environment had no effect on the nature of their movement.

The goal of the project is to create software that enhances the experience of indoor sports and exercise. Our primary focus is on improving the experience of walking, jogging, or running on a treadmill. An application that allows users wearing VR glasses, to experience various environments while walking on a treadmill was developed, making the otherwise monotonous training more enjoyable by leveraging immersive experience. The core of the application is that it synchronizes the user's movement in real time with the treadmill's speed, so the user's actual movement is reflected in the VR world. This solution offers a new type of experience in virtual reality, allowing users to explore different spaces and landscapes in a 3D environment, enabling individuals with physical limitations—who may struggle to walk outdoors—to feel closer to nature.

2 Methods

The development of the application was designed using Unity [3], a widely used game engine with strong VR capabilities. The programming logic was implemented in C#, Unity's primary scripting language, which enabled us to create interactive and responsive mechanics. To handle VR-specific interactions, we used Unity's XR Interaction Toolkit [4], which provides pre-built components for intuitive object manipulation and device integration. During the design process, we also analyzed existing VR walking simulators, to better understand technical requirements and user expectations. We aimed to avoid common falls, such as poor navigation feedback or non-intuitive interfaces, and incorporated feedback from competitor analysis to improve our design. During our project, we encountered a similar application called Octonic [5]. This is a fully developed program available on the Meta software distribution platform. Octonic is a well-crafted application that allows users to take virtual walks in various environments. However, despite receiving many positive reviews, several critical opinions surfaced that we considered important for our own project development. Most criticisms focused on the user experience, particularly navigation difficulties encountered in different environments within the program. For example, winding paths often cause discomfort for users because treadmill movement is restricted to straight lines, meaning the virtual world often does not reflect actual movement possibilities. Another example relevant to similar application development is the VR Walk: Game program [6]. In addition to using the Oculus Quest 2 headset, it employs an HTC Vive ankle sensor and the SteamVR Lighthouse tracking system to monitor walking. Like our application, the software was developed using Unity.

The technical structure of our application consists of several separate components that work together to provide a continuous and realistic VR experience. Three tracks are designed to offer users a variety of immersive environments, enhancing their virtual reality experience. The "Forest" track features a natural woodland setting, with lush greenery, providing a calming and realistic walking experience. The "Fantasy Forest" track transports users into a magical, otherworldly forest filled with vibrant colors, and enchanted elements, creating an adventurous and whimsical atmosphere. Finally, the "Rocky Valley" track features rugged, mountainous terrain with rocky cliffs and valleys. Each track is carefully crafted to cater to different preferences and provide diverse, engaging experiences for users during their virtual rehabilitation or exercise sessions.

3 Results

In the virtual environment we used a static background terrain, which plays a crucial role in ensuring that the user doesn't feel as though they are walking in an endless void while moving through the virtual world. When the player looks into the distance, a larger, static terrain follows the character, primarily showcasing the view of distant forests, mountains, and other natural elements. This solution visually creates the illusion that the user is traversing an expansive landscape, thereby enhancing the immersive experience. It has been placed lower than the original track, so the user only sees it in the distance, preventing the background from blending with the main gameplay area. To enhance immersion, we apply a light fog in our virtual environment, which serves both an aesthetic and functional purpose. The presence of the fog helps create the illusion of depth in space, contributing to a more realistic sense of the environment, while subtly obscuring distant elements beyond the horizon. This is especially important in situations where the player moves through vast open areas, as the fog amplifies the feeling that the space is truly limitless, while also hiding any lower-detail distant objects (Fig. 1).

Fig. 1. Screenshot from the application, with speed label

Our application's technical structure consists of several distinct components that work together to ensure a continuous and realistic VR experience. The core elements of the system include pre-modeled and custom-designed terrain assets, character movement synchronized with treadmill motion, and the integration of the XR Interaction Toolkit, which facilitates user interactions within the virtual environment. These components function harmoniously to provide users with a natural and seamless VR experience.

3.1 Enviroment

Our environment is built on Unity's terrain elements, enhanced with assets from the Unity Asset Store featuring diverse textures such as trees, bushes, grass, and ground patterns. These terrain tiles are dynamically and randomly placed ahead of the player as they progress, ensuring a non-repetitive and engaging experience each time. Three tracks were created, which are the following (Fig. 2):

1. Forest
2. Fantasy Forest
3. Rocky Valley

Fig. 2. Screenshot from the application, track selection

Each terrain piece has a visual mesh and a collider to detect when the player steps onto it. When triggered, a script selects a new terrain tile at random from a pool (excluding recently used ones) and places it ahead. Once all options are exhausted, the pool resets to avoid consecutive repeats. Old tiles are deleted when the player moves far enough, freeing system resources. The static background terrain plays an essential role in preventing the user from feeling like they're walking in an empty void. As the player moves, a larger background terrain follows them, showcasing distant forests, mountains, and other natural elements to create the illusion of vast, seamless environments. This background is fixed in appearance and always placed below the main game terrain to avoid visual interference. It is treated as a child object of the player, constantly updating its position to maintain consistent visuals regardless of the player's speed.

3.2 Performance Optimization

Performance is critical in VR environments. Our system constantly monitors the distance between the player and each terrain tile. If the player moves too far from a tile, it's removed from memory. This ensures that only nearby elements remain loaded, reducing memory usage and system load. We also use frustum culling and other optimization techniques to render only what is visible to the player. Objects outside the player's field of view are not processed. Additionally, Level of Detail (LOD) systems are employed, loading lower-resolution assets for distant objects, and upgrading their quality as the player approaches.

To enhance immersion, we use subtle fog effects in the virtual environment. Fog not only adds aesthetic depth but also conceals distant elements, improving performance and reinforcing a sense of vastness. Fog density can be adjusted based on the environment—denser fog for mountainous scenes, lighter fog for flat or urban areas. It can also be disabled when inappropriate for the scene.

Character movement in the VR world is continuous and synchronized with the user's physical walking on the treadmill. Movement is not controlled via a joystick but based on actual steps, providing a more natural and immersive experience. Controller-based movement is disabled entirely. Technically, this is implemented using Unity's built-in Rigidbody component, managed by custom scripts that precisely align the avatar's motion with treadmill speed. A special joystick function on the left controller allows the user to adjust the movement speed manually, which is also displayed in the player's view. When the walk ends—either by reaching a goal or manually stopping—the speed drops to zero and a statistics panel is displayed.

3.3 User System

In the main menu, users can choose a track or select a user profile. Options are navigated using the controller joystick, and selections are confirmed with the primary button. The currently highlighted option is outlined in black. Users can walk without selecting a profile, but in that case, their distance and time statistics will not be saved. Each track has its own visual button with a thumbnail for easy recognition. The top-left corner of the menu includes a button to switch to the user selection panel. All registered users are displayed in a scrollable list. A script centers the currently selected profile, giving clear visual feedback. Selecting a profile makes it active, and any walking data is saved to that user. The panel includes buttons for deleting the active user or adding a new one. When adding a user, a form appears to enter the user's name and height, which is then stored in the database and added to the user list. A final menu option (highlighted in red) allows the application to be closed.

Before starting a virtual walk, users can define their goals:

- Distance-based
- Time-based
- Infinite mode (walk freely)

This allows users to tailor their session to personal preferences, be it a short workout or a longer, relaxing walk (Fig. 3).

We use a local SQLite database to store user data without requiring internet access. Each user has a name, height, total distance walked, and total walking time, identified with a unique ID to allow duplicates in usernames. The database supports adding new users, deleting existing ones, and updating statistics after each walking session by appending new results to the existing totals.

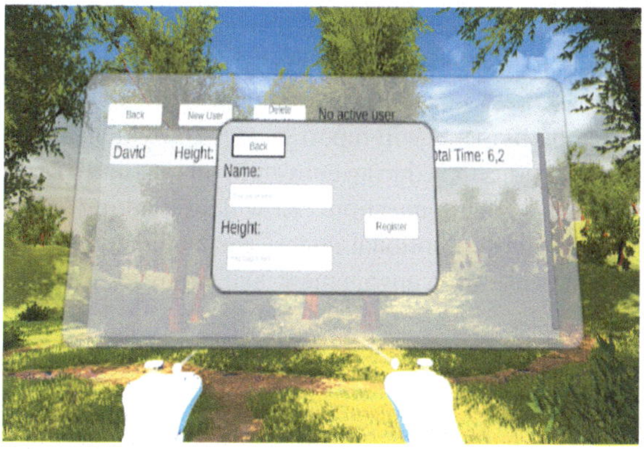

Fig. 3. Screenshot from the application, new user registration

4 Discussion

The result of our work is a VR walking simulation application that creates a seamless link between physical treadmill use and immersive virtual exploration. The user's real walking pace controls their movement in the VR world, allowing for a highly realistic and engaging experience. Our prototype offers various customizable environments that provide aesthetic and motivational value, enhancing the user's willingness to engage in regular physical activity. Compared to existing solutions, our application provides a more straightforward, user-friendly navigation system optimized for linear treadmill movement. This design decision eliminates the discomfort associated with curved virtual paths that don't match real-world walking patterns. By emphasizing accessibility and inclusivity, the application makes virtual exploration available even to those who may face physical challenges in outdoor mobility. Preliminary testing indicates improved user satisfaction and engagement, highlighting the potential of our approach for both recreational use and rehabilitative therapy.

The application has so far been tested by the development team, consisting of four individuals. These initial trials demonstrated that the system runs smoothly, with all components—virtual reality environment, treadmill integration, and user interactions—working seamlessly together to create an immersive and technically stable experience.

The next stage involves a structured exploratory study with healthy individuals. This phase is designed to assess the psychological and physiological effects of treadmill-based VR walking and running. Emotional and mood-related responses will be evaluated using self-report mood scales administered immediately after each session. To assess changes in anxiety levels, participants will complete the State-Trait Anxiety Inventory [7] – State form (STAI-S) both before and after using the application. This validated psychological tool measures temporary anxiety and provides insight into how immersive VR exercise influences short-term stress. In addition to psychological data, physiological responses will be monitored using the Polar H10 [8] heart rate sensor, which offers accurate, real-time tracking of cardiovascular activity. These combined measures will help assess whether VR-enhanced treadmill activity contributes positively to users' emotional well-being and physical engagement.

We have formulated two hypotheses for this phase of the research. The first hypothesis (H1) is that using the VR treadmill application will significantly reduce participants' state anxiety (STAI-S scores) after exercise compared to their baseline. The second hypothesis (H2) is that participants will report greater enjoyment and motivation during VR-assisted treadmill sessions than during conventional treadmill use. These hypotheses are supported by two key research questions: How does walking or running in a virtual environment affect short-term emotional states, particularly anxiety and mood? And does immersive VR increase user engagement and motivation compared to standard treadmill use?

Following the healthy participant study, the research will progress to a clinical phase in collaboration with Eötvös Loránd University in Budapest. This stage will involve individuals with neurological impairments, aiming to explore the application's potential in rehabilitation by evaluating its impact on motivation, engagement, and functional outcomes among patients recovering from neurological conditions.

5 Conclusion

The implications of this project for the Assistive Technology (AT) field are significant, particularly in advancing therapeutic applications and enhancing user experiences in rehabilitation and physical therapy. The integration of virtual reality with treadmill exercises can provide an innovative approach to rehabilitation for individuals recovering from injuries, strokes, or neurological conditions. By incorporating immersive environments, this project could help patients engage in therapy more effectively, improving adherence to rehabilitation programs and enhancing recovery outcomes. Furthermore, the dynamic synchronization of virtual landscapes with treadmill speed offers a more natural and motivating experience, potentially reducing the monotony often associated with traditional rehabilitation exercises. This enhanced engagement may lead to increased patient motivation and longer therapy sessions, which are crucial for successful rehabilitation. Additionally, the insights gained from competitor analysis and user feedback have guided the design to avoid common pitfalls, ensuring a more intuitive and user-friendly interface. Overall, this VR treadmill application represents a promising step forward in combining technology with physical therapy, offering a versatile tool that could be adapted for a wide range of therapeutic needs and settings in the future.

References

1. Winter, C., Kern, F., Gall, D., Latoschik, M.E., Pauli, P., Käthner, I.: Immersive virtual reality during gait rehabilitation increases walking speed and motivation: a usability evaluation with healthy participants and patients with multiple sclerosis and stroke. J. Neuroeng. Rehabil. **18**(1), 68 (2021). https://doi.org/10.1186/s12984-021-00848-w
2. Deutsch, J.E., et al.: Characterization and validation of virtual reality walking simulator. J. Neurol. Phys. Ther. **30**(4), 198–199 (2006). https://doi.org/10.1097/01.npt.0000281267.72948.a8
3. Unity Technologies: Unity Official Website. https://unity.com/. Accessed 10 May 2025
4. Unity Technologies: Unity XR Interaction Toolkit Manual. https://docs.unity3d.com/Packages/com.unity.xr.interaction.toolkit@3.0/manual/index.html. Accessed 10 May 2025

5. Octonic VR fitness treadmill. Octonicvr.com (n.d.). https://octonicvr.com/?srsltid=AfmBOoqu4RZ4g5MJ3s7eY2pIgRfSi2S72j8E09jRv2VrQliqkNlDls4Z. Accessed 10 May 2025
6. Bleikli, B.M., Valved, L.S.: VR walk: game. VR simulator for treadmill as a tool for constraint-based gait rehabilitation. The University of Bergen (2020)
7. Spielberger, C.D.: State-Trait Anxiety Inventory: Bibliography, 2nd edn. Consulting Psychologists Press, Palo Alto (1989)
8. Polar Electro Oy: Polar H10 Heart Rate Sensor. https://www.polar.com/hu/sensors/h10-heart-rate-sensor. Accessed 10 May 2025

Recommendation System of Virtual Reality (VR) Application for Rehabilitation Purposes

Jinat Ara[1(✉)] and Cecilia Sik-Lanyi[1,2]

[1] Department of Electrical Engineering and Information Systems, University of Pannonia, Egyetem u. 10, Veszprem 8200, Hungary
jinat.ara@mik.uni-pannon.hu

[2] Hungarian Research Network, Piarista u. 4, Budapest 1052, Hungary

Abstract. Generally, a recommendation system is considered an information aggregation system that directs appropriate information to the users. Considering the Virtual Reality (VR) application, nowadays, the use of VR applications has increased than previously for several activities, including rehabilitation purposes. Therefore, in this study, the aim is to propose and develop an effective VR application recommendation system specifically for rehabilitation purposes. The developed system performs by considering user queries and extracts the most suitable application according to the user's interest, considering different attributes of the suitable VR application. In the initial phase, the developed application can recommend 95 VR applications according to the user's interest, considering three rehabilitation purposes, including 'Cognitive', 'Physical', and 'Motion' disability. The developed application is suitable for recommending the appropriate VR application to improve the rehabilitation experience of the user.

Keywords: Virtual Reality application · Recommendation system · Rehabilitation purpose · Sentiment analysis · Keyword formulation

1 Introduction

Recommendation systems are considered tools for interacting with a large number of information sources to determine the suitable information of interest [1]. Thus, the recommendation system is acceptable and popular to the users for getting suitable and appropriate solutions/applications according to their interests. Nowadays, there are several examples of different sophisticated recommendation systems that are widely used for e-commerce platforms, streaming services, and social media platforms, for instance, Amazon.com. The advancement of these recommendation systems is prominent in the community because of their advanced functionalities and positive impact on user satisfaction. Addressing the significant impact, advanced functionalities and great user interest, nowadays, these recommendation systems inspire researchers to extend their research on recommendation systems for different new and challenging areas. On the other hand, Virtual Reality (VR) is a powerful technology that offers immersive and interactive digital environments for simulating real-world or imagined scenarios [2]. Addressing the

K. Mavrou and P. Encarnação (Eds.): AAATE 2025, *Technology for Inclusion and Participation for All: Recent Achievements and Future Directions*, pp. 166–174, 2025.
https://doi.org/10.1007/978-3-032-01628-7_21

advancement of these two advanced technologies, researchers suggested that integrating these technologies could improve the user experience and user satisfaction in the healthcare domain, especially for rehabilitation purposes [3, 4].

In recent years, the use of virtual reality (VR) in health care has gained immense popularity, which has inspired researchers to continue their investigation to determine the diverse facilities of this advanced technology [1, 2]. In general, recently, VR is being used in healthcare, particularly for rehabilitation purposes, for the assessment and therapeutic intervention of people with different disabilities such as cognitive injuries, physical injuries, temporal or permanent damage, etc. Recently, several VR applications have been developed for different rehabilitation purposes, for instance, VR-based Immersive Rehab [5] for physical therapy, VR-based SnowWorld [6] for reducing pain in burn wound care, etc. Also, VR has contributed to stroke patient rehabilitation for cognitive improvement, treating phobias, etc.

Despite these immense advancements, different VR applications have different personalized experiences, prototypes, configurations, effectiveness, and functionalities that can hinder user satisfaction and increase user dissatisfaction. For example, some applications might offer a fixed set of exercise scenarios that might fail to fulfill the user's needs. Also, some applications require extensive technical expertise to operate and navigate, which might hinder the widespread adoption in clinical settings [7]. Besides, according to the statistics of the[1] World Health Organization (WHO) and the Institute for Health Metrics and Evaluation (IHME), one-third of the world population (around 2.6 billion people) have experienced different health conditions, and 1 in 3 people can benefit from the rehabilitation/therapeutic approach. Also, in 2024, VR-based rehabilitation applications gain a large market share of around 0.83 billion USD [8]. This statistic depicts that in recent years, the demand for VR-based rehabilitation applications has gained tremendous interest, and a significant number of VR applications have been developed for rehabilitation purposes.

However, among a diverse array of VR-based rehabilitation applications, determining the appropriate VR application is challenging, and at the same time, it is an important and emerging concern for the technology researcher. Focusing on this aspect, in existing literature, a prominent number of studies have been performed focusing on the urgency of a proper VR application recommendation system for rehabilitation purposes [9–13]. Although their proposed approaches are potential, the majority of the recommendation systems focused on a specific disability type, considered limited application features, and none of the approaches considered accessibility perspectives. Therefore, considering all of these limitations, our goal is to provide an interactive solution to facilitate the rehabilitation process for people with disabilities by combining these two interactive technologies (recommendation system and VR application) and utilizing their immense potential. Moreover, this work aims to propose and develop a VR application recommendation system that considers user requirements, all the application prototypes, and accessibility perspectives to recommend the appropriate application and improve the effectiveness of the developed system. The proposed system might be beneficial for

[1] https://www.xr.health/us/blog/virtual-reality-rehabilitation/.

individuals undergoing rehabilitation for different conditions, which will facilitate finding the appropriate and most effective VR application for their rehabilitation purpose and help them accelerate their recovery.

The remainder of the paper is structured as follows: Sect. 2 provides all the methodologies from data gathering to the development of the recommendation systems, with its development strategies in detail. Section 3 provides validation results considering comparative assessment results that represent the interactivity and effectiveness of the developed system. Finally, Sect. 4 concludes the paper with a conclusion and future plan.

2 Methods

In this section, our aim is to present the methodological strategies in detail that have been followed during the development of the proposed VR application recommendation system for rehabilitation purposes. The entire development process involves four steps: i) data collection, ii) data analysis, iii) database preparation, and iv) data extraction and visualization. All of these steps are described in the following:

2.1 Data Collection

Data collection is the initial and one of the most important steps during any system development strategy. In our development process, we first selected different VR applications that were particularly developed for rehabilitation purposes for different types of disabilities. Among several disability types, in our development, we especially considered VR applications that support the rehabilitation process of people with 'Cognitive', 'Physical', and 'Motion' disabilities. All of the VR applications have been selected from the 'Google Play Store' platform, which is known as a repository of different digital applications for a diverse array of domains.

The selection was performed by setting some specific criteria, for example, we set the rehabilitation facilities for three specific purposes: Education, Therapy, and Exercise; and the supported devices are: Phone/Tablet/Laptop/Desktop. Initially, we selected 95 VR applications that fulfill all the requirements/specifications and offer rehabilitation facilities for this specific group of people with disabilities, such as 'Cognitive', 'Physical', and 'Motion' disabilities.

After selecting the appropriate VR application, we extracted all the metadata of the selected application as the 'Google Play Store' platform offers a unique page for each application by defining various information that is considered as metadata to present the application features, functionalities, and other information regarding different specifications. The metadata of each selected application was extracted using the[2] 'Google-Play-Scraper', which provides an API to extract the VR application through its application ID. After extracting all the metadata, we considered 'Application name', 'Title', 'Category', 'Application Type', 'Device Type', 'Reviews Ratio,' 'Download Ratio', 'Ratings', 'Reviews', and 'Description'.

[2] https://pypi.org/project/google-play-scraper/.

2.2 Data Analysis

In the developed system, we integrated all the extracted and chosen metadata (Application name', 'Title', 'Category', 'Application Type', 'Device Type', 'Reviews Ratio,' 'Download Ratio', 'Ratings', 'Reviews' and 'Description'), however, two types of metadata require extensive analysis in terms of their semantic aspects such as 'Reviews' and 'Description' as these metadata reflect semantic terms of the content. Therefore, 'Reviews' and 'Description' were analyzed, incorporating Artificial Intelligence (AI) techniques for sentiment classification and keyword formulation, as sentiment classification/detection helps to determine the usefulness of a particular application, and advanced keyword formulation helps to represent the actual functionalities of the application. For sentiment classification of the extracted reviews and keyword formulation from the description of the selected VR applications, we implemented Natural Language Processing (NLP) technique which is a technique under the umbrella of AI.

For sentiment classification, we used user 'Reviews', incorporating the core NLP techniques like[3] TextBlob, which is a Python library that allows us to determine the sentiment of textual data. We consider all the reviews of a particular application and analyze them to predict a final sentiment score that also represents the effectiveness status of a particular application. It performs by calculating the polarity of the words, where the polarity score ranges from [–1 to + 1]. Here, a score of + 1 indicates an entirely positive sentiment, and a score of –1 indicates an entirely negative sentiment. If the polarity score is < = –0.01, the review is considered negative; if the polarity score is > = 0.9, the review is considered positive. Besides, if the polarity score is < 0.9, then it is considered neutral sentiment. Furthermore, according to the sentiment type we determine application effectiveness status in terms of three categories: 'Accessible', 'Partially accessible' and 'Not accessible' where if the sentiment type is classified as Positive, then the application consider as Accessible application; if the sentiment type determined as Neutral, then the application consider as Partially accessible and if the sentiment type determined as Negative, then the application marked as Not accessible. A sample of user reviews with their sentiment score and type is shown in Table 1.

Table 1. Sample of user reviews with their sentiment score and sentiment type.

Sentiment classification/detection		
User Reviews	Sentiment Score	Sentiment type
Good but.. Terms are American, Faucet for Tap etc., makes it difficult to translate in time to score well if you are from the UK, also very expensive if you want to use this as it needs to be used if you want to gain any benefit from it, I can't afford it	0.7	Neutral

(*continued*)

[3] https://textblob.readthedocs.io/en/dev/.

Table 1. (*continued*)

Sentiment classification/detection		
User Reviews	Sentiment Score	Sentiment type
This app is software gore. I tried to create an account, but received the error "unable to verify email address". I'm using a valid gmail account. I went to the website (off the app) and was able to sign up. I returned to the app to login, and received an error "contact support". Deleted cache and storage and tried again, same error. No thanks. Maybe I'll give it a shot in a year or two if they're still around	0.5	Neutral
Trained Skills doesn't seem to reflect what skills are being affected by the end of the training. Everytime I try to check out a training and click the X to look for another, I end up on the Stats tab, rather than going back to where I were (Train). It is inconvenient. Lastly I think it is in poor taste that even though I pay for a fairly expensive monthly subscription, I have to pay for individual assesments (one time use) that are WAY too expensive	–0.8	Negative

Furthermore, for keyword formulation, we used the application 'Description', incorporating the[4] KeyBERT model, which is also a Python model that allows us to formulate keywords from textual data. We used N-gram 2 and all the English stop word removal functionalities to formulate the most appropriate keywords for the particular application. The top 5 keywords were selected for presenting the application information in the prediction system. Table 2 shows a sample of formulated keywords that are generated from their corresponding application description.

Table 2. Sample of formulated keywords from the application description.

Keyword formulation	
Application description	Formulated keywords (top 5)
A must have e-learning tool for 21st century learning and teaching. Based on the EduVenture outdoor learning system. EduVenture-VR is an ultimate solution for teachers to carry out VR learning in a classroom.Composing VR learning materials via the web-based authoring tools is fun and effortless.Teacher can distribute the VR content	'authoring tools, 'web based', 'learning classroom', 'eduventure outdoor', 'carry vr'

(*continued*)

[4] https://pypi.org/project/keybert/.

Table 2. (*continued*)

Keyword formulation	
Application description	Formulated keywords (top 5)
With NeuroNation scientific brain training you bring your brain day by day on trot. Whether it's weaker memory, dwindling concentration or thinking too slowly - only 15 min of training a day can make problems disappear and give your brain new momentum. Join the worldwide community of over 23 million members and treat yourself to a piece of science - right in your pocket	'increase thinking', 'training improve', 'neuronation different', 'improve memory', 'neuronation med'
Train Your Brain Did you know that you can train your brain? CogniFit makes it easier than ever to train your brain with a series of fun and engaging mental games. Our patented system takes a personalized approach helps to improve your cognitive function from anywhere, whether at home or on the go. Effective technology used by the scientific community, universities, hospitals, families, and medical centers around the world	'track brain', 'cognitive age', 'ability cognifit', 'cognitive score', 'brain exercise'

2.3 Database Preparation

After analyzing all the metadata and determining the sentiment and formulating the appropriate keywords, extracted metadata, formulated keywords, and computed sentiment scores of each VR application are stored in a database to perform the inferencing process to extract or retrieve the most suitable application according to the user's interest or user query.

2.4 Data Extraction and Visualization

To parse the information from the database according to the user's interest, we applied the tabular data representation concept. This system has been developed in the Sublime Text editor with the Python programming language. To determine the user interest, we took the user responses in terms of three questions with multiple options as follows:

Q1: What kind of application would you like to be recommended? (Cognitive/Physical/Motion)
Q2: For which purpose would you like to use the application? (Education/Therapy/Entertainment)
Q3: What kind of device would you prefer? (Phone/Tablet/Laptop/Desktop)

Upon getting the user responses regarding the questions asked, the system matched the user queries with the database information to retrieve/recommend the appropriate VR application to the user interface. The user interface was developed using the Tkinter library, which is the standard Python GUI library that allows effectively building GUI applications. Figure 1 shows the visualization of the extracted information regarding one user query. Figure 1 depicts that, according to the user query, the developed recommendation system recommends several applications with detailed information, including the application accessibility status.

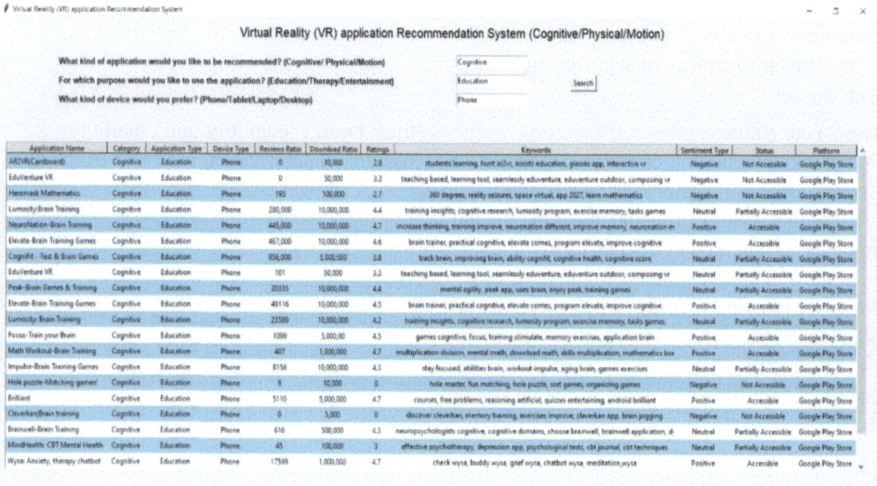

Fig. 1. User interface of the VR application recommendation system regarding user query.

3 Results

The proposed VR application recommendation system for rehabilitation has been validated by comparing it with existing similar VR application recommendation systems that have been proposed in the existing literature, considering several application prototypes or assessment functionalities as shown in Table 3. For the assessment feature, we consider four application features, such as (i) whether the recommendation system focused on multiple disability types, (ii) whether the recommendation system was developed considering rehabilitation purposes, (iii) whether the recommendation system considered accessibility aspects, and (iv) whether the recommendation system focused on the improvement of application semantic understanding. The comparative assessment result (as shown in Table 3) shows that none of the existing recommendation systems addressed all the functionalities, while our proposed and developed recommendation system addressed all the considered application functionalities during the development. It depicts that the proposed system is effective in recommending appropriate VR applications for rehabilitation purposes, which could be effective for users with disabilities and other end users.

Table 3. Comparative assessment results with existing similar approaches.

References	Multiple Disability type	Rehabilitation Purpose	Accessibility aspects of the system		Context understanding
			Sentiment score	Sentiment type	Appropriate keyword formulation
González-González et al. [6]	✓ (Yes)	✓ (Yes)	× (No)	× (No)	× (No)
Andrunyk et al. [7]	× (No)	✓ (Yes)	× (No)	× (No)	× (No)
Zhao et al. [8]	× (No)	× (No)	× (No)	× (No)	× (No)
Ge et al. [9]	× (No)	× (No)	× (No)	× (No)	× (No)
González-González [10]	× (No)	✓ (Yes)	× (No)	× (No)	× (No)
Proposed system	**✓ (Yes)**	**✓ (Yes)**	**✓ (Yes)**	**✓ (Yes)**	**✓ (Yes)**

4 Conclusion

As VR technology continues to evolve and more sophisticated applications emerge, the role of a well-designed recommendation system will maximize the potential of VR applications for rehabilitation purposes. At this initial evaluation stage of the proposed system, the outcome of the evaluation is satisfactory and has a positive impact on improving the recommendation facilities of VR applications for rehabilitation purposes for people with disabilities ('Cognitive', 'Physical', and 'Motion'). However, we are looking forward to conducting an experimental evaluation including the user testing focusing on the SUS questionnaire in our further study.

Acknowledgments. Funded by the Research Fellowship Programme (Code: 2024-2.1.1-EKÖP) of Ministry of Culture and Innovation of Hungary from the National Fund for Research, Development and Innovation.

Disclosure of Interests. There is no conflict of interest.

References

1. Resnick, P., Varian, H.R.: Recommender systems. Commun. ACM **40**(3), 56–58 (1997)
2. Rubio-Tamayo, J.L., Gertrudix Barrio, M., García García, F.: Immersive environments and virtual reality: systematic review and advances in communication, interaction and simulation. Multimodal Technol. Interact. **1**(4), 21 (2017)
3. Adamovich, S.V., Fluet, G.G., Tunik, E., Merians, A.S.: Sensorimotor training in virtual reality: a review. NeuroRehabilitation **25**(1), 29–44 (2009). https://doi.org/10.3233/NRE-2009-0497]. [FREE Full text], [Medline: 19713617]

4. Bohil, C.J., Alicea, B., Biocca, F.: Virtual reality in neuroscience research and therapy. Nat. Rev. Neurosci. **12**(12), 752–762 (2011). https://doi.org/10.1038/nrn3122] [Medline: 22048061]
5. Herrera, V., et al.: Rehab-Immersive: a framework to support the development of virtual reality applications in upper limb rehabilitation. SoftwareX, **23**, 101412 (2023)
6. Ahmadpour, N., Randall, H., Choksi, H., Gao, A., Vaughan, C., Poronnik, P.: Virtual Reality interventions for acute and chronic pain management. Int. J. Biochem. Cell Biol. **114**, 105568 (2019)
7. Kouijzer, M.M., Kip, H., Bouman, Y.H., Kelders, S.M.: Implementation of virtual reality in healthcare: a scoping review on the implementation process of virtual reality in various healthcare settings. Implement. Sci. Commun. **4**(1), 67 (2023)
8. https://www.globalgrowthinsights.com/market-reports/tele-rehabilitation-with-virtual-reality-market-112682
9. González-González, C.S., Toledo-Delgado, P.A., Muñoz-Cruz, V., Torres-Carrion, P.V.: Serious games for rehabilitation: gestural interaction in personalized gamified exercises through a recommender system. J. Biomed. Inform. **97**, 103266 (2019)
10. Andrunyk, V., Pasichnyk, V., Shestakevych, T., Antonyuk, N.: Modeling the recommender system for the synthesis of information and technology complexes for the education of students with autism. In: 2019 IEEE 14th International Conference on Computer Sciences and Information Technologies (CSIT), vol. 3, pp. 183–186. IEEE (2019, September)
11. Zhao, Z., Arya, A., Orji, R., Chan, G.: Effects of a personalized fitness recommender system using gamification and continuous player modeling: system design and long-term validation study. JMIR Serious Games **8**(4), e19968 (2020)
12. Ge, M., Persia, F., D'Auria, D.: Advanced recommender systems by exploiting social networks. In: 2019 IEEE International Conference on Humanized Computing and Communication (HCC), pp. 118–125. IEEE (2019, September)
13. González-González, C.S., Toledo-Delgado, P.A., Muñoz-Cruz, V., Hernández, M., Fardoun, H.M.: Towards a Smart System for Rehabilitation: TANGO-H Recommender System (2018)

User-Center Game Development for Older Adults

József Dávid Góczán[1] and Cecilia Sik-Lanyi[1,2](✉)

[1] University of Pannonia, 8200 Veszprem, Hungary
lanyi.cecilia@mik.uni-pannon.hu
[2] Hungarian Research Network (HUN_REN), 1052 Budapest, Hungary

Abstract. The elderly population faces numerous motor and cognitive challenges that can intensify with age. These challenges include a decline in memory and attention and a weakening of motor skills. To prevent age-related regression and improve quality of life, seeking effective development solutions is essential. Our goal was to create a developmental game in Unity that develops the motor and cognitive skills of the elderly. During the design and development of the game, those tasks and challenges were implemented, which aided in improving fine motor skills, coordination, memory, and attention. The software includes several smaller games and tasks designed to stimulate different brain functions, such as thinking, memory, problem-solving, and attention. The goal of creating this game was to provide older people with a fun and motivating tool that helps them stay active; by offering them such a game, we allow them to maintain both their motor skills and cognitive abilities enjoyably, thereby improving their quality of life. In addition to offering a motivating user experience, the game specifically targets cognitive domains such as working memory, attention, and problem-solving, as well as fine motor control. These skills were chosen based on cognitive aging literature, which highlights their relevance in maintaining independence and mental flexibility in older adults. Theoretical foundations for this design are drawn from recent studies on cognitive stimulation in aging populations and usability-centered game-based interventions.

Keywords: User-center · Inclusive design · Older adults · Game · Unity · Motor and cognitive skill

1 Introduction

The decline in motor and cognitive abilities in old age can pose significant challenges to quality of life and independent performance of daily activities for many. As time goes by, there is often a deterioration in balance and coordination, as well as a decline in memory and attention, which makes it difficult to live independently and participate in society. These changes often reinforce the other side of old age, which does not always meet the wishes and needs of the individual.

However, modern technology offers new opportunities for older people to maintain their health and well-being. The development of interactive games and applications offers

K. Mavrou and P. Encarnação (Eds.): AAATE 2025, *Technology for Inclusion and Participation for All: Recent Achievements and Future Directions*, pp. 175–180, 2025.
https://doi.org/10.1007/978-3-032-01628-7_22

solutions that can help older people to actively and entertainingly maintain their cognitive and motor abilities. Such developments are not only fun and motivating but can also be effective tools in preventing old age regression and improving the quality of everyday life. In this context, the aim of this work is to create a special development game on the Unity platform, specifically optimized for the needs and abilities of the elderly. The primary objective in the design and development of the game is to entertain and motivate elderly people, while providing an opportunity to develop their motor and cognitive skills. The game includes a variety of tasks and challenges that focus on developing fine motor skills, thinking and problem solving, as well as improving memory and attention. This initiative is not only about creating a game, but also about an innovative approach that can help elderly people stay active and healthy and prevent or slow down the decline of old age. The aim of the work is therefore to create a application that allows elderly people to maintain both their motor and cognitive skills in an enjoyable way, thus improve their quality of life and increase their self-sufficiency.

Despite promising results in prior serious games, many such solutions are not fully aligned with the usability needs of older adults. Involving users earlier in the development process—e.g., through co-design sessions or needs analysis interviews—can significantly enhance the relevance and accessibility of these tools. While this project included user feedback during testing, future iterations will incorporate participatory design from the outset.

2 Methods

The method consisted of two steps. First, based on literature research, the discovery of games with similar goals was carried out. The second step was the development of the game. Both steps are described below.

2.1 Literature Review/State of the Art

During the literature review, several serious games that focus on improving the mental and physical condition of the elderly were found. These serious games help to slow down or prevent the decline of old age in various ways. These games, as reference points, inspired and helped us to develop our application, e.g. Brain Game [1] and Neuro-therapy [2].

Software created for the elderly, such as Brain Game [1] and Neuro-therapy [2], clearly show how they can help with cognitive development and improve the quality of life. HybridPLAY [3] technology offers a new approach to games, which can be exciting for the elderly as well.

Recent literature also suggests that training programs focusing on multiple domains (e.g., attention, executive functions, and sensorimotor integration) may yield stronger cognitive benefits than single-focus interventions [4, 5]. Our game design was informed by these findings to ensure comprehensive training experience. Additional references from 2022–2024 were reviewed to better identify research gaps in current game-based cognitive training for the elderly, particularly in regard to usability adaptation and cross-platform accessibility.

2.2 Developing the Game

Choosing the right development environment was essential during the development of the project. The Unity development platform was chosen, which we considered ideal for the project due to its many beneficial features. One of the main aspects was that Unity is an extremely user-friendly and easy-to-learn environment. This greatly facilitates fast and efficient development. In addition, Unity is extremely well documented it was extremely important for the efficient resolution of problems that arose during the development process. Furthermore, another advantage of Unity is that it supports multiple platforms, such as Windows, iOS, Android, and macOS. This allows us to make the finished software widely available and reach the largest possible user base. It is important to highlight that Unity is a C++/C# language-based system that offers a wide range of functionality and options for developers. This flexibility allowed us to adapt and customize the software to the specific needs of our project. Finally, during development, it was an advantage that Unity provides the ability to easily and quickly test the already completed software using a program that can be run in the development environment. This functionality allowed us to continuously check and refine the software, which is key to a successful development process.

Throughout development, attention was paid to interface simplicity and visual clarity, with design decisions (such as color contrast, font size, button size) guided by accessibility guidelines for older users. The software architecture also supports future expansions based on user feedback and additional functionality (e.g., voice navigation, difficulty scaling).

3 Results: The Game

When starting the game, the home screen greets us. From there, the user is immediately greeted by the main menu, where several options are presented to them. They are given the opportunity to enter the settings. Here, settings that affect the entire program can be found. In this menu item, you user can change the desired language and adjust the volume. The statistics menu item is also available from the main menu, where you can track the performance of individual players. From the main menu, user can also exit the program or go to the game's menu item. Within this menu section, there are 3 types of smaller games (minigames). Before the actual start of the game, the player is greeted by a short instruction panel, where they receive a short description of the task. Only after this does the game start. After the game is over, the results are presented to the player on a panel element, and they have the opportunity to restart the game or go back to the level selection.

The user interface of the game is simple (Fig. 1): there only 4 buttons on the main menu (Games, Settings, Statisctics and Quit). The settings have 4 features (Fig. 2): choosing the language (Hungarian and English); they can adjust the music to be quieter or louder; they can install the effects; they can adjust the cursor size to be smaller or larger (more visible); and the last button is the "back" button.

Fig. 1. Main menu

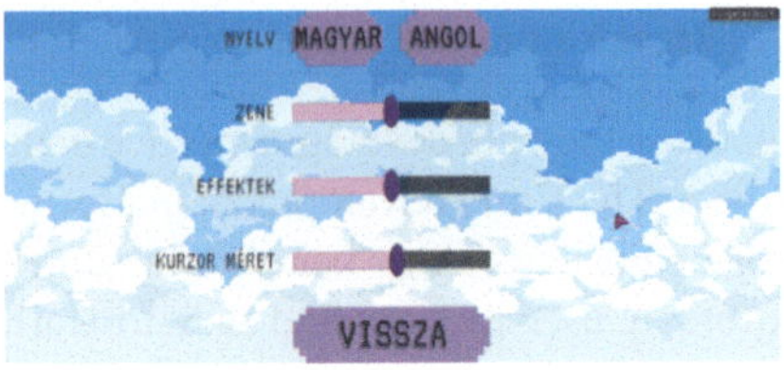

Fig. 2. Settings

The game "BugHunt" (Fig. 3) presents the player with a simple yet fun task: to catch the bugs that appear on the screen as quickly and accurately as possible. The game is available on different platforms, so it can be played on both phones and PCs, depending on which device the player prefers. In the phone version, the bugs can be caught by touching them, while in the PC version, the bugs can be caught by clicking with the left mouse button. The gameplay encourages players to react quickly and accurately, as incorrect hits hinder success, while accurate ones move forward.

The goal of the "Piano" game (Fig. 4) is for the player to accurately follow the sequence generated by the system using the piano keys that appear on the screen. Four piano keys appear in front of the player, above which are numbers showing the order in which they should be pressed. The task is to perform the task flawlessly by pressing and holding these keys in the specified order.

The "ViewPoint" game (Fig. 5) is a creative and fun challenge in which spatial visual observation and logical thinking play an important role. The game tests the ability to navigate in three-dimensional space and is fun for all ages, but at the same time requires serious concentration and attention.

Fig. 3. BugHunt game

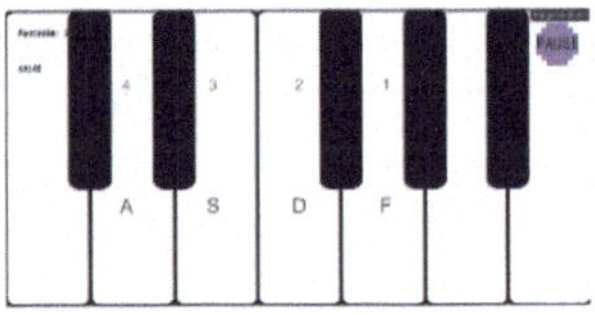

Fig. 4. Pino game

Fig. 5. ViewPoint game

4 Testing of the Game

A significant element of the project was the testing of the software in a real user environment, with the objective of providing a reliable assessment of the usability and experience of the game. In the course of the testing phase, particular emphasis was placed on the inclusion of the perspectives and experiences of individuals from diverse age demographics. The initial test subjects were selected from the older age group, primarily users over 65 years old, as the primary objective of this thesis is to develop software that is user-friendly for this demographic. Furthermore, the system was tested with a younger demographic, who are more familiar with modern technology. The purpose of this was to obtain a more extensive range of feedback and to address the diverse requirements that emerge from disparate user habits. In the course of the testing phase, particular consideration was given to the specific requirements of the older age group, including the development of an accessible and transparent user interface, as well as intuitive camera control. Concurrently, feedback from younger testers was deemed to be significant, as they expressed a preference for gameplay that was faster and more dynamic. Following the analysis of the test results from the various user groups, the game mechanics have been refined in order to provide a pleasant and enjoyable experience for all age groups. However, the needs of older users have been identified as the primary concern, and the game mechanics have been adapted to meet the needs of users in this age group, where there are conflicts of interest.

The System Usability Scale (SUS) [4] questionnaire was used to test the game. During the testing, we collected opinions from three different age groups: seniors, adults and teenagers. During the testing phase, a comprehensive set of opinions was collected from three distinct age demographics: the elderly (65+ years), adults (18–64 years), and children (10–17 years). The inclusion of participants from different age groups was intended to provide a comprehensive overview of the system's usability, thereby facilitating the acquisition of a more detailed benchmark for the evaluation of older users. An analysis of the mean scores of the responses from the various groups is presented in Table 1. These results clearly reflect the differences in the technological affinity of each age group. The lower score of the seniors was mainly due to their lower proficiency in using modern digital devices and software. Nevertheless, according to the feedback, the system structure and clear instructions were understandable and easy to follow for them, which contributed to the positive user experience. The simple navigation made it easy for them to use, and they became more confident throughout the testing process.

Table 1. SUS scores

Age groups	SUS score
Teenagers (10–17)	94
Adults (18–64)	85
Older adults) (65+)	66
Average score	**81.6**

User feedback from older participants indicated specific areas for improvement, including clearer visual cues, longer input time windows, and more feedback during interaction. Based on the findings, future iterations will include adjustable difficulty levels and training sessions to help users familiarize themselves with digital interactions. These planned improvements aim to increase usability for those who initially scored lower and reduce frustration due to unfamiliarity with touch or mouse-based interfaces.

These results are indicative of the variances in technological affinity observed among different age demographics. The lower scores attained by the older demographic were primarily attributable to their diminished proficiency in the utilisation of contemporary digital instruments and software. However, feedback indicated that the system design and clear instructions were comprehensible and straightforward, contributing to a positive user experience. The simplified navigation of the system has been found to facilitate its use, thereby enhancing the subjects' confidence during the testing process.

The increased performance levels observed in both adult and child subjects can be attributed to the incorporation of these tools into their regular routines. It was evident that children demonstrated a remarkable capacity for expeditious adaptation and an aptitude for swiftly comprehending the system's rationale. Conversely, adults expressed a favourable evaluation of the integration of simplicity and functionality. The findings indicate that the system is appropriate for various age demographics and is especially well-suited to the requirements of older users.

5 Conclusion

A motivating game was created that helps the elderly maintain their mental and fine motor skills. Although the game is understandable and easy to use for all ages, feedback from older users revealed that the structure of the system, the simple navigation and the clear instructions made it significantly easier for them to use. This was a particularly important aspect during the development, as our goal was to make the software not only fun, but also widely accessible and acceptable. The higher scores of adults and teenagers are due to the routine use of the tools. In particular, the children's quick adaptability and quick insight into the logic of the system stood out, while adults positively assessed the combination of simplicity and functionality. Overall, the results show that the system is suitable for different age groups and is particularly suitable for adapting to the needs of older adults.

This project demonstrates the feasibility and promise of serious games tailored for older adults. Future developments will integrate more user-centered design practices from the outset and focus on refining cognitive stimulation through both theoretical grounding and empirical feedback. The game will also be expanded with additional task types and multi-modal interaction options to improve inclusivity and engagement.

References

1. Gang, L.W., Jun, Z.X., Hidayat, R.: Brain game design for elderly. Int. J. Math. Comput. Sci. **18**(4), 773–779 (2023)
2. ChePa, N., Sie-Yi, L.L., Yusof, N., Yahaya, W.A.J.W., Ishak, R.: Impacts of Game-based psychotherapy intervention to elderly with memory disorder. Entertain. Comput. **44** (2023)
3. Boj, C., Díaz, D.J., Portalés, C., Casas, S.: Video games and outdoor physical activity for the elderly: applications of the HybridPLAY technology. Appl. Sci. **8**(10), 1912 (2018)
4. Brooke, J.: SUS: a quick and dirty usability scale. In: Jordan, P.W., Thomas, B., Weerdmeester, B.A., McClelland, I.L. (eds.), Usability Evaluation in Industry, pp. 189–194. Taylor & Francis, London (1996)
5. Góczán, J.D.: Developing a user-centered game for the elderly (in Hungarian: Felhasználóközpontú játék fejlesztése időseknek) (2024). BSc thesis work at the University of Pannonia, Veszprém, Hungary

Artificial Intelligence in Care and Support Ecosystems

Lessons Learnt and Hands-on Experience in Developing Personalized AI-Powered Assistive Technology Companions for Use in Patient Care Settings

Aliaksei Andrushevich[1](✉), Artem Goylo[2], Mikhail Sadouski[2], and Mikhail Kovalev[3]

[1] iHomeLab, Lucerne University of Applied Sciences and Arts, Lucerne, Switzerland
aliaksei.andrushevich@hslu.ch

[2] Lucerne, Switzerland

[3] Quintary AI, Kinira 28-303, 8011 Paphos, Cyprus

Abstract. This paper presents our experience in developing a virtual assistant tailored for use in residential care facilities, aimed at reducing the workload of facility personnel by automating the collection of patient information and addressing routine inquiries. The system integrates a wide variety of AI methods, including natural language processing (NLP), computer vision, logical inference, ontologies, and knowledge graph technology for the system to interpret user messages and generate responses, ensuring contextual and personalized interactions. The knowledge graph is populated with semantic representations of user messages and serves as a verified source of factual knowledge, enhancing the system's ability to provide accurate and reliable answers. Challenges in NLP and the integration of diverse subsystems were addressed through a hybrid approach that utilizes a specific language for unified knowledge representation, which improves explainability and addresses limitations inherent in large language models.

Keywords: Virtual Assistant · Artificial Intelligence · Large Language Models · Computer Vision · Knowledge Graph

1 Introduction

Population aging has become a critical demographic challenge that has a wide range of implications from economic side-effects to societal pressures, like increased importance of informal care [1], and healthcare concerns, such as deteriorating physical and cognitive capabilities [2]. This has recently driven active developments in the field of assistive technologies [3, 4] and technologies for ageing well [5], while recent breakthroughs in the field of Artificial Intelligence providing new promising ways of addressing mentioned challenges, with personal virtual assistants [6] recently also being driven by rapidly evolving AI technologies [7].

A. Goylo and M. Sadouski—Independent Researcher.

K. Mavrou and P. Encarnação (Eds.): AAATE 2025, *Technology for Inclusion and Participation for All: Recent Achievements and Future Directions*, pp. 183–190, 2025.
https://doi.org/10.1007/978-3-032-01628-7_23

The integration of AI has transformed healthcare by enhancing decision-making, streamlining workflows, and personalizing support for both patients and healthcare providers. The potential of AI applications is currently actively explored in the conversational interfaces for care of older adults [8] as well as in enhancing patient care quality, where virtual assistants are an important research and development direction [9, 10].

The goal of presented work was to develop a virtual assistant in order to reduce the workload of residential care facility personnel by collecting useful self-monitoring information about patients, older adults and people with disabilities and answering their (often repeated) routine questions. A comprehensive virtual assistant is required [11] to be able to converse with the user, satisfy the user's informational needs, take into account relevant context, including the environment, the user's emotional state, dialog history, as well as specific subject domain knowledge. In this paper, we outline our experience in developing such a system for use in patient care settings, specifically, at a residential care facility.

2 Methods

2.1 Underlying Technology

Satisfying the requirements for a comprehensive AI companion necessitates integration of various problem-solving methods and distinct subsystems. For this reason, our system is being developed using OSTIS, an open-source technology for designing intelligent systems. The technology allows for development of hybrid multi-agent intelligent systems based on the ontology-driven approach [12].

Such systems, called ostis-systems, consist of three components: the knowledge base (KB), the problem solver, and the interface. The basis of the KB is a hierarchical system of subject domains and ontologies. The problem solver is based on a multi-agent approach in which agents interact by specifying their actions in a shared memory, which allows for integration and combination of different problem-solving models [13]. Ostis-systems and their components are developed using an internal knowledge representation language called SC-code. The syntax of SC-code is based on graph theory and the language has basic set-theoretic semantics.

2.2 Logical Architecture of the AI-Powered Assistive Technology Companion

The architecture of the AI-powered assistive technology companion system is visualized in Fig. 1. Speech processing module solves multiple tasks:

- receiving and outputting speech signal;
- speech recognition (Whisper AI);
- speech synthesis (mimic3);
- input message processing (rule-based sentence type classification using spaCy, named entity recognition and intent recognition using Rasa's DIETClassifier, semantic analysis with LLM (ChatGPT-3.5));

- output message processing (translation from a graph representation into a natural language text (ChatGPT-3.5)).

Video signal processing module solves two tasks: emotion recognition, and object detection/recognition (YOLOv5). Dialog management module controls the dialog state and forms the context necessary to generate output messages. All components of the system are connected to the KB, which by default contains formalized subject domains and ontologies minimally required for the system to function. New information is then added to the knowledge graph by other modules during the system's lifecycle. The use of a unified language (SC-code) by all components enables their interoperability and ease of extension/scaling.

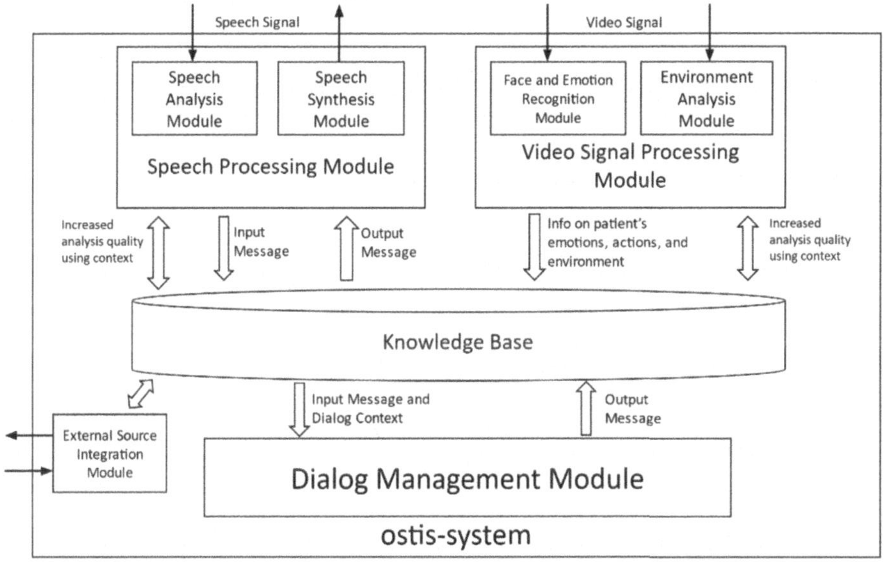

Fig. 1. Logical architecture of the AI-powered assistive technology companion.

3 Results

The main result of our work is that the user can communicate with the system on any topic by means of voice or text messages (in case of voice messages, speech recognition is used). The developed AI-powered assistive technology companion is also capable of keeping detailed dialog history, identifying user's emotions with respect to time, and collecting any information about the user available during the conversation.

3.1 Message Processing Pipeline

When the user sends the system a message, multiple agents process it to extract information, which is then added to the KB.

First, the system identifies sentence type (declarative, interrogative, imperative), as well as message topic and intent. Then, the system extracts the meaning of the message using a large language model (LLM). The prompt requests the LLM to find all keywords in the sentence and output a string that contains triples of the format <entity:relation:entity>. The string can contain any number of triples, while triples can be grouped together if they share one of their keywords. The prompt also "teaches" the LLM to differentiate between a limited set of specific relations (e.g. action, question, request) and basic set theory concepts (sets, members, membership, inclusion) needed to translate the message meaning from the intermediate representation into SC-code.

For example, the sentence *I like football* may be interpreted by the LLM as <entity_user:nrel_like:concept_football>. This string is then translated into SC-code and the resulting graph is added to the relevant segment of the KB. This way, the KB is populated with new facts about the user, which ensures learnability and allows the system to personalize responses. Figure 2 shows how the meaning of a message is added to the knowledge graph. The graph represents the semantic neighborhood of the node corresponding to the user who sent the message. As can be seen in the figure, the graph contains information about membership of the user in some relevant sets (e.g. human, user, patient) as well as the user's relations to other entities.

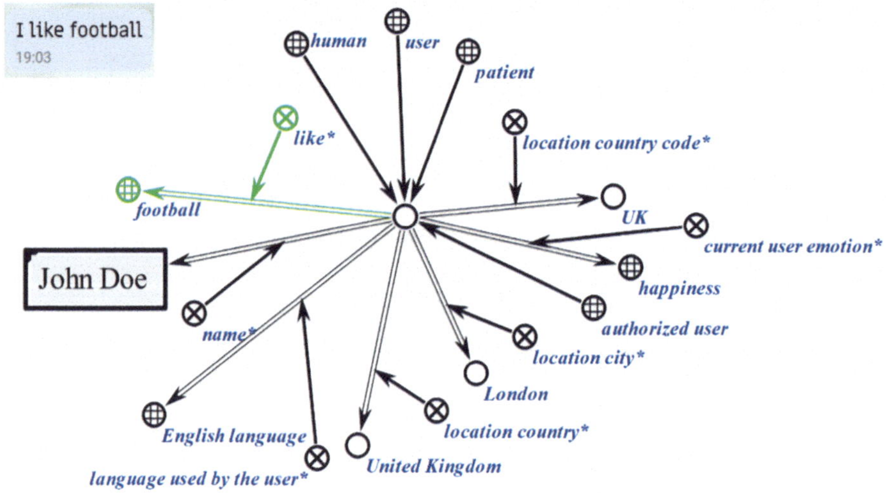

Fig. 2. Example of a message meaning being added to the KB.

3.2 Reply Generation

Once the message has been interpreted, the dialog management module forms the context necessary to generate a reply based on a set of input parameters that is provided by all other system components (message meaning, topic, intent, type; user's current emotion, objects detected on the user's scene, other fragments of the KB). The system is capable of choosing which parameters are necessary. For example, if the message type is interrogative, it will trigger a search in the KB for the required information, which will be added to the context when found, i.e. the KB serves as a source of factual knowledge for the LLM. Subgraphs from the KB are linearized into strings containing pseudo-natural language triples. The context is then added to the reply generation prompt, along with two messages immediately preceding the target message. The reply generated by the LLM is then sent to the user as a voice message via speech synthesis submodule.

3.3 Ensuring Adaptability and Robustness

The system's adaptability is facilitated by the KB, which serves as the source of verified information about the user, the user's environment, as well as facts in contextually relevant subject domains, all of which can be situationally utilized to enhance the user experience.

The state of the KB updates with each message sent by the user. Apart from information extracted from the user's messages, the KB is also populated with outputs from the computer vision submodule. The submodule's artificial neural network (ANN) models are trained [14] to recognize faces and 7 emotions (anger, disgust, fear, happiness, sadness, surprise, and neutral), as well as 80 classes of objects and their colors. The KB provides the problem solver with useful context, which can have numerous applications, such as including information from the computer vision module into the LLM prompt when the user asks a question about their environment, or changing the topic of the conversation, if the current topic led to the user having an undesirable emotional state, as well as leveraging the knowledge about the user's environment for assistance – e.g., helping the user in finding a misplaced item.

To enhance trustworthiness and robustness, the system can operate under human supervision. Authorized residential care facility personnel can observe the system's replies and intervene, if necessary, by modifying replies in real time via a special control panel. Reply modifications are saved in the KB and are taken into account when responding to similar user requests. This feature can be used as a method of training the system to respond in a particular way consistently in sensitive scenarios.

4 Discussion

Developing a complex intelligent system with many integrations presents a unique set of challenges. The main concerns are validity, reliability and significance [15] of used natural language processing methods, since information in the knowledge graph is used by the problem solver for various tasks, and the KB is populated using information extracted from the user's messages.

The original idea was to use rule-based syntactic and semantic analysis to ensure complete explainability of responses [16]. However, in our case the rule-based analysis was not sufficiently robust because any failure during an intermediate step would cause the whole pipeline to fail.

Instead, an LLM-based approach was chosen, which has its own set of related issues [17]. The translation of message meaning into intermediate representation can be unreliable due to LLM hallucinations [18, 19], while the semantic analysis in general can be quite shallow [20]. Moreover, the authors are not aware of the reliable way to ensure that the LLM respects a specific fragment of the whole context (e.g. the user's emotion), which underutilizes the potential of the KB.

Despite the shortcomings of this approach, it actually enhanced the robustness of the developed system by allowing it to parse message meanings at least up to a certain baseline level of acceptability (even if not entirely correct) and provide a reply to any message by falling back on simple contextless response generation by the LLM in case of a fault at any stage of the processing. We plan to address aforementioned issues by implementing a more comprehensive context management system [21] that would distinguish between different types of contexts (dialog context, user context, and global context) and that would have specific agents for managing information within those contexts, including fact verification so that unverified information from user context can be merged with the verified part of the KB.

Finally, there is a question of system usability and the way end users interact with the system, which could prove difficult for elderly patients housed at residential care facilities. Adoption of assistive technologies [5] and especially AI systems by older adults [11, 22] is a delicate matter, in which a balance should be found between perceived usefulness and usability, with particular attention paid to socioemotional aspects of technology use [23], especially because certain designs for assistive technology might be unpopular [24]. However, Nicenboim et al. [25] presents a challenge to the stereotype (older adults are passive and frail) and proposes to design assistive technology that leverage resourcefulness of the older adults and their existing competencies.

To streamline the process of interacting with the system, we rejected a traditional GUI in favor of a SILK-interface (Speech, Image, Language, Knowledge), in which users interact with a virtual avatar representing the system by talking to it as if having a regular conversation with a real person, thereby immersing the users in a familiar setting that only requires their general conversational skills to navigate.

5 Conclusion

We believe that the most potentially impactful result lies in a new ways hybrid approaches (that integrate various problem-solving methods) may enhance explainability, which is an important concern in the assistive technology and healthcare domain, especially considering the fact that chatbot-like assistive technologies at the current stage rely mostly on ANN-based solutions such as LLMs that notoriously lack explainability. Integration of LLMs with a verified source of factual knowledge (in our case – a KB containing a knowledge graph) helps to mitigate well-established issues of LLMs.

Since the AI-powered assistive technology companion, described here, is constantly under development, future work will focus on addressing the problems outlined in the Discussion section and conducting more extensive trials to validate the system's performance and identify more areas for future improvements.

Acknowledgements. The authors would like to express their gratitude to the colleagues from Quintary AI for their valuable thoughts and critical discussions concerning the scope of this article.

Disclosure of Interests. The authors Artem Goylo, Mikhail Sadouski, and Mikhail Kovalev worked on the project discussed in this article during its development in their capacity as employees of Quintary AI.

References

1. Mulvenna, M.D., Nugent, C.D., Moelaert, F., Craig, D., Dröes, R.M., Bengtsson, J.E.: Supporting people with dementia using pervasive healthcare technologies. In: Mulvenna, M.D., Nugent, C.D. (eds.) Supporting People with Dementia Using Pervasive Health Technologies, LNCS, pp. 3–14. Springer, London (2010). https://doi.org/10.1007/978-1-84882-551-2_1
2. Pignatti, F., Rozzini, R., Trabucchi, M.: Physical activity and cognitive decline in elderly persons. Arch. Intern. Med. **162**(3), 361–362 (2002). https://doi.org/10.1001/archinte.162.3.361
3. Pramod, D.: Assistive Technology for Elderly People: State of the Art Review and Future Research Agenda. https://ssrn.com/abstract=3955842
4. Paice, A., Biallas, M., Andrushevich, A.: Assistive and inclusive technology design for people with disabilities (special needs). In: Malakhatka, E., Wiberg, M. (eds.) Human-Technology Interaction. Springer Series in Adaptive Environments, LNCS. Springer, Cham (2025). https://doi.org/10.1007/978-3-031-78357-9_11
5. Gambo, I., et al.: Technology adoption review for ageing well: analysis of technical solutions. Front. Public Health **11** (2023). https://doi.org/10.3389/fpubh.2023.1169192
6. Velciu, M., Kaczmarek, M., Marzan, M., Teunissen, B., Andrushevich, A., Spiru, L.: Virtual assistant for healthy aging: benefits perceived by elders. In: Proceedings of the 4th International Conference on Neuroscience and Cognitive Brain Information. Presented at: BRAININFO '19, June 30–July 4, pp. 12–15. Rome, Italy (2019)
7. Donadello, I., Dragoni, M.: AI-enabled persuasive personal health assistant. Soc. Netw. Anal. Min. **12**, 106 (2022). https://doi.org/10.1007/s13278-022-00935-3
8. Biallas, M., Lu, G., Stieger, M., Moist, R., Andrushevich, A., Paice, A.: Exploring the potential of conversational interfaces for care of older adults: insights from stakeholder workshops. Stud. Health Technol. Inform. **306**, 63–69 (2023). https://doi.org/10.3233/shti230597
9. Samala, A.D., Rawas, S.: Generative AI as virtual healthcare assistant for enhancing patient care quality. Int. J. Online Biomed. Eng. (iJOE) **20**(5), 174–187 (2024). https://doi.org/10.3991/ijoe.v20i05.45937
10. Alowais, S.A., Alghamdi, S.S., Alsuhebany, N., et al.: Revolutionizing healthcare: the role of artificial intelligence in clinical practice. BMC Med. Educ. **23**, 689 (2023). https://doi.org/10.1186/s12909-023-04698-z
11. Kaczmarek, M., Velciu, M., Spiru, L., Bujnowski, A., Andrushevich, A., Birrer, E. Methodology for assessing end-user requirements in the Ella4Life project: elders' perspectives about self-monitoring. In: 41st Annual International Conference of the IEEE Engineering in Medicine and Biology Society (EMBC), pp. 1409–1412 (2019)
12. Golenkov, V., Gulyakina, N., Shunkevich, D.: Open technology for ontological design, production and operation of semantically compatible hybrid intelligent computer systems (In Russian). Bestprint, Minsk (2021)
13. Shunkevich, D.: Ontology-based design of hybrid problem solvers. In: Golenkov, V., Krasnoproshin, V., Golovko, V., Shunkevich, D. (eds.) Open Semantic Technologies for Intelligent Systems. OSTIS 2021. CCIS, vol. 1625, pp. 101–131. Springer, Cham. (2022).https://doi.org/10.1007/978-3-031-15882-7
14. Canal, F.Z., et al.: A survey on facial emotion recognition techniques: a state-of-the-art literature review. Inf. Sci. **582**, 593–617 (2022). https://doi.org/10.1016/j.ins.2021.10.005

15. Riezler, S., Hagmann, M.: Validity, Reliability, and Significance: Empirical Methods for NLP and Data Science. Springer, Cham (2022). https://doi.org/10.1007/978-3-031-02183-1
16. De Silva, D.I., Gallage, I.S.: The role of syntax and semantics in rule-based translation: a comprehensive review. In: 2024 International Conference on Information and Communication Technology for Development for Africa (ICT4DA), Bahir Dar, Ethiopia, pp. 229–234 (2024). https://doi.org/10.1109/ICT4DA62874.2024.10777198
17. Myers, D., Mohawesh, R., Chellaboina, V.I.: Foundation and large language models: fundamentals, challenges, opportunities, and social impacts. Cluster Comput. **27**, 1–26 (2024). https://doi.org/10.1007/s10586-023-04203-7
18. Perković, G., Drobnjak, A., Botički, I.: Hallucinations in LLMs: understanding and addressing challenges. In: 47th MIPRO ICT and Electronics Convention (MIPRO), Opatija, Croatia, 2024, pp. 2084–2088 (2024). https://doi.org/10.1109/MIPRO60963.2024.10569238
19. Huang, L., et al.: A survey on hallucination in large language models: principles, taxonomy, challenges, and open questions. ACM Trans. Off. Inf. Syst. **43**(2), 1–55 (2024). https://doi.org/10.1145/3703155
20. Cambria, E., Poria, S., Gelbukh, A., Thelwall, M.: Sentiment analysis is a big suitcase. IEEE Intell. Syst. **32**(6), 74–80 (2017). https://doi.org/10.1109/MIS.2017.4531228
21. Hu, P., Indulska, J., Robinson, R.: An autonomic context management system for pervasive computing. In: Sixth Annual IEEE International Conference on Pervasive Computing and Communications (PerCom), Hong Kong, China, pp. 213–223 (2008). https://doi.org/10.1109/PERCOM.2008.56
22. Bults, M., van Leersum, C.M., Olthuis, T.J.J., et al.: Acceptance of a digital assistant (anne4care) for older adult immigrants living with dementia: qualitative descriptive study. JMIR Aging **7**, e50219 (2024)
23. Spinelli, G., Micocci, M., Ajovalasit, M.: Behaviours: older adults' behavioural strategies in the adoption of new technology-based products: the effects of ageing and the promising application of smart materials for the design of smart products. In: Tsekleves, E., Cooper, R. (eds.) Design for Health, pp. 358–374. Routledge, Abingdon-on-Thames, UK (2017). https://doi.org/10.4324/9781315576619-22
24. Khakurel, J., Knutas, A., Melkas, H., Penzenstadler, B., Porras, J.: Crafting usable quantified self-wearable technologies for older adult. In: Ahram, T. (ed.) Advances in Human Factors in Wearable Technologies and Game Design. AHFE 2018, LNCS, vol. 795, pp. 75–87. Springer, Cham (2019). https://doi.org/10.1007/978-3-319-94619-1_8
25. Nicenboim, I., Kitazaki, M., Kihara, T., Marin, A.T., Havranek, M.: Connected resources: a novel approach in designing technologies for older people. In: Extended Abstracts of the 2018 CHI Conference on Human Factors in Computing Systems (CHI EA '18), pp. 1–4. Association for Computing Machinery, New York (2018) https://doi.org/10.1145/3170427.3186527

A Novel AI-Powered Approach for Improving Inclusion and Participation of Unpaid Caregivers in a Digital Carer Support Community

Rohullah Khalid[1(✉)], Elaheh Homayounvala[1], Joe Legate[2], Claire Cook[2], Karim Ouazzane[1], and Paul Calcraft[2]

[1] AI and Data Science Research Group, School of Computing and Digital Media, London Metropolitan University, London, UK
k.rohullah@londonmet.ac.uk
[2] Mobilise Care Limited, London, UK

Abstract. Unpaid carers provide support to individuals who cannot manage independently due to disability, age, illness, mental health conditions, or addiction [1]. They often face significant challenges balancing their caregiving duties while managing their own well-being. In 2021/22, the value of unpaid care across the UK was estimated at £184 billion annually, surpassing the NHS budget for all four nations [2]. Mobilise Care Limited, a digital carer support organisation, empowers carers through the Mobilise Digital Carer Support System (MDCSS)—a platform comprising two main tools: the Carer's Assessment Tool and the Mobilise Assistant Chat. While Carer's Assessment Tool delivers personalised support through guided phone conversations and follow-up via SMS or WhatsApp, Mobilise Assistant Chat offers AI-driven responses to carers' digital queries. This research introduces the Enhanced Carer Digital Support System (E-CDSS)—an upgrade to the existing MDCSS—which integrates peer-generated community knowledge from a growing carers network into its digital infrastructure to improve inclusion and participation of carers in the digital carer support community.. Using supervised machine learning [3] and Retrieval-Augmented Generation (RAG) [4], E-CDSS classifies and embeds helpful community discussions with 93% accuracy, enriching the knowledge base behind both Carer Assessment Tool and Mobilise Assistant Chat. Preliminary findings suggest increased carer satisfaction, reinforcing the role of digital innovation in reducing social isolation and enhancing accessibility, inclusion, and quality of support within the carer community.

Keywords: Personalised AI · Unpaid Carers · Community Engagement · Assistive Technology · Digital health innovation

K. Mavrou and P. Encarnação (Eds.): AAATE 2025, *Technology for Inclusion and Participation for All: Recent Achievements and Future Directions*, pp. 191–198, 2025.
https://doi.org/10.1007/978-3-032-01628-7_24

1 Introduction

Unpaid carers (caregivers) in the UK provide unpaid support to a family member, partner, or friend who is unable to manage independently due to illness, disability, mental health conditions, old age, or addiction [1]. These carers face significant challenges in balancing their caregiving responsibilities with their own well-being and daily life. With an estimated 5.8 million unpaid carers across the UK [5], they represent the backbone of informal care provision. In 2021/22, the economic value of their contribution was estimated at £184 billion—surpassing the combined NHS budget for all four UK nations [2]. This highlights both their vital role in the healthcare system and the urgent need for innovative support mechanisms.

The Mobilise Carer Digital Support System (MCDSS), developed through collaboration between Mobilise, London Metropolitan University, and UK Research and Innovation, delivers personalised information and support by leveraging individual carer profiles and responding to real-time needs. The Mobilise Carer Digital Support System (MCDSS) comprises two core components: the Carer's Assessment Tool, which assesses carers' needs through guided phone conversations and delivers tailored support plans via SMS or WhatsApp; and the Mobilise Assistant Chat, which responds to carers' digital queries. While both tools are built on Mobilise's central knowledge base, they often fall short in offering practical insights drawn from lived experiences, fostering emotional engagement, and facilitating two-way discussion on applying caregiving strategies. Online Health Communities (OHCs) allow carers and others in similar situations to connect, exchange information, share experiences, and provide mutual emotional support [6]. Recognising the value of such peer-driven support, this study focuses on the Mobilise Community—a growing digital platform with around 9,500 registered carers—where members discuss day-to-day care-related issues. This research presents a novel enhancement to MCDS: extracting and embedding useful community-generated threads into the system's knowledge base to improve the personalisation, contextual relevance, and emotional resonance of its responses. By integrating peer insights into both the Carer's Assessment Tool and Mobilise Assistant Chat, E-CDSS improves inclusion, accessibility, and the quality of informational support, while fostering stronger community engagement. Preliminary findings indicate increased user satisfaction and extended support reach, demonstrating strong potential for innovation in digital health.

To contextualise this work, Sect. 2 of this paper reviews the related work and background, discussing existing research on unpaid carers and digital health tools. Section 3 presents the methodology, detailing the AI-powered tools and community integration approach used. Section 4 provides a quantitative and qualitative analysis of the results, including model performance and user feedback. Section 5 discusses the implications of the findings and the impact on carer's engagement and well-being. Finally, Sect. 6 concludes the paper, offering suggestions for future work and further innovations in carer support systems.

2 Related Work

Unpaid carers play a vital role in healthcare systems, yet they often remain under-supported and face significant emotional, financial, and social burdens [2]. Digital interventions have emerged as promising tools to alleviate these pressures, particularly through Online Health Communities (OHCs), which enable carers to exchange information and receive emotional support. Participation in OHCs has been associated with reduced isolation and improved mental well-being [6]. These platforms foster peer support and shared lived experiences, empowering carers to manage complex caregiving challenges more effectively. However, OHCs also present limitations: community-generated content is typically siloed within the platform and can be difficult to search or surface in response to specific needs. These challenges limit the accessibility and broader utility of valuable peer knowledge outside the immediate community environment. In parallel, artificial intelligence (AI) is increasingly used in digital health to provide personalised and scalable support. Generative models such as GPT have been deployed for mental health support and education, demonstrating high engagement and perceived usefulness [7,8]. Recent research indicates that integrating peer-generated content into AI systems can enhance the contextual richness and empathy of responses. For instance, Sharma et al. (2022) developed an AI-in-the-loop agent that provided real-time feedback to peer supporters in mental health conversations, leading to a 19.6% increase in conversational empathy [9]. Similarly, Lee et al. (2024) found that responses generated by large language models were consistently rated as more empathetic than human-written responses across various life scenarios [10]. Retrieval-Augmented Generation (RAG) [4] has emerged as a promising architecture that combines structured language models with unstructured, real-world data. It enables systems to retrieve relevant contextual documents (e.g., community discussions) and use them to ground model outputs. This approach has shown strong performance in domains requiring factual accuracy and contextual reasoning, however, its application in caregiving contexts remains limited. Machine learning has also been utilized to identify factors contributing to burnout and emotional exhaustion among healthcare providers, which are pertinent to understanding carers' stress levels [11]. However, few systems integrate these capabilities into live support tools, and there is minimal research on how community-generated knowledge can be dynamically filtered, stored, and reused to inform AI-driven responses for carer support Despite growing interest in both community-based health interventions and AI-powered support tools, there is a lack of integrated systems that combine peer-generated knowledge with RAG models to deliver personalised, contextually aware assistance to unpaid carers. Furthermore, empirical evaluation of such systems in real-world care settings remains scarce.

3 Methodology

MCDSS (Fig. 1) uses RAG [4] approach, which uses the OpenAI GPT-4o API. The system retrieves data stored in a vector database to generate contextualised responses tailored to the carers' needs. In the latest feature enhancement,

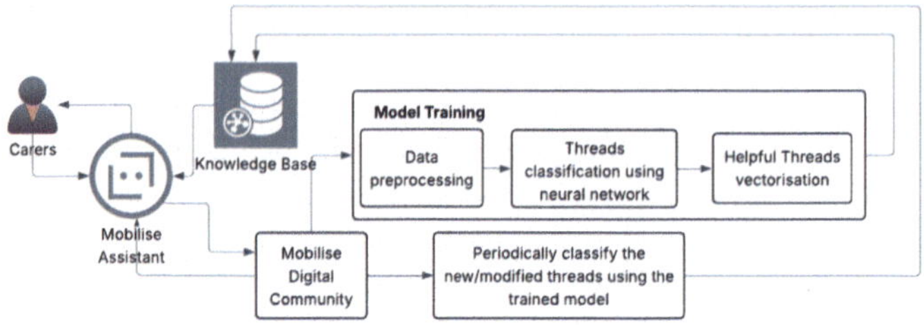

Fig. 1. Conceptual design

community threads (collection of posts and comments) were collected from the Mobilise Community platform using API. These internal discussions on caregiving topics were initially classified using a semi-supervised model [12]. with labels from the Mobilise Carer Support Team, who rated threads on a five-level Likert scale from "Very Helpful" to "Not Helpful." The dataset of 1709 threads was then binarised into "Helpful" and "Unhelpful" for model training. Subsequently, a supervised neural network with two layers (256 and 16 neurons) adopted for improved performance [3], incorporating textual features and metadata such as likes and comments. The final pipeline integrates helpful threads into the RAG [4] system by vectorising and storing them in the vector database, enabling the tool to provide carers with both core content and peer-generated community discussions. This entire process is automated to update regularly, ensuring the tool reflects evolving community insights, which are crucial for addressing niche, practical concerns and reinforcing the platform's role in fostering connected, responsive support.

4 Quantitative and Qualitative Results

The model reached 93% accuracy after 70 training epochs (Fig. 2), demonstrating strong generalisability and effective capture of linguistic and contextual features indicating helpfulness in caregiving content.

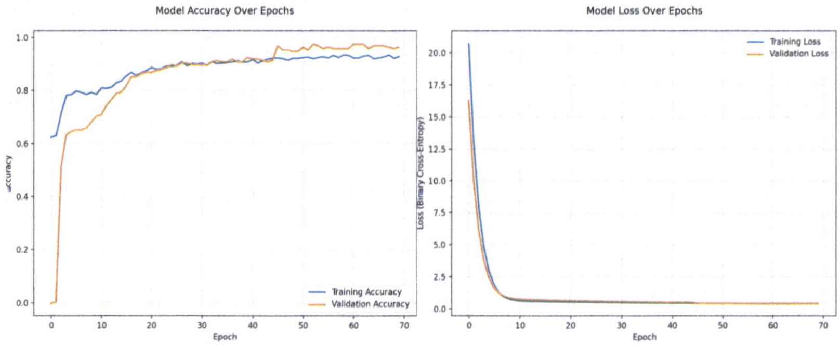

Fig. 2. Training and validation accuracy and loss curves

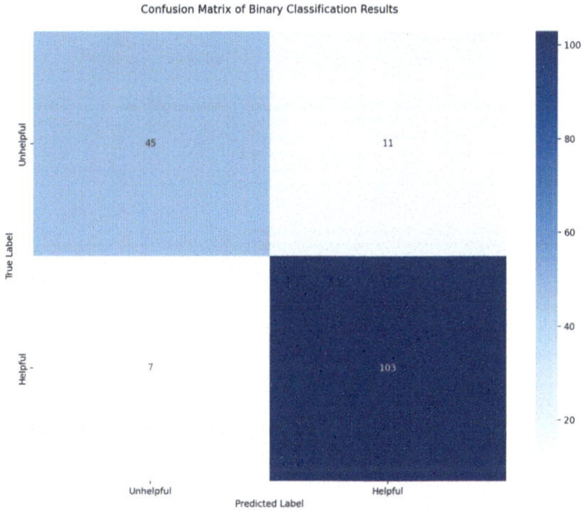

Fig. 3. Confusion matrix of the model

Figure 2 shows a steady decline in training and validation loss, indicating effective convergence without overfitting. The close alignment of the curves supports the model's robustness and reliability.

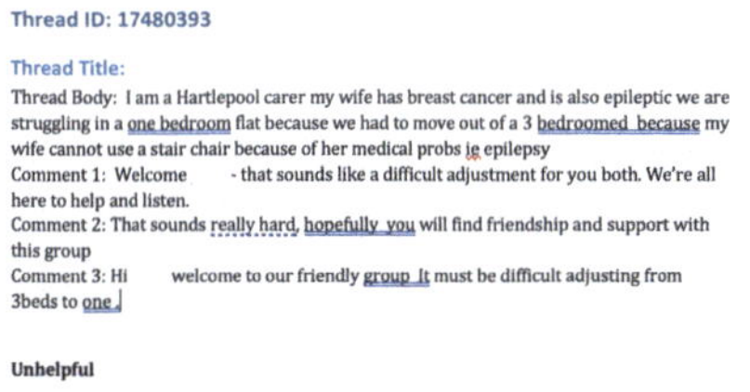

Thread ID: 17480393

Thread Title:

Thread Body: I am a Hartlepool carer my wife has breast cancer and is also epileptic we are struggling in a one bedroom flat because we had to move out of a 3 bedroomed because my wife cannot use a stair chair because of her medical probs ie epilepsy

Comment 1: Welcome - that sounds like a difficult adjustment for you both. We're all here to help and listen.

Comment 2: That sounds really hard, hopefully you will find friendship and support with this group

Comment 3: Hi welcome to our friendly group It must be difficult adjusting from 3beds to one.

Unhelpful

(a) Unhelpful

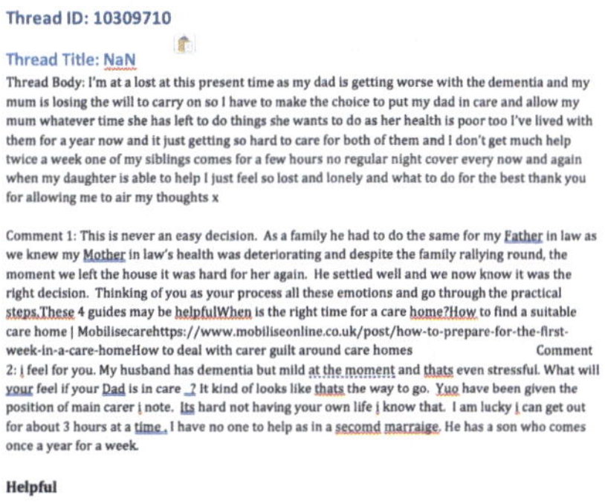

Thread ID: 10309710

Thread Title: NaN

Thread Body: I'm at a lost at this present time as my dad is getting worse with the dementia and my mum is losing the will to carry on so I have to make the choice to put my dad in care and allow my mum whatever time she has left to do things she wants to do as her health is poor too I've lived with them for a year now and it just getting so hard to care for both of them and I don't get much help twice a week one of my siblings comes for a few hours no regular night cover every now and again when my daughter is able to help I just feel so lost and lonely and what to do for the best thank you for allowing me to air my thoughts x

Comment 1: This is never an easy decision. As a family he had to do the same for my Father in law as we knew my Mother in law's health was deteriorating and despite the family rallying round, the moment we left the house it was hard for her again. He settled well and we now know it was the right decision. Thinking of you as your process all these emotions and go through the practical steps.These 4 guides may be helpfulWhen is the right time for a care home?How to find a suitable care home | Mobilisecarehttps://www.mobiliseonline.co.uk/post/how-to-prepare-for-the-first-week-in-a-care-homeHow to deal with carer guilt around care homes Comment 2: i feel for you. My husband has dementia but mild at the moment and thats even stressful. What will your feel if your Dad is in care ? It kind of looks like thats the way to go. Yuo have been given the position of main carer i note. Its hard not having your own life i know that. I am lucky i can get out for about 3 hours at a time . I have no one to help as in a secomd marraige. He has a son who comes once a year for a week.

Helpful

(b) Helpful

Fig. 4. Threads predicted as Unhelpful and Helpful

The confusion matrix (Fig. 3) shows the model correctly classified 45 helpful and 103 unhelpful posts, with 7 helpful and 11 unhelpful misclassifications, demonstrating balanced performance and high precision in identifying unhelpful content to maintain a reliable knowledge base. Figure 4 provides qualitative examples: Fig. 4a depicts an unhelpful thread lacking actionable support, while Fig. 4b shows a helpful thread offering emotional and practical advice relevant to carers. These illustrate the model's real-world classification effectiveness.

No	Question (user query)	Answer from Assistant's staging environment	Is it helpful, relevant, and coherent	Are you happy with the empathy	Comment
23	How can I get Mum to drink more	Encouraging someone to drink more can be challenging, especially if they're resistant. Here are a few suggestions: Offer a variety of drinks: Try different types of beverages like tea, coffee, squash, or banana milkshake. Sometimes, a change in taste can stimulate the desire to drink. Make it fun: You could try making tasty ice lollies to get extra fluids in. This can be especially helpful during warmer days. Use soups and porridges: These are easy to consume and can be a good way to slip both calories and extra fluid into the diet. Use hydration aids: Some carers have found products like jelly drops helpful. These are jelly sweets that help with hydration. Try a gentle approach: Instead of insisting, leave the drink within their reach and let them decide when to drink. Remember, it's important to respect their autonomy and not to force them to drink if they don't want to. For more tips and shared experiences from other carers, you can check out this discussion on the Mobilise community.	Yes, yes, yes	Yes	Includes a brilliant community link but doesn't include a link to our great piece of content. Has it swapped one for the other? I like the wording "For more tips and shared experiences from other carers, you can check out this discussion on the Mobilise community". If it could also specifically mention the topic that would be even better. E.g. "for more tips on getting your mum to drink more, you can check out...."

Fig. 5. Feedback from carer support expert

Feedback from carer support experts (Fig. 5) confirmed the system's practical value, noting improvements in content relevance, workflow efficiency, and user engagement. These results validate the AI classification's effectiveness and support its integration into the Mobilise platform, enhancing personalisation, engagement, and scalable intelligent support.

5 Discussions and Impacts

This study shows that integrating AI-powered community knowledge into MCDSS positively affects carers and key stakeholders. As shown in Fig. 6, referrals to relevant community threads have increased engagement, enhancing carers' sense of inclusion and support.

The feature also benefits carer support staff by streamlining workflows, providing quicker access to relevant content, aiding empathetic responses, and summarising queries. This reduces time for personalised support, improves communication quality, and eases staff onboarding. For local authorities, the system offers better responsiveness through aggregated insights, aiding policy and support planning.

From an organisational perspective, Mobilise gains strategic advantages such as higher user retention, improved service delivery, and operational scalability. The centralised knowledge base aids staff training, content creation, and decision-making. Data-driven metrics also help Mobilise demonstrate its impact to funders and partners, strengthening its leadership in digital carer support.

Integrating peer-generated knowledge into AI systems is a promising advance in digital health. By grounding AI responses in real-life experiences, the system enhances relevance, trust, and engagement. This approach has potential applications in mental health, chronic illness management, and other OHCs where lived experience is crucial.

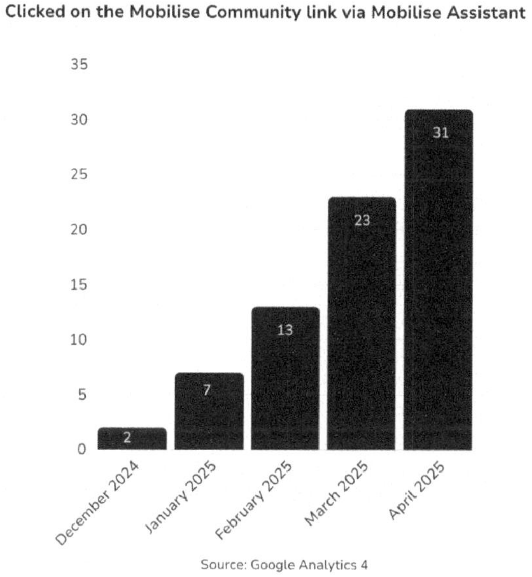

Fig. 6. User engagement

6 Conclusion and Future Work

This research presents a novel AI-powered approach that integrates personalised support with community-derived knowledge to enhance digital assistance for unpaid carers. By classifying and embedding helpful peer-generated content from a digital caregiving community into the Mobilise Carer Digital Support System (MCDSS), the system delivers richer and more contextually relevant responses. This integration improves informational accuracy while fostering inclusion, empathy, and shared experience among carers. Quantitative evaluation showed 93% accuracy in classifying helpful threads, with strong convergence and balanced performance. Qualitative testing with carers and experts indicated higher satisfaction, better problem-solving, and stronger engagement. Mobilise staff also benefit through faster responses, smoother onboarding, and insights from user behaviour.

These findings confirm the value of integrating machine learning with lived experience to enhance digital care platforms. Future work will focus on scaling support, refining engagement, and expanding multilingual and culturally sensitive features for broader inclusion.

Intellectual Property Statement

All proprietary systems and methods described in this paper, including the Mobilise Carer Digital Support System (MCDSS), Carer's Assessment Tool, and

References

1. Carers Trust: Unpaid Carers (2021). https://carers.org/downloads/resources-pdfs/working-for-carers/unpaid-carers.pdf
2. NHS England: Commitment to Carers. NHS England (2020). https://www.england.nhs.uk/ourwork/people/carers
3. Mitchell, T.M.: Machine Learning. McGraw-Hill, New York (1997)
4. Lewis, P., et al.: Retrieval-augmented generation for knowledge-intensive NLP tasks. Advances in Neural Information Processing Systems, vol. 33, pp. 9459–9474 (2020). https://arxiv.org/abs/2005.11401
5. Carers UK. https://www.carersuk.org
6. Kamalpour, M., et al.: Online health communities, contributions to caregivers and resilience of older adults. Health Soc. Care Commun. **29**(6), 1861–1873. (2021). https://doi.org/10.1111/hsc.13247.
7. Inkster, B., Sarda, S., Subramanian, V.: An empathy-driven, conversational artificial intelligence agent (Wysa) for digital mental well-being: Real-world data evaluation. JMIR MHealth UHealth **6**(11) (2018). https://mhealth.jmir.org/2018/11/e12106/
8. Zeng, Y., et al.: Empathic conversational agent platform designs and mental health outcomes. JMIR Ment. Health **11**(1) (2024). https://mental.jmir.org/2024/1/e58974
9. Sharma, A., Lin, I.W., Miner, A.S., Atkins, D.C., Althoff, T.: Human-AI collaboration enables more empathic conversations in text-based peer-to-peer mental health support. arXiv preprint: arXiv:2203.15144 (2022)
10. Lee, Y.K., Suh, J., Zhan, H., Li, J.J., Ong, D.C.: Large language models produce responses perceived to be empathic. arXiv preprint: arXiv:2403.18148 (2024)
11. Van Zyl-Cillié, M.M., Bührmann, J.H., Blignaut, A.J., Demirtas, D., Coetzee, S.K.: A machine learning model to predict the risk factors causing feelings of burnout and emotional exhaustion amongst nursing staff in South Africa. BMC Health Serv. Res. 24(1), 1665 (2024). https://doi.org/10.1186/s12913-024-12184-5.
12. Chapelle, O., Scholkopf, B., Zien, A.: Semi-Supervised Learning. MIT Press, Cambridge (2006). https://mitpress.mit.edu/9780262033589/semi-supervised-learning/

About Assistive Technology Stack Supporting Active and Healthy Aging Based on Personalization and Application of Sensor Measurements

Aliaksei Andrushevich[1(✉)] and Iosif Vojteshenko[2]

[1] iHomeLab, Lucerne University of Applied Sciences and Arts, Horw, Switzerland
aliaksei.andrushevich@hslu.ch

[2] Faculty of Applied Mathematics and Computer Science, Minsk, Belarus

Abstract. This article describes the development of an assistive sociotechnical system aimed at supporting active and healthy aging based on personalization and application of sensor measurements. In this work we rely on both historical and real-time data coming from environmental, behavioral and biomedical measures. Further, this data is analyzed to highlight patterns and to identify undesirable situations based on deviations from recognized patterns. This way the user's state and lifestyle trends can be determined based on data from several sensors located in and around the home. Our results promise to increase the safety of aging populations via assistive technology stack integration with personal AI-assistants that have started to fit into everyday life and modern households.

Keywords: Active and healthy ageing · single-board computers · wearable sensors · ambient sensors

1 Introduction/Background

The integration of assistive technologies (AT) leveraging sensor measurements and personalization is pivotal in promoting active and healthy aging. Recent scientific advancements have focused on developing comprehensive AT stacks that combine Internet of Things (IoT) devices, artificial intelligence (AI), and user-centered designs to support the elderly in maintaining independence and well-being.

Wearable and ambient sensors are central to modern AT solutions. For instance, the iCare system utilizes wireless body sensors and smartphones to monitor the health of elderly individuals, providing real-time data and emergency alerts to caregivers [1]. Similarly, the AGAPE platform employs smart bands and inertial measurement units to track physical activity and deliver personalized interventions through a mobile health assistant [2].

Artificial intelligence enhances the capability of AT by enabling complex activity recognition and decision-making. A notable example is the AI-based system that combines IoT-enabled ambient sensors with large language models to track and interpret the daily activities of elderly users, facilitating timely assistance and interaction [3].

K. Mavrou and P. Encarnação (Eds.): AAATE 2025, *Technology for Inclusion and Participation for All: Recent Achievements and Future Directions*, pp. 199–206, 2025.
https://doi.org/10.1007/978-3-032-01628-7_25

Personalization in AT is further exemplified by virtual caregiver systems that adapt to the unique needs of elderly users. A rule-based mobile chatbot has been developed to monitor mental and physical health, providing tailored advice and support based on user responses [4].

Advancements in wearable technology, such as soft exoskeletons, offer personalized mobility support. These devices utilize soft actuators and sensors to adapt to the user's movements, enhancing comfort and reducing the risk of injury [5].

Remote patient monitoring (RPM) systems [6] enable continuous health tracking outside clinical settings. By collecting physiological data through peripheral devices, RPM facilitates early detection of health issues and reduces hospital visits, thereby supporting the autonomy of elderly individuals.

With declining birth rates in developed countries, the proportion of elderly people in society is increasing, while the proportion of working-age people is decreasing. Thus, with the aging population in developed countries, the number of micro-households is increasing significantly. Older people often need assistance and accompaniment in maintaining health, moving freely and in case of unwanted accidents. Whereas in the past IT-based assistive devices were predominantly seen from a medical and health perspective, nowadays they are increasingly being used at a much earlier stage and thus contribute to comfort and safety. They are evolving from problem-oriented "stand-alone" products into integrated daily assistants that can easily fit into everyday life and home environments.

In this regard, [7–9] identified the following application areas of assistive technologies for solving challenges of elderly individuals living alone:

- informational assistance (easy availability of all necessary information),
- intelligent situational behavior (the environment should recognize typical patterns of behavior and offer appropriate assistance),
- prediction of undesirable events (recognizing such situations based on behavioral and physiological indicators and applying preventive measures),
- recognizing and responding to undesirable situations,
- security (protection against intrusions using authorization and authentication mechanisms),
- privacy (minimization of interference in personal life).

2 Methods

To meet these needs, a sociotechnical system can rely on both historical and real-time data. Two groups of data can be distinguished. Environmental data comes from environmental sensors such as temperature, motion, humidity, light, noise, etc. Data about human behavior and condition comes from biomedical sensors. The sociotechnical system may also include feedback devices for multimedia (like visual and voice) notification of various events.

Data from these sensors is processed locally using edge computing devices such as single board computers, reducing reliance on cloud services and ensuring functionality even during internet outages.

In terms of modeling such systems, three categories of indicators can be distinguished. The physical aspects include both environmental conditions and human health parameters. Due to the continuous nature of this category, the system dynamics method combined with stochastic modeling [10] to account for the role of randomness seems to be the most natural. On the other hand, discrete-event modeling [11] allows to account for emerging events such as falls. Finally, spontaneous user behavior is best modeled using the agent-based method [12, 13].

Machine learning algorithms analyze the data to establish baseline behavior patterns, flagging anomalies such as missed meals, irregular sleep, or sudden falls. A risk-scoring system evaluates multiple factors—such as inactivity combined with abnormal vital signs—to determine the severity of alerts. Among many undesirable situations, we particularly emphasize fall detection. One way to recognize falls while preserving privacy is to use a wearable acceleration and atmospheric pressure sensor simultaneously. In case of unexpected acceleration, the system reads the pressure data twice and draws a conclusion about the human body position [7].

3 Composition of the Hardware Part

The hardware part of the proposed hardware-software complex [14] includes:

1. Single-board computer with a memory card. Its choice is conditioned by its low cost.
2. Web camera 4 Mpix.
3. Volibel Intelligent Lock Tuya X9s" lock with RFID reader
4. Ethernet cable to connect the lock and single-board computer to Wi-Fi.
5. Motion detector, based on it you can turn on/off the light.
6. "Smart" lamp with a dimmer, the switching on of which depends on the motion detector.
7. Light sensor depending on the time of day and brightness of the smart lamp.
8. GY-21 module based on HTU21 temperature and humidity sensor.
9. Device controllers.

4 Software Subsystems

4.1 Application for Lighting Control in a Dwelling

Functional requirements for the application: when entering a room, the light shall light up in response to movement, and 10 s after leaving the room, the light shall turn off. The brightness of the lights shall be adjusted to account for the level of street lighting entering the dwelling. In addition, the lighting between 11pm and 6am shall operate in night mode, with reduced brightness [15].

4.2 An Application for Controlling the Entrance to the Premises ("Smart Lock")

The main part of the application is an experimental face recognition system. It is developed and functions based on dlib, cmake, OpenCV, face-recognition and streamlit libraries. The cross-platform package manager conda is used to manage the environment. The current version of the application follows standard biometric authentication procedures and algorithms and does not provide full protection against tampering by

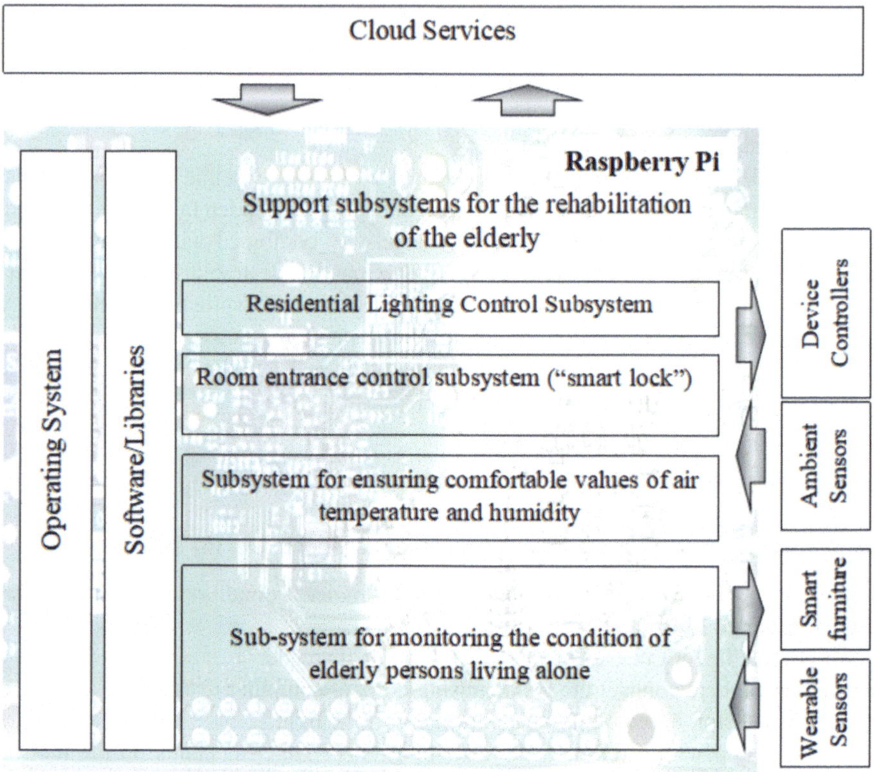

Fig. 1. The overall system architecture

substituting the image of the user's face read from the camera. In order to strengthen the biometric authentication procedure, it is supposed to use one of the Face Liveness Detection methods, providing verification of the presented identifier (face image) for belonging to a "live" user. Detailed consideration of the above two subsystems is given in [14].

4.3 Application for Ensuring Comfortable Values of Air Temperature and Humidity

When connecting the sensor, it is necessary to determine which interface it operates on. The sensor we are using operates on the interface (or rather bus) I2C (IIC, Inter-Integrated Circuit). This interface has 4 pins: VIN (input voltage), GND (ground) and two bidirectional lines SDA (data line) and SCL (clock line) for data transmission. To get readings from this sensor on a Raspberry Pi we need to connect the VIN, GND, SDA, SCL outputs of the sensor to GPIO pins 1, 9, 3, 5 of the board respectively. After, it is necessary to connect to the computer and use the raspi-config utility to enable the I2C interface and reboot the board. In general, to get readings from this sensor it is necessary to write certain bytes by the device address to the I2C bus (0xF3 for temperature and

0xF5 for relative humidity) using special kernel functions. For a detailed description of the application, see [16].

4.4 Sub-System for Monitoring the Condition of Elderly People Living Alone

The sub-system for monitoring elderly individuals living alone [17] integrates advanced IoT sensors, artificial intelligence, and remote communication technologies to ensure safety while preserving independence. The system employs [18–20] a network of wearable and environmental sensors to track daily activities, detect emergencies, and alert caregivers or emergency services when necessary.

Wearable devices such as smartwatches or medical alert bracelets monitor vital signs including heart rate, blood oxygen levels, and movement patterns, with built-in fall detection and emergency buttons for immediate assistance. Ambient sensors, strategically placed throughout the home, include motion detectors that use passive infrared (PIR) and millimeter-wave radar to track movement without intrusive cameras, ensuring privacy [21]. Door and window sensors help detect unusual inactivity, while smart plugs monitor the use of essential appliances [22] like refrigerators and kettles to confirm daily routines. Pressure-sensitive mats on beds or chairs identify prolonged inactivity [23], and environmental sensors track air quality, temperature, and humidity to maintain safe living conditions.

The alert system operates on a tiered response model [24]. Minor deviations from normal behavior trigger written notifications to family members. More concerning patterns, such as extended inactivity, prompt automated calls to designated caregivers. Critical emergencies, like detected falls or medical distress, immediately notify emergency services while enabling two-way audio communication through smart speakers for real-time assessment.

5 Results

The developed prototypes [7, 14, 15, 16] are based on inexpensive hardware and mainly use freely distributed and open-source software. The prototypes have undergone initial testing among users and have been proven to be functional.

The proposed condition monitoring sub-system for elderly people living alone represents a balanced approach to elderly care, combining technological innovation with socio-economic considerations. By enhancing safety, reducing healthcare costs, and supporting independent living, it offers benefits for individuals and society. Addressing challenges such as affordability, privacy, and reliability will be crucial to maximizing its positive impact.

Future developments could include deeper integration with telemedicine, personal virtual assistance and adaptive self-learning behavior modeling to further personalized care and healthy lifestyle coaching and nudging.

6 Discussion

The implementation of described systems has significant societal and economic implications. On a positive note, these technologies enable elderly individuals to maintain independence for longer periods, reducing the need for costly nursing home admissions.

By detecting health issues early—such as strokes or cardiac events, the system can prevent severe complications, lowering hospitalization rates and associated healthcare expenses. This proactive approach not only improves quality of life but also alleviates the financial burden on families and public health systems.

Family caregivers benefit from reduced stress, as remote monitoring provides reassurance without requiring constant physical checks. Additionally, the growing demand for elderly care technology creates new job opportunities in installation, maintenance, and remote monitoring services, contributing to economic growth. Municipalities can integrate these systems with smart city infrastructure, improving emergency response times and optimizing resource allocation in elder care.

However, challenges remain in ensuring equitable access. Low-income seniors may struggle with the costs of such systems or lack the technical literacy to use them effectively, necessitating government or insurance subsidies for broader adoption. Privacy concerns may also deter some individuals, requiring transparent policies to build trust. False alarms, though minimized by AI refinement, could strain emergency services if not properly managed.

In the long term, widespread adoption of elderly monitoring systems supports aging-in-place initiatives, easing pressure on overburdened care facilities. The mental health benefits for both seniors and their families—knowing that help is readily available—further enhance societal well-being. Economically, the shift from reactive emergency care to preventive monitoring leads to more efficient healthcare spending, benefiting both individuals and public institutions.

7 Conclusion/Implications for the AT Field

Modern wireless communication technologies allow for a reliable, easy to install, and inexpensive assistive technology infrastructure for data transmission. However, a central issue to home assistive sociotechnical systems remains the collection and storage of data on user behavior, highlighting patterns and identifying undesirable situations based on deviations from them. The user's state can be determined based on data from several sensors located in and around the home [25]. Our results promise to increase the safety of aging populations via assistive technology stack integration with daily AI-assistants that have started to fit into everyday life and modern households.

Acknowledgement. The authors express their gratitude to the students of the Faculty of Applied Mathematics and Computer Science of the Belarusian State University of different years: Emelyanova Y. A., Vardak A. S., Protasenia D. A., who took an active part in the development of system components presented in this work.

References

1. Lv, Z., Xia, F., Wu, G., Yao, L., Chen, Z.: iCare: a mobile health monitoring system for the elderly. In: 2010 IEEE/ACM Int'l Conference on Green Computing and Communications & Int'l Conference on Cyber, Physical and Social Computing, Hangzhou, China, pp. 699–705, 2010. https://doi.org/10.1109/GreenCom-CPSCom.2010.84

2. AGAPE Project. (2024). Empowering Active and Healthy Ageing: Integrating IoT and Wearable Technologies for Personalised Interventions. Appl. Sci. 2024, 14(11), 4789, https://doi.org/10.3390/app14114789
3. Sun, Y., Ortiz, J.: An AI-based system utilizing IoT-enabled ambient sensors and LLMs for complex activity tracking. Acad. J. Sci. Technol. **11**(3), 277–281 (2024). https://doi.org/10.54097/dj2pt496
4. Miura, C.H., et al.: Assisting personalized healthcare of elderly people: developing a rule-based virtual caregiver system using mobile chatbot. Sensors **22**(10), 3829 (2022). https://doi.org/10.3390/s22103829
5. Canali, S., Schiaffonati, V., Aliverti, A.: Challenges and recommendations for wearable devices in digital health: data quality, interoperability, health equity, fairness. PLOS Digit. Health **1**(10), e0000104 (2022). https://doi.org/10.1371/journal.pdig.0000104
6. Boikanyo, K., Zungeru, A.M., Sigweni, B., Yahya, A., Lebekwe, C.: Remote patient monitoring systems: applications, architecture, and challenges. Sci. Afr. **20** (2023). https://doi.org/10.1016/j.sciaf.2023.e01638
7. Andrushevich, A.: ZigBee/IEEE 802.15.4 technologies in ambient assisted living applications. In: Andrushevich, A., Kistler, R., Bieri, M., Klapproth, A. (eds.) 3rd European ZigBee Developers' Conference (EuZDC), Munich, Germany, pp. 5–30 (2009)
8. Takagi, M., Kato, Y., Suzuki, N., Maeno, J.: Analysis of problems related to the elderly living alone and development of an IT based support system. Adv. Soc. Sci. Res. J. **10**(1), 113–120 (2023). https://doi.org/10.14738/assrj.101.13788
9. Bouaziz, G., Brulin, D., Campo, E.: Technological solutions for social isolation monitoring of the elderly: a survey of selected projects from academia and industry. Sensors **22**(22), 8802 (2022). https://doi.org/10.3390/s22228802
10. Chen, N.: Stochastic Methods for Modeling and Predicting Complex Dynamical Systems. Springer, Cham (2025). https://doi.org/10.1007/978-3-031-81924-7
11. Wainer, G.: Discrete-Event Modeling and Simulation: A Practitioner's Approach (Computational Analysis, Synthesis, and Design of Dynamic Systems) (2009)
12. Lee, Y.S., Malkawi, A.M.: Simulating multiple occupant behaviors in buildings: an agent-based modeling approach. Energy Buildings **69**, 407–416 (2014). ISSN 0378-7788, https://doi.org/10.1016/j.enbuild.2013.11.020
13. Cai, L., Liu, B., Yu, J., et al.: Human behaviors modeling in multi-agent virtual environment. Multimed. Tools Appl. **76**, 5851–5871 (2017). https://doi.org/10.1007/s11042-015-2547-z
14. Andrushevich, A.A., Vardak, A.S., Vojteshenko, I.S.: Prototype of hardware and software complex for control of "smart home" applications. In: Information Systems and Technologies: Proceedings of the International Scientific Congress on Informatics. In 3 parts. Part 2, Rep. of Belarus, Minsk, 27–28 October 2022. - Minsk: Belarusian State University, pp. 231–237, 2022
15. Andrushevich, A.A., Vojteshenko, I.S., Emelianova, O.J.: About simulation environment for modeling the Internet of Things components and applications using visual flow-based programming tools and cloud services. Informatics **19**(2), 100–116 (2022). (In Russ.), https://doi.org/10.37661/1816-0301-2022-19-2-100-116
16. Andrushevich, A., Vojteshenko, I., Protasenia, D.: Prototyping the monitoring applications in public buildings using edge computing paradigm. In: 16th International Conference on Human System Interaction (HSI), Paris, France, pp. 1–6, 2024. https://doi.org/10.1109/HSI61632.2024.10613556
17. Watanabe, S., Asano, M., Nakazawa, T., Ando, R., Tasaki, T., Aoki, H.: Possibilities of simple IoT system for monitoring elderly people living alone. In: 2017 IEEE 6th Global Conference on Consumer Electronics (GCCE), Nagoya, Japan, pp. 1–4, 2017. https://doi.org/10.1109/GCCE.2017.8229239

18. Cornacchia, M., Ozcan, K., Zheng, Y., Velipasalar, S.: A survey on activity detection and classification using wearable sensors. IEEE Sens. J. **17**(2), 386–403 (2017). https://doi.org/10.1109/JSEN.2016.2628346
19. Kantoch, E., Augustyniak, P., Markiewicz, M., Prusak, D.: Monitoring activities of daily living based on wearable wireless body sensor network. In: 2014 36th Annual International Conference of the IEEE Engineering in Medicine and Biology Society, Chicago, IL, USA, 2014, pp. 586–589, 2014. https://doi.org/10.1109/EMBC.2014.6943659
20. Kaczmarek, M., Bujnowski, A., Wtorek, J., Polinski, A.: Multimodal platform for continuous monitoring of the elderly and disabled. J. Med. Imaging Health Inform. **2**(1), 56–63 (2012). https://doi.org/10.1166/jmihi.2012.1061
21. Wang, Y., Wu, Y., Zhang, S., et al.: Smart house system for safety of elderly living alone based on camera and PIR sensor. Artif. Life Robot. **29**, 43–54 (2024). https://doi.org/10.1007/s10015-023-00932-5
22. Augustyniak, P., Kantoch, E.: Turning domestic appliances into a sensor network for monitoring of activities of daily living. J. Med. Imaging Health Inform. **5**(8), 1662–1667 (2015). https://doi.org/10.1166/jmihi.2015.1627
23. Salice, F., de Vellis, M., Gwyn Barbieri, A., Masciadri, A., Comai, S.: Nonintrusive monitoring and detection of sitting and lying persons: a technological review. IEEE Access **12**, 178875–178897 (2024). https://doi.org/10.1109/ACCESS.2024.3508836
24. Sprogis, S.K., Smith, D.: Tiered Rapid Response Systems. In: DeVita, M.A. (eds.) Textbook of Rapid Response Systems. Springer, Cham (2024). https://doi.org/10.1007/978-3-031-67951-3_14
25. Biallas, M.: Living Safely and Actively in and Around the Home: Four Applied Examples from Avatars and Ambient Cubes to Active Walkers / M. Biallas [et al.] // Safe at Home with Assistive Technology / I. Kollak. — Springer, 2017

Artificial Intelligence in Higher Education: Advancing Inclusive Environments, Pedagogical Approaches, and Assistive Technologies

AI Research: Where are the Disabled Researchers and the Voices of Disabled People

Marion Hersh(✉)

University of Glasgow, Glasgow G12 8LT, Scotland
marion.hersh@glasgow.ac.uk

Abstract. The paper discusses both artificial intelligence (AI)'s potential in assistive technology and to improve accessibility and its use in control and surveillance and the disadvantages to disabled people from the bias in AI algorithms. This makes the involvement of disabled people in AI research particularly important, but a lack of literature in this area was found. There is some literature on equity, diversity and inclusion in AI, with gender the main characteristic investigated and very limited reference to disability, and a few papers on AI co-design research with non-academic disabled researchers. There is also some work on making computer science education more accessible to disabled students. Several approaches to designing databases to avoid digital discrimination against disabled people were also identified. The work highlights the importance of disabled researchers being involved in the development of new technologies, particularly where there is potential for discriminatory impacts, as well as in assistive technology development. It also highlights the need for empirical research on disabled researchers in AI and other technical areas.

Keywords: AI research · disabled researchers · EDI · justice · bias · discrimination

1 Introduction

Artificial intelligence (AI) has a wide range of potential applications including in technologies to support disabled people e.g. [1] and improve accessibility e.g. [2]. It has significant and growing economic importance with estimated market size of 293.8 billion euros in 2024 and growth rate of 13.5% [3] However, concerns have also been raised, for instance about ethical issues and privacy [4, 5] and facial and speech recognition systems not working well for some (groups of) disabled people [6]. Bias in AI algorithms is considered to be common and to be caused by a variety of factors, including under representation of minority groups. This can lead to disadvantages resulting from discriminatory predictions based on gender or minority group status [7]. Bias in AI systems has been found to disadvantage women, people of colour and gender minorities, but less attention has been given to disabled people [8].

A healthcare algorithm used with millions of patients in the USA has been found to have significant racial bias: Black patients need to have more significant illness than

K. Mavrou and P. Encarnação (Eds.): AAATE 2025, *Technology for Inclusion and Participation for All: Recent Achievements and Future Directions*, pp. 209–216, 2025.
https://doi.org/10.1007/978-3-032-01628-7_26

White patients for a given score and consequently also to receive the same level of care [9]. This bias is due do the algorithm predicting healthcare costs and unequal access to healthcare reducing spending on Black patients. AI video-interpreting systems are being marketed to recommend candidates for follow-up interviews based on analysis of speech patterns, tone of voice, facial movements and other indicators from remote video interviews [8]. The report also cites studies which show the lack of scientific evidence of the link between these factors and job competences and that these systems lead to discrimination against many groups of disabled people. This discrimination particularly affects those whose facial expressions and tone of voice are affected by their impairments.

There is also evidence of AI being used for surveillance and control, for instance by employers to manage workers and set workloads, performance targets and wage rates [8]. This can lead to arbitrary dismissal where there are no legal protections against it or arbitrarily increase workloads and reduce workers' ability to manage their work. Algorithmic wage control and AI enabled platforms can lead to very low wages and unpredictable incomes, as well as the risk of being removed from the platform without warning and finding it difficult to reopen an account. A study reported in [8] found that automation would have the greatest impacts on low-waged workers, which includes many disabled workers. AI technologies are also having an increasing role in the targeting of immigrant populations in the USA, including through the use of facial recognition technologies in an 'unprecedented surveillance infrastructure'.

The importance of AI and its potential opportunities, challenges and very serious threats make it essential that disabled people are listened to, included in the discussions and participate in the research. This is potentially also a multi-disciplinary research growth area with significant opportunities for grant funding and high impact publications. Therefore, the inclusion or exclusion of disabled researchers could have significant positive or negative impacts on their career development, while at the same time raising ethical issues and potentially removing funding from other less problematical areas.

It is probably useful to state that I am a disabled academic and researcher in biomedical engineering. While I do not currently work directly on AI, I supervise students who do.

2 Methods

The study was carried out to investigate disabled people's involvement in AI, particularly as researchers, the barriers they experienced and how they tried to overcome them. The initial literature survey involved Google Scholar, my university database and references from other papers. The initial search terms were various combinations of disabled, disability or disabilities with AI and research/er(s). They were subsequently extended to equity, diversity and inclusion/EDI and AI research/er and survey, AI and researcher(s). The search was then extended to the Pubmed and IEEE xplore databases with the title/abstract combination searched on Pubmed and abstracts only on IEEE xplore. Searches involved the terms 'researcher' with 'AI' or 'artificial intelligence' or 'machine learning' and 'disabled' or 'disabilities' or 'neurodivergent'. Though less directly relevant, the last term was replaced in some searches by 'minority'. There were no restrictions by year, language or type of literature. The low number of even tangentially relevant results obtained led to a decision not to explore other databases.

3 Results

Most of the papers discussed below were obtained from the searches of Google Scholar, my university's databases and references in the literature obtained. As an illustration, the results of the IEEE xplore surveys are summarised. A full text survey with the terms 'researcher', 'disability' and 'artificial intelligence' produced 5057 results, with nothing considered even tangentially relevant in the first 75. The use of full text surveys was therefore considered inefficient, the remaining items were not examined and subsequent surveys considered abstracts only. The following searches were carried outl: 'disabled', 'researcher' and 'AI': six results; 'disabled', 'researcher' and 'artificial intelligence': seven results; 'disabilities', 'artificial intelligence' and 'researcher': 18 results; 'neurodivergent', 'researcher' and 'artificial intelligence': no results; 'minority', 'researcher' and 'artificial intelligence': seven results; 'disabilities', 'researcher' and 'machine learning': 'minority', 'researcher' and 'machine learning': 81 results and 'disabilities' and 'student' and 'computer science': 37 results. There were no results on disabled researchers in AI, 13 on education and seven on other possibly relevant topics.

Several papers on equity, diversity and inclusion (EDI) in AI and the use of AI for EDI were identified. Diversity has been found to be of great benefit to industry with studies reported in [10] finding that diverse companies reported 2.3 times higher cash flow per employee, inclusive teams in high diversity environments had 30% better performance and companies with higher than average diversity and engagement had 46-58% better financial performance than those with lower than average diversity and engagement. Studies reported in [11] found that diverse teams increased the likelihood of biases in AI systems being recognised and addressed and could identify additional sources of bias in data or algorithms. They also determined that involving marginalised communities increased the likelihood of acceptance by those groups. A scoping review of 45 papers on AI and disability showed the prevalence of a narrow medical model and ableist perspectives and the potential of AI to both improve healthcare and perpetuate disability bias and discrimination [12]. It stressed the importance of the social model of disability, accessibility, usability and inclusivity, but did not discuss the need for disabled AI researchers.

The main focus in studies on EDI in AI has been gender and there is limited mention of disability. For instance, only four of the 48 studies reviewed in [11] considered disability and only one neurodivergence. The problems experienced by gender classifiers in recognising non-binary genders are particularly relevant to autistic people who have a higher percentage of non-binary genders than the general population. Globally women have less than one third of data science jobs [7]. The percentages of disabled women and disabled Black data scientists are likely to be less than a tenth of this.

There are a small number of papers on co-design research involving disabled non-academic researchers. People with learning disabilities were involved as co-investigators in the analysis of survey data with more than 1500 responses, including the outputs of machine learning based structural topic modeling of open question responses [13]. There has been some discussion of the perspectives of disabled people in responsible research and innovation and the importance of accessibility to avoid barriers [14]. One of the few papers on AI where all the authors stated they are disabled discusses the ways in which the use of data leads to exclusion and disadvantages to disabled people [15]. They present four different 'visions' to change this.

There are several papers on designing databases which avoid digital discrimination against disabled people resulting from automated decision making based on AI, as they fall outside the patterns in high volume data [16]. They include a 'Disability Wikibase' which was developed by a project led by two disabled researchers and involved obtaining feedback from organisations led by Black and Indigenous people and people of colour and local, national and transnational disabled people's organisations [17]. Disability rights country reports produced by Disability Rights Promotion International were the first documents used to train the AI algorithm for knowledge acquisition as part of an aim to adopt a human rights perspective. An online structure for collecting data from disabled people who are remunerated for it has been discussed as a way of making AI datasets more inclusive [18]. The authors highlighted the need for the process to be motivating, safe and accessible. Motivation could include the provision of appropriate remuneration and safety should include taking account of privacy issues. Other important factors are accessibility of all stages of data contribution to people with a wide range of different impairments and respectful treatment of disabled participants, including through providing sign language options to Deaf ones. Another approach involved the use of profiles of accommodations to be taken into account to ensure a fair assessment of, for instance, job candidates without identifying specific impairments, but unfortunately referred to them as 'limitations' despite discussing the importance of appropriate language use [16].

A review paper on ethics in the AI literature and disabled people noted the lack of engagement with disabled people [19], but not the lack of involvement of disabled researchers. An AI-enabled framework has been developed for suggested use in decision making, development and assessment of AI-enabled assistive technology [20]. However, none of the 11 researchers involved in the development indicated that they were disabled in their brief biographies and the participatory design framework does not seem to have involved any disabled researchers.

The main, though not the sole, educational pathways to AI research are through computing science and engineering. A review of a decade of computer science education research and initiatives found that disability was underresearched, with papers from before 2017 not specifying student disability status and only 2-5% of subsequent papers doing this where applicable [21]. A paper on barriers to African American and Hispanic disabled K12 students found that they were more likely to be placed in special education and that this generally reduced the opportunities open to them, including the opportunity to study computer science [22].

When teaching Deaf students (who communicate in sign language) technical subjects, it is frequently necessary to develop agreed signs for technical terms. An early and ongoing project (not identified in the searches) is the British Sign Language glossaries of curriculum project which has developed nearly 2000 signs and sign language definitions in science, technology, engineering and mathematics [23]. There is strong involvement of Deaf scientists and teachers and two of the four-person organising team are Deaf. An Argentinean Sign Language/Spanish dictionary of computing science terms has been developed and evaluated as part of a Systems Analysis Techniques degree programme course [24]. A paper on teaching hearing impaired students software engineering and programming noted the importance of communication issues related to both spoken language and sign language [25]. It makes a number of recommendations for improving communication and motivation. Students with specific learning disabilities and attention deficit conditions are often overlooked in widening access to K12 computing science courses [26]. Permitting the use of screen readers to read questions and responses, checking for understanding and illustrating if-then statements and inequalities have been recommended. A paper by a disabled researcher has stated the need for greater support to enable disabled students to obtain appropriate work integrated learning placements and positive outcomes [27]. Greater attention should also be paid to disabled students in designing placement programmes and support structures. AccessCSforAll aims to increase the successful participation of disabled students in high school computer science courses by developing teacher capacity and an accessible version of computer science principles, and acting as a US resource for K12 computer science education [28].

4 Discussion and Conclusions

AI is being integrated into all aspects of life. It has the potential to improve accessibility and participation and overcome barriers experienced disabled people and other minority groups. It can also be used to increase surveillance and control and may result in additional sources of discrimination, including through AI decision algorithms based on non-disabled norms.

Several papers have investigated approaches to developing more inclusive AI databases which meet the needs of disabled people, but further work is still required in this area. However, they do not consider in sufficient detail whether their work will perpetuate rather than overcome discrimination or give AI approaches an undeserved legitimisation. There is a small body of research on EDI in AI research, but very limited work on involving disabled researchers, the barriers they experience and the solutions to overcoming them. Gender is the main characteristic considered and the results are likely to also be relevant to disabled women, with minimal attention to disabled researchers including those with intersectional characteristics. Disabled people are seen as potential beneficiaries of a wide range of technologies based on AI and the importance of involving and consulting them has been recognised. However, the potential for disabled people to have an active role in AI development, for instance as researchers and developers of new technologies based on AI or researchers on ethical and other aspects of AI, has rarely been recognised.

There is some, though limited, work on computing science and engineering education and disabled students, including the barriers to African American and Hispanic disabled students. This is important, as they are two of the main educational routes to AI research. However, with a few exceptions, there has been no involvement of disabled researchers and teachers and much of the work has a medical model perspective. While there is some work on co-production and co-design with disabled people, there still seems to be limited recognition of the potential of disabled people to become researchers and developers of AI and other technologies.

The limited research on disabled researchers in AI and the data on the low percentages of women make it highly likely that the percentage of AI researchers who are disabled is very small. The percentage of disabled people in a position to affect decision making on the applications of AI is likely to be even smaller, but greater involvement of disabled people in AI research should result in at least some developments that benefit them. However, this will not resolve the wider issues related to AI. This is part of a much larger debate on the role of technology in society, which is beyond the remit of this paper. However, it raises questions of the need for greater regulation of the applications of AI and how decision making about developments in AI can be moved away from powerful vested interests and disabled people have a much greater role.

The lack of research on disabled AI researchers shows the need for empirical research, including interviews and surveys, of disabled researchers working in AI and related fields, as well as of computing science and engineering students and potential AI researchers. This should cover their experiences as researchers or students and their attitudes to the potential, advantages and disadvantages of AI, both in general and specifically for disabled people. The results could also be used to develop a framework for increasing the participation of disabled workers in AI. This will probably require a change in attitudes to disabled people. This should include a rejection of deficit based medical model perspectives for social model approaches and increased understanding of the barriers experienced by disabled people and the need to overcome them. It should also include recognition of their right and ability to contribute to society and that for some disabled people this ability involves the potential to carry out AI and other technical research.

Acknowledgements. I would like to thank our funders, UKRI with support from the British Academy, under grant reference ES/X008444/1 and Fenella Watson of EDICa for helpful comments and suggestions.

References

1. Almufareh, M.F., Kausar, S., Humayun, M., Tehsin, S.: A conceptual model for inclusive technology: advancing disability inclusion through artificial intelligence. J. Disabil. Res. **3**(1), 20230060 (2024)
2. Shuford, J.: Contribution of artificial Intelligence in improving accessibility for individuals with disabilities. J. Knowl. Learn. Sci. Tech. **2**(2), 421–433 (2023)
3. Özbilgin, M.F., Erbil, C., Gundogdu, N.: Developing artificial intelligence (AI) with equality, diversity, and Inclusion (EDI): towards interdisciplinary co-design with scholars and practitioners of EDI and AI. In: AI and Diversity in a Datafied World of Work: Will the Future of Work be Inclusive?, pp. 249–265 Emerald Publishing Limited (2025)
4. Hartmann, K.V., Primc, N., Rubeis, G.: Lost in translation? Conceptions of privacy and independence in the technical development of AI-based AAL. Med. Health Care Philos. **26**(1), 99–110 (2023)
5. Stahl, B.C., Wright, D.: Ethics and privacy in AI and big data: implementing responsible research and innovation. IEEE Secur. Priv. **6**(3), 26–33 (2018)
6. Guo, A., Kamar, E., Vaughan, J.W., Wallach, H., Morris, M.R.: Toward fairness in AI for people with disabilities SBG@ a research roadmap. ACM SIGACCESS Accessibility and Cmputing **2**(125), 1 (2020)

7. de Hond, A.A., van Buchem, M.M., Hernandez-Boussard, T.: Picture a data scientist: a call to action for increasing diversity, equity, and inclusion in the age of AI. J. Am. Med. Inform. Assoc. **29**(12), 2178–2181 (2022)
8. Crawford, K., Dobbe, R., Dryer, T., Fried, G., Green, B., Kaziunas, E., et al.: Disability, bias, and AI. AI Now 2019 Report. https://ainowinstitute.org/wp-content/uploads/2023/04/AI_Now_2019_Report.pdf. Accessed 14 May 2025
9. Obermeyer, Z., Powers, B., Vogeli, C., Mullainathan, S.: Dissecting racial bias in an algorithm used to manage the health of populations. Science **366**(6464), 447–453 (2019)
10. Aleem, S., Ahmed, F.: Practicing equity diversity inclusion (EDI) in software development teams: a systematic literature survey. IEEE Access (2023)
11. Shams, R.A., Zowghi, D., Bano, M.: AI and the quest for diversity and inclusion: a systematic literature review. AI Ethics 1–28 (2023
12. El Morr, C., Kundi, B., Mobeen, F., Taleghani, S., El-Lahib, Y., Gorman, R.: AI and disability: a systematic scoping review. Health Inform. J. **30**(3), 14604582241285744 (2024)
13. Chapko, D., Andrés Pérez Rothstein, P.A., Emeh, L., Frumiento, P., Kennedy, D., McNicholas, D., et al.: Supporting remote survey data analysis by co-researchers with learning disabilities through inclusive and creative practices and data science approaches. In: Proceedings of the ACM Designing Interactive System Conference, pp. 1668–1681 (2021)
14. Monteleone, R.: Forgotten publics: considering disabled perspectives in responsible research and innovation. J. Responsib. Innov. **7**(sup1), 84–91 (2020)
15. Newman-Griffis, D., Swenor, B., Valdez, R., Mason, G.: Disability data futures: achievable imaginaries for AI and disability data justice. arXiv preprint arXiv:2411.03885 (2024)
16. Buyl, M., Cociancig, C., Frattone, C., Roekens, N.: Tackling algorithmic disability discrimination in the hiring process. In: Proceedings of the 2022 ACM Conference on Fairness, Accountability, and Transparency, pp. 1071–1082 (2022)
17. Gorman, R.: Disability data justice from the ground up: a practice-led, participatory co-design approach to building an AI search engine and data repository for local, national, and transnational disability organizations. Crit. Stud. Int. Interdiscip. J. **18**(1), 28–43 (2024)
18. Park, J.S., Bragg, D., Kamar, E., Morris, M.R.: Designing an online infrastructure for collecting AI data from people with disabilities. In: Proceedings of the 2021 ACM Conference on Fairness, Accountability, and Transparency, pp. 52–63 (2021)
19. Lillywhite, A., Wolbring, G.: Coverage of ethics within the artificial intelligence and machine learning academic literature: the case of disabled people. Assist. Technol. **37** (2021)
20. Silvera-Tawil, D., Higgins, L., Packer, K., Bayor, A.A., Walker, J.G., Li, J., et al.: AI-enabled AT framework: a principles-based approach to emerging assistive technology. Disabil. Rehabil. Assist. Technol. **20**(1), 1–20 (2025)
21. Smith, J.M., McGill, M.M., Decker, A.: an examination of empirical evidence produced by a decade of K-12 computer science education research. IEEE Front. (2023)
22. Ashford-Hanserd, S., Singh, S., Muoneke, A., Eaglin, P.: Teachers' perceptions of barriers influencing African American and Hispanic students with disabilities' participation in k-12 computer science education. In: 2021 Conference on Research in Equitable and Sustained Participation in Engineering, Computing, and Technology (RESPECT), pp. 1–4 (2021)
23. About the British Sign Language Glossaries of Curriculum Terms. https://www.ssc.education.ed.ac.uk/BSL/about.html. Accessed 14 May 2025
24. Fachal, A.S., Abásolo, M.J., Sanz, C.V.: Dictionary of computer terms in LSA with operational signs proposed by and for hearing-impaired students. IEEE Revista Iberoamericana de Tecnologias del Aprendizaje **16**(4), 372–381 (2021)
25. Distante, D., Huang, S.: Challenges and lessons learned in teaching software engineering and programming to hearing-impaired students. In: 20th Conference on Software Engineering Education & Training (CSEET'07), pp. 344–354 (2007)

26. Wille, S., Century, J., Pike, M.: Computer science principles (CSP) and students with learning differences. In: 2016 Research on Equity and Sustained Participation in Engineering, Computing, and Technology (RESPECT), pp. 1–8 (2016)
27. Boye, T.: Accessibility of work-integrated learning in engineering, IT, and computer science for students with disabilities. In: 2022 Conference on Research in Equitable and Sustained Participation in Engineering, Computing, and Technology (RESPECT), pp. 109–113 (2022)
28. Ladner, R.E., Stefik, A., Blaser, B.: Addressing disability in CS for all. In: 2019 Research on Equity and Sustained Participation in Engineering, Computing, and Technology (RESPECT), p. 1 (2019)

Comparative Study on Gen-AI Models to Write Alt Text

Milena Risi[1,2](✉)

[1] University of Rome "Tor Vergata", Rome, 00173 Rome, Italy
milena.risi@students.uniroma2.eu
[2] Zanichelli Publisher, Irnerio. 34, 40126 Bologna, Italy

Abstract. Every day, an educational publishing house produces hundreds of thousands of images for e-books. Ensuring that these images are accessible to all students, including those with visual impairments, is not only an educational necessity but also a legal obligation under the European Accessibility Act (EAA). This paper presents an experiment that integrates Artificial Intelligence (AI) into the automatic generation of alternative text (alt text) for editorial images. The AI system is capable of analyzing visual content, interpreting its meaning, and generating coherent, context-aware alt texts that contribute to a more inclusive learning experience. Through a comparative analysis of various AI models, this paper highlights key benefits such as scalability, speed, and consistency in the production of alt text—without compromising semantic quality. However, the study also emphasizes the importance of maintaining a human in the loop approach. Human oversight remains essential. AI can streamline the process, but human expertise is critical to achieving truly accessible and meaningful educational content. The experiment thus represents a replicable best practice, supporting publishers in both regulatory compliance and social responsibility.

Keywords: Alt Text · AI · Educational Accessible Publishing

1 Introduction

With the growing availability of generative artificial intelligence (Gen-AI) models and the increasing attention surrounding their capabilities, this study investigates the potential of AI as a tool for enhancing accessibility in educational publishing, with a specific focus on the automatic generation of alternative text (alt text) for images.

Zanichelli, an Italian educational publishing house committed to digital accessibility for individuals with disabilities, has been exploring various applications of AI technologies. This article presents a pilot experiment designed to evaluate the effectiveness of different Gen-AI models in generating alt text. Given that a single e-book can contain hundreds or even thousands of images, and considering the high number of titles published annually, the total volume of images requiring alt text is substantial.

The study compares the performance of multiple Gen-AI models in producing accurate and contextually appropriate alt text. The research builds on a prior study that

K. Mavrou and P. Encarnação (Eds.): AAATE 2025, *Technology for Inclusion and Participation for All: Recent Achievements and Future Directions*, pp. 217–223, 2025.
https://doi.org/10.1007/978-3-032-01628-7_27

established a framework for evaluation, including a set of assessment criteria and defined image categories [1]. Based on these foundations, several models were tested under varying parameters to optimize output quality and identify the most promising approaches for practical implementation in educational publishing.

2 Scaling Accessibility: A Comparative Study of AI Models for Alt Text Generation

The AI models evaluated for alt text generation are all multimodal, meaning they are capable of processing multiple types of data simultaneously, including text, images, audio, and video. This capability allows them to understand and generate accurate, contextually relevant descriptions for visual content, thereby enhancing accessibility and interaction with various media formats [2]. The models assessed in this study include GPT-4, GPT-4s, GPT-4 with Vision (GPT-4V), Gemini 1.0, GPT-4o, and Claude.

2.1 Prompt Structure

Prompt was multi-modal [3], based on three main components:

IMAGE: The image from which the alt text needed to be generated was provided.

CONTEXT: Its context, typically the surrounding text of the image, including the preceding paragraph and caption.

INSTRUCTIONS: Finally, instructions were given to address the previously mentioned evaluation criteria [1] (e.g., identify detail from general to specific, identify image category and subcategory, etc.).

2.2 First Phase: Three AI Models Tested

The first study compares three AI models—Gemini 1.0, GPT-4, and GPT-4S—in their ability to generate alt text and long descriptions, following specific instructions.

Experimental Setup: The experiment was conducted by evaluating the models with identical prompts. For GPT-S, which had the same prompt but was provided with external resources to leverage during the process, including information on how to write alt text or guidelines for creating alt text. In this first step, no parameters (top p, frequency penalty, top k or temperature) were modified to generate alt text.

The prompt provided to each model was based on guidelines [1] inspired by accessibility standards [4–8] elaborated based on the criteria of the WCAG 2.2 [9]. The context was derived from e-books across six subjects—three STEM fields (Chemistry, Physics, Mathematics) and three humanities subjects (Geography, Art, English). For the last experiment the comparative study had also other subjects like Italian Literature, History and Science).

Results. This first study demonstrates the varying capabilities of AI models in automating alt text generation for educational purposes. GPT-S and GPT-4 offer robust frameworks for generating accessible content, but both models have areas where further refinement is needed—particularly in the consistency of image categorization and context-sensitive learning. As these technologies evolve, their potential to revolutionize content accessibility will only grow, particularly in the educational sector where clarity, accuracy, and cognitive load management are paramount. Gemini 1.0 tended to be overly verbose, often including background information or contextual details that were not requested. This sometimes detracted from clarity and efficiency, especially in tasks requiring concise output.

2.3 Second Phase: Error Identification and Solution Development

This phase compared two models—, GPT-4V, and GPT-S—, it is concentrated on common errors and explored the use of secondary prompts to address these issues. The structure was identical to the previous study: the input was the same for both models, and as before, it included context, image, and updated instructions. In this phase, the objective was to generate a secondary prompt to resolve problems identified in the initial output. For instance, if an alt text was generated with an incorrect image category, the second prompt was designed to correct it. The same approach was applied to issues such as character count, repetition of information, and other relevant concerns. Common issues observed include:

Alt Text Length: Descriptions occasionally exceeded or fell short of the required character limits, affecting clarity and compliance [1, 4–8]. With the second prompt, outputs became more consistent in length and better aligned with character constraints, though occasional deviations still occurred depending on image complexity.

Information Repetition: The models often produced redundant content, repeating the same idea or echoing context already provided outside the image [1, 4–8]. The second prompt helped reduce unnecessary repetition by narrowing the focus of the description, but full elimination of redundancy was not always achieved.

Exercise Misinterpretation: In educational images, particularly those embedded in exercises, the models frequently attempted to solve the exercise rather than simply describe the visual. This was often triggered by surrounding text containing instructional cues such as "solve" or "find," leading to unintended answers being included in the alt text [1]. The second prompt, which emphasized instructions like "do not solve the exercise," partially mitigated this behavior. However, confusion still occurred, especially when the prompt was too dense or when the task structure was ambiguous, indicating the need for clearer task separation.

Graph Interpretation: Models struggled to accurately describe visual data in charts and graphs [10], often missing key elements such as axis labels, trends, or numeric values, and sometimes hallucinating data. Even with a second prompt explicitly requesting structured, detailed descriptions, both models continued to produce unreliable outputs, suggesting a deeper limitation in interpreting structured visual information.

Comics: GPT-4 and GPT-4S showed difficulty in interpreting comic-style images, especially in correctly identifying speakers and interlocutors. While they understood the need to translate speech bubbles, they often failed to accurately represent the scene or dialogue structure—an essential requirement in educational comics focused on language learning. The second prompt improved scene coherence slightly, but outputs still lacked sufficient clarity to be fully usable.

Face Recognition: Neither GPT-4 nor GPT-S could reliably detect or describe faces. Even with a follow-up prompt, the models remained unable to identify facial features or individuals, likely due to built-in restrictions aimed at protecting user privacy.

Spelling and Punctuation: Both models frequently missed transcribing textual details such as blank spaces, underscores, or special characters [1], especially in visually dense images with overlapping text. The second prompt, while more targeted, only partially improved performance, indicating persistent limitations in handling fine-grained visual text elements.

Across categories—exercise misinterpretation, graphs, comics, face recognition, and textual precision—the second prompt provided partial improvements but did not fully resolve the issues. In complex or ambiguous cases, models continued to produce inaccurate or incomplete outputs. Moreover, long or instruction-heavy prompts often led to overlooked directives, highlighting the current limits of prompt engineering and pointing to deeper architectural or training-related constraints in multimodal models.

2.4 Third Phase: Contrast and Saturation

The hypothesis was that increasing contrast and reducing saturation would make key elements—such as axes and data points—more distinguishable. This analysis focused primarily on charts and graphs, where visual clarity is critical for accurate interpretation. Despite these adjustments, the results remained suboptimal. The models showed only slight improvements in identifying key elements, and the time spent on pre-processing did not yield enough benefit to justify the effort. This aligns with findings from the study "Vision Models Are Blind" (10), which highlights that current AI models trained to interpret images still lack the precision needed for tasks involving numerical or spatial recognition.

2.5 Fourth Phase: Testing GPT-4o Model and Parameter Tuning

In the forth phase, GPT-4V and GPT-4o were tested by varying the top-p parameter, with values set to 0.3 and 0.8, while other parameters—such as frequency penalty and temperature—were kept at zero. This setup enabled a focused evaluation of how changes in the top-p value influenced the models' performance in image analysis and text generation tasks. By isolating the top-p parameter, the experiment aimed to assess its effect on text variability, minimizing interference from factors like repetition or novelty in the generation process.

Enhanced Prompt: From the first to the forth phase, the prompt has undoubtedly evolved. The techniques of prompt engineering have been refined, enhancing the understanding of the model's behavior and increasingly guiding the study towards the targeted objective with greater precision. Consequently, the instructions in the forth phase exhibit a higher degree of detail compared to previous iterations, allowing for the identification of distinct blocks. For the forth phase, the structure of the instructions was divided into:

Role Identity: Defining the model as a generator of clear and concise alt text.

Instructions (Dos & Don'ts): Guidelines to ensure adherence to specific criteria and avoid common pitfalls [1, 4–8].

Output Format: Standardization of the result to maintain consistency and clarity.

Role and Objective Reminder: At the end, reinforcing the model's role and the goal of the request.

This structured approach aligns with established practices in AI and machine learning literature. For example, Radford et al. [11] document the significance of role definition in enhancing model performance, emphasizing the benefits of clear role assignment for achieving specific outputs. Furthermore, research by Brown et al. [12] highlights the efficacy of structured prompts in guiding models to produce relevant and accurate results. These studies support the effectiveness of clearly defined roles, specific instructions, and standardized output formats in optimizing AI-generated responses.

Results: GPT-4o is optimized for general multimodal tasks and human-computer interaction, and it benefits from a lower top-p value (0.3). This allows the model to generate more precise and focused responses, maintaining a good level of consistency and avoiding overly broad explorations that might become too generic or scattered. A lower top-p helps GPT-4o provide more accurate and targeted descriptions, optimizing its ability to handle a variety of tasks and interactions without getting lost in unnecessary details. Conversely, GPT-4V, designed to handle multimodal tasks involving both text and images, benefits from a higher top-p value (0.8). It is important to note that GPT-4o with a top-p value of 0.3 has demonstrated superior performance compared to GPT-4V with a top-p of 0.8.

2.6 Fifth Phase: Testing Claude and Parameter Tuning

Having observed that lowering the top-p value yields better results, we further reduced it and also tested Claude. When comparing GPT-4o with Claude using a fixed top-p of 0.1, Claude provided detailed and nuanced alt text, often demonstrating a solid understanding of image structure. However, it tended to include excessive details, such as repeating obvious information and adding unnecessary summary sentences. This verbosity affected its overall effectiveness. Additionally, instances of hallucinations—incorrect or fabricated information—were observed.

A critical point with this result was that the prompt, designed and evolved with and for GPT models, was based on the functioning of GPT models due to prior experiments. As a result, it did not perform as effectively with Claude, suggesting the need for model-specific prompt adjustments.

3 Necessity of Human Intervention

The assertion that GPT-4o represents progress towards AGI does not imply that it is flawless in its capabilities or that it can replicate the full range of human abilities, at least for the moment. While it proves very useful in improving accessibility to editorial content, technology cannot be relied upon exclusively. Despite numerous advancements, certain risks such as hallucinations, biases, or inaccuracies remain, and the delivery of accurate and truthful information continues to be a top priority. In particular, educational content requires special attention and cannot be fully delegated to AI models.

AI models are capable of handling various tasks, such as generating alt text and processing large volumes of data. While these models can quickly describe images, they are not without flaws. In fact, human intervention may be needed to correct imperfections produced by AI. Techniques like Retrieval-Augmented Generation (RAG) have shown promising results in improving the quality of AI-generated outputs [13], particularly in complex tasks that require significant cognitive effort. And also, AI may struggle to distinguish between decorative and informative images [1], or in identifying subtle nuances to describe, such as in the case of complex descriptions. In this regard, reviewer feedback is critical to validate the alt text generated, ensuring that it is both accurate and relevant.

4 Conclusion

Model performance varied significantly across the board. GPT-S excelled at integrating external resources but struggled with image categorization [1]. GPT-4V and GPT-4o produced clear and coherent descriptions, though they occasionally omitted important details [1]. Gemini demonstrated strong multimodal capabilities but tended to generate verbose and sometimes off-topic descriptions. Finally, Claude and GPT-4 delivered general descriptions but showed limitations when handling complex educational content.

The integration of AI into educational publishing, particularly through the generation of alt text, holds significant potential for improving accessibility, time efficiency, and cost-effectiveness. Adopting a "human-in-the-loop" approach is essential to validate AI-generated outputs and ensure their accuracy, even during the phase preceding alt text generation, when it is necessary to distinguish between informative and decorative images. Human feedback can greatly improve the quality of AI-generated alt text, reducing the risks of biases and errors. The importance of human oversight is clear: while AI can support many tasks, it cannot replace the nuanced understanding and expert judgment provided by human intervention.

Ultimately, AI offers tremendous potential, but its benefits can only be fully realized when combined with human intervention. Efforts to improve accessibility and equity in AI systems must be ongoing, ensuring that AI tools are not only effective but also fair for all users, including those with disabilities.

References

1. Risi, M.: Accessible Images: AI-Generated Alt Text in Educational Publishing. Logos: Journal of the World Book Community, pp. 49–68 (2025)
2. Chen, H., et al.: Multi-modal generative AI: multi-modal llm, diffusion and beyond, arXiv preprint arXiv:2409.14993, 1–23 (2024)
3. Guo, Z., Jin, T., Zhao, Z.: Multimodal prompt learning with missing modalities for sentiment analysis and emotion recognition. arXiv preprint arXiv:2407.05374, 1–11 (2024)

4. Ganner, J., Mrva-Montoya, A., Duncan, K., Park, M.: Books Without Barriers: A Practical Guide to Inclusive Publishing. Institute of Professional Editors and Australian Publishers Association, pp. 1–218 (2023)
5. Lewis, V.: How to Write Alt Text and Image Descriptions for the Visually Impaired. Perkins School for the Blind. https://www.perkins.org/resource/how-write-alt-text-and-image-descriptions-visually-impaired/. Accessed 22 Apr 2024
6. DIAGRAM Center: Image Description Guidelines. http://diagramcenter.org/table-of-contents-2.html. Accessed 24 Apr 2024
7. WebAIM: Techniques for Writing Alt Text. https://webaim.org/techniques/alttext/. Accessed 12 Jan 2024
8. De Gruyter: Accessible Manuscript Preparation. A Guide to Alt Text. https://degruyter-live-craftcms-assets.s3.amazonaws.com/assets/DeGruyter_Accessible_manuscript_preparation_2023.pdf. Accessed 02 Mar 2024
9. W3C: Web Content Accessibility Guidelines (WCAG) 2.2. https://www.w3.org/TR/WCAG22/. Accessed 13 Dec 2024
10. Rahmanzadehgervi, P., Bolton, L., Taesiri, M.R., Nguyen, A.T.: Vision language models are blind. In: Proceedings of the Asian Conference on Computer Vision, pp. 18–34 (2024)
11. Radford, A., Narasimhan, K., Salimans, T., Sutskever, I.: Improving language understanding by generative pre-training. OpenAI Blog 1–12 (2018)
12. Brown, T.B., et al.: Language Models are Few-Shot Learners. arXiv preprint arXiv:2005.14165, 1–75 (2020)
13. Lewis, P., et al.: Retrieval-augmented generation for knowledge-intensive NLP tasks. Adv. Neural. Inf. Process. Syst. **33**, 9459–9474 (2020)
14. O'Grady, E.: Why AI fairness conversations must include disabled people. Harvard Gazette. https://news.harvard.edu/gazette/story/2024/04/why-ai-fairness-conversations-must-include-disabled-people/. Accessed 22 Apr 2024

Evaluating Large Multimodal Models for Inclusive Mathematics Education: Addressing Dyslexia and Dyscalculia in Higher Education

Giorgia Nieddu[1], Branislav Gerazov[2], Silvio Marcello Pagliara[1](✉), Luigi Antioco Zurru[1], and Maria Cristina Carrisi[1,2]

[1] UNICA, 09127 Cagliari, Italy
silviom.pagliara@unica.it
[2] FEEIT, Ss Cyril and Methodius University in Skopje, 1000 Skopje, Republic of North Macedonia

Abstract. This study investigates the potential of Large Multimodal Models (LMMs) in generating educational materials tailored to students with learning disorders, specifically dyslexia and dyscalculia, in higher education mathematics courses. By analyzing outputs from ChatGPT o1, Llama 3.1, Phi4, DeepSeek r1, and Claude Sonnet 3.5, this research evaluates the effectiveness of AI-generated learning resources in maintaining both mathematical correctness and pedagogical usability. Using a teacher-centered prompting approach, we request model-generated adaptations for problem-solving lessons on limits and derivatives. Expert evaluators assess the outputs through a Likert scale to determine their alignment with inclusive teaching strategies. The findings provide insights into the strengths and limitations of LMMs in fostering accessible and pedagogically sound mathematics education while emphasizing the need for AI-human collaboration in instructional design.

Keywords: Learning Disorders · Mathematics Education · Generative AI · Dyslexia · Dyscalculia · Special Pedagogy

1 Introduction

The integration of Artificial Intelligence (AI) in higher education has introduced new possibilities for improving accessibility, particularly for students with learning disorders (LDs) such as dyslexia and dyscalculia. In mathematics education, these students face unique challenges that extend beyond traditional learning difficulties. Dyslexic learners often struggle with symbol recognition, equation structuring, and notation processing, while dyscalculic learners experience difficulties with number sense, sequential reasoning, and spatial organization of mathematical operations [1–5]. As AI-powered Large Multimodal Models (LMMs) gain traction in educational settings, their potential to generate adaptive and inclusive learning materials warrants systematic investigation. The

K. Mavrou and P. Encarnação (Eds.): AAATE 2025, *Technology for Inclusion and Participation for All: Recent Achievements and Future Directions*, pp. 224–231, 2025.
https://doi.org/10.1007/978-3-032-01628-7_28

Italian educational context provides a particularly relevant framework for this investigation, having embraced inclusive education principles for over four decades. Recent research has demonstrated a concordance between Italian school practices and Universal Design for Learning (UDL) principles, suggesting that principles of universality and inclusive educational design are deeply embedded in Italian school culture [6]. As Mura [6] notes, the Italian system has made commendable progress in promoting inclusion and catering to diverse student needs, with emphasis on recognizing each student's uniqueness and fostering collaborative learning environments. This cultural foundation creates an inclusive educational environment that serves as an important backdrop for evaluating AI-generated materials. Building on this tradition, Italy's National Plan on School Digital Literacy (PNSD) recognizes that educational innovation through digital technologies requires a reformulation of traditional teaching-learning methods to include pedagogical accessibility and inclusion. As D'Alonzo [7] emphasizes, differentiated instruction for inclusion requires methodological strategies and activities that recognize the heterogeneity of learners. This pedagogical tradition aligns with contemporary efforts to integrate technology into inclusive practices while acknowledging that areas requiring attention and improvement remain [6]. Despite their growing application, LMMs have not been explicitly designed for pedagogical adaptation, raising questions about their effectiveness in creating cognitively accessible and mathematically sound educational resources. Most AI-based learning tools focus on direct answer generation, potentially bypassing the cognitive engagement necessary for deep learning. However, when properly guided, AI can function as a pedagogical scaffold, assisting educators in adapting instructional materials to align with evidence-based learning strategies. As Pavone [8] notes, inclusive education requires pedagogical approaches that consider disability within a broader framework of individual differences and educational participation. This study builds on the pedagogical framework proposed in prior research on AI-enhanced inclusive STEM education by specifically evaluating the capabilities of five state-of-the-art LMMs in generating accessible mathematics content. Our research represents the first systematic investigation of how these AI systems can successfully create inclusive mathematics resources for higher education, addressing a critical gap in both the AI and inclusive education literature. The investigation focuses on a comparative analysis of model-generated educational materials for students with dyslexia and dyscalculia, examining whether these AI systems can successfully create inclusive mathematics resources that maintain the delicate balance between accessibility and mathematical rigor required in higher education contexts.

2 Methodology

This study employs an experimental approach grounded in inclusive pedagogical principles to assess the effectiveness of LMMs in generating mathematics content that aligns with inclusive teaching strategies. The research design draws inspiration from self-assessment frameworks used in inclusive education contexts, particularly the ENTELIS framework, which has proven valuable for evaluating digital inclusion practices in educational settings. The methodology is structured around a teacher-centered prompting method, reflecting the reality that educators remain central to the design and implementation of inclusive learning materials, even when assisted by AI technologies. The research framework acknowledges that effective inclusive education requires not merely the adaptation of existing materials, but a fundamental reconceptualization of how mathematical concepts are presented and explored. This aligns with the Italian pedagogical tradition of viewing inclusion as a process that benefits all learners, not just those with identified disabilities. In this context, the evaluation of AI-generated materials must consider both their technical accuracy and their potential to support diverse learning pathways. To ensure a rigorous evaluation, we focus on two key mathematical concepts: Limits and Derivatives. These topics were selected for their conceptual importance in higher education mathematics and the specific challenges they present to students with learning disorders. The abstract nature of limits and the symbolic manipulation required for derivatives create particular barriers for students with dyslexia and dyscalculia, making them ideal test cases for evaluating AI assistance in inclusive material generation. The experimental design involves structured prompts that simulate real-world instructional challenges educators face when designing inclusive lesson plans. Each prompt was carefully crafted to elicit responses that would reveal the models' understanding of both mathematical content and inclusive pedagogical principles. The queries request models to generate either new lesson content optimized for students with LDs or adaptations of existing educational materials to enhance accessibility for dyslexic and dyscalculic students. This dual approach allows for assessment of both creative generation and adaptive modification capabilities. Five LMMs were selected for evaluation based on their availability and reported capabilities: ChatGPT o1, Llama 3.1, Phi4, DeepSeek r1, and Claude Sonnet 3.5. Each model generates five outputs per topic, resulting in a total of 50 AI-generated resources. This sample size provides sufficient data for identifying patterns while remaining manageable for detailed expert evaluation. The prompting protocol emphasized pedagogical considerations, explicitly requesting adaptations that maintain mathematical rigor while enhancing accessibility. The evaluation framework employs a panel of five experts comprising three mathematics educators and two special education specialists. This interdisciplinary composition ensures that assessments consider both mathematical integrity and inclusive pedagogical effectiveness. Evaluators assess outputs based on two primary dimensions using a Likert scale ranging from 1 to 5. Mathematical correctness examines whether definitions and explanations are mathematically sound and whether analogies and simplifications maintain conceptual integrity. Usability for students with LDs evaluates whether content aligns with established pedagogical strategies for dyslexia and dyscalculia and whether modifications effectively improve accessibility without reducing cognitive engagement. The blind review process ensures objectivity, with raters assessing AI-generated materials independently before aggregating findings. This methodological approach reflects best practices in educational research while acknowledging the unique challenges of evaluating AI-generated content for inclusive education purposes. The integration of both quantitative ratings and qualitative analysis of specific examples provides a comprehensive understanding of model capabilities and limitations.

3 Results

The evaluation reveals notable variations across AI models in their ability to generate inclusive learning materials, with implications that extend beyond simple performance metrics to fundamental questions about the nature of inclusive mathematics education in the AI era. The initial quantitative analysis shows that the majority of AI-generated outputs maintain accurate mathematical definitions and procedural explanations, achieving an average correctness rating of 4.2 out of 5. However, this apparent success masks significant variations and concerning patterns that emerge upon closer examination. In terms of mathematical content, the analysis reveals critical gaps in how models approach formal mathematical concepts. Notably, only Llama 3.1 explicitly mentions mathematical properties such as linearity, while none of the models—including Llama—introduce the formal definition of a limit or even suggest that such formalization could be introduced progressively in the learning process. This absence represents a fundamental misunderstanding of mathematical pedagogy, where the journey from intuitive understanding to formal definition constitutes a crucial developmental pathway, particularly for students with learning disorders who benefit from structured, systematic approaches. The treatment of limit concepts reveals further pedagogical concerns. Most models work with limits that are not actually reached, presenting a conceptually problematic approach. The exception is Gemini, which explicitly highlights the distinction between a function's value and its limit in its learning objectives, demonstrating a more nuanced understanding of the conceptual challenges students face. However, all models introduce the term "limit" prematurely, before students have acquired any conceptual foundation, often using functions that are too complex or involve non-trivial indeterminate forms such as 0/0. This approach contradicts established principles of inclusive mathematics education, where conceptual understanding must precede formal manipulation. Regarding pedagogical usability, the effectiveness of AI-generated materials varies significantly across models and approaches. Claude Sonnet 3.5 and DeepSeek r1 demonstrate the highest alignment with inclusive teaching principles, incorporating clear step-by-step explanations, structured layouts, and multimodal representations. These features align with best practices for supporting dyslexic and dyscalculic learners, who benefit from multiple pathways to understanding and clear organizational structures. However, even these higher-performing models exhibit limitations in addressing core accessibility needs, particularly when text-heavy explanations overload dyslexic students or when numerical structuring lacks sufficient scaffolding for dyscalculic learners. The analysis of problem types and structures reveals a fundamental misunderstanding of what constitutes meaningful mathematical problem-solving in inclusive contexts. The proposed problems rarely represent true mathematical challenges; instead, they consist of loosely connected exercises that fail to engage students in authentic mathematical thinking. While all models claim to adapt problems by breaking them into smaller components, this adaptation proves largely superficial. Many outputs provide problem statements followed by implicit answers or introduce out-of-context functions that confuse rather than clarify. Only Gemini and Phi4 manage to provide at least one genuine problem each that requires student engagement and mathematical reasoning. The articulation of learning objectives across models reveals a concerning lack of pedagogical sophistication. Most models present generic objectives that could apply to any mathematical topic, failing to address the specific conceptual challenges of limits and derivatives for students with learning disorders. Llama 3.1 distinguishes itself by articulating more precise objectives, while Gemini provides superior pedagogical justification by explicitly referencing the Concrete-Pictorial-Abstract methodology, a well-established approach in special education mathematics. An interesting pattern emerges in how models differentiate between dyslexia and dyscalculia adaptations. Sequence-based problems appear more frequently in responses to dyscalculia prompts, with DeepSeek and Phi4 particularly favouring this approach, possibly reflecting an assumption that sequences are more manageable for students with number-processing difficulties. However, this assumption lacks empirical support and may reflect stereotypical rather than evidence-based understanding of dyscalculia. The use of real-life analogies represents another area of mixed success. While common across all models, these analogies often lack meaningful mathematical connections, serving more as superficial attempts at engagement rather than genuine conceptual bridges. More critically, none of the models effectively explore the teacher's role during learning activities, a significant omission given that inclusive education depends heavily on skilled educator mediation. The absence of guidance for teachers on how to facilitate activities, respond to student difficulties, or adapt materials in real-time represents a fundamental limitation in the models' understanding of inclusive pedagogy. Phi4's approach of creating dyslexia-specific adaptations, while well-intentioned, risks reducing classroom inclusivity by creating separate materials rather than universally accessible ones. This approach contradicts contemporary inclusive education principles that favor universal design over segregated adaptations. In contrast, models like Gemini that provide teacher tips for supporting diverse learners without creating differentiated materials better align with inclusive pedagogical principles. The quality disparity between outputs for dyslexia versus other conditions suggests that models may have been trained on datasets with uneven representation of different learning disorders. Outputs targeting dyslexia show generally lower quality and less sophisticated adaptations, possibly reflecting broader societal misconceptions about this condition. Additionally, occasional mathematical inaccuracies observed across several models raise concerns about reliability, particularly given that students with learning disorders may have greater difficulty identifying and correcting such errors independently.

4 Discussion

The findings highlight both the potential and current limitations of AI in generating accessible mathematics materials, revealing a complex landscape where technological capability intersects with pedagogical understanding in ways that challenge our assumptions about both AI and inclusive education. While LMMs demonstrate competence as potential assistive tools, their effectiveness depends critically on careful human guidance to ensure alignment with inclusive learning strategies. This dependency reflects deeper questions about the nature of mathematical knowledge, pedagogical expertise, and the irreducibly human aspects of inclusive education. From a pedagogical standpoint, the results reinforce that AI should not and cannot replace the cognitive effort required for learning but must function as a carefully designed scaffold. The distinction between models that simply provide direct solutions and those that emphasize structured problem-solving, guided reasoning, and multimodal adaptations proves crucial for supporting students with dyslexia and dyscalculia. This distinction aligns with D'Alonzo's [7] emphasis on differentiation that maintains high expectations while providing necessary support, a balance that current AI models struggle to achieve consistently. The findings resonate with the established concordance between Italian educational practices and UDL principles, as identified by Mura [6]. The Italian tradition of recognizing student uniqueness and fostering collaborative learning environments provides a valuable framework for evaluating AI contributions to inclusive education. However, the AI models' limitations in capturing this nuanced understanding suggest that while Italian schools have successfully internalized inclusive principles through decades of practice, translating this cultural knowledge into AI systems remains challenging. The gap between Italian inclusive education culture and AI-generated materials underscores that inclusive education emerges from complex historical, cultural, and pedagogical processes that cannot be easily codified into algorithmic responses. The universal absence of formal mathematical definitions across all models represents more than a technical limitation; it reveals a fundamental misunderstanding of mathematical learning trajectories. In higher education mathematics, the progression from intuitive understanding to formal definition constitutes not merely a pedagogical convention but a crucial cognitive journey. For students with learning disorders, who often benefit from explicit, structured approaches, this progression provides essential scaffolding. The models' failure to recognize or support this progression suggests they lack deep understanding of mathematical epistemology and pedagogy. The superficial nature of problem decomposition observed across most models indicates that current LMMs lack deep comprehension of the cognitive processes involved in mathematical problem-solving for students with learning disorders. True adaptation requires more than breaking problems into steps; it demands reconceptualization of how mathematical relationships are presented, explored, and internalized. The rare instances of genuine problem-solving opportunities underscore this limitation while pointing toward what more sophisticated AI support might achieve. The variation in pedagogical effectiveness between models suggests that training data and model architecture significantly influence educational output quality. Models performing well, such as Claude Sonnet 3.5 and DeepSeek r1, appear to better integrate multimodal representations and structured explanations, key elements identified in special education mathematics research. However, even these higher-performing models fail to address the dynamic, responsive nature of inclusive teaching, where educators continuously adjust their approaches based on student responses and emerging understanding. The absence of teacher guidance in most outputs represents a critical oversight that reflects a fundamental misunderstanding of inclusive education. As Pavone [8] emphasizes, inclusion is not achieved through materials alone but through the complex interplay of resources, pedagogical approaches, and skilled educator mediation. The models' failure to acknowledge or support this educator role suggests they conceptualize inclusive education as a product rather than a process, a static adaptation rather than a dynamic, responsive practice. These results must be understood within the broader context of AI development and educational practice. Current LMMs were not designed specifically for educational purposes, let alone for the nuanced requirements of inclusive mathematics education. Their training data likely contains limited examples of high-quality inclusive educational materials, and their optimization targets general language tasks rather than pedagogical effectiveness. These limitations suggest that significant improvements will require not just larger models or more data, but fundamental reconceptualization of how AI systems can support educational processes. The implications extend beyond technical considerations to ethical and professional questions about the role of AI in inclusive education. If AI systems consistently produce materials that appear inclusive but lack deep pedagogical grounding, they risk undermining rather than supporting inclusive education efforts. Educators using such materials without sufficient expertise to evaluate and modify them might inadvertently provide inadequate support to students with learning disorders. This risk underscores the importance of maintaining AI as a tool for augmenting rather than replacing human expertise in inclusive education.

5 Conclusion

This study provides the first systematic evaluation of LMMs as tools for generating mathematics content tailored to students with dyslexia and dyscalculia in higher education. The investigation reveals that while AI models demonstrate mathematical accuracy and attempt inclusive adaptations, they require substantial human oversight to ensure genuine accessibility and pedagogical effectiveness. Current-generation LMMs show technical competence but struggle with the deeper pedagogical adaptations necessary for truly inclusive education, particularly in their universal failure to introduce formal mathematical definitions, superficial problem adaptations, and absence of teacher guidance. The contrast between Italy's deeply embedded inclusive education culture [6] and the AI models' surface-level adaptations highlights the challenge of translating decades of pedagogical wisdom into computational systems. While Claude Sonnet 3.5 and DeepSeek r1 show superior alignment with inclusive principles, even these models fail to capture the dynamic, responsive nature of inclusive teaching that characterizes successful educational practice. Future research should focus on developing specialized training datasets with high-quality inclusive materials, creating hybrid systems that combine AI generation with expert knowledge bases, and conducting user studies with students with learning disorders. Standardized benchmarks for evaluating AI-generated inclusive content would enable systematic improvement of these systems. For practice, educators should view AI-generated materials as initial drafts requiring substantial revision, supplement outputs with formal definitions and teacher guides, and prioritize models demonstrating stronger pedagogical alignment. The focus must remain on fostering genuine mathematical problem-solving rather than procedural exercises, ensuring accessibility enhances rather than diminishes cognitive engagement. This research contributes to broader conversations about technology's role in equitable education. As AI systems become prevalent in educational contexts, understanding their capabilities and limitations in supporting diverse learners becomes crucial. While AI represents a powerful tool for educational innovation, its effective implementation requires careful integration with human expertise and pedagogical wisdom, building upon established foundations of inclusive practice. AI should be understood not as a solution to be implemented but as a tool to be wielded wisely in service of genuinely inclusive education.

Acknowledgement. As concerns dr. Silvio Marcello Pagliara, this work was produced during his research activity under financial support under the National Recovery and Resilience Plan (NRRP), Mission 4 Component 2 Investment 1.5 - Call for tender No.3277 published on December 30, 2021 by the Italian Ministry of University and Research (MUR) funded by the European Union – NextGenerationEU. Project Code ECS0000038 – Project Title eINS Ecosystem of Innovation for Next Generation Sardinia – CUP F53C22000430001 - Grant Assignment Decree No. 1056 adopted on June 23, 2022 by the Italian Ministry of Ministry of University and Research (MUR).

References

1. Lee, U., et al.: Few-shot is enough: exploring ChatGPT prompt engineering method for automatic question generation in English education. Educ. Inf. Technol. **29**, 11483–11515 (2024). https://doi.org/10.1007/s10639-023-12249-8
2. Shalev, R.S.: Developmental dyscalculia. J. Child Neurol. **19**, 765–771 (2004). https://doi.org/10.1177/08830738040190100601
3. Trott, C.: Dyscalculia. In: Neurodiversity in Higher Education: Positive Responses to Specific Learning Differences, pp. 125–148. Wiley (2009)
4. Trott, C.: Dyscalculia in higher education: systems, support and student strategies. In: Chinn, S. (ed.) The Routledge International Handbook of Dyscalculia and Mathematical Learning Difficulties, pp. 406–419. Routledge (2015)
5. Perkin, G., Croft, T.: The dyslexic student and mathematics in higher education. Dyslexia **13**, 193–210 (2007). https://doi.org/10.1002/dys.334
6. Mura, A.: Diversità e inclusione. Prospettive di cittadinanza tra processi storico-culturali e questioni aperte. FrancoAngeli, Milano (2016)
7. D'Alonzo, L.: La differenziazione didattica per l'inclusione. Metodi, strategie, attività. Erickson, Trento (2016)
8. Pavone, M.: L'inclusione educativa. Indicazioni pedagogiche per la disabilità, Erickson, Trento (2014)

Assistive Technologies Outcomes and Impact Measurement

Satisfaction with and Perceived Effectiveness of Assistive Technologies: Experiences of People with Disabilities and Older Adults in the Hague

Tsjitske M. Haanstra(✉) and Luc de Witte

Research Group Technology for Healthcare, Center of Expertise Health Innovation, The Hague University of Applied Sciences, The Hague, The Netherlands
t.m.haanstra@hhs.nl

Abstract. Despite the increasing global demand for assistive technologies (AT), many individuals still face significant obstacles in accessing appropriate devices and services. This study investigates the experiences of people with disabilities and older adults in The Hague, the Netherlands, regarding access to and use of AT. Using a cross-sectional survey (n = 192), we examined satisfaction, perceived effectiveness, and barriers in obtaining AT. Results show that while most users perceive AT as beneficial for their daily functioning and quality of life, substantial challenges persist. These include bureaucratic application procedures, limited empathy and coordination among professionals, and unclear or insufficient reimbursement. Notably, 55% of participants reported having paid for at least one device themselves. Logistic regression analyses revealed that higher income, cohabitation, and having communication or mobility limitations were positively associated with greater satisfaction and perceived effectiveness, whereas cognitive problems were negatively associated. The findings suggest that socioeconomic and demographic factors strongly influence AT outcomes. To ensure equitable access, we recommend simplifying provision procedures, improving interprofessional collaboration, and involving users more directly in system design and decision-making. A more inclusive, user-centered approach is essential to enhance the accessibility and impact of assistive technologies.

Keywords: assistive technology · functional limitations · user experience · satisfaction · inclusive care · accessibility · provision process

1 Introduction

The global demand for assistive products is rapidly increasing. According to a large-scale study conducted by the World Health Organization (WHO) and UNICEF (2022) [1], an estimated 2.5 billion people currently require assistive products, a number projected to reach 3.5 billion by 2050. This trend is largely driven by population ageing and the rising prevalence of chronic conditions [2, 3]. As societies age and the burden of disease shifts, access to effective and appropriate assistive technology becomes increasingly essential to ensure independent functioning and participation in daily life.

K. Mavrou and P. Encarnação (Eds.): AAATE 2025, *Technology for Inclusion and Participation for All: Recent Achievements and Future Directions*, pp. 235–244, 2025.
https://doi.org/10.1007/978-3-032-01628-7_29

Assistive technology refers to the systems and services related to the provision of assistive products—devices and tools designed to maintain or improve an individual's functioning and independence, thereby enhancing overall well-being [1]. Examples include wheelchairs, hearing aids, spectacles, pill organizers, and memory aids. These products play a critical role in supporting autonomy, particularly among older adults and individuals with disabilities, by enabling mobility, communication, and the performance of daily activities [4–7].

Despite the recognized benefits of assistive products, access remains a challenge for many. Research has increasingly turned to the lived experiences of users, yet a significant gap persists in our understanding of how people navigate the processes of acquiring and using assistive products [8–11]. Existing studies suggest that individuals often face barriers such as limited information, complex application procedures, and restricted product availability. These challenges are shaped not only by system-level inefficiencies but also by social and economic factors, including language barriers, digital literacy, and affordability [12, 13]. In addition, the rapidly evolving landscape of digital assistive technologies introduces new forms of inequality for those who lack access to or knowledge of such innovations.

In the Dutch context, assistive product provision is embedded within a complex policy framework that spans municipal social support (Wmo), health insurance systems, and national disability programs. While this infrastructure aims to promote equity, evidence on its effectiveness in meeting users' needs—especially among socially and ethnically diverse groups—remains limited. Moreover, most studies to date focus on specific categories of users or devices, without addressing the broader diversity of assistive product users in urban environments.

This study aims to explore the experiences of individuals with disabilities and older adults in The Hague with regard to the acquisition and use of assistive products that support mobility and activities of daily living (ADL). In addition, we examine how users evaluate the assistive products they receive, particularly in terms of their satisfaction and the extent to which these products address the problems for which they were intended.

2 Methods

2.1 Study Design

This study employed a quantitative, cross-sectional survey design. It collected data on the experiences of older adults and individuals with functional limitations living in The Hague regarding the acquisition and use of assistive products.

2.2 Study Population

The study population consisted of adults (18 years and older) with functional limitations and/or older adults who currently use or express a need for assistive products. Participants were recruited through a range of local networks and organizations, including Voorall (a local advocacy group), municipal social support teams (Wmo), healthcare providers, assistive product suppliers, educational institutions, and local and social media platforms.

2.3 Procedure

Data collection took place between January and June 2024. Participants received information through invitation letters, flyers, or advertisements disseminated via the aforementioned channels. They completed a structured, anonymous questionnaire either online or on paper. For participants without internet access or with limited digital skills, a paper version was made available. In cases where additional support was needed, a face-to-face interview was offered. The questionnaire was preceded by an information and consent form explaining the purpose of the study and the use of the data.

2.4 Measurement Instruments

The questionnaire consisted of selected items from several validated instruments: the Rapid Assistive Technology Assessment Tool [14], the KWAZO questionnaire [15], the Individually Prioritised Problem Assessment [16], and the Dutch version of the Quebec User Evaluation of Satisfaction with Assistive Technology [17]. These instruments measure aspects such as functional limitations, satisfaction with the assistive product and the provision process, perceived needs, and the effectiveness of the product. To allow for more nuanced insights, several open-ended questions were included to capture participants' explanations in cases of dissatisfaction. In addition, the questionnaire collected background characteristics of the participants, the types of assistive products used, and the functional limitations they experienced in daily life. The draft version of the questionnaire was pilot-tested with several members of the target group to assess clarity and readability, resulting in minor revisions prior to full deployment.

2.5 Data Analysis

Descriptive statistics were used to summarize participant characteristics and distribution of key variables, including frequencies, means, and standard deviations. In addition to exploratory analysis, inferential statistical analyses were conducted to examine factors associated with satisfaction and perceived effectiveness of assistive products. Binary logistic regression analysis with backward elimination was used to identify predictors of satisfaction and perceived effectiveness. The initial model included all potential predictors: education level, income, gender, age, employment status, living situation, reimbursement or financing type, perceived limitations, and the total number of assistive products used. Variables were removed stepwise based on statistical significance until only the most relevant predictors remained. A significance threshold of $p < 0.05$ was used. Model performance was evaluated using the Nagelkerke R2, which indicates the proportion of variance explained by the model. All quantitative analyses were performed using IBM SPSS Statistics version 27. Responses to open-ended questions were analyzed thematically using Atlas.ti.

2.6 Ethical Considerations

The study protocol was submitted to the Medical Ethics Review Committee of Leiden, The Hague, Delft (METC LDD), which concluded that the study did not fall within the scope of the Dutch Medical Research Involving Human Subjects Act (WMO). All participants provided informed consent by completing a consent form prior to participation.

3 Results

3.1 Description of the Study Population

The study included a total of 192 participants: 129 individuals aged between 18 and 66, and 63 individuals aged 67 and older. Among these, 89% reported using at least one assistive product. The remaining 11% indicated a need for an assistive product but were not currently using one. These respondents did not complete the questions regarding the assistive product provision process, financing, product effectiveness, or satisfaction, and were therefore excluded from analyses addressing those topics. Table 1 presents the characteristics of the total study population.

Table 1. Characteristics of the study population

	Total population (n = 192)	<67 years (N = 129)	≥67 years (N = 63)
Sex			
% Female	59.4%	54.9%	66.7%
Employment*			
In paid employment	38.5%	55.8%	3.2%
Self-employed	7.8%	10.0%	3.2%
Volunteer activities	7.8%	6.2%	11.1%
Disability benefits	25.0%	20.9%	33.3%
Unemployed	1%	0.8%	0.2%
Economically inactive (including retirement)	38.5%	18.6%	79.4%
Yearly income			
- <€25.000	38.4%	38.5%	38.1%
- ≥€25.000	61.6%	61.5%	61.9%
Household composition			
- single person household	40.2%	32.5%	55.6%
- multi person household**	59.8%	67.5%	44.4%

* multiple answers possible.
** cohabiting with a partner/spouse, children, other relatives, friends or others.

3.2 Functional Limitations in Daily Life

Participants reported a wide range of functional limitations in their daily lives (Table 2). The most commonly reported limitation was difficulty with vision, followed by mobility and hearing limitations. Younger participants (<67 years) more frequently reported communication difficulties, whereas older participants (≥67 years) more often reported hearing and mobility issues. Limitations in memory and self-care were reported in both age groups, although with varying prevalence.

Table 2. Limitations of daily living experienced by participants

Item (% at least some difficulties)	Total population (n = 192)	<67 years old (N = 129)	≥67 years old (N = 63)
Cognitive function	35.4%	32.0%	42.9%
Mobility	60.9%	56.3%	69.8%
Communication	18.2%	21.9%	11.1%
Vision	70.3%	66.4%	77.8%
Hearing	37.5%	28.1%	57.1%
Selfcare	38.0%	35.9%	41.3%

3.3 Assistive Products, Acquisition and Financing

The most commonly used assistive products among participants were glasses, hearing aids, wheelchairs, rollators, and crutches. Assistive products were primarily used to support communication, leisure activities, household tasks, and work or study. Of the assistive products mentioned by participants, 55% were self-funded, 29% were covered by health insurance, and 20% were financed by the municipality. A small number of products were funded through the Dutch Employee Insurance Agency (UWV) or through other sources. Participants primarily obtained their assistive products from private medical aid suppliers (31%), opticians (26%), and audiologist (23%). Among those who received some form of reimbursement, 53% considered it sufficient, 21% considered it insufficient, and the remaining participants could not recall whether the reimbursement had been adequate.

3.4 Satisfaction with Assistive Products

To assess satisfaction participants completed the D-QUEST, which includes eight items measuring satisfaction with characteristics of the assistive device (Table 3), and four items measuring satisfaction with the service (Table 4). In general, participants reported being satisfied to very satisfied with their assistive product. The average total satisfaction score for the assistive product (D-QUEST) was 32 on a scale from 8 to 40. Detailed item-level responses are presented in Table 3. The average satisfaction score for the service delivery process was 16.5 on a scale from 4 to 20. As shown in Table 4, respondents were more likely to report being "somewhat satisfied," "not satisfied," or "very dissatisfied" with the provision process compared to their satisfaction with the product itself.

Table 3. D-Quest scores assistive device

Dquest – assistive device	% responding 'somewhat satisfied' or less	% responding 'quite satisfied' or 'very satisfied'
Dimensions	17.0%	83.0%
Weight	17.5%	82.5%
Adjustments	36.3%	63.7%
Safety	15.2%	84.8%
Durability	22.2%	77.8%
Simplicity of use	12.9%	87.1%
Comfort	16.4%	83.6%
Effectiveness	9.4%	90.6%
General satisfaction with the characteristics of the assistive device	11.7%	88.3%

Table 4. D-Quest scores services

Dquest - services	% responding 'somewhat satisfied' or less	% responding 'quite satisfied' or 'very satisfied'
Service delivery	35.7%	64.3%
Repairs and service of the device	42.7%	57.3%
Professionalism of service	28.7%	71.3%
Follow-up service	36.3%	63.7%
General satisfaction with services	31.0%	69.0%

3.5 Qualitative Insights into Dissatisfaction

Participants who expressed dissatisfaction with the assistive product or the provision process were invited to elaborate through an open-ended question. These qualitative responses were thematically analyzed and revealed four main themes:

1. **Process-related factors**: Participants reported long waiting times and excessive bureaucracy when applying for assistive products.

2. **Human factors**: Several respondents indicated that professionals involved in the provision process did not listen adequately, lacked empathy, or failed to coordinate effectively with each other.
3. **Errors**: Some assistive products did not meet the user's needs or were deemed inappropriate. In a few cases, participants attributed the mismatch to an incorrect medical diagnosis.
4. **Reimbursement and financing**: Many participants found the reimbursement system unclear or expressed dissatisfaction with the level of coverage received—or lack thereof.

3.6 Needs and Perceived Effectiveness

A total of 24% of participants indicated a need for additional or alternative assistive products beyond those they currently possessed. Regarding perceived effectiveness, 88% of participants considered their assistive product (very) effective in addressing the specific problems it was intended to solve. Additionally, 87% reported that the product helped them greatly or completely in resuming desired daily activities. Most respondents (94%) felt the product was worth the effort and cost, and 78% stated they could no longer do without it.

3.7 Factors Associated with Satisfaction and Effectiveness

Logistic regression analyses with backward selection identified several variables associated with satisfaction and perceived effectiveness. Perceived effectiveness in addressing the specific problems for which the assistive product was intended was positively associated with higher income, cohabitation, mobility and communication limitations, sufficient reimbursement, employment, and female gender. In contrast, cognitive limitations were negatively associated with effectiveness. This model explained 43% of the variance (Nagelkerke R2 = 0.430). Effectiveness in regaining desired activities was positively associated with communication limitations and cohabitation, explaining 23% of the variance (R2 = 0.230). Satisfaction with the assistive product was positively associated with cohabitation, the total number of assistive products used, and a positive experience during the application and delivery process. The model explained 14.5% of the variance (R2 = 0.145). Satisfaction with the provision process and service was associated with a positive application and delivery experience, hearing and memory limitations, cohabitation, and higher income. This model explained 50.2% of the variance (R2 = 0.502). These findings suggest that both socioeconomic resources and the presence of functional limitations are linked to greater perceived effectiveness and satisfaction with assistive products. In particular, cohabitation, higher income, and communication limitations emerged as consistent predictors across multiple models.

4 Conclusion, Discussion, and Recommendations

This study reveals that assistive devices are generally perceived as effective and that many users are satisfied with both the device itself and the service surrounding its provision. A majority of users indicated that their assistive device helped them perform daily activities and was worth the investment. However, there is still room for improvement—particularly in the provision process and service delivery. Open-ended responses indicate that long and bureaucratic procedures, lack of empathy and coordination among professionals, provision errors, and dissatisfaction or confusion about reimbursement are key drivers of dissatisfaction. Notably, 55% of users reported having paid for their device themselves, suggesting a gap in accessible funding mechanisms. Socioeconomic factors such as higher income and cohabitation, as well as certain types of functional limitations (e.g., communication impairments), were associated with greater satisfaction and perceived effectiveness. These findings point to structural inequalities that may affect access to and the impact of assistive technologies. These outcomes are consistent with previous research. For example, Danemayer et al. [13] found socioeconomic, social and demographic barriers to use mobility assistive products in the UK and De Witte and Van der Vaart [14] raise a call for action to address inequalities in access to AT.

This study reveals that while assistive technologies (AT) are generally perceived as effective and beneficial in supporting daily functioning and quality of life, significant challenges remain in the service provision process. Many users expressed satisfaction with the assistive device itself, yet dissatisfaction often stemmed from bureaucratic application procedures, poor interprofessional coordination, unclear or inadequate reimbursement, and a lack of empathy from service providers. Notably, 55% of participants reported having paid for at least one assistive device out of pocket, indicating gaps in the accessibility of funding mechanisms.

Socioeconomic factors such as higher income and cohabitation, as well as specific functional limitations like communication impairments, were positively associated with perceived effectiveness and satisfaction. Cognitive limitations were linked to lower satisfaction. These findings suggest inequalities in access to and benefit from assistive technologies, aligning with previous research. For instance, Danemayer et al. [13] identified demographic, social, and economic barriers that hinder access to mobility assistive products in the UK. Similarly, Federici et al. [18] emphasized the role of the service delivery process (SDP) in influencing satisfaction and abandonment, highlighting the need for user-centered provision systems.

Further support for these findings comes from a systematic review by Larsson Ranada and Lidström [19], which concluded that the service delivery process as a whole contributes to satisfaction with and usability of assistive devices in relation to achieving desired participation outcomes. The authors advocate a client-centred approach throughout the process, which was found to be a key factor in ensuring both effective provision and satisfied users.

Recommendations

Based on our findings, we propose the following recommendations to enhance the accessibility and effectiveness of assistive device provision:

1. **Simplify and streamline the provision process.** Current application procedures are often experienced as overly complex and time-consuming. A more user-friendly system could increase access to necessary devices.

2. **Improve communication and coordination.** Multidisciplinary collaboration between healthcare professionals, suppliers, and policymakers should be promoted to better align provision with user needs.
3. **Enhance training and awareness.** Professionals involved in the assessment and provision of assistive technologies should be trained in empathic communication and gain a better understanding of how assistive devices impact users' daily lives.
4. **Address socioeconomic disparities.** Since income and social support are associated with better outcomes, efforts should be made to reduce financial and structural barriers for users with fewer resources.
5. **Involve users in system redesign.** Including people with lived experience in the evaluation and development of provision policies could ensure that services are more responsive to actual needs.
6. **Move towards integrated access points.** A centralized and independent application office for assistive devices, regardless of the funding route (municipality, health insurer, or employer), could make the system more navigable.

Improving the provision system requires sustained collaboration among municipalities, health insurers, assistive device providers, and users themselves. Only through a coordinated, inclusive, and user-centered approach can we ensure that assistive technologies fulfill their intended function: enabling people to participate fully in society.

References

1. Global Report on Assistive Technology, 1st edn. World Health Organization, Geneva (2022)
2. Hwang, J., Kim, K.H., Hwang, J.G., Jun, S., Yu, J., Lee, C.: Technological opportunity analysis: assistive technology for blind and visually impaired people. Sustainability **12**(20), 8689 (2020). https://doi.org/10.3390/su12208689
3. Dahler, A.M., Rasmussen, D.M., Andersen, P.T.: Meanings and experiences of assistive technologies in everyday lives of older citizens: a meta-interpretive review. Disabil. Rehabil. Assist. Technol. **11**(8), 619–629 (2016). https://doi.org/10.3109/17483107.2016.1151950
4. Stumbo, N.J., Martin, J.K., Hedrick, B.N.: Assistive technology: Impact on education, employment, and independence of individuals with physical disabilities. J. Vocat. Rehabil. **30**(2), 99–110 (2009). https://doi.org/10.3233/JVR-2009-0456
5. Muncert, E.S., Bickford, S.A., Guzic, B.L., Demuth, B.R., Bapat, A.R., Roberts, J.B.: Enhancing the quality of life and preserving independence for target needs populations through integration of assistive technology devices. Telemed. E-Health **17**(6), 478–483 (2011). https://doi.org/10.1089/tmj.2010.0206
6. Kenigsberg, P.-A., et al.: Assistive technologies to address capabilities of people with dementia: from research to practice. Dementia **18**(4), 1568–1595 (2019). https://doi.org/10.1177/1471301217714093
7. Lauriks, S., Meiland, F., Osté, J.P., Hertogh, C., Dröes, R.-M.: Effects of Assistive Home Technology on quality of life and falls of people with dementia and job satisfaction of caregivers: Results from a pilot randomized controlled trial. Assist. Technol. **32**(5), 243–250 (2020). https://doi.org/10.1080/10400435.2018.1531952
8. Steel, E.J., Layton, N.A., Foster, M.M., Bennett, S.: Challenges of user-centred assistive technology provision in Australia: shopping without a prescription. Disabil. Rehabil. Assist. Technol. **11**(3), 235–240 (2016). https://doi.org/10.3109/17483107.2014.941953

9. Manship, S., Hatzidimitriadou, E., Moore, J., Stein, M., Towse, D., Smith, R.: The experiences and perceptions of health-care professionals regarding assistive technology training: A systematic review. Assist. Technol. **36**(2), 123–146 (2024). https://doi.org/10.1080/10400435.2023.2219294
10. Nijs, S., Maes, B.: Assistive technology for persons with profound intellectual disability: a European survey on attitudes and beliefs. Disabil. Rehabil. Assist. Technol. **16**(5), 497–504 (2021). https://doi.org/10.1080/17483107.2019.1668973
11. Wang, R.H., Zdaniuk, N., Durocher, E., Wilson, M.G.: Policymaker and stakeholder perspectives on access to assistive technologies in Canada: challenges and proposed solutions for enhancing equitable access. Disabil. Rehabil. Assist. Technol. **17**(1), 61–73 (2022). https://doi.org/10.1080/17483107.2020.1765033
12. De Witte, L., Van Der Vaart, R.: Inequities in access to assistive technology: a call for action. Lancet Public Health **10**(1), e4–e5 (2025). https://doi.org/10.1016/S2468-2667(24)00270-6
13. Danemayer, J., Bloomberg, M., Mills, A., Holloway, C., Hussein, S.: Demographic, socioeconomic, and social barriers to use of mobility assistive products: a multistate analysis of the English Longitudinal Study of Ageing. Lancet Public Health **10**(1), e20–e28 (2025). https://doi.org/10.1016/S2468-2667(24)00243-3
14. Pryor, W., Liem, N.: Measuring need and unmet need for assistive technology the rapid Assistive Technology Assessment (rATA) tool for national representative survey enumeration: a manual. WHO (n.d.). 20201021-rata-enumerator-manual-final.pdf (who.int)
15. Dijcks, B.P.J., Wessels, R.D., De Vlieger, S.L.M., Post, M.W.M.: KWAZO, a new instrument to assess the quality of service delivery in assistive technology provision. Disabil. Rehabil. **28**(15), 909–914 (2006). https://doi.org/10.1080/09638280500301527
16. Wessels, R., et al.: IPPA: individually prioritised problem assessment. Technol. Disabil. **14**(3), 141–145 (2002). https://doi.org/10.3233/TAD-2002-14310
17. Wessels, R.D., Witte, L.P.D.: Reliability and validity of the Dutch version of QUEST 2.0 with users of various types of assistive devices. Disabil. Rehabil. **25**(6), 267–272 (2003). https://doi.org/10.1080/0963828021000031197
18. Federici, S., Borsci, S.: Providing assistive technology in Italy: the perceived delivery process quality as affecting abandonment. Disabil. Rehabil. Assist. Technol. **11**(1), 22–31 (2016). https://doi.org/10.3109/17483107.2014.930191
19. Larsson Ranada, Å., Lidström, H.: Satisfaction with assistive technology device in relation to the service delivery process—a systematic review. Assist. Technol. **31**(2), 82–97 (2019). https://doi.org/10.1080/10400435.2017.1367737

Shades of Grey in the Use, Underuse and Non-use of Assistive Products: Exploring the Ways Practitioners and Researchers Articulate Assistive Product Utilization

Turið Barbara Jacobsen[1](✉) and Natasha Layton[2,3]

[1] University of the Faroe Islands, 100 Tórshavn, Faroe Islands
tubaja@gmail.com

[2] Rehabilitation, Ageing and Independent Living Research Centre, Monash University, Frankston, Australia

[3] Australian Rehabilitation and Assistive Technology Association (ARATA), Beaumaris, Australia

Abstract. How might we describe 'states of use' for technology (AT)? In the AT field, uptake of or adherence to assistive products and the opposite state, abandonment of assistive products, are considered key outcome metrics.

In this paper, two occupational therapists practicing and researching in very different contexts reflect upon definitions of assistive technology 'use'. This iterative method firstly draws together definitions sourced from the various literature reviews informing i) a study recently completed in the Faroe Islands, and ii) a study currently underway in Australia. The two study methods are then presented and contrasted. The results demonstrate the ways in which operational definitions of use and non-use influence data analysis. The discussion considers the international directives which can assist practitioners and researchers to use consistent and meaningful definitions to more fully understand AT use, underuse and non-use. In conclusion, delivering on WHO UNICEF Global Report on AT Recommendation 6 *Invest in data and evidence based policy* requires comparable evidence and is supported by studies designed according to agreed definitions and standardized terminology.

Keywords: Use · Non-use · Abandonment · Adherence · Assistive products

1 Introduction/Background

In the assistive technology field, uptake of/adherence to assistive products and the opposite state, abandonment of assistive products, are considered key outcome metrics. Governments are particularly interested in uptake and retention as markers of appropriate expenditure. While many policy contexts count 'outputs' in terms of assistive products and services provided, measuring the ongoing impact of these outputs is more complex [1].

K. Mavrou and P. Encarnação (Eds.): AAATE 2025, *Technology for Inclusion and Participation for All: Recent Achievements and Future Directions*, pp. 245–256, 2025.
https://doi.org/10.1007/978-3-032-01628-7_30

Scholarly writings over several decades have tackled how to define ‘abandonment’ and the literature outlined below comprises seven frequently cited studies pertaining to ‘abandonment’ between 1993–2019 were identified and their terminology and definitions considered. These were identified through separate literature reviews informing studies run by each author, as described further in the Methods section. Several early and influential studies offer varying definitions of abandonment and statistics on abandonment rates [2–4]. Subsequent studies interrogated these definitions, grading the extent of non-use [5] and offered contrasting interpretations of the extent of non-use [6]. Several large studies interrogate abandonment for entire populations within countries and social care contexts for example national health users in Italy [7, 8] and expanded discussion of the produce – user interface, including whether non-use and under-use is in fact negative as AT users may have functional changes which render the AT no longer appropriate or needed. Others have interrogated aspects of user – product – environment which impact abandonment, for example the landmark study by Sugawara et al. which identified a range of personal factors, intervention-related factors, product-related factors or environmental factors [9].

Beyond the AT field of research, the concept of patient nonadherence has been studied much more broadly, for example related to medication adherence. Tuazon et al. conducted a conceptual review of adherence to assistive device use among older adults using synonyms for acceptance, adherence and compliance [10].

A complexity of clinical factors may impact the extent of use of AT. As functional status changes, AT use may no longer be indicated or indeed may be contraindicated, as supported by the recently developed WHO International Classification of Health Interventions which includes health intervention *UAI.PJ.ZZ: Teaching a person to function after the removal of equipment, products and technologies* [11]. It is also true that people living with complex functioning limitations may be in the position of having AT which supports some outcomes but have unmet needs associated with other products [12].

People who do not have the opportunity to use AT are clearly not in a position to underuse or abandon it, and Danemeyer et al. constructed an indicator based on the assumption that having mobility limitations and not having an assistive device indicates an unmet need [13] using data from the English Longitudinal Study of Ageing. The consideration of unmet need as an additional metric is an important development and responds to the context of global AT supply [14] and the global policy focus upon current unmet need. Significantly the recent WHO UNICEF Global Report on AT which signposts the direction for AT research mentions the term abandonment five times, compared with 24 mentions of unmet need [15], indicating the importance of this metric. Figure 1 illustrates some of this changing terminology over time, drawing on Table 1, which lists operational definitions of abandonment or non-use across these seminal studies.

The authors will use examples of current research studies underway in the Faroe Islands and Australia to:

a) Illustrate the factors which impact abandonment based on Sugawara et al. [9]
b) Problematize and question whether these impacts of these factors are in fact abandonment, or may be viewed in other ‘shades of grey’
c) Explore a continuum of use/underuse/non-use

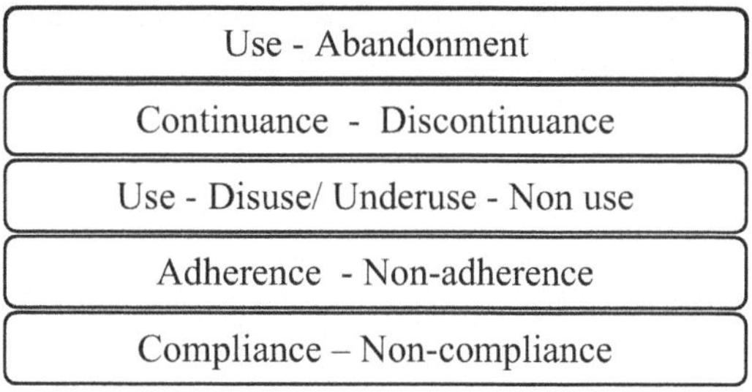

Fig. 1. Different terminology over time

Table 1. Operational definitions used in known studies.

Study	Operational definitions used
Phillips, B., & Zhao, H. (1993) [2]	• Abandonment defined as nonuse of a device type or category at the time of the survey
Riemer-Reiss & Wacker (2000) [3]	• AT continuance/discontinuance measured by 'are you still using your AT?'
Wessels, R., Dijcks, B., Soede, M., Gelderblom, G., & de Witte, L. (2003) [4]	• No use at all, no full-time use, non-use at the time of the study • No frequent use, low average use, no correct use, non-use for the activities it was prescribed for
Wielandt, McKenna,Tooth & Strong (2006) [5]	• Level of AT use using five-point Likert scale
Dijcks, Witte, Gelderblom, Wessels, Soede (2006) [6]	• Non-use or use, with use defined as: used regularly; ever used? • Actual use compared to expected use (more, about as much, less)
Federici & Borci (2016) [8] Federici, Meloni & Borsci (2016) [7]	• Abandonment as an umbrella term for any non-use (positive or negative) • AT disuse
Tuazon, Jahan & Jutai (2019) [10]	• Patient non-adherence/adherence (compliance)

2 Methods

We firstly summarize the two studies from which data is drawn and offer reflexive statements regarding the similarities and differences between them. We then present our data sources in the form of Table 2, which summarizes the methods used across both studies. For data analysis, Table 3 identifies the subset of data which will inform this current study, that is a) illuminate shades of grey in the use, underuse and non-use of assistive products; and b) explore the ways practitioners and researchers articulate assistive product utilization.

Study A: Author TBJ is an occupational therapist and Master in Health Science and guest researcher at the University of the Faroe Islands, Faculty of Health Sciences. In the Faroe Islands an AT center provides AT services for Faroe Islanders of all ages. The study was initiated by the researcher and funded by the Faroese Health Insurance Fund. Recruitment: the AT center informed that they could not provide contact with AP users due to General Data Protection Regulation (GDPR). Therefore, health practitioners who had contact with a AP users asked permission to give the researcher contact information, information about the study was posted on social media and placed in public places etc. resulting in 101 participants in total. This study aimed to investigate how the at service was organized by asking the AP users which institutions/professionals assessed; selected AP; and implemented; user satisfaction and use/non-use. All Faroese individuals using AP were included. A study-specific questionnaire administered face-to-face was used to collect data.

Study B: Author NL is an occupational therapist and senior research fellow with the Rehabilitation, Ageing and Independent Living Research Centre at Monash University, Australia. The Australian government has commissioned a 2 year longitudinal study (2024–2026) into the effectiveness of the National GEAT (goods, equipment and assistive technology) program, which provides assistive products to community dwelling older Australians. The sample size at baseline is N = 23,379. This study aimed to establish the longitudinal usage and effectiveness of assistive technology (AT) for older Australians by measuring functioning, health service utilization, quality of life, goal achievement, assistive technology (AT) useability and satisfaction as well as any differences by prescriber type. A survey questionnaire was conducted over the telephone at baseline, 3, 6, 12, 18 and 24 months. The subset of data which informs this paper is the 6 month data (n = 6,370) (27%).

2.1 Reflexivity

As authors we recognize the complexity of the AT ecosystem, envisioned as interrelationships between AT users (people), assistive products, personnel who may provide them, provision systems for access, and policy which underpins the health and community care/welfare systems within which we both, as occupational therapists, operate. Both studies were established independently. While both use survey and interview instruments both have used different outcome measures.

Our studies have some differences and some similarities. The physical, cultural and policy contexts are very different: we operate in different hemispheres, and the variety of regions in Australia (from dense urban to very remote) differs to Faroe Islands (more or less dense urban to smaller villages). In Australia, aged care services are separate to those for younger people utilizing AT and therefore the Australian study focusses on older persons only. In contrast the Faroese AT services are for all ages. Sample sizes are very different, and while one study (Australia) is commissioned by government and had access to government-collected information; the other study (Faroe Islands) was initiated by the researcher and was subject to privacy constraints. Faroese participants may have had their AT for some time when the survey was run, and the data on AT use by Australian participants was captured after 6 months of the provision date.

Our studies also had many similarities, aiming to understand AT use for people living in the community. Both studies are undertaking with ethical approval from universities, and investigate government funded AT provision. Both settings are high income countries, with occupational therapists and allied health practitioners as the central workforce.

The shared endeavor of understanding AT use and comparing data has led us to compare our results and to critically appraise the ways in which our methods, and indeed the vocabulary we have to work with – has enabled or created barriers to mutual learnings across these two studies.

2.2 Data Analysis

In terms of demographics, data was extracted for all older persons (excluding the younger Faroese participants).

3 Results

3.1 Demographics

The Australian study enrolled N = 23,379 older persons. The subset of data examines for this study is the 6 month cohort, who were provided with their AP for 6 months n = 6,370 (27%). They used between 1–19 products, and the average number of assistive products per person is 2.3. The Faroese study has N = 73 older persons. They used between 1–10 products, and the average number of assistive products per person is 4.5.

Table 4 summarizes the type of AT used in both studies. Some similarities and some differences are noted int the comparison of data from both studies, see Fig. 2. Self-care is similar, whereas double the Australian cohort used AT for domestic life. Communication and information management was also different, with nearly 22% of Australians receiving, compared with 8% Faroese participants. This is likely due to different prescribing policies, as Australia does provide devices such as IPads, whereas this is just provided to children in the Faroe Islands. Personal alarms are also in both Australia and the Faroe Islands, however these products were not included in the Faroese study. The lower number of mobility products in Australia is likely to be due to the policy of providing mobility products on discharge from hospital, or through post-acute care programs immediately after hospital discharge.

3.2 Usage of AT

Figure 2 shows the percentage of AT by category, either in use, use infrequently, or not used. The rates of use are high, ranging between 67% and 100%. Rates of use, infrequent use or non-use are comparable across the two study populations and across product categories.

Qualitative data illustrating the factors which impact abandonment

Qualitative data was only available from the Faroese study and illustrated over half of factors identified by Sugawara.

Table 2. Contrasting methods used across studies

Dimension	Faroe Study	Australia Study
Demographics	• Community dwellers • All ages • Living region	• Over 65 years or over 55 years for Indigenous Australians • Indigenous status/cultural background • Living region
Health Service Utilisation	Receives help	Career support; doctor visits; allied health visits; hospital visits; new or changed medications
Health-related quality of life	-	EQ5D
Functional problems	12 functioning areas	LifeCurve functioning scale – 19 questions
Possible diagnosis	Health condition	Health condition
Assistive products	1–12	1–20 coded with ISO 9999
Service delivery	Who assessed for and delivered the AP Who implemented the AP	Who assessed for the AP Local provider/supplier implemented
Longevity of AP	Use of and AP that the participant has	Longitudinal study across baseline, 3, 6, 12, 18 and 24 month touchpoints
Usage of AP	For each AP: In use [yes/no] If yes then 1 = daily, 2 ≥ once/week, 3 ≥ once/month, 4 ≤ once/month, 5 = never Used infrequently and not used [*free text*]	For each AP in use: In use [yes/no] Used infrequently [*free text*] If not used then [*Free text]*: a) It didn't help with the functional problem b) It is no longer required because the functional problem has changed or resolved c) It is no longer required as I found a different way to manage
Service satisfaction	KWAZO	Likert scale service satisfaction
User satisfaction with AP related service	QUEST service subscale (4 questions)	PIADS adapted

Table 3. Subset of data for the current comparison

Dimension	Faroe Study	Australia Study
Usage of AP	1. In use = daily, once/week; 2. Used infrequently ≥ once/month? 3. Not used	1.In use = daily, once/week; 2. Used infrequently ≥ once/month? 3. Not used
Qualitative data: Factors identified as impacting abandonment [5]	Qualitative data	-

Table 4. Type of AT

Functional category of assistive product (ICF)	Faroe (survey)		Australia (6 month survey)	
Managing body functions	n = 62	(18%)	*n* = 988	(5.9%)
Self-care	n = 113	(32%)	*n = 4,876*	(29.4%)
Mobility	n = 94	(27%)	*n = 1,323*	*(8.0%)*
Domestic life	n = 50	(14%)	*n = 5,523*	*(33.3%)*
Communication and information management	n = 30	(8%)	*n = 3,635*	*(21.9%)*
Home modifications	n = 5	(1%)	*n* = 235	(1.4%)
Therapy products	0		*n = 29*	*(0.2%)*
TOTALS	354	100%	16,609	100%

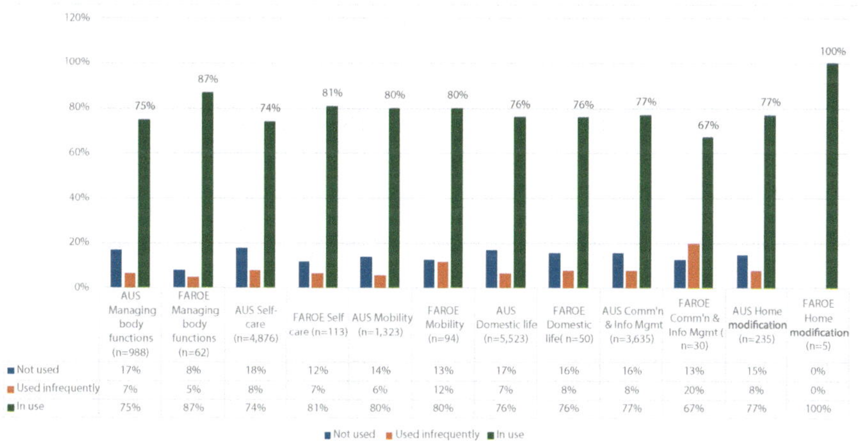

Fig. 2. Comparing use, infrequent use and non-use across Faroese and Australian data

Table 5. Reasons for non-use

Factors identified as impacting abandonment Sugawara et al. [9]	Faroe data	Examples
Personal factors		
Health condition improved or worsened	✓	Has not used the 2 rollators for a year due to worsened Parkinson
The user did not get adapted to the device or did not accept it	✓	Does not care to use the bath brush
The user was ashamed of using the device	X	
A new disease or injury prevented the use of the device	X	
The user lost or replaced the device	X	
Intervention-related factors		
The device was of inadequate size	✓	Bath stool too big for shower The wheel-chair too heavy for the family
The device did not meet user's needs	✓	Can't use the stocking applier due to no function in the left hand
The device was inadequate to the intended functions	✓	Gets too exhausted to use the rollator for shopping, reduced pulmonary function and barriers outside
The use was suspended by medical orders	X	
Using the device lead to discomfort or pain	✓	Doesn't use the activity chair, because it is uncomfortable to sit on
The user could not wear the device	X	
The user did not know how to use the device	✓	Had received the day planner without instruction, has had it for 4–5 years
Product-related factors	=	
The device was worn out or broken	✓	The charger for the Daisy player has been broken for some time
The device was under maintenance	X	
Environmental factors		
The user felt discriminated when using the device	X	
The user faced barriers to using the device in or out of home	✓	Too many barriers to use the rollator outside Too cramped inside to use the rollator

Other reasons for non-use/underuse emerged from data analysis and could be categorized as intervention related, product related or environmental factors. Intervention-related factors included situations where the need decided by someone else, for example '*A family member had decided that she needed the shower stool*' or where the device was not installed or was incorrectly installed, for example:

The way the grab bar is placed makes it impossible to use.

The rail/grab bar has not been installed (got it several months ago).

In an example of a situation where the person does not need it any more for reasons related to intervention (rather than personal factors), one participant, *'has received a stair-lift from the AT center and can go to the toilet upstairs and does not deed the commode chair downstairs'*.

Product related factors related to the device's functions making it difficult to use,

"I don't use the Amigo HD much, too small screen, too sensitive so the colors change too much"

Two expansions to environmental factors were identified. Environmental factors related to seasonal weather, for example:

"I use the rollator daily in the summer for walks, not every month in the winter due to challenging weather".

"I use the rollator when the weather is favorable"

Also, environmental factors related to considerations of others:

"I only use the rollator outside together with family, they will not let me use it alone"

Describing use, underuse and non-use of assistive products

Figure 1 describes a range of terms commonly used to describe the status of AT use. Many of these terms are synonyms for each other, meaning that data could be coded under *many* terms. For example, the following quote could be categorized as discontinuance, a decrease in adherence, or underuse:

"I use the rollator less now due to reduced pulmonary function. Edges etc. make it difficult to get in and out with the rollator"

In other examples of non-use, abandonment, non-compliance or non-adherence, participants stated:

"I do not know why I got this wheelchair, I have never used it"

"My daughter thought I needed the shower stool. I do not use it."

"I have had the rollator for several years. I do not need it. I support myself to furniture."

"I do not use the wheelchair. I prefer to use my legs".

The results of these studies suggest the following graded set of terms situated along a continuum of use/non-use (Fig. 2). The measurement of ranking of reasons for underuse and non-use, we suggest, could follow a revised set of factors based on Table 5.

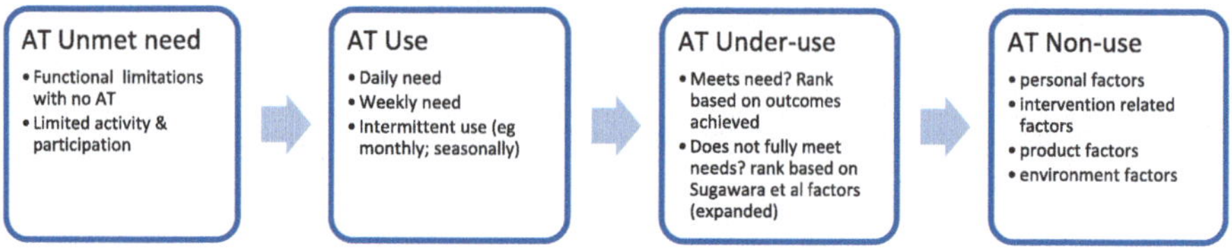

Fig. 3. Continuum of AT unmet need/AT use/AT underuse/AT non-use

4 Discussion

This paper has compared data from two studies, run using different methods and in different hemispheres. In terms of the data across studies, the high rates of use were remarkably consistent, as were the much lower rates of infrequent use, and of non-use. As described above, different services, systems and policy contexts across two hemispheres are likely to explain differences seen in the two datasets.

The ranking of use/non-use was collapsed to achieve comparable data across the studies into a 3 point scale (In use = daily, once/week; 2 = Used infrequently ≥ once month; 3. Not used). We suggest this is not the ideal grouping for studies of usage, as it is overly broad, proposing instead the continuum in Fig. 3.

This brings us to our initial determination – that guidance as to standard terms and concepts would improve the feasibility of running studies in which data can be pooled and compared. Qualitative data is invaluable in understanding reasons for use and non-use, and we suggest mixed methods studies are vital when investigating AT use. An available taxonomy of factors from Sugawara [9] was used to analyze the Faroese data. Of these 9 resonated with the Faroese data while 7 did not, and 2 extra Environmental factors, 4 extra Intervention related factors and 1 extra Product related factor emerged from the data.

As authors, we have proposed a tentative continuum from AT unmet need through to meet need. Our proposal acknowledges that use may be daily, weekly, intermittently or seasonally, such as for assistive products used for recreation or for differing weather conditions. Our data suggests that under-use can only be understood in relation to the person's context and goals. Narratives of under-use and non-use explain whether this is positive, such as a person no longer requiring support from the AP; or negative, such as being unable to utilize AT to achieve the outcome.

5 Limitations

This paper has identified the need for a graded set of terms situated along a continuum of use/non-use, and informed a draft set of terms through the author's clinical experience; data from several current studies, as well as known studies from the assistive technology literature base. The datasets of the study were matched for age, but not on other dimensions. Matching for region (that is, excluding the dense urban centres as well as the very, very remote settings int the Australian study) may have provided more comparable contexts with the regional living situation of most Faroese people. Given the multiple other contextual factors at play, it was not felt this would strengthen the comparison significantly, and we note the limitation this presents. Other limitations include the lack of a systematic review of the assistive technology literature; or systematic searches to identify and understand 'adherence' terminology used in similar interventions which may be subject to non-use, such as prescription of pharmaceuticals. These were beyond the scope of the enquiry.

6 Conclusion/Implications for the AT Field

The WHO/UNICEF Global Report Recommendation 6 proposes we *Invest in data and evidence-based policy* [15]. Accurate terminology has been identified as crucial for the AT sector for many reasons [16–18]. In some instances, non-use is indicated, and it is vital that research studies are structured to capture this nuance [11]. Clear definitions of when assistive product may be abandoned versus under-used, no longer required or a range of other subtleties, could significantly inform the evidence base and support understanding by policymakers and others of the ongoing, reflexive and multifactorial realities of AT use. We hope this paper with its grounded, practitioner-based and emergent approach to the issue of gradations in definition, has in a small way informed critical knowledge gaps in assistive technology outcome measurement.

Acknowledgments. Both authors undertook writing of this paper, and secondary data analysis, in their own time without funding. The Australian study was funded by the Australian Government Department of Health and Aged Care. The Faroese study was funded by the Faroese Health Insurance Fund.

Disclosure of Interests. The authors have no competing interests to declare that are relevant to the content of this article.

References

1. De Witte, L., Steel, E., Gupta, S., et al.: Assistive technology provision: towards an international framework for assuring availability and accessibility of affordable high quality assistive technology. Disabil. Rehabil. Assist. Technol. **13**(5), 467–472 (2018)
2. Phillips, B., Zhao, H.: Predictors of assistive technology abandonment. Assist. Technol. **5**, 36–45 (1993)
3. Riemer-Reiss, M., Wacker, R.: Factors associated with assistive technology discontinuance among individuals with disabilities. J. Rehabil. **66**, 44–50 (2000)
4. Wessels, R., Dijcks, B., Soede, M., et al.: Non-use of provided assistive technology devices, a literature overview. Technol. Disabil. **15**(4), 231–238 (2003)
5. Wieland, T., McKenna, K., Tooth, L., et al.: Factors that predict the post-discharge of use of recommended assistive technology (AT). Disabil. Rehabil. **1**(1–2), 29–40 (2006)

6. Dijcks, B.P., Witte, L.P., Gelderblom, G.J., et al.: Non-use of assistive technology in the Netherlands: a non-issue? Disabil. Rehabil. Assist. Technol. **1**(1–2), 97–102 (2006)
7. Federici, S., Meloni, F., Borsci, S.: The abandonment of assistive technology in Italy: a survey of National health Service users. Eur. J. Phys. Rehabil. Med. **52**(4), 516–526 (2016)
8. Federici, S., Borsci, S.: Providing assistive technology in Italy: the perceived delivery process quality as affecting abandonment. Disabil. Rehabil. Assist. Technol. **11**(1), 22–31 (2016)
9. Sugawara, A.T., Ramos, V.D., Alfieri, F.M., et al.: Abandonment of assistive products: assessing abandonment levels and factors that impact on it. Disabil. Rehabil. Assist. Technol. **13**(7), 716–723 (2018)
10. Tuazon, J.R., Jahan, A., Jutai, J.W.: Understanding adherence to assistive devices among older adults: a conceptual review. Disabil. Rehabil. Assist. Technol. **14**(5), 424–433 (2019)
11. World Health Organization. International Classification of Health Interventions (2018)
12. Layton, N., Wilson, E., Colgan, S., et al.: The Equipping Inclusion Studies: Assistive Technology Use and Outcomes in Victoria. Deakin University, Melbourne (2010)
13. Danemayer, J., Bloomberg, M., Mills, A., et al.: Demographic, socioeconomic, and social barriers to use of mobility assistive products: a multistate analysis of the English Longitudinal Study of Ageing. Lancet Public Health **10**(1), e20–e28 (2025)
14. AT Scale. The Case for Investing in Assistive technology (2021)
15. WHO & UNICEF. Global Report on Assistive Technology. Geneva: WHO & UNICEF (2022)
16. Global Alliance of AT Organisations. GAATO AT Outcomes and Impact: A Global Grand Challenge. GAATO, Geneva, Switzerland (2022)
17. Bell, D., Layton, N., Smith, R., et al.: Developing a holistic process to measure assistive technology outcomes. Assistive Technol. Outcomes Benefits **18**(Spring 2024) (2024)
18. Elsaesser, L.-J., Layton, N., Scherer, M., et al.: Standard terminology is critical to advancing rehabilitation and assistive technology: a call to action. Disab. Rehabil. Assistive Technol. 1–3 (2022)

A Full Lifespan Comparison of Virtual Reality Based Cognition Testing on Acceptance, Effectiveness and Satisfaction

Soma Zsebi[1,2](✉), Patrícia Szabó[3,4], Cecilia Sik-Lanyi[3,4], and Renáta Cserjési[2]

[1] Doctoral School of Psychology, Eötvös Loránd University, Budapest, Hungary
zsebi.soma@ppk.elte.hu

[2] Department of Psychology, Eötvös Loránd University, Budapest, Hungary

[3] University of Pannonia, Egyetem u. 10, 8200 Veszprem, Hungary

[4] Hungarian Research Network, Piarista u. 4, 1052 Budapest, Hungary

Abstract. Virtual Reality (VR) technology can pose a good alternative in cognitive testing to the original paper-and-pencil based tasks; however, the acceptance and user experience is often left unattended when it comes to developing cognition measuring tasks. In order to solely measure the effects induced by the VR device or environment, we created a 1:1 copy of a classical cognitive psychological test, the Corsi-Block Tapping Task, and digitalized it into a VR environment. We investigated if different age groups, from children to elderly, showed different rates on mood, emotion, VR experience and user experience questionnaires and if these results impacted their performance. Our pilot results indicate that age is decreasing factor on performance, however, VR makes no difference in that matter.

Keywords: Virtual Reality · Cognitive testing · Corsi-Block Tapping Task

1 Introduction

1.1 Cognitive Testing in Virtual Reality

Virtual Reality (VR) based cognitive testing has been becoming more and more popular in the last 10 years both in experimental and clinical settings [1–3]. Whilst posing a very attractive alternative to classical paper-and-pencil based testing, only a handful of VR based cognition measuring tasks have quantitatively measured their participants' acceptance and satisfaction of the tasks [4]. One of the challenges for that is the constant new developments of VR based cognitive tests, which may seem desirable first, however, it hinders the generalizability and comparability of such tasks [5]. Whilst both classical paper-and-pencil tasks and VR based tasks can be useful, the latter poses questions about its reliability as stimuli rich VR environments can distract participants, change cybersickness levels and alter cognitive performance [6–8]. Another potential problem might come from the varying age of participants as different age groups, especially after the age of 60, might perform differently in a VR based task compared to one another [9–11].

K. Mavrou and P. Encarnação (Eds.): AAATE 2025, *Technology for Inclusion and Participation for All: Recent Achievements and Future Directions*, pp. 257–264, 2025.
https://doi.org/10.1007/978-3-032-01628-7_31

This poses a turmoil in VR based testing as such tasks have been shown to be useful in elderly populations, whilst also producing different results when compared to younger populations [12–14]. Because of these potential problems, comparing classical tests to newly developed VR tasks is desired, however, current research hardly ever does that. In order to do so, implementations of already known and validated cognitive tests to VR environments could be a solution as these tasks have normative data, supporting theoretical backgrounds and known age-related differences. If an already existing, validated and reliable cognitive test was implemented to VR, then any observed difference between the performance in the original and the VR version of the task could be solely attributed to the VR environment as the task would be the same. This would allow researchers to examine the effectiveness of VR more directly while excluding any potential task related differences in performance.

1.2 The Corsi-Block Tapping Task

The Corsi-Block Tapping Task (CBTT) is one such widely known and validated neuropsychological task which measures cognitive functions related to the medial temporal lobe (MTL), especially the hippocampus [15, 16]. Originally intended to measure spatial working memory, it has since been expanded as a more overall measurement of the MTL assessing spatial and temporal information as well as their combined, but still unique, piece of memory, spatiotemporal memory [17, 18]. In the task, nine $3 \times 3 \times 3$ cm cubes are placed over a wooden 28×21 cm board in a preorganized manner [16]. The task consists of two parts, the so-called forward and backward conditions. In the forward condition, the experimenter taps the blocks in ascending sequences of quantity, starting with tapping two random blocks after one another. The participant then must reproduce the exact sequence of the blocks, meaning they ought to tap the same blocks in the same order as the experimenter did; if they manage to do so, the experimenter taps a new sequence of randoms blocks increased by one more block each time. The task is done until the participants fails to reproduce the sequence either by tapping a wrong block or messing up the order of the blocks. In the backwards condition, the experimenter taps the sequences of blocks in the same order, however, the participant must reproduce it in a reverse order, from the last block to the first one.

Whilst the CBTT was originally intended to be a general cognitive test, research ever since its development has pointed out its effectiveness for measuring age related cognitive decline [19–21]. This ability is probably due to fact that the underlying neural structure activated during task administration is the MTL, which is a good predictor of age-related cognitive decline as well as has the ability to predict probabilities of different dementia types' occurrences [22–24]. This allows the CBTT to measure and predict cognitive decline related to age and its validity and reliability makes it a good candidate for VR implementation to examine potential VR related differences in testing procedures whilst excluding task related differences.

1.3 Our Development of the VR Corsi Task and the Present Study

In order to test the effectiveness of VR in cognitive testing, we created a 1:1 replica of the original Corsi-Block Tapping Task, called VR Corsi, and implemented it to a

VR environment. The VR replica completely followed the measures of the original task declared in the validation article by Kessels et al. (2000) [16]. With this software, we were able to directly test the effectiveness of the VR device because if it is as usable as original/classical cognitive tasks, we should observe similar results in the VR Corsi task as in the CBTT.

Since we expected to see potential VR related differences in performance, we also measured participants' pre- and post-emotional states, as well as their subjective experience with the VR device, previous VR experience, level of immersion and cybersickness symptoms. We expected to observe no significant difference between the CBTT and the VR Corsi, with the small distinctions being explainable with the aforementioned factors.

2 Methods

2.1 Participants

Using GPower version 3.1.9.7., we calculated the optimal number of participants to be 54 for group comparison (t-test: two dependent means; Tail(s): Two; Effect size d: 0.5; α err prob: 0.05; β err prob: 0.95), which would let us observe if there was any big difference between the performance of two tasks. Participants were recruited from online ads, leaflets and direct approaches to the acquaintances of the experimenters. Participants were recruited from ages 14 to 90 in order to provide a homogenous age pool. Participants had to be free from all neurological and psychiatric diseases as well as had to have normal or corrected-to-normal vision. Applying these criteria resulted in excluding 5 applicants ultimately creating a sample of 49 participants (Mean age = 32.673 years, SD = 23.494; 11 male, 38 female). Participants were not given anything for their participation.

2.2 Materials

An Oculus Quest 2 VR headset was utilized, connected to a laptop through a USB-C cable. The laptop operated on a Windows 11 operating system, running the Oculus application for the VR Corsi task. The VR screen was casted to the laptop through QuestLink enabling the experimenter to observe and supervise the participant. The VR environment consisted of a typical children's room with some toys (e.g. wooden cars, teddy bears) laying around and a wooden table in the middle of the room with the Corsi-Block Tapping Task on it. Participants could navigate around the room using the controllers on the joysticks, however, they couldn't interact with anything but the CBTT. Participants had hands animated as the controllers mimicking the feeling of using their own hands.

For measuring emotional states, the Self-Assessment Mannequin (MAN) was applied in which participants had to indicate how much they felt three different moods on 1–5 Likert-scales (Happiness, Excitement, Feeling of Control) and the Positive and Negative Affective Scale (PANAS) in which participants had to indicate on 1–5 Likert-scales how much they felt 20 different feelings at the moment of answering. We also asked participants to indicate how much previous VR experience they had on a 1–5 Likert-scale, regardless of in what occasion they had previously used a VR device. We also created a short, 5 item questionnaire measuring the subjective effectiveness and attitude towards VR (titled Subjective Impression Scale) in which participants had to indicate on 1–5 Likert-scales how much they agreed with each statement (see appendix 1). We also created a short cybersickness questionnaire in which participants had to indicate on 1–4 Likert-scales how much they felt certain physical symptoms during testing (see appendix 2). Lastly, participants had to indicate on a simple 1–7 Likert-scale how immersed they felt during testing. In all questionnaires, the higher number indicated a more positive score and none of the tools had any reverse scales.

2.3 Procedure

Participants all signed an informed consent before beginning the data collection procedure. Participants indicated their age, sex, highest level of education and previous VR experience. After, the pre-questionnaires took place in which participants filled out the MAN and PANAS questionnaires (titled as pre-MAN and pre-PANAS). Once finished, participants did the classical CBTT with half of them beginning with the forward and the other half with the backward condition. After finishing, participants went on to the VR Corsi testing. Participants were informed about the usage of the VR device and were given time and encouraged to roam around before starting testing. In the VR Corsi, the blocks highlighted after one another in white color to present the sequence, as there was no experimenter in the environment to tap the blocks. Participants had to reproduce the sequences using the joystick to point to a certain block and pressing the "select" button on the corresponding joystick to mark a block. Once again, half of the participants began with the forward and the other half began with the backward condition. Once done, participants again filled out the MAN and PANAS questionnaires (titled post-MAN and post-PANAS), after which they completed our Subjective Impression Scale and Cybersickness Questionnaire. Lastly, they indicated their level of immersion. The whole procedure took about 30 min per participant.

2.4 Data Analysis

The longest successfully recalled sequence in each task was recorder as performance indicators (CBTT Forward; CBTT Backward; VR Corsi Forward; VR Corsi Backward) and was used for comparison between the tasks. The raw SAM scores were used for analysis in both cases (pre- and post). The PANAS questionnaire doesn't have a total score but rather calculates two subscores as half of the feelings represent positive, whilst the other half represent negative feelings; total scores are calculated for these two subscales by summing the scores given to of each feeling in both subscales. Negative- and positive affective change scores were calculated by subtracting the pre-PANAS scores from the post-PANAS scores, respective to the subcales. These scores indicated how much the positive and negative feelings of our participants changed between the beginning and the end of the data collection process. The Subjective Impression Scale and Cybersickness Questionnaire scores were calculated by summing all the scores given to each item. Unfortunately, the Cybersickness Questionnaire was only later added to our repertoire, for which reason we didn't have this questionnaire's data on 20 participants.

Nevertheless, we still decided to include this variable in our analysis as complementary, exploratory score.

3 Results

We used Jasp 0.19.1. software for statistical analysis. Since we intended to test non-difference between the classical and the VR test scores, we compared the CBTT Forward and VR Corsi Forward as well as the CBTT Backward and VR Corsi Backward performance indicator scores in two separate Bayesian paired samples t-tests measuring probability of the null hypothesis. Descriptive data about the performance values are presented in Table 1.

Table 1. Descriptive data of the performance values in the four conditions.

Condition	Mean	Std. Deviation
CBTT Forward	5.204	1.000
CBTT Backward	4.918	1.134
VR Corsi Forward	4.776	1.403
VR Corsi Backward	4.592	1.306

The forward condition comparison resulted in a very small, anecdotal probability ($BF_{01} = 0.290$, error % = 6.951×10^{-7}); whilst the backward condition resulted in a slightly bigger but still negligible result ($BF_{01} = 1.538$, error % = 0.030). For the correlation analysis, we applied two separate Spearman's correlations as pairwise normality was violated in case of the backward condition comparison; and because the performance indicator values can only be integers making them discrete data. Correlation between the forward conditions resulted in a significant, positive, moderate correlation (rho = 0.560, $p < 0.001$); whilst correlation between the backward conditions resulted in a similarly significant, positive, moderate correlation (rho = 0.460, $p < 0.001$).

We also conducted four different linear regression models in which we analyzed the predictive value of age on performance in all four conditions. Results suggested a significant results in all conditions (all $p < 0.001$) (see Table 2 for results). Interestingly, though, cybersickness didn't moderate either of the VR conditions suggesting a nominal effect of VR induced physical symptoms on cognitive performance (all $p > 0.05$).

Furthermore, we found that age also significantly predicted the negative affective change score, although the effect size was relatively low ($R^2 = 0.130$; $\beta = 0.360$, $p = 0.011$), indicating that the older the participant, the more likely they are going to experience negative emotions during VR testing. Not unexpectedly, cybersickness had a significant predictive effect on negative affective change score ($R^2 = 0.328$, $\beta = -0.573$, $p = 0.001$), but did not show any inverse prediction on the positive affective change score ($p > 0.05$).

To our surprise, the level of immersion did not have a significant predictive effect on either of the VR conditions' performance values, neither on the positive and negative

Table 2. Values of the linear regression models.

Condition	R^2 value	β value
CBTT Forward	0.272	−0.521
CBTT Backward	0.318	−0.563
VR Corsi Forward	0.253	−0.503
VR Corsi Backward	0.426	−0.652

affective change scores, as well as it didn't predict cybersickness scores (all $p > 0.05$). We also did not find any significant predictive value of SAM scores as well as didn't find moderating effects of them on either of the affective change scores.

4 Discussion

In the present study, we examined the effects of VR based cognitive testing using an already known neuropsychological task, the Corsi-Block Tapping Task (CBTT), for comparison between classical and VR based testing. Previous studies employing VR environments usually created their own tasks which hindered generalizability and comparability between classical and VR based cognitive testing. This difference could perhaps be bridged by implementing already existing cognitive tasks into VR environments which then would allow researchers to directly examine the potentially altering effects of VR without the task related differences. For this purpose, we implemented the CBTT into a VR environment, called VR Corsi, as a 1:1 replica and examined differences between the original CBTT and the VR Corsi.

Our results indicated that both the forward and backward conditions differed in the CBTT and the VR Corsi, with participants providing lower scores on both conditions in the VR Corsi task. This result suggests that participants are indeed affected by the VR environment, even in completely identical tasks. This result is partially refined by the correlational results providing positive, moderate correlations between the CBTT and the VR Corsi in both conditions. However, in such a completely identical 1:1 replica, one would probably expect a high correlation between the two tasks if VR really had no effect on performance. For this reason, we conclude that VR environments do have a performance altering effect, which cannot be attributed to the different tasks employed in classical testing settings and in VR environments, but solely to the VR device itself.

Whilst we concluded that there is an altering effect of VR, we failed to provide an explanation for why that difference might be, as neither emotional changes, nor cybersickness, nor the level of immersion provided moderating effects on performance. We also managed to prove that the already known age related difference in performance can be detected in a VR environment as well, however, the predictive value of age was low-moderate in all conditions, proposing the need to find other predictive variables as well. Interestingly, age apparently did not differ between classical and VR testing, suggesting that the elderly can perhaps use VR just as effectively as the young. Differences also couldn't be differences in perceived subjective effectiveness.

Virtual Reality environments do make a difference in cognitive testing and such environments cannot be considered the same or approximately-the-same as real world environments. This difference has to be taken into account when employing VR based cognitive testing, however, the reason for this difference is unclear at the moment. Future studies may employ more complex emotional testing, perhaps even using psychophysiological measurements, and more reliable cybersickness data as this variable in our study was rather incomplete.

References

1. Parsons, T.D., Rizzo, A.A., Bamattre, J., Brennan, J.: Virtual reality in neuropsychological assessment: a review of the literature. Annu. Rev. CyberTherapy Telemed. **5**, 163 (2007)
2. Pieri, L., Tosi, G., Romano, D.: Virtual reality technology in neuropsychological testing: a systematic review. J. Neuropsychol. **17**, 382–399 (2023)
3. Huygelier, H., Schraepen, B., Van Ee, R., Vanden Abeele, V., Gillebert, C.R.: Perceived realism and the role of contextual cues in virtual reality environments: Implications for cognitive performance. Sci. Rep. **9**, 4519 (2019)
4. Arlati, S., et al.: A virtual reality-based cognitive stimulation for the elderly: a usability and feasibility study. J. Alzheimers Dis. **80**, 1025–1034 (2021)
5. Neguţ, A., Matu, S.A., Sava, F.A., David, D.: Task difficulty of virtual reality-based assessment tools compared to classical paper-and-pencil or computerized measures: a meta-analytic approach. Comput. Hum. Behav. **54**, 414–424 (2016)
6. Georgopoulou, E.N., et al.: Computer-based cognitive training vs. paper-and-pencil training for language and cognitive deficits in Greek patients with mild Alzheimer's disease: a preliminary study. Healthcare **11**(3), 443 (2023)
7. Zhou, C., Bryan, C.L., Wang, E., Artan, N.S., Dong, Z.: Cognitive distraction to improve cybersickness in virtual reality environment. In: Proceedings of the IEEE International Conference Mobile Ad Hoc Sensor System Workshops (MASSW), pp. 72–76 (2019)
8. Gamito, P., Oliveira, J., Alves, C., Santos, N., Coelho, C., Brito, R.: Virtual reality-based cognitive stimulation to improve cognitive functioning in community elderly: a controlled study. Cyberpsychol. Behav. Soc. Netw. **23**(3), 150–156 (2020)
9. Tan, N.C., et al.: Age-related performance in using a fully immersive and automated virtual reality system to assess cognitive function. Front. Psychol. **13**, 847590 (2022)
10. Guo, Y., et al.: Age effect analysis of different gender groups in spatial ability test based on virtual reality technology. Front. Psychol. **15**, 1494048 (2025)
11. Oliveira, C.R., et al.: Neuropsychological assessment of older adults with virtual reality: association of age, schooling, and general cognitive status. Front. Psychol. **9**, 1085 (2018)
12. Yun, S.J., et al.: Cognitive training using fully immersive, enriched environment virtual reality for patients with mild cognitive impairment and mild dementia: feasibility and usability study. JMIR Serious Games **8**(4), e18127 (2020)
13. Moon, H.J., Choi, Y.R., Lee, S.K.: Effects of virtual reality cognitive rehabilitation program on cognitive function, physical function and depression in the elders with dementia. J. Int. Acad. Phys. Ther. Res. **5**(2), 730–737 (2014)
14. Park, J.S., Jung, Y.J., Lee, G.: Virtual reality-based cognitive–motor rehabilitation in older adults with mild cognitive impairment: a randomized controlled study on motivation and cognitive function. Healthcare **8**(3), 335 (2020)
15. Corsi, P.M.: Human memory and the medial temporal region of the brain (1972)

16. Kessels, R.P., Van Zandvoort, M.J., Postma, A., Kappelle, L.J., De Haan, E.H.: The Corsi block-tapping task: standardization and normative data. Appl. Neuropsychol. **7**(4), 252–258 (2000)
17. Ferreira, C.T., Verin, M., Pillon, B., Levy, R., Dubois, B., Agid, Y.: Spatio-temporal working memory and frontal lesions in man. Cortex **34**(1), 83–98 (1998)
18. Zimmer, H.D., Speiser, H.R., Seidler, B.: Spatio-temporal working-memory and short-term object-location tasks use different memory mechanisms. Acta Psychol. **114**(1), 41–65 (2003)
19. Guariglia, C.C.: Spatial working memory in Alzheimer's disease: a study using the Corsi block-tapping test. Dement. Neuropsychol. **1**(4), 392–395 (2007)
20. Saggino, A., Balsamo, M., Grieco, A., Cerbone, M.R., Raviele, N.N.: Corsi's Block-Tapping Task: standardization and location in factor space with the WAIS–R for two normal samples of older adults. Percept. Mot. Skills **98**(3), 840–848 (2004)
21. Saikia, B., Tripathi, R.: Executive functions, processing speed, and memory performance: untangling the age-related effects. J. Psychiatry Spectr. **3**(1), 12–19 (2024)
22. Jaroudi, W., Garami, J., Garrido, S., Hornberger, M., Keri, S., Moustafa, A.A.: Factors underlying cognitive decline in old age and Alzheimer's disease: the role of the hippocampus. Rev. Neurosci. **28**(7), 705–714 (2017)
23. Bettio, L.E., Rajendran, L., Gil-Mohapel, J.: The effects of aging in the hippocampus and cognitive decline. Neurosci. Biobehav. Rev. **79**, 66–86 (2017)
24. Mosconi, L., et al.: Hippocampal hypometabolism predicts cognitive decline from normal aging. Neurobiol. Aging. Aging **29**(5), 676–692 (2008)

From End Users to Principal Communicators: Ensuring Deaf Agency in Technologies

Amir Zuccalà[1] and Abraham Glasser[2](✉)

[1] Sapienza University, Via Salaria 113, 00198 Rome, Italy
amir.zuccala@uniroma1.it
[2] Gallaudet University, 800 Florida Ave NE, Washington, DC 20002, USA
abraham.glasser@gallaudet.edu

Abstract. The increasing use of artificial intelligence (AI) in sign language translation is often framed as a breakthrough in accessibility. However, the top-down approach in AI-driven sign language technologies largely positions Deaf individuals as passive stakeholders rather than active decision-makers. This leads to the proliferation of projects such as signing gloves and avatars, often developed without meaningful Deaf involvement. While international regulatory frameworks, including the European AI Act 2024/1689 and the 2023 U.S. President's Executive Order on AI, attempt to address AI ethics, they fail to consider the specific challenges of AI-driven sign language translation and its broader implications for Deaf empowerment. Furthermore, discussions around AI and Deaf communities remain narrowly focused on sign language translation, overlooking AI's potential to support Deaf education, language acquisition, and broader accessibility. This presentation critiques these gaps, drawing on literature and qualitative data from social media discussions and questionnaires, and introduces a Deaf-led approach to AI governance. The #DeafSafeAI initiative, developed by the Coalition on Sign Language Equity in Technology (Co-SET), is presented as a framework for ensuring ethical AI development that centers Deaf perspectives, advocating for a shift from technological solutionism to true inclusivity.

Keywords: Sign Language AI · Deaf Agency · #DeafSafeAI

1 Introduction

The rapid proliferation of AI-driven sign language translation tools exemplifies a broader trend of technological solutionism, where innovations are celebrated without adequately considering their real-world impact on the communities they aim to serve [1]. Research and development in sign language technologies frequently lack meaningful Deaf involvement, resulting in systemic biases that shape both research agendas and practical applications [2–4].

This technological enthusiasm aligns with a top-down accessibility model in which Deaf individuals are primarily regarded as passive stakeholders and *end users* of prepackaged "inclusion solutions," rather than active contributors shaping the design and

K. Mavrou and P. Encarnação (Eds.): AAATE 2025, *Technology for Inclusion and Participation for All: Recent Achievements and Future Directions*, pp. 265–270, 2025.
https://doi.org/10.1007/978-3-032-01628-7_32

implementation of these technologies [5]. The prevailing focus remains on translating into sign languages rather than genuinely engaging with Deaf people's voices, proposals, and expectations. Consequently, projects such as signing gloves and AI-generated avatars continue to emerge, often without meaningful Deaf participation.

While international regulatory frameworks, including the European AI Act 2024/1689 and the 2023 U.S. President's Executive Order on AI, strive to establish ethical AI governance, they fail to address the specific challenges associated with AI in sign language translation. Additionally, discussions surrounding AI and Deaf communities remain narrowly centered on sign language translation, overlooking the broader potential of AI to support Deaf empowerment in areas such as education, language acquisition, and self-advocacy.

This paper highlights these critical gaps and advocates for a Deaf-led, ethical approach to AI governance in sign language and related technologies. Drawing on relevant literature and qualitative data collected through questionnaires and social media group discussions, this work aims to reframe the discourse on AI and accessibility, ensuring that Deaf perspectives are central to the development and deployment of these technologies.

2 Avatars, Biases and the Quest for Deaf Agency

Much of the international literature on sign-language translation interfaces has been driven by a strong social-utility motivation, envisioning these systems primarily as accessibility tools for Deaf users alongside their scientific and technical development aims. However, this field presents numerous challenges in practice [6].

At the same time, a growing critical perspective has highlighted systemic biases in avatar–based research: many projects fail to account for the complexity and variability of sign languages, the inherently multimodal nature of signed communication, the ways in which datasets and sign samples are constructed, and—crucially—the perspectives, skills, and lived experience of Deaf signers, who too rarely participate as equal partners in development [7]. Scholars have begun to interrogate the very assumptions underpinning AI research in sign language, calling for decolonizing approaches that prevent Deaf communities from being further marginalized by technology [8]. These voices try to move"from discourses around access and assistiveness towards agency and self-determination" [9].

Despite this critique, many avatar projects continue to perpetuate ableist—or, more specifically, audist—assumptions about Deaf people's roles as passive recipients rather than active co-creators. Moreover, contemporary discourses tend to reduce the Deaf experience—and the needs of Deaf individuals—almost exclusively to the domain of translation, particularly sign language translation. This narrow framing fuels the current hype surrounding the development of signing avatars. At the same time, there is a noticeable lack of attention to how AI might be integrated into various aspects of everyday life for Deaf people—for instance, through text generation tools—or how they might benefit from broader applications of AI beyond sign language interfaces. Equally absent is a systematic effort to provide accessible training that would empower Deaf individuals to understand, use, and critically engage with AI technologies to their full potential.

In response to this institutional gap, Deaf associations have stepped in to address the need for accessible education. They organize courses and workshops designed specifically for Deaf participants, offering information on how AI systems work, their risks and benefits, and practical strategies for navigating and taking advantage of these emerging technologies. These initiatives are crucial in fostering digital autonomy and ensuring that Deaf individuals are not left behind in the AI revolution.

The World Federation of the Deaf (WFD) and the World Association of Sign Language Interpreters (WASLI) in 2018[1] have jointly issued a statement to raise awareness about the implications of AI on sign languages and interpreting. More recently, the WFD has also launched a call for experts to establish an Ad Hoc Group tasked with developing a comprehensive position on AI and national sign languages. This initiative reflects the growing concern within the international Deaf community about the impact of AI on linguistic rights and access to information.

Similarly, during the "AI and Interpretation" Conference hosted by the Directorate-General for Interpretation of the European Commission[2], the European Union of the Deaf (EUD) stressed that AI currently lacks the linguistic depth and cultural nuance necessary for accurate and context-sensitive national sign language interpretation. While AI-generated sign language tools may hold potential, they also pose serious risks of inaccuracy, which could ultimately jeopardize Deaf people's right to access essential information in a reliable and culturally appropriate manner. EUD is also working on building a strong ethical and policy framework concerning the use of Artificial Intelligence in Sign Languages and related Deaf technologies. The same caution applied to the use of AI in spoken language interpretation must also be extended to sign language. Without this, the drive for cost efficiency risks undermining accessibility, compromising not only the quality of communication but also the fundamental rights of Deaf individuals.

Further, other organizations and governmental entities have released statements relating to AI and sign languages; including but not limited to the British Deaf Association[3], the Council of Europe[4], and the Government of Canada[5]. Several of these include calls to action for further research and evaluation of these technologies for the future of accessible AI.

Feeling the need to fill all these gaps, Co-SET developed Guidelines to ensure that Deaf perspectives on the use of automated or automatic interpreting via artificial intelligence (AI) are thoroughly integrated into all policy recommendations. Deaf perspectives are inherently intersectional, representing all stakeholder groups (end users, qualified practitioners, vendors and buyers) and diverse communities within and across identity groups, including ethnicity, nationality, Deafblind and other disabilities.

[1] https://wfdeaf.org/news/resources/wfd-wasli-statement-use-signing-avatars/.

[2] https://eud.eu/ensuring-equity-in-ai-for-sign-language-interpretation-euds-advocacy-at-eu-conference/.

[3] https://bda.org.uk/discussion-paper-on-ai/.

[4] https://www.coe.int/en/web/artificial-intelligence/the-framework-convention-on-artificial-intelligence.

[5] https://accessible.canada.ca/creating-accessibility-standards/asc-62-accessible-equitable-artificial-intelligence-systems.

3 #DeafSafeAI Guidelines

Co-SET, composed of leaders from the Deaf community and the sign language interpreting field, hosted webinars with over 350 participants discussing Deaf communities and perspectives on ethics, privacy, risks, benefits, and more regarding sign language AI technologies. These discourses were analyzed and a detailed report of the sociotechnical system was written (in both English and ASL)[6].

These, among with other activities by CoSET and the SAFE-AI Task Force[7] have led to a publication of evidence-based guidelines on AI and interpreting services. These guidelines include (1) End-User Autonomy, (2) Evidence of Improving Safety and Wellbeing for End-Users, (3) Transparency of AI Quality for the General Public, and (4) Accountability by Companies for Errors and Harm to End-Users.[8]

These four principles are drawn from ethical, professional practices of high and low resource languages, and are intended to drive legal protections and promote innovations in fairness and equity in design and delivery so that all can benefit from the potential of AI interpreting products. This Guidance covers the full range of AI-generated, AI-assisted, and Human-supported AI interpreting products and services.

CoSET has also worked to disseminate these guidelines and surrounding topics across a breadth of different disciplines and venues. Working to increase awareness and encourage all stakeholders to ensure that AI continues to develop safely, fairly, and ethically, CoSET has presented to the language industry, interpreting agencies, academic research conferences, government organizations, private companies, and beyond.

4 Preliminary Results from Case Studies

To demonstrate the relevance and urgency of developing transparent and effective guidelines on AI and sign language technologies, we present some preliminary findings from ongoing PhD dissertation work, conducted by one of this paper's authors, within the Italian Deaf community. This case study highlights the critical need for participatory frameworks and inclusive policymaking. The study adopts a mixed-methods ethnographic approach, drawing on several sources: direct involvement in technical working groups with governmental bodies and private organizations in Rome (2023–2024), observation of a digital skills and AI training course designed for unemployed Deaf participants, monitoring of social media discussions focused on Deafness (such as the Tecnologia ed elettronica in LIS and Gruppo sordi italiani Facebook groups), preliminary results from a survey of interpreting students, and ten semi-structured interviews with Deaf leaders active in Deaf-led organizations, aged 25–50 [10, 11].

Through these observations, a stark imbalance is found in how technology is discussed and understood within the community. For instance, online conversations in Facebook groups are primarily concerned with practical tools; devices, subtitles, internet connectivity, electric vehicles, and communication apps—while features such as AI integrations in platforms like Meta's WhatsApp receive only superficial attention, often

6 https://safeaitf.org/deafsafeai/.

7 https://safeaitf.org/.

8 https://safeaitf.org/guidance/.

limited to concerns about data privacy. Broader ethical questions, including those relating to social justice, surveillance, or participatory design, are largely absent from these exchanges.

Similarly, participants in the digital skills course exhibited limited digital literacy, especially with regard to computer use, and basic understanding of telecommunication tools. This gap restricts their capacity to critically engage with the societal and ethical implications of AI, including debates about the replacement of human interpreters with signing avatars.

Preliminary results from a survey of early-stage interpreting students indicate general optimism about future collaboration between humans and AI in sign language translation. However, these future professionals show limited awareness of ethical risks and practical challenges associated with replacing human interpreters with avatars, particularly in sensitive or high-stakes contexts. This situation reveals a significant lack of accessible, critical information for Deaf communities, preventing them from fully participating in conversations about technological innovation that directly affect their lives. Additionally, we observe the rise of a form of technological solutionism, often driven by external fascination with sign languages and the symbolic promise of signing avatars. Despite growing evidence of the limitations and risks of such systems [5, 7], research and development continue to push forward with little regard for co-design or community validation.

In institutional contexts, the Italian Deaf Association (ENS) has frequently encountered top-down proposals for signing-avatar systems, often presented as ready-made accessibility solutions. These technologies, however, are still underdeveloped and unsuitable for use in critical domains such as emergency communication or healthcare. Yet they are promoted enthusiastically, buoyed by substantial funding opportunities—often from European sources—which contribute to the momentum of AI deployment, regardless of readiness or community acceptance.

Even the 2021 legal recognition (Law 69/2021) of Italian Sign Language, while significant, has not led to the establishment of Deaf-led research centers or dedicated spaces for Deaf epistemologies, such as Deaf Studies programs. Instead, institutional responses have focus on expanding interpreter training programs, reinforcing a translation-based model of accessibility. This reflects a broader social narrative in which Deaf individuals are positioned as passive recipients of information rather than active agents shaping their own communicative environments, where access is privileged over participation [12].

Consequently, decades of advocacy by Deaf communities for the recognition of national sign languages and the development of services rooted in their lived experiences risk being diluted into generalized, top-down narratives of accessibility—ones that fail to acknowledge how Deaf individuals themselves give meaning to their experiences and articulate pathways for reducing social injustice.

5 Conclusion

While AI and sign language technologies in general have tremendous potential to improve, innovate, and add to the current AT landscape, uninformed and unethical use would contribute to increased discrimination along with many ethical issues, risks, and opportunities for Deaf individuals. This work contributes to a larger discussion concerning AI frameworks that balance the interests of all stakeholders to ensure that AI continues to develop safely, fairly, and ethically.

References

1. Morozov, E.: To Save Everything, Click Here: The Folly of Technological Solutionism. PublicAffairs, New York (2014)
2. De Meulder, M.: Is "good enough" good enough? Ethical and responsible development of sign language technologies. In: Proceedings of the 1st International Workshop on Automatic Translation for Signed and Spoken Languages (AT4SSL), pp. 12–22. Association for Machine Translation in the Americas, Virtual (2021)
3. Desai, A., De Meulder, M., Hochgesang, J.A., Kocab, A., Lu, A.X.: Systemic biases in sign language AI research: a deaf-led call to reevaluate research agendas. In: Efthimiou, E., Fotinea, S.-E., Hanke, T., Hochgesang, J.A., Mesch, J., Schulder, M. (eds.) Proceedings of the LREC-COLING 2024 11th Workshop on the Representation and Processing of Sign Languages: Evaluation of Sign Language Resources. pp. 54–65. ELRA and ICCL, Torino, Italia (2024)
4. Fox, N., Woll, B., Cormier, K.: Best practices for sign language technology research. Univers. Access Inf. Soc. (2023). https://doi.org/10.1007/s10209-023-01039-1
5. De Meulder, M.: Deaf in AI: AI language technologies and the erosion of linguistic rights. arXiv:2505.02519 (2025)
6. Esselink, L., et al.: Exploring automatic text-to-sign translation in a healthcare setting. Univers. Access Inf. Soc. **23**, 35–57 (2024). https://doi.org/10.1007/s10209-023-01042-6
7. Desai, A., de Meulder, M., Hochgesang, J.A., Kocab, A., Lu, A.X.: Systemic biases in sign language AI research: a deaf-led call to reevaluate research agendas (2024)
8. Angelini, R.: Contrasting technologists' and activists' positions on signing avatars. In: Extended Abstracts of the 2023 CHI Conference on Human Factors in Computing Systems, pp. 1–6. Association for Computing Machinery, New York (2023)
9. Angelini, R., Spiel, K., De Meulder, M.: Speculating deaf tech: reimagining technologies centering deaf people. In: Proceedings of the 2025 CHI Conference on Human Factors in Computing Systems. Association for Computing Machinery, New York (2025)
10. Zuccala', A.: Put the interpreter in your pocket. Sign language technologies from accessibility to empowerment. Studi Sociol. **1**, 75–86 (2025). https://doi.org/10.26350/000309_000216
11. Zuccala', A., Di Renzo, A.: Avatar segnanti tra rappresentazioni grafiche e sociali. In: Avsar, B., et al. (eds.) Innovamenti. spazi e percorsi per una ricerca multidisciplinare, Siena, pp. 191–198 (2025)
12. Sauma, J.F.: Performing normal: deafness, intersectionality and academic exhaustion. In: Fagan Robinson, K., Carew, M.T., Groce, N.E. (eds.) Inaccessible Access: Rethinking Disability Inclusion in Academic Knowledge. Rutgers University Press, New Jersey (2025)

Implementation of a Web Platform to Improve Residential Autonomy: Usability and Self-determination for Persons with Physical Disability

Claude Vincent[1,2(✉)], Frederic S. Dumont[2], Marie-Eve Schmouth[2], Martin Caouette[3], Normand Boucher[2], and Eliane Begnanhi[2]

[1] Université Laval, Québec, QC G1V 0A6, Canada
claude.vincent@rea.ulaval.ca

[2] Centre Interdisciplinaire de Recherche en Réadaptation Et Intégration Sociale (Cirris), Québec, QC G1M 2S8, Canada

[3] Université du Québec à Trois-Rivières, Trois-Rivières, QC G8Z 4M3, Canada

Abstract. AppLOAD is an innovative solution proposed by a community organization to address the problems of finding accessible housing and aims at the self-determination of their members. No studies exist measuring residential autonomy for individuals with physical disabilities in 2019. Objectives were 1) to document the usability of the integrated web platform, i.e. effectiveness, efficiency and satisfaction of the platform), and 2 to explore the self-determination of people with disabilities in relation to residential autonomy. A mixed methodology based on a non-experimental cross-sectional design was adopted. Three questionnaires were developed to capture usability and self-determination. Two participants indicated that effectiveness was poor (6/12, 6/15) as self-determination (10/15, 8/20) and satisfaction was moderate (3.5/5, 3.4/5). Later interviews with conceptors revealed improvements and solutions.

Keywords: effectiveness · satisfaction · accessible housing platform

1 Introduction

The United Nations Habitat Assembly (UN-HABITAT) defines access to safe, secure, habitable and affordable housing as a fundamental right. People with disabilities and people with reduced mobility struggle to find housing on their own and are often referred to housing by a third party (e.g. a social worker). People with disabilities, like the general population, want to choose their own place of residence, that is, a place that meets their needs [1].

In the field of physical disability, very few scientific studies have documented residential autonomy in physical disability. Wright, Colley and Kendall [2] recently published a systematic review on alternative housing solutions for individuals who have

K. Mavrou and P. Encarnação (Eds.): AAATE 2025, *Technology for Inclusion and Participation for All: Recent Achievements and Future Directions*, pp. 271–279, 2025.
https://doi.org/10.1007/978-3-032-01628-7_33

suffered a brain injury or spinal cord injury. The type of housing has been shown to be associated with several important elements. For example, those in a family-like environment (home or housing) have greater community integration and participation, greater independence, psychosocial well-being and quality of life compared to structured environments, such as a health facility. A family-like environment is therefore desirable but is not always an option for people with physical or cognitive disabilities, due to the physically inaccessible or unaffordable nature of private housing.

AppLOAD (https://www.appload.ca) is a French-Canadian innovative solution proposed by a community organization to address the problems of finding accessible housing (e.g. accessible apartment) and aims at the self-determination of their members, to the greater Quebec City area. In 2020, for the community organization, it was more than relevant to collaborate with a research project to ensure they better met the need for residential autonomy of people with physical disabilities. We have received funding from the Social Sciences and Humanities Research Council of Canada for a partnership project [5].

Objectives were: 1) to document the usability of the integrated web platform, i.e. effectiveness, efficiency and satisfaction of the platform, and 2) to explore the impact of the use of the integrated web platform on the self-determination of people with disabilities in relation to residential autonomy. We have not identified any other platforms of this kind in 2019. Updating housing records, and deciding when homes are rented or become vacant, remain major challenges.

2 Methods

2.1 Research Design

A mixed methodology based on a non-experimental cross-sectional design, of the intervention-evaluation type, was preferred.

2.2 Intervention

The intervention involved the use of the AppLoad platform by individuals with physical disabilities, housing developers and aid organizations offering a range of services. The AppLOAD platform is structured into five sections: Self-Evaluation, Situate Yourself, Moving, Find out more, and Contact. The **Self-Evaluation section** (see Fig. 1A) allows users to assess their level of autonomy with ease, acting as a filter on the offers presented. The participant is prompted to first select his or her diagnosis (cerebral palsy, head injury, osteoporosis, amputation, etc.), which suggests an autonomy rate based on what is known about this condition, where 100% is Completely autonomous and 0% where the person is not autonomous at all. The participant can then adjust his autonomy rate according to his situation. The **Situate Yourself section** (see Fig. 1B) enables people with disabilities to locate a home in the regions around Quebec City. By clicking on the number, a technical sheet appears (see Fig. 2) to show the type of dwelling (e.g. apartment, bungalow), number of rooms, parking and contact person, building accessibility, dwelling accessibility, and technical aids accessibility (e.g. manual wheelchair).

A

B

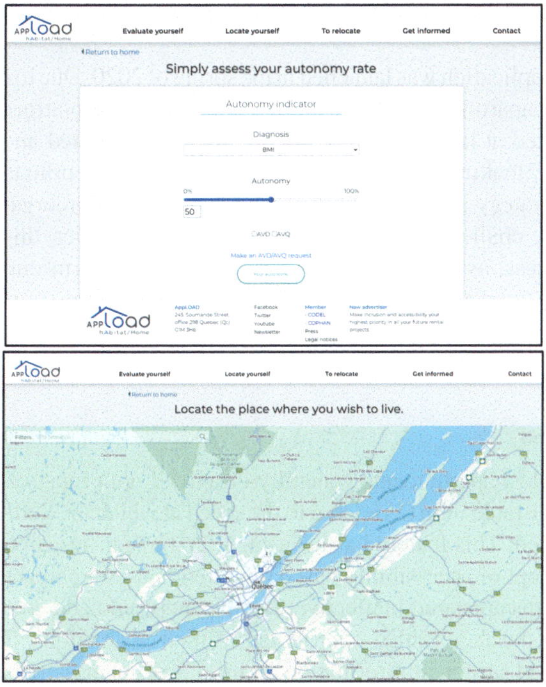

Fig. 1. Self-evaluation section (https://www.appload.ca/evaluer) and Situate yourself (https://www.appload.ca/situer) sections. Available in French, but here in English using the Google Translate plugin.

2.3 Measurements

There were three questionnaires: effectiveness (6 questions; no, moderate, yes), self-determination (5 questions; 1-strongly disagree to 5-strongly agree) and satisfaction towards the five sections of AppLOAD, adapted from the QUEST 2 [3] (12 questions about the technology and services, 1-not satisfy at all to 5-very satisfy). Participants for the interviews were contacted to explore technological aspects, AppLOAD services, and strategies to promote its use.

2.4 Recruitment

An invitation link to participate in the research was available on AppLOAD. A student then contacted each participant to obtain their consent and to invite them to complete questionnaires. Since this was a usability study, recruiting 5 users with diverse characteristics was deemed sufficient to gather relevant insights for the developer. Given the circumstances (Covid-19 pandemic and the impossibility of visiting homes), it was decided to recruit two staff members from the community organization at the end of the project. The director and the project manager agreed to be interviewed by the principal investigator.

3 Results

The AppLOAD application was launched in the spring of 2020. Due to delays in gathering data on accessible apartments caused by the lockdown, the web platform was not formally or widely promoted at launch. Instead, the organization adopted an iterative approach to bug resolution, making the web platform available online primarily to collect user feedback. This strategy also allowed the organization time to reorganize its operations in response to the challenges posed by the pandemic. However, this context hindered the research process, as few individuals used the web-platform during the six-month recruitment period, and even fewer consented to participate in the study. Two participants were recruited: P1 was a 44-year-old man without physical disability and P2 a 25-year-old woman with a physical disability.

3.1 Effectiveness

As shown in Table 1, participants generally disagreed with most of the questionnaire items. The highest-rated item (1) related to the speed at which the Self-evaluation section could be completed. This feature has been suggested by our research team to help personalize the concept of accessibility based on individual user characteristics. However, the remaining items appeared ineffective. Participants could add comments which provide insight into the factors that contributed to their evaluations. They reported that the platform lacked sufficient information on most apartment listings, which prevented them from contacting landlords. Furthermore, none of the participants found a suitable apartment using the platform, indicating that the intended data collection on actual accessibility features did not occur. Thus, they answered non-applicable (NA).

Table 1. Effectiveness of the AppLOAD platform in facilitating access to accessible housing for persons with disabilities (n = 2)

Items	*1 = No, 2 = moderate, 3 = Yes*	P1	P2
1.	Is the Capability Profile questionnaire completed quickly? (e.g. login, find link)	2	3
2.	Did the associated services facilitate the move (e.g. moving assistance, reference bank for home help (ADL/ADL attendants), electronic lease signature)?	1	1
3.	Were the housing opportunities offered relevant to you?	1	1
4.	Were the features of the units sufficient to complete your search (main entrance, intercom, indoor/outdoor parking, hallway, 3 ½, 4 ½, elevator, nearby services (e.g. pharmacy, school, church…)	2	1
5.	After visiting the property virtually or in person, was the property really accessible?	NA	NA
6.	Did the platform enable you to choose your own accommodation?	NA	1
	Maximum total score on /18	6/12	6/15

3.2 Self-Determination

Use a different scale than the previous tables, ranging from 1 to 5. Two distinct response profiles emerge for the self-determination scores presented in Table 2. P1, the person without physical disabilities consistently rated items 3 or higher, whereas the participant with disabilities (P2) rated all items 3 or lower. This contrast suggests that, for the participant with disabilities, the platform did not meet its objectives in supporting self-determination. As Participant 2 stated in their comment, *"The site must be improved in order to function properly."*

Table 2. Personal outcomes (self-determination) associated with the use of the AppLOAD platform in facilitating access to accessible housing for persons with disabilities (n = 2)

Items	*1 = strongly disagree to 5 = strongly agree*	P1	P2
1.	This platform has allowed me to feel more in control of choosing my next living environment	3	2
2.	Using this platform makes me feel proud of myself	NA	2
3.	This platform helped me plan my move	3	1
4.	Using this platform has made me feel more independent	4	3
	Comment: What were the main benefits for you of using this platform?	-	See text above
	Maximum score on /20	10/15	8/20

3.3 Satisfaction

Table 3 shows that the satisfaction levels reported by both evaluators were relatively similar, apart from item 8. Several aspects of the platform were rated positively, including ease of access, ease of use, visual appearance, and the "Self-evaluation" section. However, satisfaction decreased for both participants when it came to the information available on the platform. Their comments further highlighted these shortcomings, revealing key weaknesses that influenced their satisfaction (e.g. item 7: "*I could not find an apartment*", item 9: "*Some filters do not work*").

Table 3. Satisfaction questionnaire regarding the AppLOAD that aimed at facilitating access to accessible housing for persons with disabilities (ESAT; n = 2)

Items	*1 = Not satisfied at all to 5 = Very satisfied*	P1	P2
1.	**TECHNOLOGY**		
2.	Ease of access to the platform (e.g. login, to find link)	5	5
3.	Self-evaluation section	3	4
4.	Ease of use	4	4
5.	Information contained on the platform	2	2
6.	Visual appearance (e.g. text size, contrasts)	5	4
7.	Geographical coverage	NA	4
8.	Efficiency (find a home that meets your needs)	NA	2
9.	Platform configuration (e.g. layout, tabs)	2	4
	SERVICES		
9.	Filters, technical sheets (Situate yourself section)	NA	1
10.	Availability of moving services	NA	4
11.	Computer support	NA	NA
12.	Professional service to help find accommodation or related services	NA	NA
	Total mean score on /5	3.5	3.4

3.4 Technological Aspects of the Web Platform

This section summarizes what emerged from two interviews. They were done 1 year after the launch of the web platform. The first is the organization's director and the second is the project manager. Interviews covered the technological aspects of the platform, namely items 2, 4, 5, 6, and 8 of the satisfaction questionnaire.

"Self-evaluation" Section - The project manager indicated that the range of disabilities supported by the site has been expanded. Initially focused on physical disabilities, the platform is now being adapted to include visual, auditory, and cognitive impairments. The director noted that feedback from the research team led to improvements in this section, notably by allowing users to manually adjust their perceived level of autonomy.

"Situate Yourself" Section - The project manager reported that they encountered several technological barriers when attempting to create a direct link with another database. A web developer was hired to address this issue. Consequently, the latest problems related to linking the database to apartment listings have been resolved. As a result, the listings will now include more information and photos (see Fig. 2).

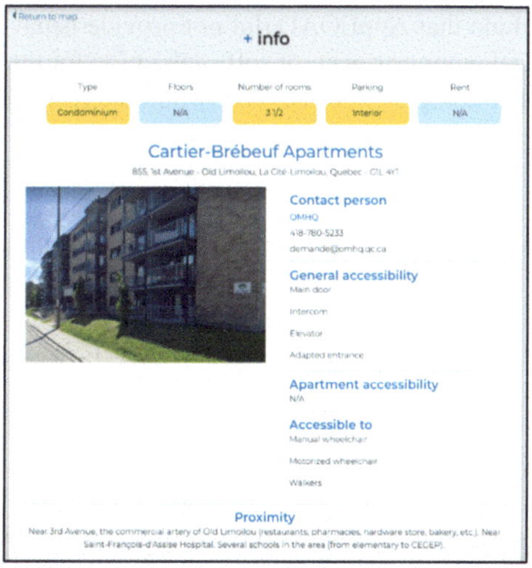

Fig. 2. Technical sheet from the "Situate yourself section" (available in French, but here in English using the Google Translate plugin).

Geographic Coverage - With respect to geographic coverage, the first participant stated that it is currently focused on the city of Québec. However, the broader goal was to serve the area covered by CAPVISH, which includes the Capitale-Nationale region.

"Find out More" section - The development of this section was not without challenges. According to the project manager, the issues that had previously prevented regular updates have now been resolved. Both participants highlighted the significant contribution of two law students who provided legal information regarding housing rights. One participant noted that this section stemmed from a desire to educate the public on their rights, after having supported individuals in court in defense of those rights.

Users' Accessibility - Regarding this dimension, the first participant emphasized that a key improvement was making the website accessible to individuals with visual or auditory impairments. The site now includes a reading mode that simplifies the display, and allows users to adjust font type, size, color, and spacing. The second participant mentioned more technical aspects, such as the ability to filter and select the type of information users are seeking, with these filters being easily accessible during map navigation, as well as the addition of a glossary.

4 Discussion

Through this research project, we witnessed several difficulties and barriers that prevented us from achieving some of our research objectives. Data collected nonetheless allowed for the documentation of the four self-determination variables for people with motor disabilities concerning their residential autonomy [4]. We were able to observe that the "Self-evaluation" tab on AppLOAD allows for the documentation of several types of capabilities (9 diagnoses). Since three of the variables could not be optimized at that time, we can conclude that AppLOAD did not provide behavioral autonomy, self-regulation, psychological empowerment, or self-realization for the platform's users [4]. However, we believe that doing this project now would increase the self-determination results. The research report is also available with more details [5].

It should be noted that the ambition of this web platform is significant, since CAPVISH had no long-term financing commitments. It is something that is rarely seen, primarily due to the scale of work involved in creating and updating apartment listings. Additionally, the technical challenge of linking several databases and services was ambitious. We learned that an even more ambitious platform than APPLOAD had been set up in Australia. When submitting this article to AAATE, we were referred to a new platform in Australia, the Housing Hub (see Fig. 3). We would be very interested to know if their site enables self-determination for people with disabilities. It says that they have supported over 1300 people to achieve a home and living goal.

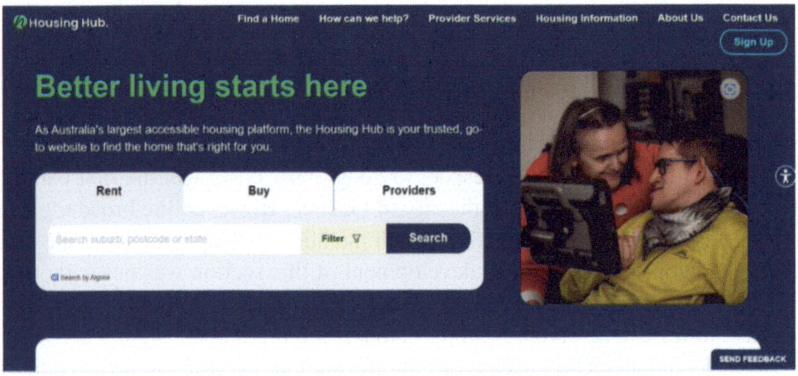

Fig. 3. Capture of Housing Hub website (Housing Hub) in 2025.

5 Conclusion/Implications for the AT Field

A platform that lets you find accommodation for yourself remains a technological innovation. Updating housing records, and finding when homes are rented or become vacant, remain major challenges. The platform is now in use, and we hope that people can find accommodation for themselves (https://www.appload.ca). One can visit the web platform today and appreciate the progress made.

Funding and Ethic Statements. The research project was funded by the Social Sciences and Humanities Research Council (892-2020-0048) and approved by the CIUSSS-CN ethics committee (#2021-2107).

References

1. Boucher, N., Miller, W.C., Smith, M.E.: Housing and attendant services: cornerstone of community reintegration after spinal cord injury. In: Eng, J.J., Teasell, R.W., Miller, W.C. (eds.) (2014)

2. Wright, C., Colley, J., Kendall, E.: Exploring the efficacy of housing alternatives for adults with an acquired brain or spinal injury: a systematic review. Brain Impairment **21**(2), 125–153 (2020)
3. Demers, L., Weiss-Lambrou, R., Ska, B.: The Quebec user evaluation of satisfaction with assistive technology (QUEST 2.0): an overview and recent progress. Technol. Disabil. **14**, 101–105 (2002)
4. Wehmeyer, M.L.: A functional model of self-determination: describing development and implementing instruction. Focus Autism Other Dev. Disabil. **14**(1), 53–62 (1999)
5. Vincent, C., Dumont, F.S., Schmouth, M.-E., Caouette, M., Boucher, N., Begnanhi, E.: Residential autonomy, accessible housing and self-determination for people with physical disabilities. Cirris Research Report Archives (2024). https://www.cirris.ulaval.ca/en/produits/residential-autonomy-and-self-determination-for-people-with-physical-disabilities/

Enhancing Mobility and Quality of Life: A Longitudinal Study on the Impact of Customized Wheelchairs in India

Rajendra Prasad, Suvlaxmi Gurumayum(✉), and Rohini Gaur

Kara Medical Foundation, Patna, India
hello@kara.foundation

Abstract. Access to appropriate mobility aids is crucial for individuals with locomotor impairments, yet many in India lack access to well-fitted assistive devices. This study evaluates the long-term impact of customized wheelchairs on mobility, independence, and overall quality of life. A longitudinal study was conducted with 43 participants, assessing outcomes three months and two years after receiving customized wheelchairs. Data were collected using the WHOQOL, FEW, and PIADS scales, with paired t-tests performed to determine statistical significance. Results indicate sustained improvements in mobility, psychosocial well-being, and daily functioning, with a 12.5% increase in FEW scores and notable gains in quality of life and independence. Despite these benefits, challenges such as accessibility to repairs and affordability persist. The findings highlight the need for better support structures, including repair services, peer mentorship programs, and policy initiatives to enhance affordability. Addressing these barriers could ensure that individuals with disabilities experience long-term benefits from customized mobility solutions, ultimately fostering greater independence and inclusion.

Keywords: Customized Wheelchairs · Assistive Technology · Mobility · Quality of Life · India

1 Introduction

Mobility is fundamental to independence and participation in daily life. However, for millions of people in India with mobility impairments, access to appropriate assistive devices remains a challenge. According to the 2011 Census of India, approximately 26.8 million people have disabilities, with locomotor impairments being among the most common. Despite this, the availability of mobility aids, particularly customized wheelchairs, is limited. Research indicates that while many individuals could benefit from mobility devices, only a small percentage receive them, and an even smaller number obtain one that meets their specific needs [1].

Existing literature highlights the importance of aligning wheelchair provision with the World Health Organization's (WHO) guidelines. The WHO emphasizes a user-centered approach that includes assessment, prescription, fitting, training, and follow-up. Studies have demonstrated that adherence to these guidelines leads to improved

K. Mavrou and P. Encarnação (Eds.): AAATE 2025, *Technology for Inclusion and Participation for All: Recent Achievements and Future Directions*, pp. 280–287, 2025.
https://doi.org/10.1007/978-3-032-01628-7_34

outcomes such as reduced pressure sores, increased mobility, and greater participation in daily activities. However, these guidelines are often inconsistently implemented in resource-constrained settings [1].

Mukherjee and Samanta [2] argue that the widespread practice of wheelchair donation—often under the guise of charity—can be counterproductive within community-based rehabilitation (CBR) contexts. Their work critiques the efficacy of such donations, highlighting that uncoordinated or inappropriate wheelchair distribution can result in a mismatch between device and user needs, especially regarding environmental suitability and postural support. This mismatch may cause secondary health complications and decreased mobility. Further compounding these challenges is the lack of user training in safe wheelchair use and maintenance. In addition, reliance on charitable donations may undermine local service ecosystems, stifle local manufacturing and repair capabilities, and perpetuate social stigma and dependency. These issues underscore the need for a paradigm shift from donation-driven models to sustainable, locally integrated wheelchair provision systems aligned with WHO guidelines [2].

Adding further depth to the rural Indian context, Gupta et al. [3] explore how persons with disabilities in rural communities experience various dimensions of invisibility. Their findings emphasize that beyond assistive technology, systemic barriers rooted in poverty, social stigma, and infrastructural exclusion shape daily realities. The study identifies five domains of invisibility: physical (due to inaccessible infrastructure), economic (lack of employment and skills training), social (stigma and exclusion), political (absence from decision-making), and cultural (deep-rooted misconceptions about disability). These intersecting factors limit participation, reinforce marginalization, and hinder the effectiveness of even well-intentioned interventions.

Moreover, Gupta et al. highlight gaps in the implementation of disability-related services and policies in rural India. Bureaucratic inefficiencies, lack of awareness, and intersectional disadvantages linked to gender, caste, and socioeconomic status all compound the challenges faced by persons with disabilities. Importantly, their study calls for more inclusive, community-driven solutions that center the voices and experiences of disabled individuals, especially in rural settings where invisibility is most acute [3].

Despite their advantages, customized wheelchairs remain inaccessible to many due to financial constraints, lack of awareness, and limited distribution channels. Current data suggests that only about 16% of individuals requiring mobility devices actually own one. Startups such as NeoMotion have introduced innovative solutions, yet widespread adoption continues to be a challenge.

2 Methodology

2.1 Study Design and Background

This study builds upon a previously published pilot project that evaluated the feasibility and effectiveness of teleassessment-based wheelchair provision in rural India [1]. That study focused on short-term (3-month) outcomes and established that remote assessments could lead to effective wheelchair customization and improved user satisfaction.

The current research extends that work through a mixed-methods, longitudinal follow-up, assessing the same cohort (n = 43) two years after wheelchair provision to evaluate long term impacts on mobility, psychosocial well-being, and quality of life.

2.2 Participants

Participants were 43 adults (41 males, 2 females) with locomotor impairments who were enrolled in the original teleassessment study [1]. Inclusion criteria for that study were:

- Verified need for a wheelchair
- Access to a smartphone and caregiver to support telehealth assessments
- Functional upper limbs and ability to travel to a regional assistive technology (AT) center
- Willingness to participate in training and follow-up assessments

Exclusion criteria included children, individuals with severe spasticity or who were bedridden, and those without reliable mobile technology. The same cohort was retained for this two-year follow-up, ensuring continuity of evaluation.

2.3 Teleassessment Protocol

As outlined in the initial study [1], the assessment process followed WHO-TAP guidelines and was adapted for rural delivery via telehealth. Assessment involved 1–3 structured video calls that covered:

- Socioeconomic and environmental context (e.g., housing conditions, terrain, mobility needs)
- Functional and neurological assessment of impairments
- Remote guidance for measuring seat width, back height, and footrest length, using caregiver support
- Wheelchair selection based on individual use patterns, home layout, and mobility range

Participants received their customized wheelchairs at a designated AT center, where peer mentors provided operational training. All beneficiaries were trained to use and maintain their wheelchairs independently.

2.4 Data Collection

At the two-year follow-up, participants were re-evaluated using the same validated instruments:

- WHOQOL-BREF: To assess overall quality of life across physical, psychological, social, and environmental domains
- FEW (Functioning Every Day with a Wheelchair), Version 2.0: Administered to track changes in functional mobility since the 3-month baseline
- PIADS (Psychosocial Impact of Assistive Devices Scale): Used to assess psychological well-being, adaptability, and independence

3 Results

The findings demonstrate a significant positive impact of customized wheelchairs on mobility, independence, and overall well-being.

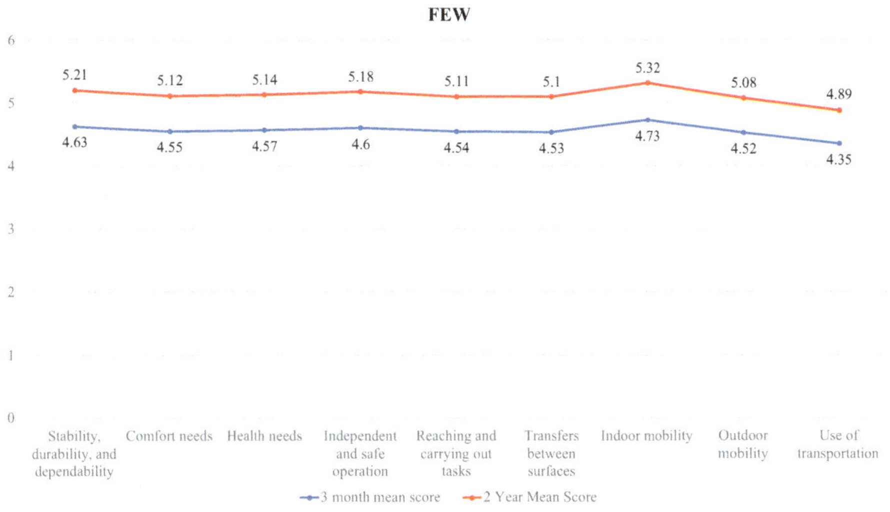

Fig. 1. FEW (Functioning in Everyday Life with Wheelchair Scale)

As shown above in Fig. 1, Two years post-distribution, FEW scores reflected increased user satisfaction, stability, and independence compared to the three-month evaluation. The average FEW score increased by 12.5%, with 81.4% of participants (35 out of 43) reporting enhanced mobility. Initially, 18.6% (8 out of 43) had scores below 50/60, but this figure dropped to 7% (3 out of 43) after two years, highlighting improved adaptability.

Notable improvements include:

- Item 4 (Operate): Ease of use increased from 86% to 93%.
- Item 10 (Transportability): Users of rigid wheelchairs reported improved transportability, reflecting better environmental adaptation and user experience.
- Para-athletes maintained high satisfaction levels, demonstrating that training and prior experience contribute to long-term usability.

While most participants continued using their original wheelchairs, seven required repairs, such as front wheel replacements, six repaired it and kept using it and one participant had to replace their wheelchair entirely due to irreparable damage.

As shown above in Fig. 2, Post-assessment scores indicated significant improvements in multiple life domains:

- Health satisfaction (M = 4.29)
- Daily activities (M = 4.55)
- Personal relationships (M = 4.81)

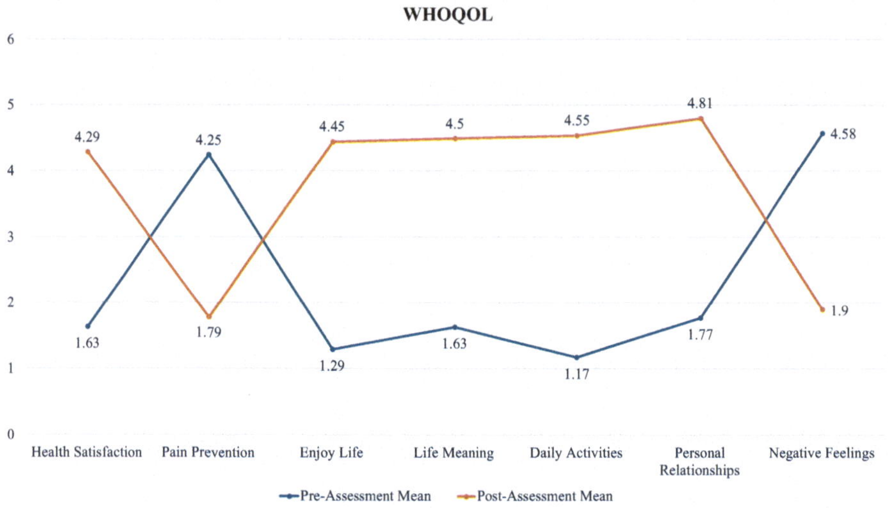

Fig. 2. WHOQOL (World Health Organization Quality of Life Scale)

- Reduction in physical pain (M = 1.79) and emotional distress (M = 1.90)

These results emphasize the broader health and social benefits of customized wheelchair use beyond mobility enhancement.

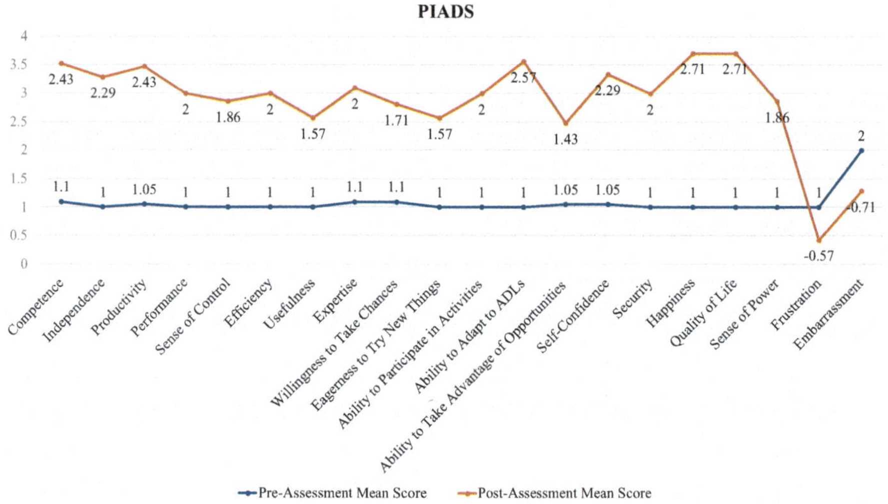

Fig. 3. PIADS (Psychosocial Impact of Assistive Devices Scale)

As shown above in Fig. 3, Post-assessment PIADS scores revealed substantial improvements:

- Competence (+2.43), Independence (+2.43), and Quality of Life (+2.71)
- Adaptability to daily activities increased to 2.57
- Reduction in frustration (-0.57) and embarrassment (−0.71)

Although gains were noted across all domains, willingness to take risks (+1.71) and eagerness to try new activities (+1.57) saw relatively lower increases, suggesting some users may still experience psychological barriers to full social participation.

4 Discussion

The findings of this longitudinal study affirm the long-term benefits of customized wheelchair provision on mobility, psychosocial well-being, and quality of life. Consistent improvements across all three measurement instruments (FEW, WHOQOL, and PIADS) suggest that tailored mobility solutions not only enhance day-to-day functioning but also positively influence psychological and emotional outcomes.

The FEW scale improvements between the 3-month and 2-year follow-up underscore sustained user adaptation and increased satisfaction over time. Particularly notable were gains in indoor mobility, safe operation, and health-related comfort, highlighting that appropriate fitting and training result in improved usability in both home and community settings. These results align with earlier findings from Griggs (2024), who emphasized that user satisfaction with manual wheelchairs (MWCs) is heavily influenced by fit, comfort, maneuverability, and terrain adaptability [1].

Our PIADS results further demonstrate the psychological value of proper assistive device provision. Competence, independence, and quality of life all saw substantial gains, reflecting a deeper shift in perceived self-efficacy and autonomy. These findings extend the conclusions of Mukherjee and Samanta (2005), who cautioned that inappropriate, charity-based wheelchair donations may foster dependency and psychological disengagement [2]. In contrast, our data supports a more empowering, user-centered model of provision, consistent with WHO guidelines and a rights-based approach to disability services.

The WHOQOL outcomes reflect the broader impacts of mobility on emotional and social dimensions. Participants reported substantial reductions in pain and negative feelings, and dramatic improvements in their engagement in daily activities, relationships, and life meaning. These shifts mirror observations by Gupta et al. (2020), who documented how disability-related social and cultural invisibility undermines participation in rural India [3]. Our findings suggest that when mobility needs are met appropriately, these barriers can begin to be dismantled.

Importantly, the study also captured substantial social and economic transformation among participants. All individuals were able to resume income-generating or developmental activities, such as reopening local shops, attending vocational training, or returning to college. Para-athlete participants resumed sports training and reported participating in tournaments—an outcome made possible through improved outdoor mobility and safe transport. These shifts highlight how mobility is foundational not only to physical independence but also to economic reintegration and future planning.

A notable consequence of this transition was the redistribution of household caregiving responsibilities. Prior to the intervention, women—often mothers or wives—were the default caregivers, bearing the dual burden of economic provision and physical assistance. Post-intervention, participants were able to contribute financially and function more independently, reducing caregiver strain and shifting family dynamics toward shared responsibility. This is a significant yet often underreported outcome of assistive technology interventions, particularly in patriarchal and economically disadvantaged rural contexts.

Participants also re-entered public life. Many reported attending religious events, community gatherings, and awareness drives, which not only enriched their social identity but also helped raise local consciousness about mobility-related disabilities. This aligns with the goals of community-based rehabilitation (CBR) and echoes the empowerment themes discussed by Gupta et al. [3].

From a socioeconomic perspective, 83.7% of participants were below the poverty line, yet they demonstrated sustained use of and adaptation to their wheelchairs. Repairs were successfully managed at local workshops, indicating that the intervention model was not only technically sustainable but also economically viable in low-resource settings. This stands in contrast to Griggs' (2024) findings in the UK, where dissatisfaction often stemmed from maintenance barriers and lack of modular design [1].

Additionally, our study reinforces Griggs' call for improvements in customisation, pressure relief, and user training. The satisfaction gains seen in our cohort are likely attributable to the teleassessment model, which allowed for detailed measurement, user education, and environmental tailoring. This contrasts with common MWC dissatisfaction in the UK, where poor fit and lack of training were key barriers [1]. Furthermore, Griggs' emphasis on the need for lighter wheelchairs and better environmental compatibility is echoed in our results: foldable, terrain-appropriate wheelchairs showed higher FEW and PIADS scores, particularly in users who required outdoor mobility.

While much of the existing literature on wheelchair provision has focused on high-income or urban contexts, our study fills a critical gap by demonstrating the feasibility, sustainability, and multi-dimensional impact of teleassessment-based, customized wheelchair provision in rural low-resource settings. The strong outcomes reported after two years reinforce the sustainability of this approach—not only in terms of device functionality but also in restoring dignity, enabling participation, and supporting systemic inclusion.

5 Conclusion

This study provides compelling evidence that customized wheelchair provision—delivered through a teleassessment model—can lead to lasting improvements in mobility, independence, psychosocial well-being, and overall quality of life for persons with disabilities in rural India. The integration of user-centered design, individualized assessment, and peer-supported training proved critical to the success of the intervention.

In contrast to traditional donation-based models critiqued by Mukherjee and Samanta [2], this approach empowers users and facilitates true participation—not just physical movement. Our results also validate and extend insights from Griggs [1], emphasizing that comfort, adjustability, fit, and training are universal priorities across cultural and geographic contexts.

Going forward, we recommend:

Expansion of teleassessment-based wheelchair services in low-resource and remote settings.

Integration of pressure relief technologies and modular customization options to further enhance comfort and reduce secondary complications.

Structured caregiver training and follow-up support to ensure proper use, maintenance, and sustainability.

Policy advocacy to support local repair networks and funding mechanisms for customized mobility solutions.

Future research should explore comparative cost-effectiveness, integration with community-based rehabilitation (CBR) systems, and broader psychosocial impacts across genders, age groups, and disability types. Ultimately, meeting the mobility needs of persons with disabilities must go beyond distribution—it requires systems that are accessible, responsive, and grounded in dignity and autonomy.

References

1. Griggs, K.E.: Wheelchair satisfaction and recommended improvements of manual wheelchairs in the UK. Disabil. Rehabil. Assist. Technol. **20**(1), 163–170 (2024). https://doi.org/10.1080/17483107.2024.2359473
2. Mukherjee, G., Samanta, A.: Wheelchair charity: a useless benevolence in community-based rehabilitation. Disabil. Rehabil. **27**(10), 591–596 (2005). https://doi.org/10.1080/09638280400018387
3. Gupta, S., de Witte, L.P., Meershoek, A.: Dimensions of invisibility: insights into the daily realities of persons with disabilities living in rural communities in India. Disabil. Soc. **36**(8), 1285–1307 (2020). https://doi.org/10.1080/09687599.2020.1788509
4. Geilen, B., Witte, L., Norman, G., George, C.: Quality of wheelchair services as perceived by users in rural Bangalore district, India: a cross-sectional survey. Disabil. Rehabil. Assist. Technol. **17**, 1–9 (2020). https://doi.org/10.1080/17483107.2020.1839576
5. Prasad, R., Sinha, V.K., Gurumayum, S., Gaur, R., Nayar, S.: Implementation of tele-assessment and delivery of wheelchairs in low/middle-income countries. In: 18th AAATE Conference 2023, pp. 1–12. AAATE, Paris (2023)

Development of a Value Appraisal Tool for Implementing Technology in Care for Older Adults

Rosalie van der Vaart(✉) and Luc P. De Witte

Technology for Health Care Research Group, The Hague University of Applied Sciences, The Hague 2521EP, Netherlands
r.vandervaart@hhs.nl

Abstract. At present, innovation via technology is vital for sustainable, high-quality care. Yet, implementing such innovations is complex and requires careful decision-making. This study, conducted in collaboration with eleven care organizations, aimed to develop a value appraisal tool to support informed implementation of care technology. The tool was created through stakeholder analysis, literature review, checklist development, and testing. Stakeholders included clients, informal caregivers, care professionals, care teams, and organizational managers. Aspects related to technology, financial considerations, and regulatory requirements were also included. Key values, needs, and preconditions for each stakeholder were identified and translated into a comprehensive checklist. This was tested via scoping reviews and interviews for six technologies, e.g. medication dispensers and lifestyle monitoring. The current tool offers a structured approach to assess the value of technologies, balancing evidence with the specific needs and capacities of organizations. By actively involving diverse stakeholders and identifying critical success factors, it supports the sustainable implementation of care technology. This approach has significant implications for advancing the adoption of technologies in healthcare, as part of the broad field of assistive technology, emphasizing the need for inclusivity, contextual adaptation, and systematic evaluation in care innovation.

Keywords: Care technology · Care for older adults · Value appraisal · Implementation

1 Introduction

Care innovation involving the use of assistive technology and care technology has become indispensable in care, and also in care for older adults, both in the Netherlands and internationally [1, 2]. Aging populations and the associated rising demand and complexity of care pose significant challenges. Workforce shortages in a shrinking labor market exacerbate these issues [3]. Care innovation is essential to address these challenges while maintaining or improving care quality. However, implementing

K. Mavrou and P. Encarnação (Eds.): AAATE 2025, *Technology for Inclusion and Participation for All: Recent Achievements and Future Directions*, pp. 288–295, 2025.
https://doi.org/10.1007/978-3-032-01628-7_35

technological innovations is a complex process, requiring significant time, planning, resources, and perseverance [4]. Making well-informed decisions about which technologies to implement, involving whom, and how to integrate them into routine care is critical. These are multifaceted decisions that require balancing the interests of various stakeholders, including clients, caregivers, informal caregivers, care management, and health insurers [5].

To better understand and navigate the complexities of implementing care technologies, the NASSS (Non-adoption, Abandonment, Scale-up, Spread, and Sustainability) framework provides a valuable lens. This framework identifies key domains influencing the adoption and long-term use of technology, including the nature of the condition being addressed, the technology itself, the value proposition, the workforce and organizational context, the wider system, and the process of embedding and adapting the innovation over time [5, 6]. By applying the NASSS framework on the context of care for older adults, organizations can anticipate on relevant potential barriers and facilitators in the decision-making process.

This project, conducted in collaboration with a partnership of eleven care organizations, aimed to develop a value appraisal tool for care technology in care for older adults, using the NASSS framework as a basis. This tool should support health care organizations to collaboratively assess new technologies or to increase the scale up of existing technologies. The tool is intended to be used reflexively, to guide conversations and decision making in teams and organizations. It incorporates scientific knowledge on acceptance, impact and outcomes of assistive and care technologies, practical experience from care organizations, and the diverse needs of stakeholders, to facilitate sound choices regarding the adoption of care technologies within the care for older adults.

2 Methods

A project team consisting of researchers and innovation managers and project leaders from four out of the eleven care organizations was formed. For the appraisal tool four steps were taken: (1) Stakeholder analysis: the relevant stakeholders' perspectives to be included in the tool were based on the NASSS framework and further identified through discussions with the project team. (2) Gathering values: a thorough review of existing literature and expert opinions were used to determine the values, needs, and contextual considerations relevant to each stakeholder group. This resulted in a comprehensive list of outcomes. (3) Tool development: the comprehensive list was refined into a practical tool, shaped like a checklist, through discussions with the project team. (4) Testing: the tool was tested using six technologies with high potential in care for older adults, namely medication dispensers, lifestyle monitoring, video consulting, day structuring tools, electronic home access management and smart incontinence materials. For each technology, a scoping review of scientific and grey literature was conducted, and interviews with innovation managers and care professionals were held on their experiences with the care technology. These combined insights were tailored to the context of each organization through discussions with the project team. The gathered information was synthesized and used to develop a final version of the value appraisal tool.

3 Results

3.1 Stakeholder Analysis

For the stakeholder analysis, the NASSS framework was used as a basis. Together with the project team, the different people, roles and parties involved were mapped within the setting of care for older adults. The following key stakeholders were identified: clients, informal caregivers, care professionals (intramural and extramural), professional teams, and care organizations and management. Additionally, aspects related to the technology, financial considerations, and regulatory requirements were included.

3.2 Gathering Values

To gather values for each group of stakeholders, a literature search was done, including a search for other value appraisal tools. Also, conversations with project leaders within the participation organizations were held. This provided insights into the needs, values, and contextual requirements of each stakeholder group. The identified aspects included user attitudes, satisfaction, usability, training and support needs, effects, inclusiveness, privacy and safety considerations, tailoring or adaptability, impact on care delivery, organizational readiness, leadership, financial and regulatory aspects, and sustainability.

3.3 Tool Development

The list of values was translated into a checklist in which for each value specific questions were posed to gain insight into that specific value. This resulted in a list of three to seven values per stakeholder, represented in six to eleven questions. For example, questions regarding clients focused on usability (What do clients need to do to utilize the care technology?), effects (What is the influence of the care technology on clients regarding independence or autonomy, health problems, quality of life, etc.?), required support and training (Do clients need specific skills to use the care technology? Do clients need training or support to acquire these skills?), inclusiveness (What are in- and exclusion criteria for use? Are there any language barriers?), and intrusiveness (What are points of attention regarding privacy and GDPR?). Questions regarding the health care organization focused on vision (Does the CT align with the organizations' vision, Is there urgency to start with the CT and on what time line?), leadership (How does the organizations' leadership support the implementation and adoption of the CT?), organizational capacity (How does the CT align with other projects or changes within the organization. Is there capacity to monitor the CT's effects and progress?) and readiness for change (Is there sufficient capacity for change? (e.g. project teams, staff involvement, time)). Appendix A provides an overview of the complete value appraisal tool.

3.4 Testing

The checklist was tested using six innovations: electronic home access management, day structuring tools, video consultation, lifestyle monitoring, medication dispensers, and smart incontinence materials. For each technology, a scoping review of scientific and grey literature was conducted, and interviews with innovation managers and care professionals were held on their experiences with the technology. This process provided a comprehensive overview of positive and negative aspects of each technology and preconditions for implementation, from the viewpoint of all included stakeholders. Based on discussions within the project team, the appraisal tool was refined, and individual reports were created for each technology to support decision-making in the respective organizations.

4 Discussion

This project represents a significant step towards a systematic approach to innovation and care improvement within organizations delivering care for older adults. The value appraisal tool provides a structured method to assess the added value and needed preconditions of care technologies and to select promising innovations by leveraging existing scientific knowledge and practical experience within the context of the participating organizations. It supports deliberate considerations on the technology's target group, intended use, place in the care process and outcome monitoring needed.

A key strength of this approach is the integration of the NASSS (Non-adoption, Abandonment, Scale-up, Spread, and Sustainability) framework. NASSS informed the identification of the stakeholders and the values used within the appraisal tool, by highlighting critical domains such as the characteristics of the technology, inclusiveness for clients, organizational readiness, and broader systemic influences. This way, we aimed to ensure that the appraisal tool captures both the scientific evidence base and the specific needs and conditions of each organization. By explicitly considering these dimensions, the process enables organizations to anticipate potential barriers and facilitators to adoption, and to make better-informed, context-sensitive decisions. By carefully weighing the pros and cons of each technology, it helps prevent premature implementation without adequate preparation.

This structured approach facilitates the identification of success factors and preconditions, contributing to more effective and sustainable integration of technologies. A central principle is the active involvement of diverse stakeholders—including care staff, management, clients, and informal caregivers—which fosters support and increases the likelihood of successful implementation. Completing the value appraisal tool forms the basis for the discussion on the innovation or technology. In this discussion, the aspects that determine a 'go' or 'no go' to start (or scale up) the innovation are discussed. During these discussions, the preconditions that need to be met are clearly defined. Depending on the outcome, a decision can be made to conduct further research or a pilot to assess whether and how these preconditions can be met, and the priority that is given to the specific technology in relation to other relevant innovations. The next challenge lies in embedding this tool into organizations' internal practices and broader collaborative networks.

4.1 Implications for the Field of Assistive Technology

Care technology in care for older adults intersects with assistive technology by focusing on enhancing care delivery and patient outcomes, while complementing the independence and quality of life improvements facilitated by assistive tools in general. Our approach in the development of a value appraisal tool for care technologies underscores the value of comprehensive assessment as a means to enhance the alignment between technological innovation and practical care delivery. It emphasizes the critical role of stakeholder involvement and contextual adaptation in achieving sustainable implementation. Future efforts should focus on refining and scaling this tool across diverse care contexts and fostering collaboration between end-users, care organizations, technology developers, and researchers. This will support the development and dissemination of assistive technologies that are not only innovative but also meaningful, practical, and widely adopted in care settings.

Appendix

Appendix a: Health Technology Value Appraisal Tool

Stakeholders & values	Focus area / key questions *CT = Care Technology
Client	
User satisfaction	How do clients experience the CT?
	How do clients perceive the new way of working and care delivery with the CT?
Usability	How user-friendly is the CT for clients?
	What do clients need to do to use the CT, both during initial setup and regular use (e.g., charging)?
Skills and support	Do clients need specific skills to use the CT?
	Is training or support needed for clients? If so, what kind?
Effects	What is the impact of CT on clients? (e.g., independence/autonomy, symptoms, safety, quality of life, etc.)
Inclusiveness	What are the inclusion and exclusion criteria for using the CT? Are there any language barriers?
Intrusiveness (privacy, GDPR)	What are the key privacy and GDPR considerations for clients?
Informal caregiver	
(Family)Values	What is the attitude of the client's social network towards the CT?
Usability	How user-friendly is the CT for informal caregivers?

(*continued*)

(*continued*)

Stakeholders & values	**Focus area / key questions** *CT = Care Technology
	What do informal caregivers need to do to use the CT, both during initial setup and regular use (e.g., charging)?
Skills and support	Do informal caregivers need specific skills to use the CT?
	Is training or support needed for informal caregivers? If so, what kind?
Effects	What is the impact of CT on informal caregivers? (e.g., workload, task support, reduced stress, etc.)
Collaboration with formal care	Does the CT contribute to the involvement of informal caregivers? If so, how?
Recognition and appreciation	Does the technology include features that recognize and support the role of informal caregivers?
Privacy	How do informal caregivers perceive the impact of CT on their own privacy and that of the client?
Healthcare professional	
Attitude	Is there motivation, trust, and intention to use the CT?
Usability	How do healthcare professionals experience the usability of the CT?
	Is the method of care delivery with CT perceived as comfortable?
	What do healthcare professionals need to do to use the CT, both during initial setup and regular use (e.g., charging)?
Skills and support	Do healthcare professionals need specific skills to use the CT?
	Is training or support needed for healthcare professionals? If so, what kind?
Work processes	What are the most significant changes in work processes when using the CT?
	How do healthcare professionals perceive the associated way of working?
Effects	What are the effects of CT on healthcare professionals? (e.g., autonomy, safety, time savings, support, workload, etc.)
Team	
Attitude	Is the team open to a different way of working? Is there a willingness to change?
Work processes	What are the biggest changes in team workflows when working with the CT?
	How does the team experience the associated way of working?

(*continued*)

(*continued*)

Stakeholders & values	**Focus area / key questions** *CT = Care Technology
Strong leadership	To what extent can team leaders/managers guide their staff through the transition to the new way of working?
Tailoring/adaptability	Can the CT be customized to meet the specific needs of the team?
Implementation	What is needed to implement the CT in the team? (e.g., time and resources)
Technology	
Safety	Does the CT meet legal safety requirements (e.g., NEN/ISO certifications)?
	Are there safety concerns regarding the connection between the CT and the organization's systems?
	What other security considerations exist? What is required for the CT's security?
Sustainability	How sustainable is the CT and its use? (e.g., duration of use, energy requirements, reuse possibilities, cost savings, etc.)
Readiness	Can/should the CT be modified to fit this organization's needs?
	What is needed for system management after implementation?
	What system modifications are required to integrate CT, and what is their impact?
Health care organization	
Vision	Does the CT align with the organization's vision?
	Is there clarity on what the organization aims to achieve with the CT?
	Is there urgency to start with the CT, and on what timeline?
Leadership	How does the organization's leadership support the implementation and adoption of the CT?
Organizational capacity	How does the CT align with other projects or changes within the organization?
	How is collaboration structured between different disciplines (both within and outside the organization)?
	Is there capacity to monitor the CT's effects and progress?
	Are there best practices and previous experiences within the organization?
Readiness for change	Is there sufficient capacity for change? (e.g., project teams, staff involvement, time)

(*continued*)

(continued)

Stakeholders & values	**Focus area / key questions** *CT = Care Technology
	What support is needed for healthcare professionals during and after implementation? Can this be provided?
	What are the risks, and how can the organization mitigate them?
Finances, Tax Considerations, and Policies & Regulations	
Policy and legislation	Does the CT align with policies from the Ministry of Health, healthcare offices, and insurers?
	What relevant legislation exists, and what key considerations arise?
Financial considerations	What is the required one-time investment?
	What are the operational costs?
	What are the financial benefits of using the CT?
Reimbursement mechanisms	To what extent is the CT reimbursed (e.g., by health insurers, long-term care offices, or the Social Support Act)?
	Are there subsidy opportunities?
Medical device regulations	What are the applicable regulations regarding the CT as a medical device, and what is their impact?

References

1. Global strategy on digital health 2020–2025. World Health Organization, Geneva (2021)
2. Thoma-Lürken, T., Bleijlevens, M.H.C., Lexis, M.A.S., Hamers, J.P.H., de Witte, L.P.: An overview of potential labor-saving and quality-improving innovations in long-term care for older people. J. Am. Med. Dir. Assoc. **16**(6), 482–489 (2015). https://doi.org/10.1016/j.jamda.2014.12.017
3. Global strategy on human resources for health: workforce 2030. World Health Organization, Geneva (2016)
4. Ko, M., Wagner, L., Spetz, J. Nursing home implementation of health information technology: review of the literature finds inadequate investment in preparation, infrastructure, and training. Inquiry **55**, 46958018778902 (2018). https://doi.org/10.1177/0046958018778902
5. Greenhalgh, T., et al.: Beyond adoption: a new framework for theorizing and evaluating non-adoption, abandonment, and challenges to the scale-up, spread, and sustainability of health and care technologies. J. Med. Internet Res. **19**(11), e367 (2017). https://doi.org/10.2196/jmir.8775
6. Greenhalgh, T., Abimbola, S. The NASSS framework - a synthesis of multiple theories of technology implementation. Stud. Health Technol. Inf. **263**, 193–204 (2019). https://doi.org/10.3233/SCTI190123

My Assistive Technology Outcomes Framework: Tools for Assistive Technology Users and the Practitioners that Support Them

Natasha Layton[1,2(✉)], Meredith Prain[3], Diane Bell[4], Melissa Noonan[5], and Libby Callaway[1,2,6]

[1] Rehabilitation, Ageing and Independent Living Research Centre, Monash University, Frankston, Australia
natasha@natashalayton.com.au
[2] Australian Rehabilitation and Assistive Technology Association (ARATA), Beaumaris, Australia
[3] Centre of Excellence – Deafblind, Melbourne, Able Australia, Australia
[4] Business School, Stellenbosch University, Bellville, Cape Town, South Africa
[5] Limbs 4 Life Inc., Mt Waverley, Australia
[6] Occupational Therapy Department, Monash University, Frankston, Australia

Abstract. Which outcomes should we measure to demonstrate the transformative power of assistive technology (AT)? The answer may differ according to whether the question is being asked by an AT provider or funder; by practitioners involved with a certain type of AT such as assistive products for vision, or prosthetics; or - as represented in the World Health Organization's People-Centred 5P Assistive Technology Model - by the person central to the AT process being the AT user themselves. This paper presents My Assistive Technology Outcomes Framework, an AT outcome measure co-produced with AT users. My Assistive Technology Outcomes Framework captures data regarding the supports, service delivery pathways and customer experience of AT users. It also measures the outcomes, costs and cost benefits, and human rights realised or not yet realised with assistive products and services. Evidence for the Framework will be reviewed and its reach across functioning areas and assistive product categories will be presented, as well as areas for future research.

Keywords: Assistive Technology · Outcomes · User Empowerment

1 Introduction

Assistive Technology AT outcomes research refers to 'systematic investigation aimed at identifying the changes that are produced by AT in the lives of users and their environments' [1] (p. 1244). Published measurement tools frequently focus on one type of AT or environmental modification, one type of disability, or one outcome area (such as satisfaction or independence). The focus is often on outcomes of interest to funders

K. Mavrou and P. Encarnação (Eds.): AAATE 2025, *Technology for Inclusion and Participation for All: Recent Achievements and Future Directions*, pp. 296–303, 2025.
https://doi.org/10.1007/978-3-032-01628-7_36

or policymakers, such as 'value for money' or 'hospitalisations avoided' [2]. Further, published measurement tools are usually designed to be administered by health care professionals and not by AT users themselves. In contrast, the My Assistive Technology Outcomes Framework (MyATOF) was developed in Australia through a co-production process with AT users and the AT sector [3, 4]. MyATOF is a mixed methods data collection framework intended for use by AT users, in collaboration with their care partners and the AT personnel that engage with them. MyATOF asks questions across six evidence-based domains: supports; outcomes; costs; rights; service delivery pathways; and customer experience [5] (see Table 1).

Table 1. My Assistive Technology Outcomes Framework: Tools, Operating Dimensions and Supporting References.

MyATOF Tool	Contents and source references
Tool A – My supports	Assistive products and environmental adaptations: ISO 9999 [6]
Tool B – My outcomes	WHO ICF Activity and participation chapters [7]
Tool C – My costs	Costs (direct and indirect) and cost offsets including social return on investment [8, 9]
Tool D – My rights	Subset of twelve Articles from UN CRPD [10]
Tool E – My AT service delivery pathway	AT service delivery steps [11]
Tool F – My customer experience	Eight aspects of customer experience [12]
Pre- and post-measurement tool	Self-report measures (difficulty; time use, satisfaction with participation) Independent measure for extent of participation [13]

Systematic collation of these data enables AT users and their supporters to articulate the outcomes sought and the supports required. The base question set (see Appendix 1) can been adapted into accessible formats to assure useability for different access needs. To accommodate user needs across their lives and across the lifecycle of their assistive products, AT users have the capacity to build multiple reports, either for various AT solutions or bundles, or at various points in time. These reports record AT use over time, track the process as capability changes and outcomes are achieved, and provide a record of the impact of AT use (and need) for both the AT user and funders of assistive products and services [3, 4].

2 Method

This paper summarizes data from the published evidence base over a decade of MyATOF development. Inclusions are drawn from the grey literature (conference proceedings (n = 9) and black literature (n = 2, peer reviewed). Exclusions are non peer-reviewed materials such as education materials, newsletter entries, blogs and reviews.

3 Results

MyATOF has been used as a pre- and post-evaluation of assistive products and related supports for seating and positioning [14]; prosthetics [15]; and to evaluate home-integrated technologies for acquired brain injury [16]. MyATOF has also been used to explore the assistive technology needs of people living with deafblindness [17–19]; chronic illness [20]; and the late effects of polio [21]. Table 2 summarizes the use cases for MyATOF according to functioning limitations through linkage with the body structure and function chapters of the WHO International Classification of Functioning, Disability and Health (WHO ICF) [22].

Table 2. Mapping MyATOF use across WHO ICF functioning chapters

WHO ICF Body Structure and Function Chapters	MyATOF Publications
Mental functions & the nervous system	[16, 21]
Sensory functions & structures Voice and speech	[17–19, 21]
Cardio-vascular, hematological, immunological and respiratory systems Digestive, metabolic and endocrine systems, e.g. ingestion Genitourinary and reproductive systems	[20, 21]
Neuromusculoskeletal and movement-related systems	[3, 4, 14, 21]
Skin and related structures	[15]

Table 3 indicates the assistive product classes for which MyATOF has been used.

Table 3. Mapping MyATOF use across ISO 9999 Classes

ISO 9999 Classes	MyATOF application
04 Assistive products for measuring, stimulating or training physiological and psychological functions	✓
06 Orthoses and prostheses	✓
09 Assistive products for self-care activities and participation in self-care	✓
12 Assistive products for activities and participation relating to personal mobility and transportation	✓

(*continued*)

Table 3. (*continued*)

ISO 9999 Classes	MyATOF application
15 Assistive products for domestic activities and participation in domestic life	✓
18 Furnishings, fixtures and other assistive products for supporting activities in indoor and outdoor human-made environments	✓
22 Assistive products for communication and information management	✓
24 Assistive products for controlling, carrying, moving and handling objects and devices	✓
27 Assistive products for controlling, adapting or measuring elements of physical environments	✓
28 Assistive products for work activities and participation in employment	✓
30 Assistive products for recreation and leisure	✓

4 Discussion and Next Steps

Emerging from the AT user community in Australia and now being explored internationally, MyATOF is intended to enable AT users and the AT personnel who support them to describe the tailored and multiple supports which deliver valued outcomes. This documentation is in a narrative format rather than formal measurement scaling. Nevertheless a strength is the linkage with agreed international benchmarks. That is, AT users and the practitioners that support them are able to document the AT is used or that may be required according to the ISO 9999 [6] and the WHO ICF activity and participation chapters [7]. Service delivery quality is reported according to best practice service provision standards [11]. The realization of human rights through AT are described in relation to the UN Convention on the Rights of Persons with Disability (CRPD) [10]. Tools within the Framework where there is not yet a standardized taxonomy available for benchmarking have used best available evidence, and these include the customer experience [12] and cost tools [8, 9].

The results demonstrate wide applicability across ISO 9999 assistive product classes and across functioning areas. The MyATOF evidence base does not yet demonstrate its use in contexts other than Australia (one high income country) and several countries within the Southern African Development Community (SADC). A Delphi study established face validity and relevance of MyATOF for people with deafblindness in Malawi, South Africa, Zambia and Zimbabwe [19], demonstrating the importance of a functioning AT ecosystem enabling AT information and access.

5 Limitations

The decade of development of MyATOF has been less than systematic, as it is shaped by available development and funding opportunities. The span of disability groups for whom it has been of interest demonstrates comprehensive applicability across functioning limitations. However, a systematic research agenda would explore applicability across ISO 9999 constructs as well as across low and middle income countries more broadly. Strengthening the primary evidence base to support the costing and customer experience tools is needed and is a current evidence gap for the sector. Finally, further researching the pre and post evaluation tool with reliability studies will provide a much-needed standardized element to the suite of Tools with in the MyATOF.

6 Conclusion and Implications for the AT Field

Translating outcome measures into formats that can be used by AT users is epistemologically just. Co-production of tools such as the MyATOF has led to tools which address valued outcomes, and which are applicable across user groups, across types of AT and across contexts. The MyATOF has illustrated that it is feasible to 'knowledge translate' international classification systems, research evidence, regulatory and service delivery principles into plain language tools which can be used by a diverse range of AT users. While it can be argued that self-assessment is insufficiently objective to replace standardized measures, we suggest capturing AT user lived experiences aligned to internationally recognized indicators, such as rights and participation, is a valuable and empowering addition to outcome measurement.

Acknowledgments. The authors would like to acknowledge the input and knowledge of the AT user community internationally over a decade of MyATOF development.

Disclosure of Interests. The authors have no competing interests to declare that are relevant to the content of this article.

Appendix 1 MyATOF Domains and Question Sets

MyATOF enables people to talk about their assistive technology and accessible environments. Assistive technology is all products and software that optimize functioning and reduce the experience of disability for people who have functioning limitations as well as the services required to deliver them [23]. This includes assistive products which are especially made or adapted for people with disabilities as well as mainstream or generally available products with features which optimize functioning and reduce the experience of disability [6],.

Tool 1 My supports: what AT and other things do I use

PROMPT: To fully document all supports, it is suggested that MyATOF users review at the range of AT listed on relevant national databases [24] to get the names for things

04	Assistive products for measuring, stimulating or training physiological and psychological functions
06	Orthoses and prostheses
09	Assistive products for self-care activities and participation in self-care
12	Assistive products for activities and participation relating to personal mobility and transportation
15	Assistive products for domestic activities and participation in domestic life
18	Furnishings, fixtures and other assistive products for supporting activities in indoor and outdoor human-made environments
22	Assistive products for communication and information management
24	Assistive products for controlling, carrying, moving and handling objects and devices
27	Assistive products for controlling, adapting or measuring elements of physical environments
28	Assistive products for work activities and participation in employment
30	Assistive products for recreation and leisure

Tool 2 My outcomes: what does my AT enable me to do?

PROMPT: the World Health Organisation tells us we participate in lots of different ways. Tick the areas you participate in. Write down what AT enables you to do these things:

- Mobility
- Manage self-care
- Communicate
- Manage general tasks and demands
- Manage your domestic (home) life
- Learn and apply knowledge (e.g. remember, write, read)
- Relate to others
- Have an educational life
- Have an economic (working, volunteering) life
- Have a civic life (being part of the community)
- Undertake recreation
- Have a spiritual life (e.g. get to worship)
- Have a political life (e.g. get to vote in person)

Tool 3 My rights: how does my AT meet my human rights. Indicate if your right is "realized" or "not yet realized" Here are some rights from the United Nations Convention on the Rights of Persons with Disabilities

- Article 9 Accessibility
- Article 19 Living independently and being included in the community
- Article 24 Education
- Article 20 Personal Mobility
- Article 21 Freedom of expression and opinion, and access to information
- Article 25 Health

- Article 26 Habilitation and Rehabilitation
- Article 27 Work and Employment
- Article 29 Participation in political and public life
- Article 30 Participation in cultural life, recreation, leisure and sport

Tool 4 My Costs: how much do my supports cost? How much do my supports save? PROMPT: thinking about how much supports cost. Also, thinking about savings, that is, money we don't have to spend because the supports are doing the trick. Please tell us about:

- The cost of the AT and getting it set up and maintained
- Think about things you don't have to spend because you have the AT e.g. another piece of equipment, or less support work hours
- Can you think of any costs you may save later on, because you have your AT right now? This might include fewer falls, or better pressure care management due to better AT now.

Tool 5 My Service Delivery Pathway: Good practice standards say that about 6 AT service steps should be followed to get the right AT set up in the right way for you. Did you get access to these AT service steps? PROMPTS:

- Initiation: Did I know where to start in getting my AT?
- Assessment: Did I get to receive info about "why" and "how" of assessment; did I contribute knowledge of my self, my environment, my goals? Was I asked "how is this going for you?"
- Trying out my AT: Could I contribute opinion and preferences if I wished to?
- Choosing the AT I want: Could I take part in the process of buying, applying for funding, talking to suppliers if I wished to?
- Getting my AT: was I involved with delivery/setup/trial as much as I wanted?
- Afterwards: could I go back and ask people questions? Did I have the opportunity to give feedback?

References

1. Fuhrer, M.J., et al.: A Framework for the conceptual modelling of assistive technology device outcomes. Disabil. Rehabil. **25**(22), 1243–1251 (2003)
2. Rist, P., et al.: Recovery from Disablement: what functional abilities do rehabilitation professionals value the most? Arch. Phys. Med. Rehabil. **89**, 1600–1606 (2008)
3. Layton, N., Volkert, A., Joyce, R.: ARATA/AFDO/OTA Breakfast forum on assistive technology outcomes plenary Session Thursday 15 November 7.30–9.00. In: Australian Assistive Technology Conference. ARATA, Melbourne (2018)
4. Layton, N., Doyle, L., Volkert, A.: My Outcomes Framework: synthesising the evidence on valued outcomes into a set of user-led and policy-relevant tools. In: Occupational Therapy Australia 28th National Conference and Exhibition. Sydney (2019)
5. Layton, N., et al.: My assistive technology outcomes framework: rights-based outcome tools for consumers to 'measure what matters'. Assistive Technol.,1–9 (2023)
6. ISO. ISO 9999 Assistive products - Classification and terminology (2022)

7. World Health Organization. International classification of functioning, disability and health. Geneva: World Health Organization; 2001; 2024 release
8. Andrich, R., Caracciolo, A.: Analysing the cost of individual assistive technology programmes. Disabil. Rehabil.: Assistive Technol., 1–28 (2007)
9. Layton, N., Gardy, M., McDonald, R.: Assistive technology costs and outcomes. In: Chambers, D., (ed.) Assistive Technology to Support Inclusive Education. International Perspectives on Inclusive Education: Emerald Publishing Limited, pp. 109-124 (2020)
10. United Nations. Convention on the rights of persons with disabilities and optional protocol. United Nations, Geneva (2006)
11. Layton, N., et al.: Guidelines for assistive technology service provision – a scoping review. Disabil. Rehabil.: Assistive Technol.,1–12 (2024)
12. de Jonge, D., et al.: Motivations and incentives: exploring assistive technology service delivery from the perspectives of multiple stakeholders. In: Jan, Y.-K., Grott, R., (eds.) RESNA 2015: New Frontiers in Assistive Technology. RESNA, Denver, CO, USA (2015)
13. Bricknell, S., Madden R., (eds.) Exploring the measurement of participation. Meeting of heads of who collaborating centres for the classification of diseases, Brisbane (2002)
14. Layton, N., Thomson, C., Noonan, M.: Consumer-defined outcomes: the assistive technology outcomes framework applied to wheeled mobility and seating. Oceania seating symposium, Melbourne (2019)
15. Layton, N., Noonan, M., O'Connor, J.: My assistive technology outcomes framework – a collaborative tool for AT users and their practitioners. AOPA, Melbourne (2019)
16. Layton, N., Le Cerf, C.: Measuring what matters: a descriptive participatory case evaluation of a tool for measuring outcomes of assistive technology after acquired brain injury. Brain Impairment. **24**(2), 137–147 (2023)
17. Tellefson, C., et al.: Supporting inclusion and participation: Issues for Australians who are deafblind in accessing telecommunication devices. In: 'Living with Equipment in 2011' MASS Conference; 24–25 August 2011. Queensland Health, Brisbane (2011)
18. Bell, D., Prain, M., Layton, N.: Measuring outcomes of assistive technology use for deafblindness. In: 11th National Deafblind Conference; 29 June to 1 July, 2022, Fremantle, Western Australia (2022)
19. Bell, D., Prain, M., Layton, N.: Assistive technology for people with deafblindness in Southern Africa: a Delphi study exploring dimensions of impact. Disabil. Rehabil. Assist. Technol. **18**(1), 30–43 (2021)
20. Layton, N., Buchanan, R.: Owning our outcome measures: co-producing tools for change. In: Virtual Disability Research Conference, Melbourne; February; Virtual (2018)
21. Layton, N.: Best practice in providing assistive technology and environmental adaptations to the polio community. J. Rehabil. Med. **48**(8), 743 (2016)
22. World Health Organization. Rehabilitation 2030: A Call for Action. Geneva: World Health Organisation (2017)
23. WHO & UNICEF. Global Report on Assistive Technology. Geneva: WHO & UNICEF (2022)
24. E.A.S.T.I.N. European Assistive Technology Information Network nd. https://www.eastin.eu/en/searches/products/index

Theoretical Framework for Video Game Accessibility: Bridging User Experiences and Industry Data for Inclusive Design

Nadia Di Leo(✉)

Department of Humanities, University of Foggia, 71121 Foggia, Italy
nadia.dileo@unifg.it

Abstract. Video game accessibility is no longer a peripheral concern but a core requirement for equitable participation in digital play. This paper proposes a three-dimensional theoretical framework—physical, cognitive, and socio-emotional—that unifies disability-studies theory, user-generated evidence from Can I Play That?, and quantitative market data from Newzoo. A mixed-methods pipeline first scraped and embedded 2019–2024 CIPT reviews, producing six thematic clusters (Controls, UI & Navigation, Audio, Difficulty, Visual Accessibility, Subtitles). These user-level insights were then triangulated with Newzoo's global reports to reveal how accessible features correlate with broader audience reach and revenue potential. By illustrating both ethical imperatives and commercial gains, the framework provides developers, researchers, and publishers with actionable guidance for embedding accessibility from concept to post-launch. Ultimately, universal design in gaming is demonstrated to be both a moral and strategic imperative.

Keywords: Video game accessibility · Inclusive design · Disability studies · Assistive technology · Human–Computer Interaction · Market analysis

1 Introduction

Video game accessibility has emerged as a pivotal concern in the realm of Human–Computer Interaction (HCI), particularly as the gaming industry increasingly intersects with assistive technology and broader entertainment platforms (Cairns et al., 2019; von Gillern & Nash, 2023). Recent work highlights how games, when consciously designed, can benefit individuals with diverse abilities by mitigating barriers related to vision, hearing, cognition, and motor function (Deng, 2024; von Gillern & Nash, 2023). Assistive technologies integrated into gaming can enhance social participation, promote inclusive design, and foster innovation across multiple sectors (Cairns et al., 2019). Moreover, platforms such as *Can I Play That?*[1] document firsthand experiences of players who need flexible input methods, on-demand accessibility features, and community support. By bridging these user experiences with systematic industry data, stakeholders can advance both ethical and commercial imperatives for inclusion within game development.

[1] https://caniplaythat.com/.

K. Mavrou and P. Encarnação (Eds.): AAATE 2025, *Technology for Inclusion and Participation for All: Recent Achievements and Future Directions*, pp. 304–314, 2025.
https://doi.org/10.1007/978-3-032-01628-7_37

Despite notable progress, video games still pose significant challenges for players with diverse abilities. Core issues range from unreadable text, inaccessible control schemes, and minimal difficulty-adjustment settings, to broader socio-emotional barriers such as exclusion from community features (Cairns et al., 2019; Clark et al., 2023; von Gillern & Nash, 2023). As a result, many would-be gamers either abandon certain titles or require extensive third-party modifications to participate fully[1]. The lack of coherent, industry-wide standards for accessibility—particularly during early development phases—often leads to inconsistent or superficial implementation of accessibility measures (Deng, 2024). In turn, this inconsistency perpetuates social exclusion, hampering opportunities for people with disabilities to experience equitable participation and enjoyment within gaming environments (Wang et al., 2020).

This paper proposes a theoretical framework that integrates qualitative insights from *Can I Play That?* user feedback with quantitative market data provided by Newzoo[2]. Rooted in disability studies and inclusive design principles, the framework focuses on three dimensions of accessibility—physical, cognitive, and socio-emotional—thereby offering a holistic lens for understanding player needs and industry benefits. By aligning user-reported barriers (e.g., unwieldy controls, limited customizations, social exclusion) with evidence that inclusive design can boost market share and brand reputation, this paper underscores that accessibility is not merely an ethical consideration but also a strategic and financial one. Ultimately, the goal is to present a user-centric model that can serve as a resource for developers, researchers, and industry stakeholders, encouraging a shift toward universal accessibility practices in game design.

2 Background and Related Work

2.1 Historical Perspectives on Disability

Disability has long been situated within a "medical model" framework in which impairments were perceived as deficits or illnesses that individuals carry (Hogan, 2019; Adesina et al., 2021). Under this model, the primary responsibility for coping with or "correcting" the impairment fell on the individual rather than on systemic or environmental factors (Hogan, 2019). However, a growing body of work now affirms that disability should be approached through more socio-cultural lenses, shifting attention toward the ways social structures, technologies, and design choices can either disable or empower people with a range of abilities (Griggs & Moore, 2023; Wang et al., 2020).

In particular, design mismatches have been highlighted as one of the root causes of exclusion, suggesting that many "disabilities" arise at the intersection of user needs and poorly adapted environments or products (Griggs & Moore, 2023). Such perspectives echo the view that assistive technology and inclusive design can help remove disabling barriers in daily life (Adesina et al., 2021). Within this framework, accessibility is no longer regarded as a mere add-on or retrofit; instead, it becomes a proactive strategy requiring developers, industry leaders, and community members to visualize and design for human diversity from the outset (Hogan, 2019).

[2] https://newzoo.com/.

2.2 Accessible Video Game Design

Building on these socio-cultural insights, recent research addresses how video games specifically reflect—or neglect—inclusive design principles. Scholars and practitioners have noted that conventional "one-size-fits-all" interfaces, such as controllers with complex button layouts, lack of large text, or visually cluttered displays, often exclude people with diverse abilities (Clark et al., 2023; Deng, 2024; Wang et al., 2020). Consequently, the gaming community has seen an uptick in discussions about adjustable input methods, such as remappable controllers or adaptive peripherals, as well as the importance of textual clarity, color-blind modes, and subtitles (Brown & Anderson, 2021; Deng, 2024).

For instance, Clark et al. (2023) draw attention to location-based games that ignore the needs of players with physical disabilities. Likewise, Deng (2024) examines how seemingly small tweaks—like scalable fonts or providing high contrast modes—can facilitate meaningful engagement for visually impaired players. Although initiatives like "Game Accessibility Guidelines[3]" have been proposed, recent analyses suggest that inconsistent adoption and late-in-the-cycle retrofitting remain prevalent (Brown & Anderson, 2021). These gaps underscore the ongoing need for multi-dimensional accessibility frameworks that integrate physical, cognitive, and socio-emotional dimensions throughout all phases of game development (Deng, 2024; Hogan, 2019).

2.3 Market and Community Insights

Increasingly, the business case for inclusive design aligns with these socio-cultural imperatives. Newzoo's research[4] highlights that improving accessibility and diversity in gaming not only fosters inclusivity but also presents significant opportunities for publishers and developers to appeal to broader audiences, enhance engagement, and drive revenue growth. Such motivations parallel broader consumer shifts, with more players expressing willingness to invest in games that accommodate diverse needs (Brown & Anderson, 2021).

Simultaneously, feedback platforms like *Can I Play That?*[1] provide firsthand accounts of players encountering barriers related to text size, color-contrast deficiencies, limited controller customization, or lack of socio-emotional inclusion in gaming communities. Their reviews, written "by disabled gamers, for disabled gamers," highlight the importance of transparent communication about accessibility features prior to purchase and also emphasize the role of iterative community-driven testing (Brown & Anderson, 2021). As a result, these interconnected user perspectives and market forecasts not only validate the socio-cultural view of disability but also illuminate how thoughtful, early-stage inclusive design can spell economic, social, and creative benefits for the gaming industry at large.

---

[3] https://gameaccessibilityguidelines.com/full-list/.

[4] https://newzoo.com/resources/trend-reports/diversity-inclusion-report-intel-newzooo.

3 Methodology

This paper follows a mixed-methods approach that integrates qualitative user-centric insights from *Can I Play That?*[1] with quantitative market analyses from Newzoo[5]. By closely examining both firsthand user feedback on accessibility and larger-scale market forecasts, the study offers a nuanced view that bridges the micro-level of individual gaming experiences and the macro-level of industry-wide data and opportunities.

The study extracted full texts of reviews from 2019–2024 using web scraping. Keywords were extracted using OpenAI's (2024a) GPT-4o LLM model through few-shot prompting (Brown et al., 2020), with each keyword labeled as "positive," "negative," or "mixed." These keywords underwent textual embedding using OpenAI's (2024b) text-embedding-3-large model. Clustering was performed using the k-means (Bishop, 2006) algorithm with qualitative supervision, identifying six thematic macro-categories. The embeddings' dimensionality was reduced using t-SNE (Van der Maaten & Hinton, 2008) for visualization (Fig. 1). The clusters were then classified into: Controls (Fig. 2), UI and Navigation (Fig. 3), Audio, Speech and Sound (Fig. 4), Difficulty Settings (Fig. 5), Visual Accessibility (Fig. 6), and Subtitles (Fig. 7).

In parallel with the CIPT review corpus, the study leveraged key metrics and forecasts from a range of Newzoo reports. The data spanned overall market sizing, user demographics, and monetization patterns, with special attention to how inclusive design choices influence market share and in-game revenues. Recent editions of the Global Gamer Study, the PC & Console Gaming Report, and the Diversity and Inclusion in Games insights were used to capture the breadth of consumer segments, willingness to pay for accessible or inclusive titles, and how these preferences changed over time.

All relevant Newzoo statistics were organized into categories that paralleled some of the CIPT clusters. After acquiring the CIPT reviews and Newzoo forecasts, the findings were triangulated by mapping user critiques to relevant market data. In line with standard HCI approaches, iterative coding of CIPT's textual content was used to refine the categories of accessibility problems and strengths that might resonate with the data from Newzoo. Coding was continually updated through reliability checks, ensuring that if new insights emerged (for example, themes like "persistent eye-strain from overly small text"), these insights would feed directly into subsequent matching with the quantitative side. The methodology, therefore, remains close to iterative design traditions in user research, enabling a dialogue between real user experiences, theoretical guidelines in game accessibility, and industry-level economic data.

By synthesizing CIPT's user-level evidence and Newzoo's large-scale metrics, this methodology yields a panoramic view of modern accessibility challenges and market potentials in the games industry.

4 Results

In total, 448 accessibility reviews were extracted from Can I Play That?[1] (2019–2024), yielding a corpus of textual data that underwent keyword extraction, embedding, clustering, and dimensionality reduction. The analysis identified six macro-categories of accessibility features: Controls, UI and Navigation, Audio/Speech/Sound, Difficulty Settings,

[5] https://newzoo.com/resources/trend-reports/the-pc-console-gaming-report-2025.

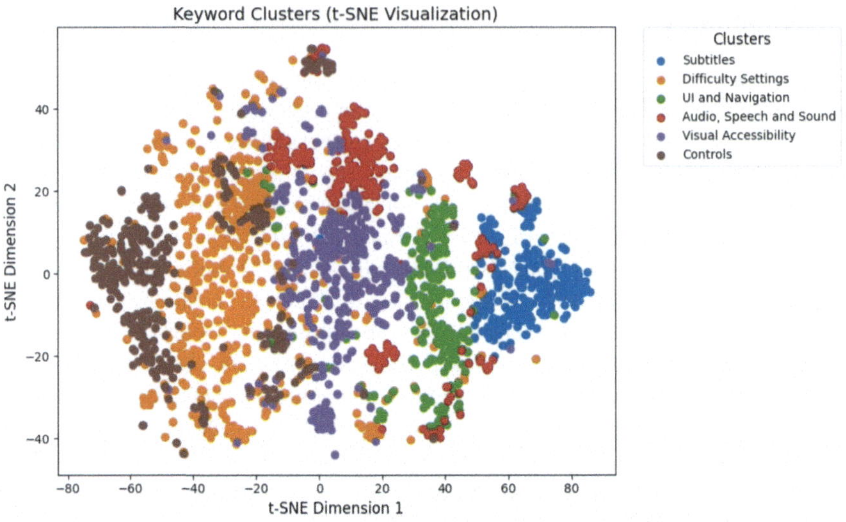

Fig. 1. Keyword Clusters (t-SNE Visualization)

Visual Accessibility, and Subtitles. These categories align with core areas that users and accessibility advocates frequently emphasize (Hassan, 2024; Moscoso, 2020). Parallel findings from consumer behavior studies on representation and diversity in gaming[6] provide context for how these accessibility features resonate with larger gamer segments and may influence whether players engage with certain titles. As stated in Newzoo's Diversity and Inclusion report[6], "While many gamers feel the market has come a long way in terms of representation, the majority feels that games should be more diverse and inclusive". Newzoo also launched, in 2022, the Gamer Sentiment Study on Diversity and Inclusion[7], based on a survey among over 4,000 gamers across the U.S. and U.K. This study higlighted that 31% of U.S. gamers have a disability, with mental health being the most reported[8].

Furthermore, according to Newzoo's Diversity and Inclusion research, 47% of all U.S. gamers do not play games they feel aren't made for them[6].

The analysis below integrates these findings with market-level data on player engagement to offer a comprehensive overview of accessibility practices.

4.1 Controls

Keywords in this cluster included "full control remapping", "motor-function presets", and "adaptive controller support". Reviews highlighted the importance of eliminating

[6] https://newzoo.com/resources/trend-reports/diversity-inclusion-report-intel-newzooo.

[7] https://raisethegame.com/news/key-highlights-from-newzoo-S-gamer-sentiment-study-on-diversity-equity-and-inclusion.

[8] https://newzoo.com/resources/blog/newzoos-gamer-sentiment-diversity-inclusion-gender-etnicity-sexual-identity-disability.

or reducing "required button holds," which can pose a significant motor barrier (Hassan, 2024). Titles that successfully implemented adaptable control schemes, such as robust remapping, toggle modes, and compatibility with devices like the Xbox Adaptive Controller, were generally rated positively. Games with mandatory motion controls showed the highest proportion of negative evaluations, consistent with prior studies that show high dropout rates among players lacking the motor abilities to engage effectively (Hassan, 2024).

These control-related features are critical in an industry context where diverse gamer demographics increasingly demand accessibility. Findings also suggest a correlation between the inclusion of diverse characters and willingness to experiment with new releases[6].

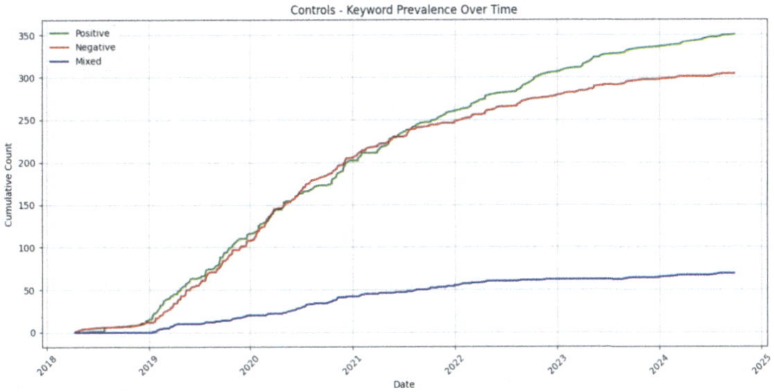

Fig. 2. Controls – Keyword prevalence over time

4.2 UI and Navigation

The second cluster encompassed textual references to "UI scaling", "colorblind filters", and "navigational aids". Reviews frequently commended games that use distinct shapes or bold outlines to convey critical information beyond color differentiation (Moscoso, 2020). This resonates with broader consumer feelings around inclusive game and character design: according to the Gamer Sentiment Study on Diversity and Inclusion[7], gamers prefer full character customization (40% in the U.S. and 32% in the U.K.), and most prefer characters that resemble them, especially in the United States.

Within the corpus, "eye tracking support" appeared infrequently, but such mentions elicited distinctly positive feedback from users with limited upper-body mobility, reaffirming that advanced technological solutions, though still niche, can open channels of play for otherwise constrained users (Hassan, 2024).

4.3 Audio, Speech, and Sound

References included "sound captioning", "speaker labels", and "visual cues for audio-based events". The frequency of positive reviews for robust audio accessibility features

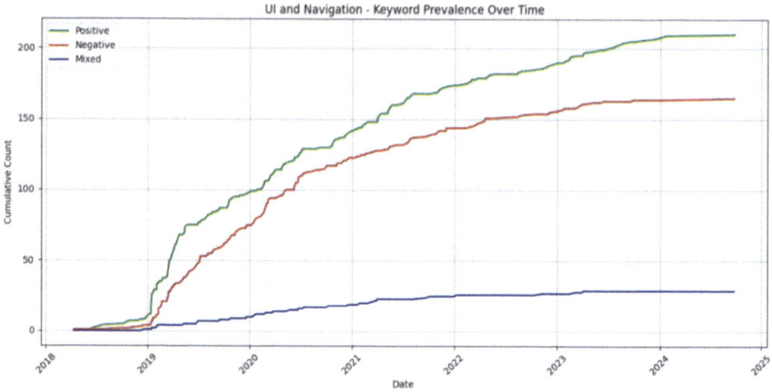

Fig. 3. UI and Navigation – Keyword Prevalence Over Time

consistently outpaced negative ones, possibly influenced by the growing awareness that partial or complete hearing loss need not present an insurmountable barrier to gameplay (Moscoso, 2020). Reviews stressed the importance of "individual volume controls" and "color-coded speaker indication", features that improved clarity in multiplayer modes. Newzoo's findings[8] highlight how globally, 64% of gamers indicate that diversity in games is important to them. This is based on the Global Gamer Study's[9] aggregate of 36 markets, indicating that diversity is meaningful for the global gaming community.

The increased share among younger gamers points to a generational difference, with younger players considering diversity in games more significant than their predecessors (Fig. 4).

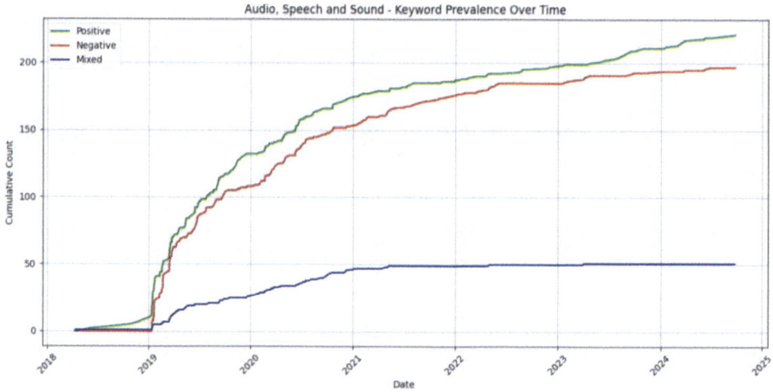

Fig. 4. Audio, Speech and Sound – Keyword Prevalence Over Time

[9] https://newzoo.com/global-gamer-study.

4.4 Difficulty Settings

A notably large volume of reviews addressed "dynamic difficulty," "assist mode," and "scalable difficulty sliders." Many players recounted that nuanced difficulty adjustment—rather than a simple "easy/normal/hard" triad—positively influenced their experience, as they could adapt challenges to suit evolving skill levels and disabilities (Hassan, 2024). Negative feedback focused on titles that locked difficulty behind progression or offered only perfunctory modifications, such as reduced enemy health without adjusting complex mechanical requirements. Moreover, consumer behavior data shows that the games market benefits when gaming is welcoming to everyone, from every generation, and to every identity[8].

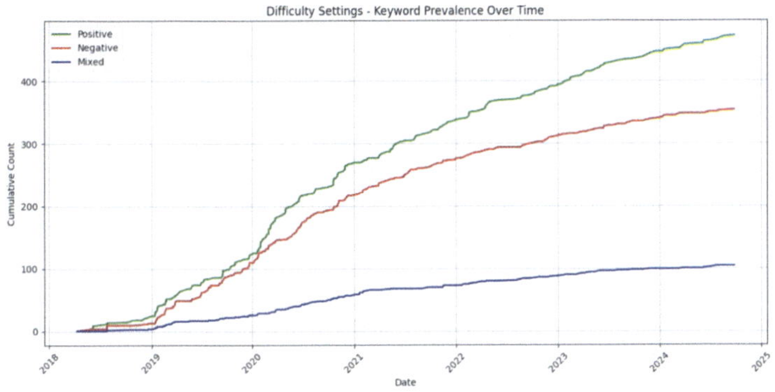

Fig. 5. Difficulty Settings – Keyword Prevalence Over Time

4.5 Visual Accessibility

Features like "text size customization" and "UI scaling" are linked to the needs of players with visual impairments, while "high contrast text" and "dyslexia-friendly fonts" support those with specific reading challenges. "Screen reader support" and "text-to-speech" capabilities are inclusive towards blind players, allowing them to navigate menus and understand dialogue.

4.6 Subtitles

Similarly, a robust system should include "adjustable subtitle size" to accommodate various visual needs and viewing distances, along with "high-contrast subtitle background" options to ensure readability across different gaming environments. Beyond basic dialogue, "environmental closed captions" and "sound effect subtitles" help create an equivalent gaming experience by conveying gameplay information traditionally relayed through audio cues. "Full closed captioning" should be consistently implemented across all game content, from "cutscene subtitles" to "ambient dialogue" and "battle commentary", ensuring no critical information is missed due to accessibility barriers.

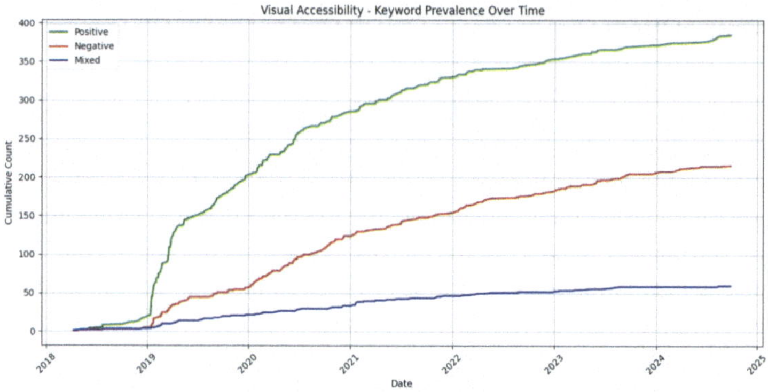

Fig. 6. Visual Accessibility – Keyword Prevalence Over Time

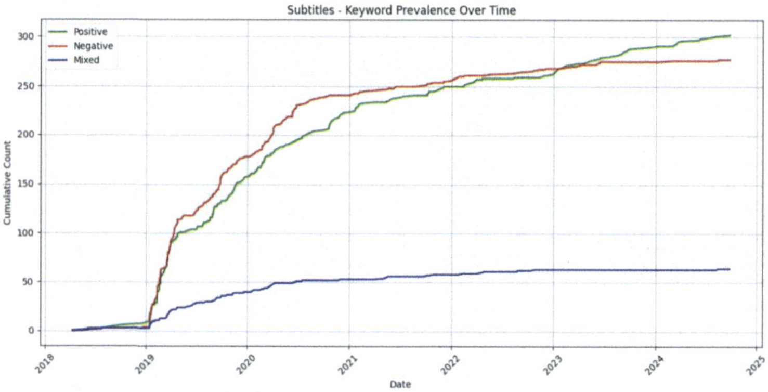

Fig. 7. Subtitles – Keyword Prevalence Over Time

4.7 Analysis of Keyword Prevalence Over Time

The most significant growth in accessibility reviews occurred during 2019–2021 across all categories, indicating increased industry and community focus on accessibility during this period.

Difficulty settings show the highest volume of reviews (roughly 450 positive, 350 negative keywords by 2024), suggesting this is a primary accessibility concern in gaming, while visual accessibility demonstrated the largest positive-negative gap (380 vs 220).

Subtitle reviews showed balanced feedback until 2023, when positive reviews increased. Audio and UI/navigation maintained modest positive leads. Control schemes showed convergence during 2020–2021, followed by slight positive advantage. Post-2022 review growth slowed, suggesting standardization of accessibility evaluation.

These patterns parallel the surge in global market awareness about inclusive design. As documented by Newzoo[6], players across multiple demographics—ranging from ethnicity, gender and disability—are more likely to invest time and money in titles that depict diverse representations and robust accessibility. Growth in the frequency of reviews post-2021 coincides with the introduction of guidelines like the Game Accessibility Guidelines and the institutional recognition of accessibility as a critical aspect of game design (Hassan, 2024; Moscoso, 2020).

Players who responded positively to robust accessibility measures on CIPT consistently highlighted an enhanced sense of inclusion, echoing broader industry insights that affordability and inclusivity can expand a title's reach. Conversely, negative reviews show the persisting frustration among players who feel marginalized by the lack of adaptive options, especially in fast-paced online titles that demand quick reflexes and complex control inputs.

5 Conclusions

This study positions accessibility as a structural principle of video-game design by weaving together disability-studies theory, the lived experiences collected by *Can I Play That?*, and the market evidence reported by Newzoo. Hardware and control schemes that respect diverse motor abilities, for example, are most effective when accompanied by user interfaces that lower cognitive load and by community features that recognise the emotional realities of disabled players. Triangulating qualitative and quantitative data shows that these same design choices have tangible commercial value: they enlarge the addressable audience, lengthen engagement times, and strengthen brand reputation. Accessibility, therefore, is revealed to be simultaneously a moral obligation and a strategic avenue for growth, and it must be incorporated from the earliest concept stage and revisited throughout the development life cycle through iterative, user-centred testing.

The research is limited by its focus on English-language reviews and its reliance on proprietary industry reports; future investigations should widen the cultural scope, employ longitudinal methodologies, and establish stronger causal links between specific accessibility interventions and revenue outcomes. Even within these constraints, the evidence is unequivocal: treating accessibility as an optional add-on perpetuates exclusion, whereas embedding it as a guiding philosophy can reshape the medium into a more inclusive, innovative, and profitable cultural force.

References

Adesina, M., Olufadewa, I.I., Oladele, R.I., Oduguwa, I., Abudu, F.R.: Historical perspective and classification of disability. Yenagoa Med. J **3**, 17–24 (2021)

Bishop, C.M.: Pattern Recognition and Machine Learning. Springer (2006)

Brown, M., Anderson, S.L.: Designing for disability: evaluating the state of accessibility design in video games. Games Culture **16**(6), 702–718 (2021)

Brown, T., et al.: Language models are few shot learners. In: Advances in Neural Information Processing Systems, vol. 33, pp. 1877–1901 (2020). https://proceedings.neurips.cc/paper/2020/file/1457c0d6bfcb4967418bfb8ac142f64aPaper.pdf

Cairns, P., Power, C., Barlet, M., Haynes, G.: Future design of accessibility in games: a design vocabulary. Int. J. Hum Comput Stud. **131**, 64–71 (2019)

Clark, M.O., Singh, A., Barbareschi, G.: Towards greater inclusion and accessibility for physically disabled players in location-based games. In: Proceedings of the ACM on Human-Computer Interaction, vol. 7, no. CHI PLAY, pp. 699–720 (2023)

Deng, Y.: Inclusive games: accessible game design for the visually impaired. Appl. Comput. Eng. **37**, 61–67 (2024)

Griggs, N., Moore, R.: Removing systemic barriers for learners with diverse identities: antiracism, universal design for learning, and Edpuzzle. J. Spec. Educ. Technol. **38**(1), 15–22 (2023)

Hassan, L.: Accessibility of games and game-based applications: a systematic literature review and mapping of future directions. New Media Soc. **26**(4), 2336–2384 (2024)

Hogan, A.J.: Moving away from the" medical model": the development and revision of the World Health Organization's Classification of Disability. Bull. Hist. Med. **93**(2), 241–269 (2019)

Manning, C.D., Raghavan, P., Schütze, H.: Introduction to Information Retrieval. Cambridge University Press (2008).https://doi.org/10.1017/CBO9780511809071

Moscoso, C.: Assassin's Creed Valhalla: Accessibility Review. Can I Play That? https://caniplaythat.com/2020/12/08/assassins-creed-valhallaaccessibility-review/. Accessed 8 Dec 2020

OpenAI. (2024a). GPT-4o API. https://openai.com

OpenAI. (2024b). text-embedding-ada-002. https://platform.openai.com/docs/guides/embeddings

Van der Maaten, L., Hinton, G.: Visualizing data using t-SNE. J. Mach. Learn. Res. **9**, 2579–2605 (2008). http://www.jmlr.org/papers/volume9/vandermaaten08a/vandermaaten08a.pdf

von Gillern, S., Nash, B.: Accessibility in video gaming: an overview and implications for English language arts education. J. Adolesc. Health. **66**(6), 382–390 (2023)

Wang, H., Kwan, M.P., Hu, M.: Social exclusion and accessibility among low-and non-low-income groups: a case study of Nanjing. China. Cities **101**, 102684 (2020)

Assistive Technology in Education: Impact of Smartphone-Based Accessibility Features for Young Adults with Visual Impairments

Rajendra Prasad, Suvlaxmi Gurumayum(✉), and Rohini Gaur

Kara Medical Foundation, Patna, India
hello@kara.foundation

Abstract. The integration of assistive technology in education is vital for enhancing independence among visually impaired students. This study examines the impact of smartphone-based accessibility tools, specifically Google TalkBack, on the quality of life and psychosocial well-being of 18 visually impaired young women (aged 20–26) living in a college hostel in Delhi, India. Participants underwent a three-month, structured digital literacy program using a "Train the Trainer" module. Pre- and post-assessments were conducted using the WHOQOL-BREF and PIADS scales. Initial results showed low confidence in using accessibility features, high frustration, and moderate emotional distress, despite frequent smartphone exposure. Following the intervention, participants reported marked improvements in mobility, life satisfaction, emotional resilience, and independence. Feelings of Frustration and embarrassment significantly declined, indicating greater digital autonomy and psychosocial wellbeing. The findings highlight the effectiveness of targeted, context-sensitive training in bridging the digital literacy gap for visually impaired users. The study underscores the importance of moving beyond device provision to sustained, user-centred assistive technology education for long-term impact.

Keywords: Talkback · smartphones · visual impairments · Assistive Technology · Quality of Life · India

1 Introduction

The 2011 Census of India reported that 2.21% of the total population has disabilities, with 19% being visually impaired. India accounts for one-fifth of the global burden of blindness, with approximately 6.2 million blind individuals and 55 million experiencing moderate to severe visual impairment [1]. Despite these statistics, only 29.16% of visually impaired individuals are part of the education system, and just 6.86% of schools in India have access to Braille books and audio content [2]. Research highlights major challenges, including inadequately trained educators, poorly designed schools, and funding issues, which further hinder inclusive education [2].

K. Mavrou and P. Encarnação (Eds.): AAATE 2025, *Technology for Inclusion and Participation for All: Recent Achievements and Future Directions*, pp. 315–322, 2025.
https://doi.org/10.1007/978-3-032-01628-7_38

Smartphone-based accessibility tools like TalkBack offer a promising solution to these challenges by enabling visually impaired students to navigate digital platforms independently. However, despite the availability of such tools, many students lack adequate training and confidence in using them effectively.

Existing literature has emphasized the transformative role of mobile-based screen readers such as TalkBack and VoiceOver in supporting visually impaired users in accessing digital interfaces. These tools help users read text messages, browse the internet, and access educational content using auditory feedback, thereby promoting autonomy and digital inclusion [4]. Research has also shown that these accessibility features contribute positively to users' mental well-being and self-confidence when appropriately utilized [5].

However, challenges remain. Studies report that even among frequent smartphone users, confidence in navigating built-in accessibility features is low, largely due to a lack of structured training and awareness [6]. Barriers include inconsistent gesture recognition, device-specific limitations, and cognitive overload due to complex menu structures [7]. In the Indian context, these challenges are exacerbated by limited digital literacy training and lack of localized instructional support for assistive technology use [8].

This study investigates the impact of a structured TalkBack training program on the quality of life and perceived independence of visually impaired students in India, using the WHOQOL and PIADS assessment tools.

2 Methodology

2.1 Study Design and Background

This study builds upon a pilot training intervention previously conducted at a residential school for visually impaired girls in Patna, Bihar. While the initial pilot assessed the feasibility of introducing TalkBack-enabled smartphones in a school setting, the current study focuses on a more advanced and sustained intervention carried out in a college hostel for visually impaired young women in Delhi, India. The design is experimental and action-based, involving a structured 3-month training period using a pre- and post-assessment framework to evaluate the impact of TalkBack training on quality of life and psychosocial well-being. The training was informed by the WHOGATE framework and used a structured "Train the Trainer" curriculum to ensure consistency and replicability.

TalkBack, part of the Android Accessibility Suite, is a built-in screen reader designed for visually impaired users to navigate smartphones. Through spoken feedback and gesture-based interaction, TalkBack enables users to access apps, input text, browse the web, and manage daily tasks without relying on vision. It transforms smartphones into essential tools for education, communication, and autonomy.

TalkBack was chosen because it comes pre-installed on Android smartphones, which are significantly more accessible and affordable in India than iOS devices. This makes Android—and TalkBack in particular—a practical and scalable solution for assistive technology interventions in low-resource settings.

2.2 Participants

The study involved 18 female participants aged 20–26 years, residing in a government-supported hostel for visually impaired students in Delhi. Of these, 3 participants had low vision, and the remaining were completely blind. All participants were familiar with smartphones conceptually but did not own one prior to the study. They were enrolled in various academic programs, ranging from undergraduate to PhD levels. Inclusion criteria:

- Residing in the targeted hostel during the intervention period
- Willingness to participate in training sessions and assessments
- No personal smartphone ownership prior to intervention
- Verbal consent to participate in the study Exclusion criteria:
- Students not residing in the selected hostel
- Individuals already using TalkBack or similar assistive technologies independently

Participants were primarily from underserved regions such as Bihar and Jharkhand. All were provided with TalkBack-enabled smartphones for the duration of the study.

2.3 Training Protocol

Training was delivered in-person through weekly group sessions, each lasting 1–2 h. The sessions were led by a trained facilitator using a structured "Training of Trainer" (TOT) module specifically developed for the study. The TOT module covered:

- Accessibility foundations: Non-visual interaction principles and introduction to TalkBack
- Gesture library: Core and advanced gestures, including three-finger navigation and text editing
- Functional app use: Messaging, email, navigation, voice assistants, YouTube, WhatsApp, and other task-based applications
- Advanced tools: Bluetooth keyboard integration, document writing (Google Docs), screen reading (EasyReader), and online ordering (Swiggy, Uber, Amazon, Zepto)
- Troubleshooting: Speech rate adjustment, notification handling, and accessibility settings optimization

Each session emphasized hands-on practice and peer learning, supplemented by a tactile workbook and printed gesture guides. Local mentors provided follow-up support between classes.

2.4 Data Collection

Assessments were conducted before the intervention (baseline) and after the 3-month training period. Two standardized instruments were used:

- WHOQOL-BREF: Assessed perceived quality of life across physical, psychological, social, and environmental domains

- PIADS (Psychosocial Impact of Assistive Devices Scale): Measured perceived competence, adaptability, and self-esteem related to technology use Data collection involved in-person administration of questionnaires, supported by trained facilitators who assisted with interpretation and navigation where necessary. Observational notes were also taken to capture qualitative insights into each participant's learning trajectory, engagement level, and functional gains.

3 Results

The findings demonstrate a significant positive impact on participants' quality of life, particularly in psychological and functional domains.

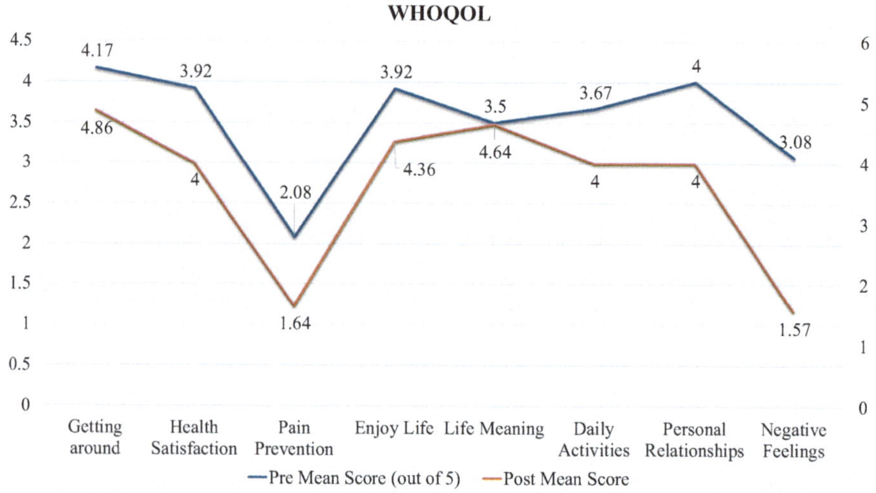

Fig. 1. WHOQOL (World Health Organisation Quality of Life Scale)

As shown above in Fig. 1, the pre-assessment WHOQOL scores indicated that participants had moderate satisfaction with their health and daily life, with notable challenges in managing physical pain (mean score: 2.08/5) and frequent negative emotional experiences (mean score: 3.08/5). Despite relatively high scores in areas such as personal relationships (4.00) and mobility (4.17), there were signs of reduced psychological well-being, as reflected in lower scores for life meaning (3.50) and enjoyment (3.92). These findings align with the initial reports of frustration, dependency, and limited confidence in handling daily tasks independently.

Following the intervention, post-assessment scores showed clear improvements across most domains. Participants reported better mobility (4.86), greater sense of life meaning (4.64), and increased enjoyment of life (4.36). There was also a substantial reduction in negative feelings (1.57) and perceived physical hindrance from pain (1.64), suggesting enhanced emotional and physical well-being. Daily activity performance and health satisfaction also improved modestly. Overall, the WHOQOL results suggest that

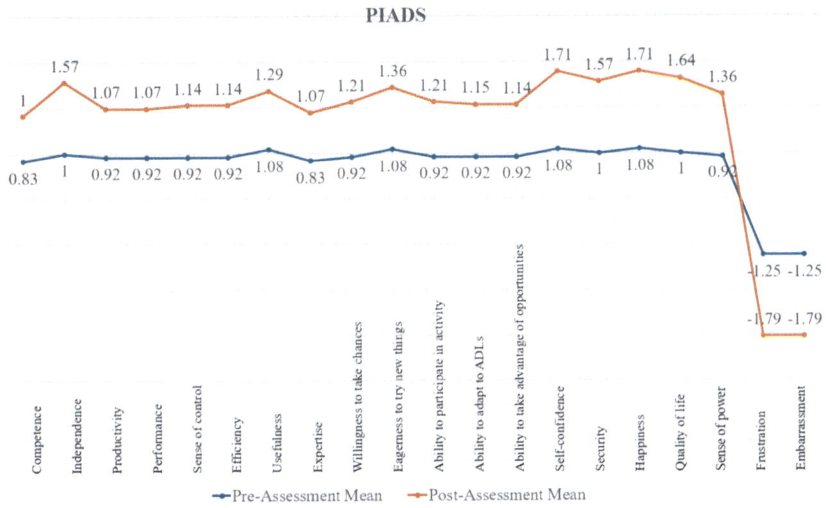

Fig. 2. PIADS (Psychosocial Impact of Assistive Devices Scale)

the digital literacy training positively impacted participants' quality of life, particularly in psychological and functional domains.

As shown above in Fig. 2, the pre-assessment results from the Psychosocial Impact of Assistive Devices Scale (PIADS) revealed that participants experienced moderate to high levels of competence, self-confidence, and productivity. However, they also reported notable feelings of frustration (mean score: 3.58/7) and embarrassment (4.08/7), suggesting a psychosocial strain associated with their limited digital independence. Scores related to independence, sense of control, and security were moderate, reflecting a lack of confidence and reliance on others for everyday digital tasks. Post-assessment scores demonstrated substantial improvement across nearly all PIADS domains. The most prominent gains were observed in independence (increased by 1.07 points), sense of control (+0.89), security (+0.90), and self-confidence (+.), indicating that the intervention significantly enhanced participants' psychological empowerment and functional autonomy. Notably, frustration and embarrassment showed large decreases (−. And −., respectively), reflecting improved comfort, ease, and reduced anxiety when using assistive technology. These positive shifts highlight the training's effectiveness in not only building technical skills but also in fostering emotional resilience, confidence, and overall psychosocial wellbeing.

4 Discussion

The findings of this study affirm the positive impact of structured TalkBack training on the psychosocial and functional well-being of young adults with visual impairments. Pre-assessment data from the WHOQOL indicated that while participants reported relatively high mobility (mean: 4.17) and interpersonal relationships (4.00), they faced significant challenges related to physical discomfort (2.08), negative emotions (3.08), and reduced psychological fulfillment, such as meaning in life (3.50) and enjoyment (3.92). These results resonate with prior research highlighting that despite frequent smartphone usage, visually impaired users often lack confidence and proficiency in using accessibility tools without guided instruction [4, 5].

Post-intervention WHOQOL scores demonstrated marked improvements in nearly all domains. Participants reported increased life satisfaction and emotional resilience, with a substantial reduction in negative feelings (from 3.08 to 1.57) and physical discomfort (from 2.08 to 1.64). Gains in life meaning (4.64) and enjoyment (4.36) further point to enhanced psychological well-being and self-efficacy resulting from the intervention. These results align with existing literature, w[here smartphone-based accessibility tools were shown to contribute to user independence and emotional empowerment when paired with structured training and support mechanisms [6, 7]. PIADS outcomes mirrored these trends. Pre-assessment scores revealed moderate-tohigh self-confidence and competence, but also reported psychosocial strains including frustration (mean: 3.58) and embarrassment (4.08). Such emotional barriers, particularly among novice users, have been widely documented, emphasizing the psychological toll of digital exclusion [5, 6]. After three months of TalkBack training, participants exhibited significant gains in core areas of independence (+1.07), sense of control (+0.89), and self-confidence (+0.71), while experiencing a sharp reduction in frustration (−.) and embarrassment (−.). These results illustrate that digital literacy training, when tailored to the needs of visually impaired individuals, can serve not only as a technical intervention but also as a form of psychosocial rehabilitation.

Importantly, the training also had direct benefits for participants' educational engagement. All participants were enrolled in government colleges under the University of Delhi, pursuing studies in the humanities. The TalkBack sessions specifically included instruction in using screen readers with Google Docs and Microsoft Word— critical tools for academic writing and documentation. Students also learned to access digital libraries such as Bookshare, enabling them to read textbooks and course materials independently. The ability to conduct web searches, participate in online assignments, and communicate more effectively with peers and faculty marked a transformative step toward inclusive learning.

From a socioeconomic perspective, the program addressed critical digital access gaps. All participants came from low-income households, primarily from Bihar and Jharkhand. While most families owned at least one smartphone—usually controlled by a male head of the household—the students themselves had no personal devices before the intervention. Despite receiving financial support through government welfare and scholarship schemes, systemic barriers had previously limited their technological autonomy. By providing personal smartphones and tailored instruction, the project enabled participants to use tools such as Google Maps, online food delivery apps, and ride-hailing platforms, significantly enhancing their functional independence beyond the academic context.

Despite these successes, the study also encountered several implementation challenges. The residential nature of the hostel meant that students followed diverse academic schedules, leading to coordination difficulties for group sessions. To mitigate this, weekly Sunday sessions were adopted and reinforced with peer-led support and selfpaced materials. Limited prior exposure to smartphones and the complete absence of personal devices posed early hurdles. Additionally, students required extended repetition and peer-assisted practice to master more complex gestures—highlighting the need for scaffolded learning over a longer term, especially when introducing advanced features like Bluetooth keyboard integration or multi-app workflows. The program's success demonstrates how smartphones—equipped with accessibility features and embedded within a structured, socially supportive learning framework— can be powerful agents of change. However, access barriers still remain, particularly in underserved regions, as highlighted by Meher and Rout, who found that nearly half of their participants reported smartphone usage as a source of distraction. In contrast, none of the participants in our study reported distraction as a barrier—likely due to the structured nature of the training and their strong academic motivations [5]. These findings collectively underscore the importance of moving beyond hardware distribution to develop localized, skills-based assistive technology programs that are responsive to the educational and socioeconomic realities of visually impaired users.

5 Conclusion

This study demonstrates that structured, context-sensitive training in smartphone accessibility tools—specifically Google's TalkBack—can significantly improve the quality of life, psychosocial well-being, and digital independence of young visually impaired women. Participants showed measurable gains in confidence, emotional resilience, mobility, and control over their digital interactions, as evidenced by improvements in both WHOQOL and PIADS scores.

Beyond technical skills, the intervention fostered psychological empowerment, reduced frustration, and addressed long-standing barriers of digital exclusion. These findings reinforce the idea that access to assistive technology must be accompanied by sustained training and support, especially for populations with historically limited exposure to digital tools.

The study also sheds light on implementation challenges in low-resource, real-world settings—such as scheduling constraints, limited device familiarity, and the need for extended guided practice. Addressing these barriers through adaptive program design, peer mentoring, and modular training resources was critical to the program's success. As India and other LMICs move toward greater digital inclusion, initiatives like this highlight the need for national policies that go beyond provisioning devices and emphasize structured, scalable training programs rooted in user empowerment. This approach can serve as a model for integrating assistive technology into inclusive education and independent living frameworks worldwide.

References

1. Senjam, S., Foster, A., Bascaran, C.: Assistive technology for visual impairment and trainers at schools for the blind in Delhi. Assist. Technol. **34**, 1–9 (2020). https://doi.org/10.1080/10400435.2020.1839144

2. Status of inclusive education for visually impaired students of India in 20th century. Int. J. Sci. Res. **9**(7), 1–4 (2020)
3. IAPB Vision Atlas: Magnitude and Projections - Countries. https://www.iapb.org/learn/vision-atlas/. Accessed 13 May 2025
4. Muhammad, W., Shireen, S.S.: Academic Use of Smartphones for Social Development of Visually Impaired Students of University of Karachi: A Study of Android Smartphone Applications by VI Students. Library Philosophy and Practice (e-journal) 3026, 1–14 (2019)
5. Meher, G., Rout, S.K.: Smartphone usages among students with visually challenged for their socio-academic life in higher education. Int. J. Finan. Manag. Res. **7**(1), 94–102 (2025). g84fbw
6. Martiniello, N., Eisenbarth, W., Lehane, C., Johnson, A., Wittich, W.: Exploring the use of smartphones and tablets among people with visual impairments: are mainstream devices replacing the use of traditional visual aids? Assist. Technol. **34**(1), 34–45 (2019). https://doi.org/10.1080/10400435.2019.1682084
7. Thomson, L.: Evaluating accessibility features across operating systems. In: 9th International Proceedings on Proceedings, pp. 101–110. TechAccess Publications, London (2014)
8. Ismaili, J., Ibrahimi, E.H.O.: Mobile learning as alternative to assistive technology devices for special needs students. Educ. Inf. Technol. **22**, 883–899 (2017)

Teachers' Evaluation of Tools for Assessing the Usability of Technology Designed for Children with Autism

Oronzo Parlangeli[1](✉), Margherita Bracci[1], Lorenzo Reina[1], Francesco Currò[1], Paola Palmitesta[1], Stefano D'Ambrosio[2], Sara Cecchetti[2], Stefano Guidi[1], and Enrica Marchigiani[1]

[1] Università di Siena, Via Roma 56, Siena, SI 53100, Italy
oronzo.parlangeli@unisi.it

[2] Università degli Studi di Bergamo, Piazzale S. Agostino 2, Bergamo, BG 24129, Italy

Abstract. The inclusion of children with autism in classroom play activities involving technology requires consideration of the usability level of these technologies. However, methods for assessing user experience are not always designed around the specific needs of children with autism and are often limited to observations made during play. The broader PRIN 2022 Family of ROBots for children with disabilities (FROB) project aims to design and implement modular playful robots for school-based interventions with children with motor disabilities, autism spectrum disorders, and complex disabilities. The present study aimed to redesign some evaluation tools by involving 41 teachers who have at least one child diagnosed with autism in their classrooms. Teachers gave their opinions on four user experience assessment tools that could involve verbal, visual, or even motor processing. The co-design activity made the assessment tools considered more suitable for the needs of children with autism, proposing an approach that integrates direct assessments with observations during play.

Keywords: Autism · Children · User Experience (UX) · Assessment tools · System Usability Scale · SUS · Teachers · Co-design · Robots for play

1 Introduction and Background

Autism Spectrum Disorder (ASD) includes a broad range of neurodevelopmental conditions that affect sensory, perceptual, and cognitive functioning to varying degrees throughout life. It is primarily characterised by persistent difficulties in social interaction and communication, along with restricted and repetitive patterns of behaviour [1, 4–6, 9, 15]. In educational contexts, several evidence-based practices (EBPs) have been developed to support the inclusion and development of children with autism. These include intensive, individualised instruction and social feedback within familiar environments, and peer-mediated interventions (PMIs) to promote learning through social

K. Mavrou and P. Encarnação (Eds.): AAATE 2025, *Technology for Inclusion and Participation for All: Recent Achievements and Future Directions*, pp. 323–331, 2025.
https://doi.org/10.1007/978-3-032-01628-7_39

interaction, offering opportunities to practice acquired skills with classmates [19]. In parallel, game-based interventions (GBIs) leverage play to enhance cognitive, emotional, and social functioning [7]. To further support inclusion, several studies have investigated the use of playful technologies—including mobile apps, virtual reality, and social robots—to facilitate communication and interaction for children with ASD [8, 13, 21]. These technologies often employ visual supports and structured tasks, which incorporate participatory and co-design frameworks that aim to empower children by recognising both their creative potential and their challenges [5, 11]. However, designing UX interventions for children with autism presents several critical challenges [5]. Individuals with ASD often display hesitant, distressed, or anxious behaviors when faced with unfamiliar or unexpected contexts, environments, or materials in UX studies. Additional difficulties arise from the use of tools and materials that impose high cognitive demands during interventions [5, 12, 14]. Therefore, it is essential to identify methods that harmonise the diverse contributions of experts, designers, and children into a unified and coherent intervention framework [5].

This study represents a preliminary phase of the broader PRIN 2022 Family of ROBots for children with disabilities (FROB) project, which aims to design and implement modular playful robots for school-based interventions with children with motor disability, autism spectrum disorders and complex disabilities [3]. The project's core goal is to create playful robots capable of adapting to children's individual preferences and potential, promoting inclusive and accessible play. In this preparatory phase, we explore the suitability and clarity of the tools intended to assess children's experiences after interaction with the robots. Specifically, we conducted interviews with teachers to evaluate a set of questionnaires and non-verbal response modalities (e.g., clapping, jumping) designed to capture children's emotional and cognitive responses. Following suggestions by Kerr and colleagues [10], the involvement of teachers in the co-design and review of assessment tools ensures that instruments are both context-sensitive and developmentally appropriate, while accounting for children's communicative and attentional profiles. The outcomes of this study will inform the design of UX evaluation tools to be used in the upcoming experimental phase, contributing to the development of inclusive, evidence-based robotic interventions for autistic children.

2 Method

2.1 Study Design and Participants

The study was conducted between December 2024 and April 2025 and involved a total of 41 teachers from five different regions of Italy. The sample consisted of 31 primary school teachers (75.6%) and 10 preschool teachers (24.4%), all of whom were female, aged between 23 and 62 years, with teaching experience ranging from 3 to 42 years.

Preschool teachers (N = 10) worked in mixed-age classrooms with children aged 3 to 6 years. Among the primary school teachers, 48.4% taught lower grades (N = 15), specifically first (12.9%), and second (35.5%) grades to children aged 6 to 7 years, while the remaining 51.6% taught upper grades (N = 16), third (12.9%), fourth (22.6%) and fifth (16.1%), to children aged 8 to 10 years.

All participating teachers had at least one student in their classroom diagnosed with autism or cognitive impairment. Specifically, 32 (78%) of the children referred by the teachers are male, 37 (90.2%) are autistic, of whom 20 (55.6%) are low-functioning and 4 (9.8%) are not diagnosed as autistic but have cognitive impairment. For greater sample consistency, the interviews with the 4 teachers who do not have children with an autism diagnosis in their class were not included in the rest of the analysis. These children were distributed across one second-grade class, one fifth-grade class, and two fourth-grade classes in the primary school. Of the total remaining sample (N = 37), 12 (29.3%) have no problems in the area of language, while all the others do, whether they are verbal or non-verbal children.

Each teacher took part in a face-to-face interview, either in-person or online, using a structured questionnaire. The aim was to co-design, in collaboration with the teachers, four user experience (UX) assessment tools: three tools already validated for typically developing children and one method developed specifically for this study.

2.2 Methodology and Instruments

Each of the four assessment tools was presented to the teachers in both its original and a modified version, the latter adapted by the researchers to better suit autistic children's needs. During the interviews, teachers were asked to respond keeping in mind a specific student, namely, the autistic child in their class. Teachers were invited to select between the original and adapted versions, propose alternative formulations, or suggest the removal of unsuitable items.

The interview track was structured into three sections:

Teacher Profile: Data collected included gender, age, teaching tenure, role (curricular or support), and school level (preschool or primary).

Child Profile: Teachers provided information about the child's gender, autism spectrum classification (high or low functioning), and any additional motor, cognitive, sensory, linguistic, or socio-communicative difficulties.

Tool Evaluation and Redesign: Teachers were asked to review and provide feedback on four UX tools.

The four UX assessment tools are described below.

Smileyometer. A pictorial tool composed of five emotive faces ranging from sad to happy, with a neutral midpoint, and labeled descriptors from "awful" to "brilliant" [17, 18], (see Fig. 1).

Fig. 1. One of the representations of the Smileyometer shown to the teachers

Adapted System Usability Scale (SUS). Originally designed by Putnam et al. [16] for children aged 7–8 and older, this ten-item scale assesses usability and learnability. Both the original and researcher-modified versions were presented to participants. Teachers had the opportunity to revise or eliminate items deemed overly complex. The items are presented in Table 1. Teachers also rated the visual layout of the 5-point Likert scale, anchored by thumbs (from strongly disagree to strongly agree), (see Fig. 2).

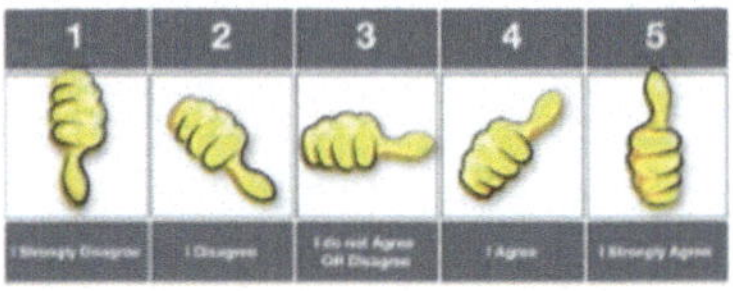

Fig. 2. Visual representation of the scale of agreement of the SUS

Relative Enjoyment Scale for Children with CVI (RESC-CVI). Adapted from Van Looy et al. [20], this scale was contextualized for the present study by comparing a robot with different types of play-based activities, for example [2]: ball games (practice play), Lego (constructive play), soft toys (symbolic play), and card games (rule-based play). Children express their preferences using a five-point Likert scale (see Fig. 3).

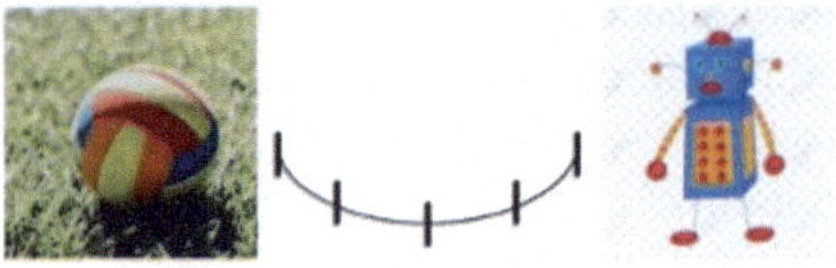

Fig. 3. Example of an item of the scale used for the study

Behavioral Observation Scale (Jumpometer and Clapometer). Developed by the authors of the present study, this instrument is thought to capture children's non-verbal behavioral expressions (e.g., clapping, jumping) as indicators of satisfaction during the robotic interaction. The method was included to provide an alternative means of collecting UX data from children with limited verbal communication abilities.

3 Results

Preferences regarding the Smileyometer, preschool educators (N = 10), 80.0% favored a 3-smile version and considered verbal descriptions unnecessary due to the images' inherent clarity. Half preferred the standard question phrasing ("How much did you enjoy playing with the robot?" or "How fun was it to play with the toy Robot?), while others proposed alternatives for younger children. First and second grade teachers (N = 14) predominantly adopted the standard question formulation (92.9%). Most educators (57.1%) deemed images self-explanatory, while 28.6% preferred embedded descriptions. The 5-smile scale was favored by nearly two-thirds of respondents (64.3%), suggesting a preference for more nuanced response options at this educational level. Upper primary teachers (grades 3–5, N = 13) similarly used standard phrasing (61.5%), with some variation in text preferences: 38.5% preferred written descriptions, 30.8% favored embedded descriptions, and 30.8% considered them unnecessary. The 3-smile version was selected by 76.9% of these educators.

The System Usability Scale (SUS) evaluation revealed varying perceptions of clarity. Among preschool teachers (N = 10), 70.0% rated the instructions as 'not very clear' and expressed a preference for the '3 thumbs' response format (50.0%) or suggested alternative formats (40.0%). In contrast, first and second grade teachers (N = 14) reported higher levels of comprehension, with 71.4% describing the instructions as 'very clear' and 85.7% favoring the '3 thumbs' scale. For teachers of grades 3–5 (N = 13), perceptions of clarity were mixed, with 53.8% rating the instructions as 'very clear' and 46.2% as 'not very clear.' Nonetheless, the '3 thumbs' format remained popular (69.2%).

Regarding the results on the formulation of the SUS statements—whether in their original version, modified version, or in the case of any changes or removals—details are provided in Table 1.

Some primary school teachers recommended the use of AAC symbols to support the wording of the questions. For example, statement 1, 'I would like to play with the toy robot a lot more,' could be accompanied by the AAC symbols for 'More' and 'Enough,' while statement 4, 'I need help to play with the toy robot,' could include the symbol for 'Help.'

The RESC-CVI methodology assessment showed that preschool educators (N = 10) unanimously deemed the RESC scale – the line with 5 thicks- not understandable, with image comparison task yielding mixed evaluations (40.0% fully comprehensible, 10.0% somewhat, 50.0% not entirely). First and second grade teachers (N = 14) were divided on scale comprehensibility (57.1% understandable), while most found the image comparison at least fairly understandable: 57.1% selected "fairly understandable", 28.6% "fully understandable", and 14.3% reported difficulties by selecting "not entirely understandable" or "not understandable". Upper primary teachers (N = 13) similarly showed divided opinions on scale comprehension (46.2% understandable), while rating image comparison task more favorably (69.2% fully comprehensible, 30.8% fairly comprehensible. Primary school teachers suggested presenting the pictures two at a time, as displaying them all at once may make comparison difficult.

The Behavioral Observation Scale revealed consistent patterns across groups. Preschool teachers (N = 10) preferred applause (50.0%) or jumping (30.0%) as indicators, with 20.0% rejecting both methods. Among first and second grade teachers (N = 14), 57.1% favored applause-based responses, 28.6% suggested jumping, and 14.3% rejected the method entirely. For grades 3–5 educators (N = 13), 46.2% selected applause as suitable and 15.4% preferred jumping, while 53.8% expressed reservations or rejection based on complexity concerns or distraction risks, indicating greater heterogeneity in assessment preferences for older student groups.

4 Discussion and Conclusion

The results suggest that simplifying response scales is essential, as teachers strongly favored a three-option format, both for the Smileyometer [18] and the SUS [16].

Particularly informative are the data related to the SUS, which show that only one statement (4) was deemed adequate in its original version. For four statements (1, 3, 9, and 10), teachers from at least two different grade levels preferred the revised version. However, the most significant finding appears to be the inadequacy of as many as five statements (2, 5, 6, 7, and 8) among those that make up the scale.

Table 1. Analysis of SUS statements. Percentages are based on four response categories: Original formulation - O1 (Option 1), Revised formulation(s) - O2 (Option 2), Suggestions (S), and Elimination (E). Bold values indicate the highest percentage for each statement.

Statements	Preschool	Grade 1–2	Grade 3–4-5
1 O1 I would like to play with the robot toy much more O2a I would like to play with the robot toy again	O1: 0% O2a: **60%** S: 40% E: 0%	O1: 14.3% O2a: **64.3%** S: 21.4% E: 0%	O1: 0% O2a: **69.2%** S: 23.1% E: 7.7%
2 O1 Playing with the robot toy was difficult	O1: 30% S: 10% E: **60%**	O1: 35.7% S: 7.1% E: **57.1%**	O1: 46.2% S: 0% E: **53.8%**
3 O1 The robot toy was easy to use O2a It was easy to use the robot toy	O1: 10% O2a: 40% S: 0% E: **50%**	O1: **35.7%** O2a: **35.7%** S: 7.1% E: 21.4%	O1: 0% O2a: **61.5%** S: 0% E: 38.5%
4 O1 I need help to play with the robot toy O2a If I want to play with the robot toy, someone has to help me	O1: **50%** O2a: 0% S: 30% E: 20%	O1: 35.7% O2a: 0% S: 14.3% E: **50%**	O1: **53.8%** O2a: 7.7% S: 30.8% E: 7.7%
5 O1 I knew what to do when playing with the robot toy O2a I knew how to play with the robot toy	O1: 0% O2a: 30% S: 0% E: **70%**	O1: 14.3% O2a: 21.4% S: 0% E: **64.3%**	O1: 23.1% O2a: 23.1% S: 7.7% E: **46.2%**
6 O1 Some things about the robot toy didn't make sense O2a Something about the robot toy I didn't understand	O1: 0% O2a: 30% S: 0% E: **70%**	O1: 14.3% O2a: 35.7% S: 0% E: **50%**	O1: 0% O2a: 30.8% S: 15.4% E: **53.8%**
7 O1 The robot toy would be easy to learn for my friends O2a Even for my friends it would be easy to learn to play with the robot toy	O1: 0% O2a: 20% S: 0% E: **80%**	O1: 7.1% O2a: 21.4% S: 0% E: **71.4%**	O1: 7.7% O2a: 23.1% S: 7.7% E: **61.5%**
8 O1 To play with the robot toy I had to do strange things O2a Sometimes, to play with the robot toy, I had to make up what I had to do	O1: 10% O2a: 30% S: 0% E: **60%**	O1: 28.6% O2a: 7.1% S: 0% E: **64.3%**	O1: 7.7% O2a: 15.4% S: 23.1% E: **53.8%**

(continued)

Table 1. (*continued*)

Statements	Preschool	Grade 1–2	Grade 3–4-5
9 O1 I was proud of myself when I played with the robot toy O2a I liked myself when I played with the robot toy O2b I was good at playing with the robot toy	O1: 20% O2a: 0% O2b: **60%** S: 0% E: 20%	O1: 0% O2a: 14.3% O2b: **57.1%** S: 7.1% E: 21.4%	O1: 15.4% O2a: 0% O2b: **46.2%** S: 23.1% E: 15.4%
10 O1 There were many things to learn to play with the robot toy O2 I had to learn many things to play with the robot toy	O1: 10% O2a: **50%** S: 0% E: 40%	O1: 28.6% O2a: **42.9%** S: 0% E: 28.6%	O1: 7.7% O2a: 30.8% S: 0% E: **61.5%**

The RESC-CVI [20] appears suitable from age six onward. Simplifying the response scale or clarifying preference directionality is recommended. In contrast, the Behavioral Observation Scale showed no age-related differences, suggesting its suitability may depend more on sensory or motor abilities than on cognitive or relational skills.

The analysis highlights key areas for improving UX assessment tools, and new versions of these tools are already under development. Most importantly, the results obtained suggest continuing the study with autistic children.

From this perspective, the greatest effort should probably be dedicated to testing a version of the SUS consisting of only five statements: statement 4 in its original version, and statements 1, 3, 9, and 10 in their revised versions, in order to evaluate its validity and factorial structure. The use of this tool, appropriately structured with reference to the abilities of autistic children, could represent a fundamental step forward toward the development of assistive technologies that truly include these users in play activities.

Acknowledgements. We sincerely thank all the teachers who kindly participated in the Project.

This work was funded by the European Union – NextGenerationEU, under the National Recovery and Resilience Plan (NRRP), Mission 4, Component 2, Investment 1.1, funding call PRIN 2022 D.D. 104 published on 2.2.2022 by the Italian Ministry of University and Research (Ministero dell'Università e della Ricerca), Project Title: FROB - Family of ROBots for children with disabilities – CUP F53D23006100006.

References

1. American Psychiatric Association: Diagnostic and Statistical Manual of Mental Disorders. 5th edn. American Psychiatric Association, Arlington (2013). https://doi.org/10.1176/appi.books.9780890425596
2. Besio, S.: What is play? In: Encarnação, P., Ray-Kaeser, S., Bianquin, N. (eds.) Guidelines for Supporting Children with Disabilities' Play, pp. 1–12. De Gruyter Open (2018). https://doi.org/10.1515/9783110613445-005
3. Bonarini, A., Besio, S.: Robot Play for All: Developing Toys and Games for Disability. Research for Development, Springer International Publishing, Cham (2022). https://doi.org/10.1007/978-3-031-05042-8

4. Bölte, S.: Social cognition in autism and ADHD. Neurosci. Biobehav. Rev. **169**, 106022 (2025). https://doi.org/10.1016/j.neubiorev.2025.106022
5. Corlu, D., Ta¸sel, , Turan, S.G., Gatos, A., Yanta¸c, A.E.: Involving autistics in user experience studies: a critical review. In: Proceedings of the 2017 Conference on Designing Interactive Systems. pp. 43–55. ACM, Edinburgh United Kingdom (2017). https://doi.org/10.1145/3064663.3064771
6. Fletcher-Watson, S., Pain, H., Hammond, S., Humphry, A., McConachie, H.: Designing for young children with autism spectrum disorder: a case study of an iPad app. Int. J. Child-Comput. Inter. **7**, 1–14 (2016). https://doi.org/10.1016/j.ijcci.2016.03.002
7. Gao, J., Song, W., Huang, D., Zhang, A., Ke, X.: The effect of game-based interventions on children and adolescents with autism spectrum disorder: a systematic review and meta-analysis. Front. Pediatr. **13**, 1498563 (2025). https://doi.org/10.3389/fped.2025.1498563
8. Guidi, S., Palmitesta, P., Bracci, M., Marchigiani, E., Parlangeli, O.: The perception of the utility of social media by caregivers of persons with autism spectrum disorders during a period of home quarantine. Interact. Comput. **35**(5), 615–627 (2023). https://doi.org/10.1093/iwc/iwad006
9. Kamp-Becker, I.: Autism spectrum disorder in ICD-11—a critical reflection of its possible impact on clinical practice and research. Mol. Psychiatry **29**(3), 633–638 (2024). https://doi.org/10.1038/s41380-023-02354-y
10. Kerr, J., Kelly, N.: Use of personas in co-designing learning experiences with teachers: an exploratory case study. Int. J. Technol. Des. Educ. **35**(1), 189–207 (2025). https://doi.org/10.1007/s10798-024-09900-x
11. Makhaeva, J., Frauenberger, C., Spiel, K.: Creating creative spaces for co-designing with autistic children: the concept of a "Handlungsspielraum". In: Proc. 14th Participatory Design Conf., vol. 1, pp. 51–60. ACM, New York (2016). https://doi.org/10.1145/2940299.2940306
12. Malinverni, L., Mora-Guiard, J., Padillo, V., Valero, L., Herv´as, A., Pares, N.: An inclusive design approach for developing video games for children with autism spectrum disorder. Comput. Hum. Behav. **71**, 535–549 (2017). https://doi.org/10.1016/j.chb.2016.01.018
13. Mazon, C., Fage, C., Sauz´eon, H.: Effectiveness and usability of technology-based interventions for children and adolescents with ASD: a systematic review of reliability, consistency, generalization and durability related to the effects of intervention. Comput. Hum. Behav. **93**, 235–251 (2019). https://doi.org/10.1016/j.chb.2018.12.001
14. Parlangeli, O., Caratozzolo, M.C., Guidi, S.: Multitasking and mentalizing machines: how the workload can have influence on the system comprehension. In: Harris, D. (ed.) Engineering Psychology and Cognitive Ergonomics, LNCS, vol. 8532, pp. 50–58. Springer, Cham (2014). https://doi.org/10.1007/978-3-319-07515-0_6
15. Pellicano, E., Den Houting, J.: Annual research review: shifting from 'normal science' to neurodiversity in autism science. Child Psychol. Psychiatry **63**(4), 381–396 (2022). https://doi.org/10.1111/jcpp.13534
16. Putnam, C., Puthenmadom, M., Cuerdo, M.A., Wang, W., Paul, N.: Adaptation of the system usability scale for user testing with children. In: Extended Abstracts of the 2020 CHI Conference on Human Factors in Computing Systems, pp. 1–7. ACM, Honolulu HI USA (2020). https://doi.org/10.1145/3334480.3382840
17. Read, J., MacFarlane, S., Casey, C.: Endurability, Engagement and Expectations: Measuring Children's Fun (2002)
18. Read, J.C.: Validating the fun toolkit: an instrument for measuring children's opinions of technology. Cogn. Tech. Work **10**(2), 119–128 (2008). https://doi.org/10.1007/s10111-007-0069-9
19. Reichow, B., Doehring, P., Volkmar, F.R. (eds.): Handbook of Evidence-Based Practices in Autism Spectrum Disorder. Springer Nature Switzerland, Cham (2025). https://doi.org/10.1007/978-3-031-78143-8

20. Van Looy, J., Houttekier, E., Nunez Castellar, E., All, A.: elative enjoyment scale for primary school children (RES-C): development and testing of reliability, validity and sensitivity. Prague (2018)
21. Velentza, A.M., Pliasa, S., Biziou, F., Fachantidis, N.: Let's play with social robots!—Improving social skills in autistic children through game activities'. In: Understanding Autism, pp. 161–172. Elsevier (2025). https://doi.org/10.1016/B978-0-443-27366-7.00026-2

Investigating the Potential of Assistive Technology Passports for Enhanced Access to Assistive Technology and the Development of an Implementation and Adoption Framework

Mohamed I. Maalim(✉) and MacLachlan Malcolm

Maynooth University, Maynooth, Co. Kildare, Ireland
Mohamed.maalim.2020@mumail.ie

Abstract. This research examines the potential of an Assistive Technology Passport to enhance AT access for individuals with disabilities, which aligns with human rights frameworks and the global movement for inclusive technology. The proposed AT Passport is envisioned as a centralised, personalised document that outlines individual AT needs, preferences, and recommendations. The research seeks to identify current barriers to AT access, clarify the passport's role in addressing these barriers, and create an implementation framework.

A four-phased, mixed-methods approach was employed. Initially, the lived experiences of individuals with disabilities and their access to assistive technology (AT) were explored using Interpretative Phenomenological Analysis to identify barriers and the potential value of an AT Passport. This qualitative phase informed a subsequent quantitative survey that gathered data on AT usage patterns and challenges. Concurrently, a Delphi study engaged expert AT users to establish consensus on key functionalities and design features for the passport, ensuring a user-centred approach.

Finally, the research synthesised qualitative and quantitative findings, incorporating the consensus from the Delphi study, to develop a comprehensive AT Passport framework. This framework provides recommendations for the passport's design, development, and implementation, emphasising integration with existing support systems and long-term sustainability. While promising, the framework's limitations necessitate further research to refine it and maximise its impact on improving AT access and fostering a more inclusive future for people with disabilities.

Keywords: Assistive Technology · Assistive Technology Passport · Accessibility · Inclusion · Mixed-Methods

1 Introduction

Assistive technology (AT) enables individuals with disabilities to lead independent and fulfilling lives, fostering their inclusion [1]. A significant disparity persists despite global efforts, with millions lacking access to the necessary AT [2]. Data reveals a substantial

K. Mavrou and P. Encarnação (Eds.): AAATE 2025, *Technology for Inclusion and Participation for All: Recent Achievements and Future Directions*, pp. 332–343, 2025.
https://doi.org/10.1007/978-3-032-01628-7_40

unmet need, with only one in ten individuals having access to the AT they require, amounting to approximately one billion underserved people worldwide [2]. This issue is particularly pronounced in low- and middle-income countries, where high costs, limited availability, and inadequate systems impede access [3].

Given these unmet needs and the importance of user-centred approaches, there is a clear need to reinvestigate and refine strategies for improving access [4–6]. To address these disparities, this research examines the potential of the AT Passport, a concept proposed to promote a more equitable and user-centred approach to AT provision [7, 8].

The AT Passport aims to enhance accessibility, raise awareness, facilitate training, and encourage user involvement by centralising and streamlining information about an individual's AT needs and history. It is envisioned as a vital tool in addressing the identified gap in effective AT access among individuals with disabilities and older persons. While its definitive structure is still being developed, the AT Passport is conceived as a central repository of user-specific data and requirements.

This research investigates the Assistive Technology Passport and situates its conceptualisation within a cohesive framework of interconnected models and principles, including the Convention on the Rights of Persons with Disabilities, the International Classification of Functioning, Disability, and Health, the Capability Approach, systems thinking, and Inclusive Design [9–17]. By adopting this comprehensive perspective, we can attain a holistic understanding of the AT landscape, develop more effective solutions, and maintain a user-centred focus [15, 18, 19]. This research draws on a user-centred, interdisciplinary approach to Assistive Technology provision [15, 19–23]. It aims to investigate the design of a sustainable and impactful AT passport and understand its potential to empower individuals with disabilities, promote inclusivity, and ensure equitable access to technology. This paper builds upon foundational research conducted as part of the author's doctoral thesis [24]. While sharing the same title and addressing the Assistive Technology Passport, this work presents a refined and focused analysis, offering a more concise perspective on the potential of an assistive technology passport in improving access to assistive technology.

2 Methodology and Research Design

To investigate the potential of AT Passports as a solution to enhance AT accessibility, especially where established systems are lacking, this research employed a pragmatic, exploratory sequential mixed methods design. Grounded in pragmatism, the study initially examined the lived experiences of individuals with disabilities through Interpretative Phenomenological Analysis. This initial qualitative phase informed a subsequent quantitative survey to gather data on AT usage patterns. Concurrently with the survey, a Delphi study engaged expert AT users to achieve consensus on essential functionalities and design. These phases followed an interconnected flow, culminating in synthesising all findings to develop a comprehensive AT Passport framework. We will introduce the design of each study phase in the findings section before presenting the results. This multi-phased approach facilitated a rich understanding, informing the development of a robust and impactful AT Passport framework.

3 The Findings

This mixed-methods study provides valuable insights into the challenges and potential solutions for enhancing access to assistive technology. By integrating qualitative and quantitative research methods, alongside expert consensus, the study offers a comprehensive understanding of the barriers to AT access and a potential framework for addressing these issues. Key findings from each phase are summarised below:

3.1 Qualitative Study

This study employed Interpretative Phenomenological Analysis to investigate the lived experiences of individuals with disabilities regarding assistive technology and the AT Passport concept. IPA focuses on understanding how individuals interpret their experiences. The analysis connected the participants' accounts with the researcher's interpretations.

As illustrated in Table 1, the analytical process involved identifying patterns in participants' initial experiential statements. This process led to the establishment of Personal Experiential Themes (PET) for each participant and culminated in Group Experiential Themes (GET) representing shared experiences within the group. GETs are not collective norms; instead, they serve as a cross-analysis tool to reveal recurring and unique patterns.

The analysis revealed key themes:

- **Striving for Human Diversity, Equity, and Social Cohesion**
- **Individual Agency and Freedom of Choice**
- **The AT Passport as a Capability-Enhancing Resource**

These themes highlight the value of AT, diverse views on access factors, and the importance of personal agency and human diversity.

Striving for Human Diversity, Equity, and Social Cohesion

This theme emphasises the desire for inclusion and recognition as valuable members of society. Participants face daily challenges navigating a non-inclusive world, citing difficulties with the built environment and web accessibility. As one participant, Leo, stated, *"I don't see myself at the edge; I am part of a greater society... People have different abilities, and that's all just a great patchwork of life."*

Beyond the built environment, participants highlighted a lack of accessibility information, indirect discrimination in employment, and the stigma sometimes associated with AT use. Physical and social barriers restrict participation, and participants stated that policies do not always benefit those they are intended to assist. "Social and Structural Discrimination" is a key personal experience contributing to this desire for diversity, equity, and social cohesion.

Individual Agency and Freedom of Choice

This theme emphasises self-identity and choice in AT adoption. Social perception significantly influences these choices; for instance, one participant, Amy, noted, *"it's more*

Table 1. Group Experiential Categories

Initial Experiential Statements	PET	GET
Daily struggles in a non-inclusive world; accessibility challenges; lack of information; discrimination; stigma; lack of awareness & disability-conscious policies; grief	*Social & Structural Discrimination*	Strive for Diversity, Equity & Social Cohesion
Need for personalised AT fitting user needs & lifestyle; Importance of advocacy skills, knowledge of rights, and support networks	*Self-identity & Choice*	Individual Agency & Freedom to Choose
AT promotes participation, is a human right, enhances ability, promotes independent living & autonomy	*Value of AT for Equal Participation*	Individual Agency & Freedom to Choose
Fragmented systems, lack of information, coordination, policy, awareness, user engagement, training, and abandonment of AT	*Social Contextual Barriers to AT Access*	AT Passport Functionalities
AT Passport facilitates transitions, provides info, and records AT journey	*AT Passport Functionalities*	AT Passport as Capability-Enhancing Resource
Accessibility standards; user experience (ease of use); user ownership; information security and privacy; and operability within the broader system	*AT Passport Usability & Acceptability*	AT Passport as Capability-Enhancing Resource

aesthetically acceptable in our society to maybe wear Apple glasses... to navigate the environment as opposed to using a cane."

Proper assessment, training, and support are essential for successful AT adoption. To prevent disappointment, AT must provide opportunities for trial before purchase. It should also meet user needs and preferences, aligning with the independent living philosophy by personalising and adapting to the user's lifestyle.

AT Passport as a Capability-Enhancing Resource

Significant issues include fragmented AT service provision, a lack of information, and inadequate transition coordination. One participant, Scott, captured the transformative power of AT: "For most people, technology makes things easier, but for someone with a disability, it makes things possible." Another participant, Mary, underscored the importance of communication, stating that AT is essential to provide a voice.

A lack of robust assistive technology (AT) policy infrastructure, limited awareness, complex maintenance pathways, the inexperience of AT personnel, and the significance of user engagement and training are all crucial factors. The AT Passport can address these issues by facilitating transitions and providing information on AT service provision pathways.

3.2 Quantitative Study

This study explored the experiences of university students with Assistive Technology throughout their educational journeys. Participants were recruited via email from the disability support service at an Irish university. After providing informed consent, the participants completed an anonymous online survey that examined demographic characteristics (gender, age, education level, and disability status) and investigated AT Utilisation, AT Knowledge and Perception, and AT Access and Integration.

Study Design and Participant Characteristics
The study included 57 students, predominantly female (59%) and aged between 18 and 23 (60%). Most were enrolled in undergraduate programmes (84%), with a significant proportion (51%) gaining access to university through the Disability Access Route to Education programme. Participants self-reported various disabilities, including specific learning difficulties (21%) and mental health conditions (17%), with some indicating multiple, co-occurring disabilities. Data were gathered using an anonymous online survey hosted on Microsoft Forms, employing Likert-type scale questions to evaluate demographic information, AT usage patterns, knowledge and perceptions of AT, and the accessibility and integration of AT within their educational experiences.

Key Observations and Findings
Most students (84%) reported using assistive technology (AT) during their education. The main reasons included accessing educational materials and facilitating communication. Assistive technology also supports participation in employment-related activities and independent living. Access to computers was the most prevalent form of AT (40%).

Students' self-reported AT Knowledge (*Mean* = 3.0), AT Confidence (*Mean* = 3.2), and perceived AT Importance (*Mean* = 3.9) were assessed.

Perceptions of educators' AT knowledge varied, with university educators viewed as more knowledgeable (Mean = 3.5) compared to secondary (Mean = 2.3) and primary educators (Mean = 1.9). Students were notably aware (Mean = 4.2) and satisfied (Mean = 4.5) with university AT support services. *A Kruskal-Wallis* test indicated significant differences in perceived AT knowledge across educator levels ($p < 0.001$).

Students with documented AT needs demonstrated a better understanding of AT ($p = 0.01$) and regarded it as more important ($p = 0.05$). Significant differences were identified in their experiences accessing AT during educational transitions ($p = 0.05$) and communicating their AT requirements ($p = 0.01$). The combined Access to AT construct also revealed significant differences ($p = 0.01$).

3.3 The Delphi Study

A Delphi study was conducted to establish expert consensus on an Assistive Technology Passport's essential features, design considerations, and security and privacy aspects. Following the CREDES guidelines, the study utilised structured questionnaires and iterative feedback to define the core elements necessary to implement the AT Passport successfully. Participants were selected based on their experience as AT users.

Study Design

The Delphi study employed a mixed-methods approach to gather expert opinions on the AT Passport. Experts were recruited from a university registry of AT users. Data were collected using Likert scales and open-ended questions through Microsoft Forms, and the results were analysed via Microsoft Excel. A consensus threshold of 80% was established, with items not achieving this level in the first round being reintroduced. Thematic analysis uncovered recurring themes in the qualitative data. Participants received information about the study and questionnaires, alongside email reminders. Descriptive statistics highlighted areas of consensus, while content analysis elicited insights from the qualitative data.

Results

The Delphi study focused on establishing consensus regarding the essential functionalities and design elements of the AT Passport. After two rounds of discussion, a final agreement was reached, incorporating 22 items into the AT Passport concept (representing 81% average agreement) while excluding five. The agreed-upon key functionalities include facilitating transitions, enabling communication, providing information, tracking the user's AT journey, and streamlining access to AT support. Design considerations emphasised user-friendliness, web accessibility, user-controlled access, and a dynamic, interoperable design. Security and privacy aspects that received strong agreement included GDPR compliance, confidentiality, secure storage and authentication, data verification, and user-controlled access.

3.4 Integrating Qualitative, Quantitative, and Expert Perspectives

Data triangulation, which involves comparing and integrating data from qualitative interviews, quantitative surveys, and expert opinions derived from the Delphi study, enhanced the validity and reliability of the findings. The analysis explored convergence, complementarity, silence, and dissonance across the data sources.

Qualitative and quantitative data both emphasised the significance of AT in enhancing accessibility and understanding individual needs, as well as the importance of personal, social, and contextual barriers. Qualitative data complemented quantitative findings by offering rich contextual details, such as lived experiences and challenges in accessing and using AT in educational settings. In contrast, quantitative data revealed a higher prevalence of AT among university students. However, quantitative data lacked insights into socio-economic factors affecting AT usage, while qualitative data overlooked specificities related to AT training. Dissonances arose regarding perceptions of the AT Passport, educator roles, and AT usage across educational stages, with qualitative data not fully supporting the higher AT usage in tertiary education; these issues were explored for potential improvements in AT design or implementation.

Key Findings

The AT Passport, a personalised digital tool that maps a user's AT requirements and history, can enhance AT access and usability by consolidating essential information and fostering communication. Research findings suggest passport-like tools improve communication, user involvement, and service access. A Delphi study identified key features, including a user-friendly design and real-time updates, further supporting this potential.

Recommendations and Framework

This research study proposes the AT Passport framework to enhance access to Assistive Technology. The framework integrates research and expert insights to promote an accessible, user-friendly, secure AT solution. The ultimate aim is to empower individuals to manage their AT needs effectively throughout their lives and various service settings.

4 AT Passport Conceptual Framework

The AT Passport framework integrates research findings and expert opinions to address gaps in Assistive Technology access. Designed with a person-centred approach, the framework is a central repository for an individual's AT journey, facilitating communication and ensuring access to necessary supports. Its design emphasises key features

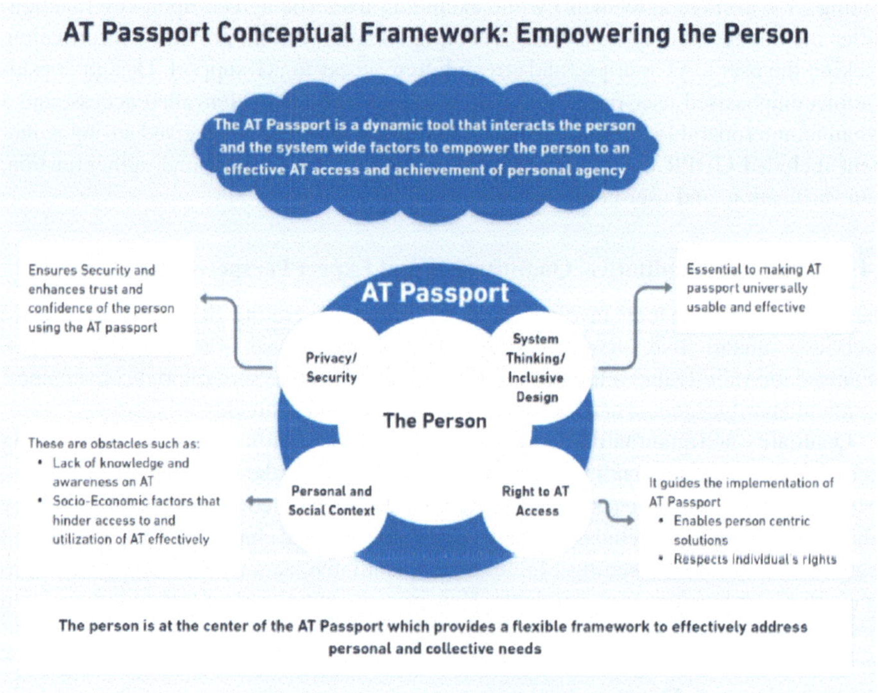

Fig. 1. AT Passport Conceptual Framework

such as intuitive use, adaptability, user-controlled access for privacy, and interoperability with existing systems, enhancing its potential as a tool for navigating AT access. Figure 1 illustrates the AT Passport Conceptual Framework, depicting it as a dynamic ecosystem centred on the individual and their needs. Furthermore, the AT Passport development framework builds on existing knowledge by incorporating inclusive and user-centred design for a versatile, secure, and engaging experience, aiming to improve AT access and usability by addressing facilitators and barriers.

5 Discussion

This study highlights the potential of Assistive Technology Passports to improve AT accessibility, particularly in contexts lacking established AT systems. A pragmatic, exploratory sequential mixed-methods design combines qualitative insights into lived experiences with quantitative data on user needs [25–28]. The study employs Interpretative Phenomenological Analysis of interviews to understand experiences and barriers to AT access, followed by a quantitative survey and a Delphi study to achieve expert consensus on essential functionalities and design features [29–31] The resulting framework addresses personal and system-wide needs by mapping AT requirements and providing a new method for optimising the AT service system.

The research identifies social barriers to AT accessibility, advocating for a multi-sectoral approach [20, 32–35]. The AT Passport is a dynamic element that optimises interactions between individuals, AT, service providers, and policies to facilitate access to AT [12, 36–39]. Key features, informed by the Delphi study, facilitate transitions, communication, AT information, user journey tracking, and streamlined support [1, 3, 4, 40]. Design emphasises a user-friendly interface, accessibility, and interoperability, while security prioritises GDPR compliance and data protection [15, 18, 41–43]. The AT Passport should facilitate transitions, empower users, and align with autonomy principles [38]. Recognising the need for adaptability and data security, the AT Passport must be dynamic and flexible [21, 23].

The culmination of this study is developing the AT Passport conceptual framework. This framework promotes the AT Passport's progression and adoption, incorporating critical functionalities and design features established through the mixed-method and Delphi studies. Designed to address the unique needs of people with disabilities, it prioritises the individual. It tackles the relationship between personal and contextual barriers, countering challenges like limited awareness and socioeconomic factors with inclusive and universal principles [14, 15, 41, 44–47]. Enriched by theoretical concepts such as human rights, the framework guides the AT Passport's design and implementation to respect individual rights and enhance AT access from the personal level up to macro-level facilitators [1, 48–51]. While focused on the AT Passport, the framework can be recontextualised for any innovative concept to operationalise system-wide AT access, potentially leading to a model that emphasises the individual at the heart of service provision and policymaking.

6 Conclusion

This study highlights the potential of AT Passports to enhance access to and utilisation of assistive technology for individuals with disabilities. The framework is developed using mixed methods and prioritises user-centred design and adaptability. The AT Passport manages needs and promotes inclusivity by placing the individual at the centre. It fosters communication among users and stakeholders, promising to improve access while requiring careful design and refinement. Further research should concentrate on its practical design and impact.

7 Limitations

This study on the AT Passport has limitations, including qualitative data from a single organisation and a quantitative phase from a single institution, which affect generalisability. The diversity of the Delphi study and the involvement of only one researcher also raise concerns. Future research should utilise diverse samples and multiple researchers.

References

1. Smith, E.M., et al.: Assistive technologies are central to the realization of the convention on the rights of persons with disabilities. Disabil. Rehabil. Assistive Technol. (2022). ahead-of-print (ahead-ofprint):1–6
2. World Health Organization U. Global report on assistive technology. Geneva (2022)
3. Assembly WH. World Health Assembly 71.8 improving access to assistive technology. Geneva (2018). https://apps.who.int/gb/ebwha/pdf_files/WHA71/A71_R8en.pdf?ua=1
4. WHO. Improving Access to Assistive Technology; Report by the secretariat Geneva. WHO (2016). https://apps.who.int/gb/ebwha/pdf_files/EB142/B142_21-en.pdf
5. WHO. A71/21: WHA Resolution on Improving Access to Assistive Technology Geneva. WHO (2018). https://apps.who.int/gb/ebwha/pdf_files/WHA71/A71_R8-en.pdf
6. du Toit, R., Keeffe, J., Jackson, J., Bell, D., Minto, H., Hoare, P.: A global public health perspective: facilitating access to assistive technology. Optom. Vis. Sci. **95**, 883–888 (2018)
7. Maalim, M.I., MacLachlan, M.: The assistive technology passport: a resource for enhancing capabilities as a result of better access to assistive technology. Societies **12**(6), 182 (2022)
8. O'Donnell, J., Long, S., Richardson, P.: Assistive technology for people with disabilities and older people a discussion paper, enable Ireland and disability federation of Ireland Dublin, Ireland (2016). http://www.enableireland.ie/sites/default/files/publication/AT%20Paper%20final%20version.pdf
9. Nations. U. Convention on the Rights of Persons with Disabilities New York. USA (2006). https://www.un.org/development/desa/disabilities/convention-on-therights-of-persons-with-disabilities/convention-on-the-rights-of-persons-with-disabilities2.html
10. World Health O. International Classification of Functioning, Disability, and Health: Children & Youth Version: ICF-CY: World Health Organization (2007). https://www.who.int/standards/classifications/international-classification-of-functioningdisability-and-health
11. Savigny, D., Adam, T.: Alliance for Health Policy and Systems Research & World Health Organization. Systems thinking for health systems strengthening / edited by Don de Savigny and Taghreed Adam. WHO (2009)

12. MacLachlan, M., Scherer, M.J.: Systems thinking for assistive technology: a commentary on the GREAT summit. Disabil. Rehabil. Assistive Technol. **13**(5), 492–496 (2018)
13. MacLachlan, M., et al.: Intersections between systems thinking and market shaping for assistive technology: the SMART thinking matrix. Int. J. Environ. Res. Public Health, 15 (2018)
14. Mitra, S.: The capability approach and disability. J. Disabil. Policy Stud. **16**, 236–247 (2006)
15. Steinfeld, E., Maisel, J.: Universal Design: Creating Inclusive Environments. Wiley, Hoboken (2012)
16. Buchanan, R., Layton, N.: Innovation in assistive technology: voice of the user. Societies **9**, 48 (2019)
17. Imrie, R., Hall, P.: Inclusive Design: Designing and Developing Accessible Environments. Taylor & Francis (2003)
18. Heidrich, R., Bassani, P.: Inclusive design-assistive technology for people with cerebral palsy. Work **41**, 4762–4766 (2012)
19. Smith, R.O., Scherer, M.J., Cooper, R., Bell, D., Hobbs, D.A., Pettersson, C., et al.: Assistive technology products: a position paper from the first global research, innovation, and education on assistive technology summit. Disabil. Rehabil. Assist. Technol. **13**, 473–485 (2018)
20. Mallin, S.S.V., de Carvalho, H.G.: Assistive technology and user-centered design: emotion as element for innovation
21. van Ommeren, A.L., Smulders, L.C., Prange-Lasonder, G.B., Buurke, J.H., Veltink, P.H., Rietman, J.S.: Assistive technology for the upper extremities after stroke: systematic review of users' needs. JMIR Rehabil. Assistive Technol. **5**(2), e10510 (2018)
22. Shore, L., Power, V., De Eyto, A., O'Sullivan, L.W.: Technology acceptance and user-centred design of assistive exoskeletons for older adults: a commentary. Robotics **7**, 3 (2018)
23. Natasha Layton, D.L.W.: AT user capability building framework; global research, education and innovation in assistive technology summit (2017). https://cdn.who.int/media/docs/default-source/assistive-technology-2/eproceedings-great2017/gs25_layton_atusercapabilitybuildingaustralia.pdf?sfvrsn=fdc8861c_7
24. Maalim, M.: Investigating the potential of assistive technology passports for enhanced access to assistive technology and the development of an implementation and adoption framework (2023)
25. Dorrington, P., Wilkinson, C., Tasker, L., Walters, A.: User-centered design method for the design of assistive switch devices to improve user experience, accessibility, and independence. J. Usability Stud. **11** (2016)
26. Woolley, C.M.: Meeting the mixed methods challenge of integration in a sociological study of structure and agency. J. Mixed Methods Res. **3**, 7–25 (2009)
27. Teddlie, C., Tashakkori, A.: Foundations of Mixed Methods Research: Integrating Quantitative and Qualitative Approaches in the Social and Behavioral Sciences. Sage (2009)
28. Tashakkori, A., Creswell, J.W.: Exploring the nature of research questions in mixed methods research. Sage Publications Sage CA: Los Angeles, CA, pp. 207–211 (2007)
29. O'Cathain, A., Murphy, E., Nicholl, J.: Why, and how, mixed methods research is undertaken in health services research in England: a mixed methods study. BMC Health Serv. Res. **7**, 1–11 (2007)
30. Smith, J.A., Osborn, M.: Interpretative phenomenological analysis as a useful methodology for research on the lived experience of pain. Br. J. Pain **9**, 41–42 (2015)
31. Nizza, I.E., Farr, J., Smith, J.A.: Achieving excellence in interpretative phenomenological analysis: four markers of high quality. Qual. Res. Psychol. **18**, 369–386 (2021)
32. Smith, J.A., Flowers, P., Larkin, M.: Interpretative Phenomenological Analysis: Theory, Method and Research: SAGE Publications (2021)
33. Yusif, S., Soar, J., Hafeez-Baig, A.: Older people, assistive technologies, and the barriers to adoption: a systematic review. Int. J. Med. Informatics **94**, 1126 (2016)

34. McNicholl, A., Casey, H., Desmond, D., Gallagher, P.: The impact of assistive technology use for students with disabilities in higher education: a systematic review. Disabil. Rehabil. Assist. Technol. **16**, 130–143 (2021)
35. Howard, J., Fisher, Z., Kemp, A.H., Lindsay, S., Tasker, L.H., Tree, J.J.: Exploring the barriers to using assistive technology for individuals with chronic conditions: a meta-synthesis review. Disabi. Rehabil. Assistive Technol. **17**(4), 390–408 (2022)
36. Boot, F.H., Owuor, J., Dinsmore, J., MacLachlan, M.: Access to assistive technology for people with intellectual disabilities: a systematic review to identify barriers and facilitators. J. Intellect. Disabil. Res. **62**(10), 900–921 (2018)
37. Youm, H.Y.: An overview of de-identification techniques and their standardization directions. IEICE Trans. Inf. Syst. **103**(7), 1448–1461 (2020)
38. O'Sullivan, K., McGrane, A., Long, S., Marshall, K., Maclachlan, M.: Using a systems thinking approach to understand teachers perceptions and use of assistive technology in the republic of Ireland. Disabil. Rehabil. Assistive Technol. **18**(5), 502–510 (2023)
39. Maalim, M., Maclachlan, M., Long, S., O'Donnell, J., Ahern, S., Gilligan, J.: Access to assistive technology: a descriptive review and application of systems-thinking approach in the conceptualization of the assistive technology passport
40. Layton, N., et al.: Opening the GATE: systems thinking from the global assistive technology alliance. Disabil. Rehabil. Assistive Technol. **15**(5), 484–490 (2020)
41. Sugawara, A.T., Ramos, V.D., Alfieri, F.M., Battistella, L.R.: Abandonment of assistive products: assessing abandonment levels and factors that impact on it. Disabil. Rehabil. Assistive Technol. **13**(7), 716–723 (2018)
42. Newell, A.: Inclusive design or assistive technology. Inclusive design: design for the whole population 172–81
43. Srivastav, A.K., Das, P., Srivastava, A.K.: Data management, security, and ethical considerations. biotech and IoT: an introduction using cloud-driven labs, pp. 133–49. Springer, Cham
44. Regulation 2016/679 of the European Parliament and of the Council of 27 April 2016 on the protection of natural persons with regard to the processing of personal data and on the free movement of such data, and repealing Directive 95/46/EC:OJ L 119, 4.5.2016, p. 1–88. 46. Harnacke, C.: Disability and Capability: Exploring the Usefulness of Martha Nussbaum's Capabilities Approach for the UN Disability Rights Convention. The Journal of law, medicine & ethics 41, 768–80
45. Borg, J., Östergren, P-O., Larsson, S., Rahman, A.S.M.A., Bari, N., Khan, A.H.M.N.: Assistive technology use is associated with reduced capability poverty: a cross-sectional study in Bangladesh. Disability and rehabilitation: Assistive technology 7, 112–21
46. Nations, U.: SDG4: Ensure inclusive and equitable quality education and promote lifelong learning opportunities for all. https://sdgs.un.org/goals/goal4
47. Abrahams, S.: How Co-creation is Enabling a More Inclusive Workforce: Disability and Adjustment Passport in: Fujistu. https://www.fujitsu.com/uk/Images/Fujitsu_Tech_for_Good_10_Jan_2020_Final_Medium_Images.pdf
48. Zhu, Y-Q, Chen, H-G.: Social media and human need satisfaction: Implications for social media marketing. Bus. Horiz. **58**, 335–45
49. Inclusive Workforce: Disability and Adjustment Passport in: Fujistu (2019). https://www.fujitsu.com/uk/Images/Fujitsu_Tech_for_Good_10_Jan_2020_Final_Medium_Images.pdf
50. Zhu, Y-Q., Chen, H-G.: Social media and human need satisfaction: implications for social media marketing. Bus. Horiz. **58**, 335–45
51. Layton, N., et al.: Assistive technology as a pillar of universal health coverage: qualitative analysis of stakeholder responses to the world health assembly resolution on assistive technology. Disabil. Rehabil. Assist. Technol. 1–7

52. de Witte, L., Steel, E., Gupta, S., Ramos, V.D., Roentgen, U.: Assistive technology provision: towards an international framework for assuring availability and accessibility of affordable high-quality assistive technology. Disabil. Rehabil. Assist. Technol. **13**, 467–72
53. Blanck, P., Flynn, E.: Routledge handbook of disability law and human rights

Method for Adapting the SUS and UTAUT-2 for Individuals with Autism, Traumatic Brain Injury and Intellectual Disability

Claude Vincent[1,2](✉), Cynthia Henriksen[2,3], Frédéric Dumont[2], Chaimaa Zemrani[2], Marie-Eve Lamontagne[1,2], Alexandra Lecours[2,4], and Ernesto Morales[1,2]

[1] School of Rehabilitation Sciences, Université Laval, Quebec City, QC, Canada
claude.vincent@rea.ulaval.ca

[2] Center of Interdisciplinary Research in Rehabilitation and Social Integration (Cirris), Quebec City, QC, Canada

[3] Réseau d'Évaluation et d'Innovation en Santé Durable (REISD), Quebec City, Canada

[4] Université du Québec à Trois-Rivières (UQTR), Trois-Rivières, QC, Canada

Abstract. The System Usability Scale (SUS) and the Unified Theory of Acceptance and Use of Technology (UTAUT-2) were adapted for a project in which individuals with level 2 autism, light or moderate brain injury, or light intellectual disability were asked to rate the usability and acceptability of a prototype for a computerized workstation training system. The objectives are (1) to detail the five steps of the co-design method and (2) to disseminate adapted versions of the SUS and UTAUT-2 for neuroatypical clients. The co-design method involved to: 1- establish a multidisciplinary advisory group, 2- develop initial drafts of the prototype, 3- conduct think-aloud usability evaluation, 4- test the prototype in clinical simulations and 5- generate a final prototype informed by workshops. The co-design method allowed us to rigorously and efficiently adapt these questionnaires and to validate them in context with the target population. The results from the adapted SUS (78.73 out of 100 = good) led to the pursuit of a computerized training system to facilitate the employability of the target population. The results from the adapted UTAUT-2 are presented for step 3. Steps 4 and 5 will be completed during work activities (assembly of parts) with the training system once manufactured.

Keywords: Usability and acceptance · Pre-employment training system · Autism · brain injury and intellectual disability

1 Introduction

The **System Usability Scale** (SUS) is a single page widely used scale to measure the perceived usability of a system or product, particularly in the field of technology and user interfaces [1]. The SUS contains 10 items, rated on a Likert scale from 1 (strongly

K. Mavrou and P. Encarnação (Eds.): AAATE 2025, *Technology for Inclusion and Participation for All: Recent Achievements and Future Directions*, pp. 344–352, 2025.
https://doi.org/10.1007/978-3-032-01628-7_41

disagree) to 5 (strongly agree). Users' answers to these questions are then converted into an overall score between 0 and 100, where a higher score indicates better usability. The advantages of the SUS include its simplicity, speed of administration, and use in a wide range of contexts [2].

The **Unified Theory of Acceptance and Use of Technology** (UTAUT-2) contains 27 items, rated on a Likert scale from 1 to 7 (completely disagree to neutral to completely agree) including participant information questions [3]. It measures the acceptance of technology considering performance expectancy, effort expectancy, social influence, hedonic motivation, facilitating conditions and behavioral intentions. The UTAUT-2 has been administered to users of banking applications [4], students using software in business school [5], nurses who use computerized medical records [6] and business administrators using resource planning systems [7].

It would be an added-value to use SUS and UTAUT-2 in work rehabilitation, in particular for computerized workstation training systems to increase employability. The SUS and the UTAUT-2 have not been validated among individuals with invisible disabilities such as autism, traumatic brain injury and intellectual disability who use specially developed information technologies such as computerized workstation training systems to increase employability. The **objectives** are (1) to detail the 5 steps of the co-design method and (2) to disseminate adapted versions of the SUS and UTAUT-2 for individuals with autism level 2, light and moderate brain injury and light intellectual disability.

2 Method

To adapt the questionnaires, the *framework for co-design of clinical tools* was followed in five steps [8].

2.1 Establish a Multidisciplinary Advisory Group

An evaluator from the *Sustainable Healthcare Innovation and Evaluation Network* (2nd author), the principal investigator (1st author) and the project coordinator (3rd author) made up the advisory group.

2.2 Develop Initial Drafts of the Prototype (i.e. Questionnaire)

In the 1st column of an EXCEL file, the evaluator positioned the original items to be adapted, and in the 2nd column, drafts items according to the readability of the target population, using the reference of the Autistic self advocacy Network [9]. The evaluator sent the EXCEL file to the PI and the coordinator. Individually, they agreed with the proposed adaptation by indicating, "OK" in the third column or proposed a new wording. This step has been done for the SUS and UTAUT-2.

2.3 Conduct Think-Aloud Usability Evaluation (i.e. Critical Appraisal)

The evaluator compiled all draft items in the 3rd column. In a face-to-face discussion, they reached an agreement on a final draft version of the items. The Likert scales presenting the level of agreement has been enhanced with familiar-colored icons as suggested by the Autistic self advocacy Network [9], for examples smiling man in green, neutral man in white, disgruntled man in red). By e-mail, the evaluator forwarded a PowerPoint version of the questionnaire SUS and UTAUT-2, respecting accessibility requirements for the target population, like one idea per line [9]. The PI and the coordinator proposed format changes.

2.4 Test the Prototype in Clinical Simulations

As planned in the initial research project, a researcher (7th author) led clinical simulations, followed by co-design sessions to gather feedback to validate [12] the instructions (training system) and the workstations. After gathering feedback, the same researcher invited everyone present to complete the SUS 'prototype'. In total, four trainees, three specialized educator and three researchers (1st, 5th and 6th author) filled the SUS. Four trainees were involved in the clinical simulation in three different settings: one with level 2 autism (Workshop ITSA), one with light chronic brain injury (Servio), one with moderate brain injury (Servio) and one with light intellectual disability (Workshop Lapierre). The clinical simulation consisted in trying the training system and the computerized workstation. Workstations were designed to help trainees assemble an electrical resistor, as pre-employment training system (see Fig. 1a). The training system provides adapted instructions on a tablet computer with accessible and well identified components to facilitate assembly. Each step appears on a PowerPoint slide (see Fig. 1b). There is always a brief written indication, a diagram with arrows and a short video of the action to be taken.

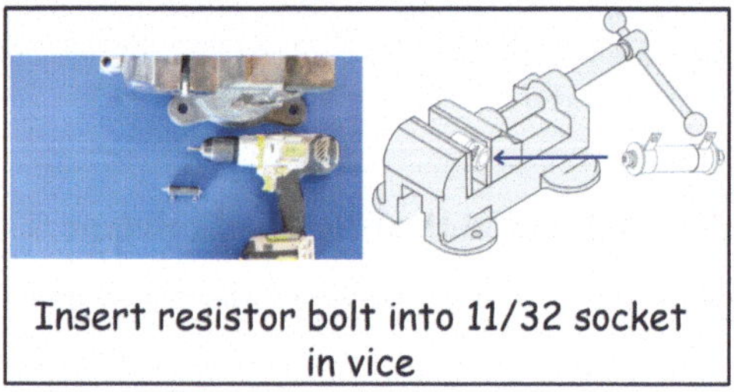

Fig. 1. a. Prototype of the workstation to help trainees assemble an electrical resistor. **b.** PowerPoint slide showing one adapted instruction to facilitate assembly.

2.5 Generate a Final Prototype Informed by Workshops

A master student (4th author) compiled 12 SUS questionnaires, including comments. To provide a score out of 100 [9], you had to invert the score of the odd-numbered items (1, 3, 5, 7, 9) to be able to add them up because they are formulated in reverse of the even-numbered items (2, 4, 6, 9, 10). Usability scores were from 72.3 to 87.5, with an overall average of 78.73 out of 100. Only one score was in the 'excellent' category (≥ 85.5 out of 100), the others being in the 'good' category (71.4 to 85.4 out of 100) [9]. A TEAMS® workshop was organized with the researchers, the coordinator and the master student to present the results, comments and the final prototype of SUS. See Table 1 for all compilation of the F-SUS.

3 Results

3.1 The 10 Modified SUS Items

1. I would very much like to use this training system.
2. I find this training system complicated for assembling a resistor.
3. I find this training system easy to use.
4. I need my special educator to use this training system.
5. The instructions and the layout of the equipment at the workstation work well for assembling the parts.
6. There are some confusing elements in this training system.
7. Other people in the shop will quickly learn to assemble parts with this training system.
8. This training system is demanding to use.
9. I'll be able to use the training system without help from anyone.
10. This training system does NOT contain the instructions needed to assemble the parts.

3.2 The Modified SUS Scale

In the SUS, all odd-numbered items must be formulated positively, while all even-numbered items must be formulated in the opposite way. To illustrate this, items 3 and 4 both assess the same variable (ease of use) but are formulated differently (see Fig. 2). The scale remains the same as in the original version, but different-colored face stickers have been added to reflect the degree of agreement. It was requested to have only one idea per page to respect the neuroatypical population [9]. The questionnaire includes 10 pages of questions and must be printed in color.

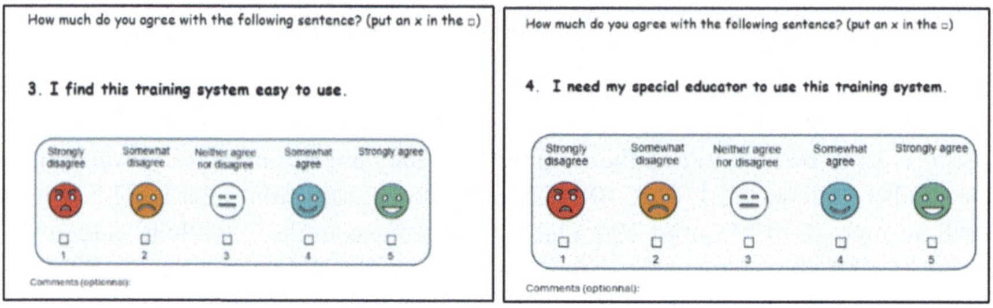

Fig. 2. Items that assess the same variable but formulated in the opposite way.

3.3 The 16 Modified UTAUT-2 Items

Eleven items have been removed from UTAUT-2 [11] as it only assessed acceptance of the training system by users in a research project. Thus, the evaluation was not required for *Social influence* (items 8, 9, 10), *Hedonic motivation* (items 15, 16, 17), *Financial value* (items 18, 19, 20), *Habit* (item 22) and *Behavioral intention* (item 25).

The 16 modified items are:

1. I find the training system useful for assembling parts the right way.
2. I assemble parts faster with the training system.
3. The training system enables me to manufacture electrical components quickly and without errors.
4. Learning to use the training system is easy for me.
5. It's simple and easy for me to understand the training system using the tablet screen.
6. I find the training system easy to use.
7. I'm getting good at using the training system.
8. I have everything I need to use the training system.
9. I have received the information I need to use the training system.
10. The training system considers the other technologies I use (e.g., tools, bins, pictograms).
11. I can get help when I have difficulty using the training system.
12. I now use the training system out of habit.
13. I use the training system to assemble microelectronic parts.
14. I intend to continue using the training system to assemble parts.
15. I will always use a training system to remind me of the steps involved in assembling parts.
16. I often use instructions for -Prepare the screwdriver and place the jig in the vice. -Insert nuts into screw. -Screw the resistor into the vice

3.4 The Modified UTAUT-2 Scale

The scale is not the same as in the original version. We've replaced "completely" by "strongly" and we've removed the word "neutral". The idea is for users to indicate their level of agreement with the training system from 1 to 7. The qualifier "strongly" is preferable because it is more concrete than "completely". The word "neutral" is removed as it remains abstract to indicate a rating of 4. We have added seven different-colored face stickers to reflect the degree of agreement with numbers 1 to 7, all positioned on an upward slope. This is more accessible for the neuroatypical population [9]. All questions measuring the same dimension are presented on the same page, as requested to have only one idea per page to respect the neuroatypical population [9]. For example, questions 1, 2 and 3 are presented together for the Performance expectancy (see Fig. 3). The questionnaire comprises 7 pages of questions and must be printed in color.

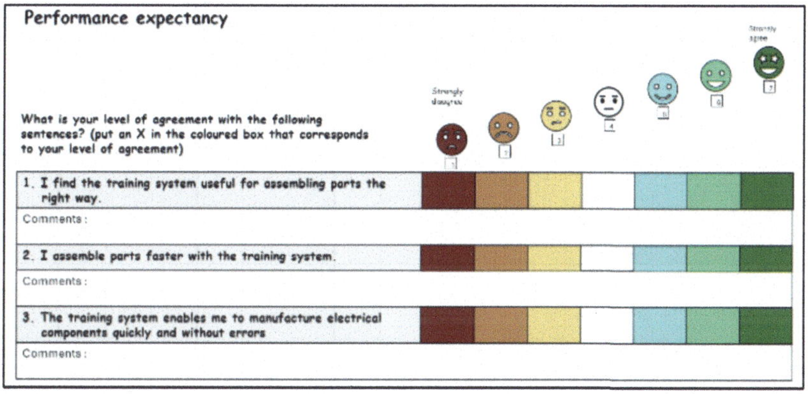

Fig. 3. Items that assess the same concept on one page

4 Discussion

The adapted SUS enabled us to conclude that the usability of the workstation training system was sufficient to proceed with the manufacture of additional workstations and purchase the required tools. The adapted UTAUT-2 will be submitted to steps 4 and 5 of co-design of clinical tools, but after 3 months of using the real computerized workstation training system by the target population.

Table 1. Compilation of the French- System Usability Scale (F-SUS) [10], adapted for neuroatypical populations, carried out with 10 participants in three settings following the use of a training system to assemble an electric resistor

	Settings												
	Servio Traumatic brain injury (L) Light vs (M) Moderate						**Workshop Lapierre** Intellectual Disability light			**Workshop ITSA** Autism level 2			
Groups: Users, Specialized educators and Researchers	U_L	S_L	S_M	R_L	R_M	U_M	U	S	R	U	S	R	Median
Age in years	28	35		46		60	21	44	40	24	31	57	37,5
Sex (F or M)	M	M		F		F	F	F	F	M	F	F	7F 3M
F-SUS – Total adjusted [1]	33	35	34	30	26	34	29	29	29	34	32	30	31,0
F-SUS – score on 100 [2]	82.5	87.5	85	75	72.3	85	72.5	72.5	72.5	85	80	75	77,5
F-SUS – Interpretation [3]	B	A	B	B	B	B	B	B	B	B	B	B	B-good
Standardised score items 3, 4, 10 (raw score on 5)													
3. *I find this training system easy to use*[5]	3 (4)	3 (4)	3 (4)	2 (3)	3 (4)	4 (5)	4 (5)	4 (5)	3 (4)	4 (5)	3 (4)	3 (4)	3 (4)
4. *I need my specialised educator to use this training system.*[6]	4 (1)	2 (3)	2 (3)	3 (2)	-[8]	1 (4)	2 (3)	2 (3)	3 (2)	2 (3)	1 (4)	2 (3)	2 (3)
10. *This training system does NOT contain the instructions needed to assemble the parts.*[7]	4 (1)	4 (1)	4 (1)	4 (1)	4 (1)	4 (1)	4 (1)	3 (2)	3 (2)	4 (1)	4 (1)	3 (2)	4 (1)
F-SUS – Mean score User n=4	x					x	x			x			83.75
F-SUS – Mean score Sp. ed. n=4		x	x					x			x		82.5
F-SUS – Mean score Reser. n=4				x	x				x			x	73.75

Note 1. Total adjusted score= SUS 1-N+SUS 2-N+SUS 3-N+SUS 4-N+…+SUS 10-N

Note 2. Score SUS on 100= Total adjusted score ×2.5

Note 3. 85.5 – 90: Excellent (A); 71.4 – 85: Good (B); 50.9 – 71: Okay (C); >50: Poor (D)

Note 4. 1= Strongly disagree, 2= Somewhat disagree, 3= Neither agree or disagree, 4= Somewhat agree, 5= Totally disagree

Note 5. Odd-numbered items (1, 3, 5, 7, 9) are weighted to obtain a score out of 100: SUS "X"-N = SUS "X" Raw - 1

Note 6. Even-numbered items (2, 4, 6, 9, 10) are weighted to obtain a score out of 100: SUS "X"-N = 5 −SUS "X" Raw

Note 7. Item 10 was modified with a negative formulation to go in the same direction as the original SUS, when the item is even.

Note 8. This participant marked Not applicable, for item 4. The raw score was therefore calculated on 9 items instead of 10.

5 Conclusion/Implications for the AT Field

The co-design method allowed us to quickly adapt the SUS and UTAUT-2 while validating them in context among people with autism, traumatic brain injury and intellectual disability. It is planned to use both questionnaires, once the workstations are operational, with special educators and trainees. Both questionnaires are available in French and in English at [13].

Regarding implication of AT fields, we therefore validated 3 different training systems, in terms of instructions and workstations. For example, the instructions for people with intellectual disabilities indicate that users should ask the specialised educator how to prepare the workstation (e.g. adjust the screwdriver, put the jig in the vice). Instructions for people with autism rely more on the direction of the arrows to carry out actions such as where to insert the bolts and the direction to turn (e.g. Figure 1b). Workstations for people with traumatic brain injury need to take greater account of fatigue, so chairs with armrests, a cane holder and a stool for putting down feet. Whereas for people with autism, workstations need to take more account types of stimulation, such as being able to adjust the color of the light and side panels to reduce visual inputs.

Funding and Ethic Statements. The research project was funded by the Ministry of Economy, Innovation and Energy (2022-2027-PSOv2a-UL-IS-63676). The project was approved by the CIUSSS-CN Re-search and Ethics Committee (#2022-2378).

References

1. Brooke, J.: System usability scale (SUS). A Retrospective. J. Usability Stud. **8**, 29–40 (2013)
2. Baumgartner, J., Ruettgers, N., Hasler, A., Sonderegger, A., Sauer, J.: Questionnaire experience and the hybrid system usability scale: using a novel concept to evaluate a new instrument. Int. J. Hum-Comput. Stud. **147**, 1–11 (2021)
3. Venkatesh, V., Thong, J.Y.L., Xu, X.: Consumer acceptance and use of information technology: extending the unified theory of acceptance and use of technology, MIS Q **36**(1), 157–178 (2012)
4. Slade, E., Williams, M., Dwivedi, Y., Piercy, N.: Exploring consumer adoption of proximity mobile payments. J. Strateg. Mark. **23**(3), 209–223 (2015)
5. Chauhan, S., Jaiswal, M.: Determinants of acceptance of ERP software training in business schools: empirical investigation using UTAUT model. Int. J. Manag. Educ. **14**(3), 248–262 (2016)
6. Maillet, É., Mathieu, L., Sicotte, C.: Modeling factors explaining the acceptance, actual use and satisfaction of nurses using an electronic patient record in acute care settings: an extension of the UTAUT. Int. J. Med. Inform. **84**(1), 36–47 (2015)
7. Keong, M., Ramayah, T., Kurnia, S., Lo, M.C. Explaining intention to use an enterprise resource planning system: an extension of the UTAUT. Bus. Strategy Ser. **13**(4), 173–180 (2012)
8. Woodward, M., Dixon-Woods, M.: How to co-design a prototype of a clinical practice tool: a framework with practical guidance and a case study. BMJ Qual. Saf. **33**, 258–270 (2024)
9. Autistic self advocacy Network. One Idea Per Line: A Guide to Making Easy Read Resources, p. 65. https://autisticadvocacy.org/wp-content/uploads/2021/07/One-Idea-Per-Line.pdf
10. Gronier, G., Baudet, A.: Psychometric evaluation of the F-SUS: creation and validation of the French version of the SUS. Int. J. Hum.-Comput. Interact. **37**(16), 1571–1582 (2021)

11. Pagé, I., et al.: UTAUT2-based questionnaire: cross-cultural adaptation to Canadian French. Disabil. Rehabil. **45**(4), 709–716 (2022)
12. Morales, E., Rousseau, J., Passini, R.: Using a co-design methodology for research on environmental gerontology. J. Gerontol. Geriat. Res. **1**(3), 100–106 (2012)
13. Vincent et al.: Adapting SUS and UTAUT-2 for neuroatypical people. Cirris research products (2025). https://cirris.ulaval.ca/en/produits/adaptation-du-sus-et-de-lutaut-2-pour-des-personnes-neuro-atypiques/

Assistive Technology and Accessibility for Cognitive Disabilities

From Challenges to Solutions: Evaluating Assistive Technologies for Students with Dyslexia in Higher Education

Melanie Schaur(✉) and Reinhard Koutny

Johannes Kepler Universtiy, Altenberger Str. 69, 4040 Linz, Austria
{melanie.schaur,reinhard.koutny}@jku.at

Abstract. The following paper examines the state of the art in Assistive Technologies (AT) for neurodiverse individuals in higher education, focusing on dyslexia as a primary example. The rationale of the study is to highlight the significant role that AT plays in supporting students with dyslexia in higher education, improving their academic performance, their overall learning experience and their ability to cope in everyday life. The primary objective of this study was to provide an overview of existing ATs and evaluate their effectiveness for neurodiverse individuals, focusing on dyslexia. This involved identifying current advances, understanding user experiences and identifying the most beneficial tools for supporting students with dyslexia in higher education. The following paper is based on a three-step approach, which includes a systematic review of the literature, a comprehensive collection of additional ATs from online sources, and an empirical assessment of the effectiveness of selected tools for dyslexic students in higher education. The conclusion identifies the use of technology to address the challenges of persons with dyslexia in higher education, as well as technology gaps and suggests future research directions for AT for neurodiverse students in higher education.

: Assistive Technology · Neurodiversity · Dyslexia · Higher Education

1 Introduction and State of the Art

Students need strong reading and writing skills to succeed academically, yet around 20% face difficulties, with 5–8% affected by dyslexia [1, 2]. Despite its prevalence, there is no universally accepted definition of dyslexia, largely due to differences in researchers' backgrounds and interpretations of its causes [2]. The International Dyslexia Association defines dyslexia as a neurobiological learning disability characterized by difficulties in word recognition, spelling and decoding, often due to phonological deficits [3]. Dyslexia can be developmental, affecting early reading acquisition, or acquired as a result of brain injury. Those with dyslexia often struggle with letter-sound associations, letter sequencing and sound recall, and may exhibit letter reversals, word omissions and other reading difficulties [1, 2, 4, 5]. The condition manifests in various ways, including visual-spatial difficulties (problems identifying letters), speech sound difficulties (issues

K. Mavrou and P. Encarnação (Eds.): AAATE 2025, *Technology for Inclusion and Participation for All: Recent Achievements and Future Directions*, pp. 355–363, 2025.
https://doi.org/10.1007/978-3-032-01628-7_42

forming and understanding sentences and spoken language), and writing difficulties. These challenges hinder the development of both written and spoken language, making it difficult for dyslexic individuals to learn additional languages. Dyslexia often occurs alongside other conditions, exacerbating learning difficulties and sometimes leading to broader issues such as school failure, behavioral problems, anxiety, or depression. These consequences can affect any dyslexic person, regardless of their intelligence, interests, or the education system they are part of [2].

Assistive Technology (AT) can be a valuable support for people with dyslexia, providing tools to enhance reading, writing, memory and organizational skills [2, 6]. While traditional reading and writing methods can be challenging, technology offers alternative strategies that are tailored to the needs of dyslexic users [2]. Tools such as spell checkers, word prediction software, text-to-speech (TTS), and speech-to-text (STT) improve spelling accuracy and writing fluency [4–6]. In addition to supporting spelling and transcription, AT can include customizable visual features such as fonts and colors, voice recognition, animated visual cues, vocabulary aids and tools that promote awareness of writing. These technologies make it easier for dyslexic individuals to engage with information and express their ideas effectively [2]. AT reduces transcription fatigue, increases motivation, and facilitates greater participation in educational and professional settings, thereby boosting self-confidence and academic engagement [1, 6].

Given its importance, this paper reviews the current state of AT for dyslexia, focusing on its applications and effectiveness in higher education.

2 Methodological Approach

The overall study adopts a three-step approach, comprising a systematic literature review (SLR) [7, 8], an extensive collection of additional ATs for individuals with dyslexia via the web [9], and an empirical evaluation of the effectiveness of selected tools for students with dyslexia in higher education [10–12].

The findings of the first step, the SLR, revealed two key insights: First, not all technological advancements have been adequately explored in existing literature, and second, evaluations of effectiveness are scarce. Additionally, the SLR revealed that a further collection of ATs needs to be done [5]. Therefore, additional ATs for students with dyslexia in higher education were collected in the second step through an extensive web search [9]. The tools found in the SLR and the tools found in the extended web-search were summarized in a spreadsheet and divided into the following categories:

Text input and writing,

Text output and reading,

Conceptual support.

The tools were then tested by the project team, who listed the advantages and disadvantages. Advantages include features such as easy login via a Google account, free or low-cost availability, compatibility with multiple devices and automatic data synchronization. Disadvantages included complex usability, high purchase costs or subscription fees, limitations to specific operating systems, or the requirement to create a separate account.

The next step of the study, the empirical assessment of the effectiveness of the collected and pre-tested tools, included three phases to evaluate digital tools designed to support individuals facing challenges in reading, writing and organizing everyday tasks. A total of twelve participants were recruited for the study through E-Mail invitations. The participants had already taken part in other parts of a bigger dyslexia study at the Johannes Kepler University and were contacted by email using contact lists. The survey period took place between October 2024 and April 2024. The first phase of this part of the study started with an initial Zoom interview, in which participants described their difficulties in reading, writing and organizing tasks. They also discussed the tools they had previously used, what had worked well and what hadn't, and their preferences for certain types of tool or usage contexts. Based on these insights, suitable tools were selected internally and tailored to each participant's individual needs. In the second phase, participants took part in on-site so-called "technical testing", during which a curated set of seven to ten tools was demonstrated. They tried the tools themselves and provided immediate qualitative feedback focused on usability, potential use cases, and personal preferences. One to two weeks later, a follow-up session was conducted either via Zoom or in person to assess the tools' longer-term usability. For the evaluation phase, participants recorded their experiences using a daily usage diary for each tool they tested. They reflected on how the tools were used, what worked well or poorly, and whether they would recommend or continue to use them in future. During a qualitative interview, participants shared insights on how and how often the tool was used, in what situations, what participants liked or disliked, and any issues encountered. They were also asked to suggest improvements, comment on unused features, and reflect on whether they would continue using the tool and why. This data was used to assess which tools are more helpful or less helpful based on user experiences and their effectiveness in overcoming dyslexia-related challenges [10–12].

3 Results

This paper focusses on the third phase of the study which consisted of an empirical part. The main aim was on the one hand to identify the key challenges experienced by students in higher education (HEI) settings in relation to reading, writing and organizing daily activities. Another major goal was to find technical solutions that support students in HEI in overcoming these challenges.

3.1 Challenges of Students with Dyslexia in Higher Education

The findings offer a nuanced insight into the specific challenges faced by individuals, the strategies they employ to overcome these challenges, and the role of ATs in their daily academic and personal lives.

One of the central findings relates to difficulties with reading comprehension. Several participants reported skipping individual words or even entire sections of text, which significantly impaired their ability to understand the content. Some also described difficulties in keeping their place within a line of text, especially when reading longer passages. This disrupts reading fluency and can lead to confusion or misinterpretation of the material. Although entertainment literature poses fewer difficulties, reading in general—especially complex or information-dense texts—remains a substantial challenge for many. Physical books are generally preferred, whereas reading on computer screens is often perceived as uncomfortable. E-book readers have not been widely adopted by participants, and experiences with audiobooks have been mixed—although it was noted that negative experiences may have been influenced by suboptimal listening conditions. Under improved circumstances, audiobooks may represent a viable alternative. Smartphones are rarely used for reading or writing tasks; thus, no specific tools or adaptations are employed in this context.

Participants also identified a number of obstacles in terms of writing. Reading and understanding source texts can be so time-consuming that it leaves limited cognitive resources for producing written responses. Line breaks were mentioned as a difficulty, though not consistently. Some participants already use tools such as a ruler to help them maintain structure when writing longer texts, for example. Interestingly, while formal writing is perceived as demanding, creative or recreational writing is experienced as less problematic. However, one participant expressed the need for support with writing as part of a continuing education course. The use of a ruler was mentioned again as a potentially helpful aid.

In contrast to the challenges associated with reading and writing, difficulties in organizing tasks were reported to be minimal. Most participants had already developed effective coping strategies. These include maintaining calendars and creating to-do lists, which help to structure daily activities and reduce mental strain. Despite these established routines, many participants reported experiencing comorbid conditions that contribute to additional stress. One commonly cited example is ADHD, which can impact concentration, time management and emotional regulation. Such coexisting conditions can exacerbate existing difficulties and must be considered when designing support systems.

Participants also shared valuable insights into their experiences with digital tools designed to assist with reading, writing or organizing tasks. It became clear across the board that tools must meet specific requirements to be perceived as helpful. Ease of use and low-threshold accessibility were particularly important to participants, who emphasized the need for tools that integrate seamlessly into daily life without drawing attention. This was particularly important to those who preferred not to stand out or disturb others. Some participants were highly concerned about cybersecurity and refused to use tools that stored data in the cloud, On the other hand, others appreciated the flexibility of accessing their information across multiple devices and saw cloud-based storage as an advantage.

3.2 Solutions for Students with Dyslexia in Higher Education

For the technical testing, tools were collected through a SLR and an extensive web search. A total of 118 tools were identified in this phase. Of these, 38 were designed to support text input and writing, 51 focused on text output and reading, and 29 provided conceptual support. For each tool, detailed information was compiled, including the tool's name, a link to access it, a description of its functionality and area of application, the type of tool, associated costs, and the compatible devices or systems.

Following an initial evaluation phase, a selection of 16 tools was shortlisted for technical testing. This final group consisted of seven tools supporting text input and writing, seven tools aimed at text output and reading, and two tools offering conceptual support.

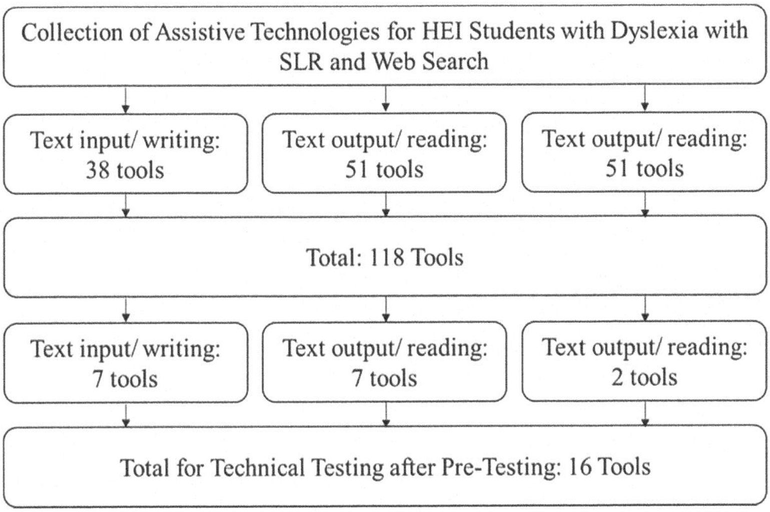

These pre-tested tools were used for the technical testing.

First of all, the participants emphasized the importance of tools being user-friendly and accessible without major technical obstacles. This low-threshold access enables seamless integration into daily academic routines without creating an additional cognitive burden. Privacy emerged as a particularly sensitive concern. Some students prefer tools that do not upload data to the cloud due to cybersecurity concerns. However, others see cloud-based solutions as beneficial, especially because they enable easy access across multiple devices. A recurring theme among the participants was the desire for tools to be discreet. One student shared that she avoids drawing attention to herself and does not want to disrupt others, suggesting that tools should operate discreetly and blend into her workflow. This desire for subtlety reflects a broader need for AT that empower users without drawing attention to their use in public.

Reading difficulties were primarily reported when engaging with complex texts, such as academic articles. Tools with text-to-speech (TTS) functionality were widely regarded as helpful, particularly when they included a feature that highlights the spoken word. This helps users follow the reading and track their progress, allowing them to revisit specific parts of the text when needed. The combination of optical character recognition (OCR) and TTS is particularly valuable in making scanned or printed documents accessible in an interactive way. Some students even considered replacing the traditional reading aid of a ruler with digital ruler functions included in certain tools to help them maintain focus when reading on screen.

The summarization feature in tools like ChatGPT or ChatPDF was particularly well-received. Rather than reading entire articles, students could generate summaries first and then decide which texts warranted further reading. This approach saves time and energy, and enables better navigation of dense academic material. However, effectively using such tools requires a certain level of familiarity and digital literacy, which not all students may have initially.

Students reported mixed experiences when it came to writing. They praised tools that offered grammar suggestions and style improvements, particularly when they provided multiple alternatives and explanations. These tools not only improved the quality of the students' writing, but also supported their learning and awareness of language. Integration with familiar platforms such as Microsoft Word or Outlook was considered a significant advantage, facilitating a more seamless and natural workflow.

Nevertheless, not all tools met expectations. For instance, the Beeline Reader was unreliable on certain websites, which limited its usefulness. Other tools, such as Helperbird, were less effective due to technical limitations, such as an inability to integrate with Safari—a browser that some students frequently switch back to out of habit. The free version of the Natural Reader was also criticized for its poor voice quality, which detracted from the user experience. Meanwhile, although tools such as Deepl Write were available, they were rarely used because alternatives such as ChatGPT and Word were considered more effective for most purposes.

3.3 Summary

The results of the empirical phase of this broader study revealed a diverse picture. Participants showed interest in the tools presented, with some already familiar to them. However, new functions for already known tools—such as DeepL's AI-powered improvement feature—were demonstrated and were largely unknown to the participants prior to the session. Following the on-site technical testing, participants selected three tools for extended testing. In the follow-up interviews, it became clear that not all three tools could be integrated into the participants' daily routines. Typically, only one to two tools were used consistently. These tools, however, were used intensively and perceived as particularly useful.

According to participants, the primary use case for ATs is reading academic texts, such as research papers. Leisure reading, by contrast, was typically reported as not being a problem. This suggests that reading difficulties are more likely to arise in contexts requiring high levels of cognitive engagement and concentration, rather than in informal or narrative reading settings.

The empirical findings also provide insight into the types of ATs that participants find helpful when it comes to overcoming the identified challenges, particularly with regard to reading and writing. When asked about their preferences regarding digital tools, all participants emphasized the need for simplicity and easy access. These tools should be user-friendly and integrate seamlessly into daily routines without requiring extensive technical knowledge or training. Several participants expressed a desire for discretion: ATs should not draw attention in public or professional settings, as some individuals felt self-conscious about being perceived as different or disruptive.

The findings also highlight the complexity of participants' needs and the importance of tailoring support tools to individual preferences and circumstances. Rather than relying on one-size-fits-all solutions, a personalized approach appears essential for fostering inclusion and supporting autonomy in academic and everyday settings.

4 Discussion

This study emphasizes the considerable potential of AT to enhance academic performance among students with dyslexia. The findings suggest that tools such as text-to-speech, speech-to-text, spellcheckers and word prediction software can meaningfully reduce cognitive load and improve essential literacy skills. By shifting the focus from decoding and transcription difficulties to content comprehension and critical thinking, AT enables students to engage more fully with academic tasks. These outcomes are consistent with prior research emphasizing the cognitive offloading benefits of AT and its role in fostering autonomy and academic confidence among learners with dyslexia.

However, despite these advantages, the study also revealed ongoing challenges regarding the accessibility and affordability of AT. While mainstream technologies have improved in recent years by integrating accessibility features, many specialized tools remain prohibitively expensive. This financial constraint particularly affects students from lower socioeconomic backgrounds, potentially exacerbating existing educational inequalities. Therefore, institutional strategies aimed at embedding AT into standard support services, such as providing access through libraries, disability offices or centralized platforms, are critical to ensuring equitable opportunities for all students.

Furthermore, variability in user experiences suggests that AT is not universally effective in a one-size-fits-all model. While some students adapted quickly and found the tools intuitive, others struggled with usability or found that the tools were incompatible with their learning styles. This highlights the importance of a more personalized and flexible approach to AT implementation, including initial training, ongoing support and tool selection that is tailored to individual preferences and needs.

The study also reveals a significant discrepancy between the availability of AT and its effective, sustained integration into daily academic routines. Although participants initially expressed interest and selected multiple tools to trial, most ultimately adopted only one for long-term use. This finding highlights the importance of providing not only access, but also fostering meaningful engagement and sustained usage through ongoing guidance, contextualized support and raising awareness.

Taken together, these findings emphasize the vital role of AT in supporting dyslexic students in higher education. They also highlight the need for more inclusive policies, better funding mechanisms and longitudinal research examining the impact of AT on learning trajectories over time. Future studies should evaluate the long-term effectiveness of AT interventions and explore strategies for seamlessly integrating them into mainstream educational environments.

5 Conclusion and Implications for the AT Field

The paper provides a comprehensive review of the current state of ATs for neurodiverse individuals, with a particular focus on dyslexia in higher education. By systematically reviewing the existing literature, collecting additional ATs through an extensive web search, and empirically evaluating the effectiveness of selected tools, the paper will highlight the significant role that AT plays in improving the academic performance, overall learning experience, and everyday coping strategies of students with dyslexia.

The findings underline that AT, such as spell-checkers, word prediction software, autocorrection, text-to-speech and speech-to-text, are essential in addressing the challenges faced by dyslexic students. These technologies not only improve literacy, but also increase motivation, boost self-esteem and facilitate smoother academic and vocational integration. However, the research also identified gaps, particularly in the evaluation of newer technologies and their long-term impact on users. Moreover, the findings revealed that while significant progress has been made in developing effective AT, there is a need for ongoing research to further explore new tools and their practical applications.

In conclusion, supporting students with dyslexia in higher education requires more than just offering tools; it necessitates an understanding of individual preferences, usability, privacy safeguarding, and the embedding of support into existing academic ecosystems. When selected and implemented thoughtfully, digital tools can play a transformative role in making academic reading and writing more accessible and less stressful for dyslexic students.

References

1. Svensson, I., et al.: Effects of assistive technology for students with reading and writing disabilities. Disabil. Rehabil. Assistive Technol. **16**(2), 196–208 (2021). https://doi.org/10.1080/17483107.2019.1646821
2. Schaur, M., Koutny, R.: Dyslexia, reading/writing disorders: assistive technology and accessibility. In: Miesenberger, K., Peňáz, P., and Kobayashi, M. (eds.) Computers Helping People with Special Needs, pp. 269–274. Springer, Cham (2024). https://doi.org/10.1007/978-3-031-62849-8_33
3. International Dyslexia Association: Definition of Dyslexia - International Dyslexia Association. https://dyslexiaida.org/definition-of-dyslexia/. Accessed 12 Apr 2024
4. Dawson, K., Antonenko, P., Lane, H., Zhu, J.: Assistive technologies to support students with dyslexia. Teach. Except. Child. **51**, 226–239 (2019). https://doi.org/10.1177/0040059918794027
5. Koutny, R., Schaur, M.: Words unleashed: a systematic literature review study on the use of current assistive technology for adults with dyslexia (2024). https://icchp.org/sites/default/files/OAC_24.pdf
6. Mossige, M., Arendal, E., Kongskov, L., Svendsen, H.B.: How do technologies meet the needs of the writer with dyslexia? An examination of functions scaffolding the transcription and proofreading in text production aimed towards researchers and practitioners in education. Dyslexia **29**, 408–425 (2023). https://doi.org/10.1002/dys.1752
7. Wetterich, C., Plänitz, E.: Systematische Literaturanalysen in den Sozialwissenschaften. Eine praxisorientierte Einführung. Verlag Barbara Budrich GmbH, Opladen, Berlin, Toronto (2021)
8. Newman, M., Gough, D.: Systematic reviews in educational research: methodology, perspectives and application. In: Zawacki-Richter, O., Kerres, M., Bedenlier, S., Bond, M., and Buntins, K. (eds.) Systematic Reviews in Educational Research, pp. 3–22. Springer, Cham (2020). https://doi.org/10.1007/978-3-658-27602-7_1
9. Munzert, S., Nyhuis, D.: Die nutzung von webdaten in den sozialwissenschaften. In: Wagemann, C., Goerres, A., and Siewert, M.B. (eds.) Handbuch Methoden der Politikwissenschaft, pp. 373–398. Springer, Cham (2020). https://doi.org/10.1007/978-3-658-16936-7

10. Birkholz, S., Bochmann, A., Schank, J.: Ethnografie und teilnehmende beobachtung. In: Wagemann, C., Goerres, A., and Siewert, M.B. (eds.) Handbuch Methoden der Politikwissenschaft, pp. 325–350. Springer, Cham (2020). https://doi.org/10.1007/978-3-658-16936-7
11. Goldberg, F., Hildebrandt, A.: Experteninterviews. In: Wagemann, C., Goerres, A., and Siewert, M.B. (eds.) Handbuch Methoden der Politikwissenschaft, pp. 267–284. Springer, Cham (2020). https://doi.org/10.1007/978-3-658-16936-7
12. Kaiser, R.: Offene interviews – von semistrukturiert bis narrativ. In: Wagemann, C., Goerres, A., and Siewert, M.B. (eds.) Handbuch Methoden der Politikwissenschaft, pp. 285–304. Springer, Cham (2020). https://doi.org/10.1007/978-3-658-16936-7

Is Easy Good Enough? Evaluation of Easy-Web-Check, a Simplified and Open Method for Testing Website Accessibility

Nadine Auer, Ronja Brauchle(✉), and Gottfried Zimmermann

Stuttgart Media University, 70569 Stuttgart, Germany
auer@hdm-stuttgart.de, ronja.brauchle@gmx.de, gzimmermann@acm.org

Abstract. Testing websites for accessibility with certified agencies is often time-consuming, complex, and costly. The Easy-Web-Check (EWC) is a simplified and open method for testing website accessibility with free tools, but it is not yet validated. This paper examines the correct use of the EWC by non-experts. To this end, a user study is conducted in which participants review the test criteria of the EWC on 2–3 websites each. In total, 24 participants have taken part, 10 completing the entire test. On average, 67% of the tasks were completed correctly, with automated testing methods demonstrating a higher number of correct assessments. The preliminary results suggest that the EWC is suitable for identifying accessibility barriers and has the potential to help non-experts improve website accessibility. However, some test criteria require more detailed explanation and examples to reduce testing errors.

Keywords: Web Accessibility · Accessibility Testing · Easy-Web-Check

1 Introduction

Web accessibility testing is becoming increasingly important, especially since various laws, such as the EU's Web Accessibility Directive and European Accessibility Act require it. However, comprehensive tests by certified agencies such as the BIK BITV test are often time-consuming, expensive and require expert knowledge [1]. The Easy-Web-Check (EWC) [2] is intended to bridge this gap. It is a simplified and open test method created by the Competence Center for Digital Accessibility at the Stuttgart Media University for checking the accessibility of websites [2]. Until now, the EWC has already been tested for its applicability [3], but has not yet been validated in the sense of correct testing results by users, especially by non-experts. The aim of this paper is to describe this validation. To this end, the paper seeks to address the following research question: Does the EWC serve as a useful instrument for the evaluation of digital accessibility on websites and the identification of barriers?

K. Mavrou and P. Encarnação (Eds.): AAATE 2025, *Technology for Inclusion and Participation for All: Recent Achievements and Future Directions*, pp. 364–371, 2025.
https://doi.org/10.1007/978-3-032-01628-7_43

2 Background

Various laws require organizations to provide barrier-free access to their websites and products. For example, the European Directive EU 2016/2102, titled „on the accessibility of the websites and mobile applications of public sector bodies", published in 2016, obliges all public authorities within the European Union to ensure the accessibility of their websites and mobile applications [4]. Furthermore, the Directive (EU) 2019/882, known as the "European Accessibility Act," published in 2019, refers to the economy, establishing specific requirements for the accessibility of digital products and services [5]. Both directives refer to the EN 301 549 [6] for specific test criteria, which, in turn, includes the Web Content Accessibility Guidelines (WCAG) 2.1 A-AA criteria [7]. The WCAG are understood as the leading international guidelines for the accessible design of websites.

In order to comply with legal obligations and guarantee accessibility, it is imperative to conduct regular accessibility testing of websites. According to Abascal et al. [8], there are three primary methods of testing websites for accessibility: automated testing, manual testing and user testing. The necessity of user tests is predicated on the premise that they provide concrete feedback from real users. Nevertheless, user tests are very time-consuming and complex to carry out [9]. Manual testing procedures employ lists of criteria that refer to accessibility guidelines. A range of established test procedures is available for performing such an accessibility assessment. These include comprehensive tests such as the BIK BITV test [11], the BIK WCAG test (web) and the WCAG-EM [10]. All these tests have in common, that they require a significant investment of effort, are costly and are typically conducted by experts [1, 12]. Automated tools involve the analysis of code by automated means and offer the advantage of saving time; however, they are unable to identify all types of barriers [13, 14]. Consequently, the utilization of these tools alone is insufficient for conducting a comprehensive accessibility audit [8, 13]. Instead, they should be incorporated into a broader testing framework, complementing manual testing procedures [8, 13].

Because the aforementioned approaches are both time-consuming and expensive, and consequently impractical for regular accessibility testing, there is an imperative for alternative methodologies. The EWC is a simplified and open test method created by the Competence Center for Digital Accessibility at the Stuttgart Media University for checking the accessibility of websites [2]. It is based on the accessibility criteria of EN 301 549, but reduced to the most common barriers [2], resulting in 81 test criteria. According to Zimmermann et al. [2], 52 of the 94 criteria from EN 301 549 v3.2.1 Table A.1 relating to embedded web content are covered by the EWC, 6 of which partially. Therefore, the EWC is not intended to be a complete accessibility test, but to help identifying barriers on websites easily even by non-experts [2]. For this purpose, the EWC uses automatic tools and manual checks, where all the tools used are free.

The EWC has already been tested by Brauchle [3], who conducted a qualitative study with 5 participants to analyze the comprehensibility of the test steps and the application of the test tools involved. The results show that without expert knowledge in accessibility, the EWC can be used efficiently to find many barriers on websites in a short period of time, with a step-by-step guide being essential as an aid. Testing errors were mainly made in subjective and exceptional cases. However, the EWC has not yet been validated with regard to whether users (including non-experts) correctly apply the test criteria contained in the EWC.

3 Methods

In order to validate the EWC, a user study is conducted in which participants apply the test criteria of the EWC on various websites and report the test result per website (fulfilled (without error), not fulfilled (with error), not applicable). The participants' answers are then reviewed for correctness in order to analyze how many and which test criteria were correctly tested by the participants by a confusion matrix.

Before starting the study, the EWC is revised. The EWC criteria are revised in terms of their relevance based on the regular accessibility test report of Germany [15]. In the report, websites are checked for accessibility according to the EN 301 549, with the criteria being assessed as "fulfilled", "not fulfilled" or "not applicable" [15]. EWC test criteria mapped to the EN 301 549 test criteria that were assessed as "not applicable" for more than 60% of the websites tested are considered irrelevant and are removed from the EWC to shorten the application in the user study. Finally, the EWC is extended by 12 easily testable test criteria that are applicable in general and not yet included. Additionally, the test criteria are divided into test criteria for individual pages and the entire website, from which only the criteria for individual pages are analyzed in the study – excluding test criteria like the availability of a declaration on accessibility. With these changes, the EWC that is validated in the study consists of 79 test criteria covering 42 test criteria of EN 301 549 v.3.2.1 Table A1, 5 of which partially.

To find suited websites for the user study, websites from accessibility test reports of the Competence Center for Digital Accessibility are examined, as well as other websites that have been evaluated with the EWC before. For each criterion, 2–3 of the websites are selected. If possible, at least one website where the test criterion is fulfilled and one where it is not fulfilled is selected per criterion. A third website is optionally added, with a random assessment of fulfilled, not fulfilled or not applicable, to break the pattern that one website is always fulfilled and one is not. For data collection, an online survey is created in which the EWC criteria are described individually and participants are asked to evaluate the respective criteria for the 2–3 websites. At the end of the survey, participants are asked about their experience before the study in testing websites for accessibility.

4 Results

A total of 24 participants took part in the study, 10 of whom completed the survey in its entirety; the remaining 14 participants either ceased or paused their participation at various points during the study and did not complete it until the end. The website tests carried out up to the point of discontinuation are still evaluated, as full participation in the study is not a mandatory requirement for participants. The participants who completed the entire study included non-experts in the field of website accessibility testing ($n = 1$) and individuals with initial experience ($n = 9$).

In general, 67% of the test tasks were correctly answered by the participants (Accuracy). A more detailed analysis of the testing errors enables the analysis of a confusion matrix for the assessments and, therefore, a distinction between true and false positives and negatives (see Table 1). Note that test criteria that are not fulfilled (with error) in the study are labeled as "positives", as they are findings of errors by the participants. Test criteria that are fulfilled (i.e. those without an error) are labeled as "negatives". That means, that "true positives" are test criteria with error as the actual condition and also assessed as not fulfilled by the participants.

Table 1. Confusion matrix for the assessments of the study.

	Assessed as with error (assessed positives)	Assessed as without error (assessed negatives)	Assessed as not applicable (assessed n/a)
Actually with error (positives)	688 (true positives)	356 (false negatives)	87 (false n/a)
Actually without error (negatives)	259 (false positives)	951 (true negatives)	91 (false n/a)
Actually not applicable (n/a)	56 (false positives)	48 (false negatives)	142 (true n/a)

The analysis of the confusion matrix facilitates the evaluation of various key performance indicators, which can be utilized to ascertain the accuracy of models. These indicators include accuracy, which denotes the percentage of participants' assessments that are correct (67%). In this context, discrepancies are particularly evident between the testing tools that were used. As shown in Table 2 the best performing testing tools are Siteimprove Accessibility Checker (90%), Bookmarklet "Trigger keyboard shortcuts" (88%) and WCAG Color Contrast Checker (81%). Conversely, the testing tools that performed most inadequately were the Bookmarklet "Lang" (17%), the Colour Contrast Analyser (53%), and the manual test of error messages (51%). These findings already indicate a difference between automatic and manual testing tools, which can be further emphasized by the other performance indicators.

The accuracy of the assessments and some additional indicators of the confusion matrix are outlined in Table 2, which is divided according to the different testing tools of the EWC.

The True Positive Rate (TPR) is a metric used to describe the percentage of errors on the websites that were correctly identified. The overall TPR for the EWC is 0.61, indicating that 61% of the errors were identified by the participants. Conversely, this means that 39% of the errors went unrecognized, of which 31% were classified as false negatives (without error) and 8% as false n/a. Furthermore, a clear distinction emerges among the various testing tools in terms of their TPR (see Table 2).

The True Negative Rate (TNR) is a metric that quantifies the percentage of assessments accurately identified as without error, with a value of 73% for the overall EWC. However, it is important to note that 20% of actually negative assessments are erroneously

Table 2. Performance Indicators separated in the different testing tools of the Easy-Web-Check

Tools	Automatic or manual	Accuracy	True Positive Rate (TPR)	True Negative Rate (TNR)	False Positive Rate (FPR)
Bookmarklet Trigger Down-Events	automatic	0.67	No positives	0.67	0.33
Bookmarklet Trigger keyboard shortcuts	automatic	0.88	0.83	0.92	0.08
Siteimprove Accessibility Checker	automatic	0.90	0.89	0.92	0.07
WCAG Color Contrast Checker	automatic	0.81	0.75	0.92	0.08
Bookmarklet content structured	semi-automatic	0.65	0.61	0.75	0.30
Bookmarklet Images	semi-automatic	0.65	0.75	0.57	0.37
Bookmarklet Lang	semi-automatic	0.17	0.17	-	-
Bookmarklet Lists	semi-automatic	0.55	0.37	0.73	0.28
Bookmarklet Tables	semi-automatic	0.61	0.44	0.65	0.10
Bookmarklet text spacing	semi-automatic	0.69	0.17	0.96	0.04
Colour Contrast Analyser	semi-automatic	0.53	0.38	0.83	0.17
Landmarks	semi-automatic	0.58	0.33	0.71	0.29
taba11y/Inspector	semi-automatic	0.63	0.62	0.63	0.37
Web Developer	semi-automatic	0.67	0.42	0.92	0.08
Error messages	manual	0.51	0.54	0.42	0.26
Keyboard	manual	0.65	0.47	0.78	0.17
Video and audio files	manual	0.58	0.53	0.41	0.24
Video player	manual	0.67	0.52	0.73	0.13
Visual inspection	manual	0.54	0.44	0.56	0.15
Total		0.67	0.61	0.73	0.20

classified as with errors, a phenomenon that is also reflected in the False Positive Rate (FPR). The FPR denotes the number of assessments that are incorrectly identified as errors, despite being fulfilled. As demonstrated in Table 2, discrepancies between the testing tools are observable in both TNR and FPR; however, these variations are less pronounced in comparison to those observed in Accuracy and TPR.

As demonstrated in Table 2, the testing tools can be categorized according to their utilization. For instance, there are testing tools that evaluate a website and generate results automatically, thereby necessitating only the transfer of results to the EWC. Manual testing tools describe test steps that users must perform, check and interpret independently. The integration of both approaches in semi-automatic tools, for instance, through highlighting website structures, still necessitates manual checking and interpretation.

The investigation of the testing tools utilizing this grouping exposes substantial disparities in several performance indicators. For instance, a t-test discloses significant variations between the automatic and manual testing tools with regard to Accuracy ($p = 0.007$), TPR ($p = 0.003$) and TNR ($p = 0.01$). Additionally, a significant disparity was identified between the automatic and semi-automatic testing tools in terms of Accuracy ($p = 0.005$) and TPR ($p = 0.0002$). However, the differences in TNR between semi-automatic and manual testing tools were found to be marginally non-significant ($p = 0.051$). The FPR exhibited no significant differences among the three groups. The visualization of the groups per performance indicator is also summarized in Fig. 1.

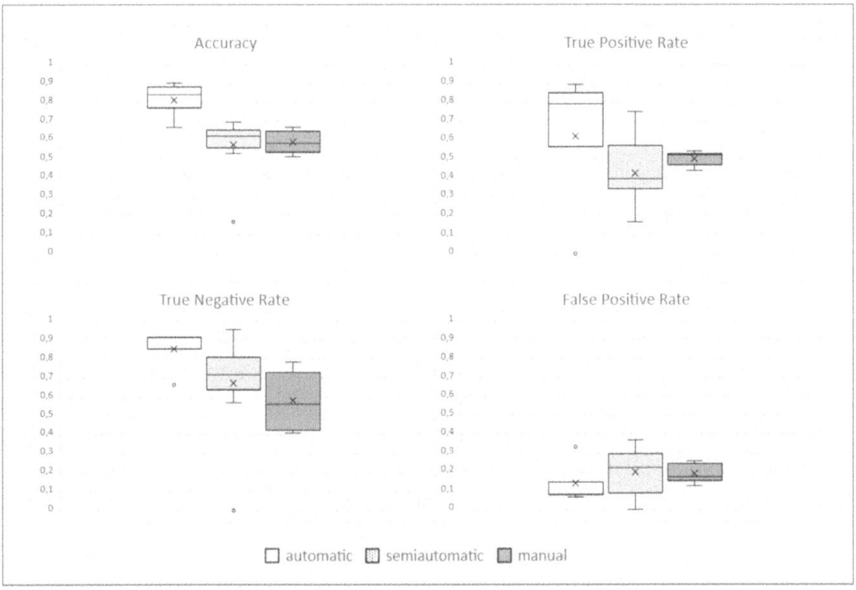

Fig. 1. Comparison of automatic, semi-automatic and manual testing tools in the performance indicators Accuracy, True Positive Rate, True Negative Rate, and False Positive Rate.

5 Discussion

The preliminary findings of the study indicate that the EWC is a useful tool for evaluating digital accessibility on websites and identifying barriers, although far away from completeness, with 61% of barriers correctly identified by participants and 73% of fulfilled criteria correctly assessed. On the downside, 39% of barriers were missed. However, a closer look at the various barriers on the websites reveals that only 4% were not identified by all participants. Therefore, the EWC can be considered a tool that helps detect more than half of the errors on websites, thus serving as an assistance to improve website accessibility even for non-experts.

Nevertheless, the analysis demonstrates significant variations between the testing tools utilized for testing. The automatic test tools performed significantly better than the manual tools for Accuracy, TPR and TNR, and also better than the semi-automatic tools for Accuracy and TPR. This indicates that the application problems of the EWC are mainly due to the fact that the participants do not understand the test criteria sufficiently or have too little prior knowledge, as the manual and semi-automatic test tools rely on autonomous testing and subjective judgment of the results. In such cases, the provision of more detailed instructions for the test and additional examples could assist the participants, leading to better results.

These results are also consistent with the findings of Brauchle [3], whose study also identified testing errors primarily in the subjectively rated test criteria.

6 Conclusion

The findings of the study demonstrate that the EWC is indeed a tool that has the potential to assist non-experts with limited or no prior knowledge in testing websites for accessibility and thus enhancing the accessibility of websites. However, it is important to note, that certain test criteria, particularly manual tests, are more challenging to test, resulting in an overall error rate of 33%. There is a need to improve the EWC, especially for more subjective test criteria, through the provision of illustrative examples and more detailed descriptions.

At the moment we are working on developing a web application for the utilization of the EWC.[1] This application has the potential, to support especially non-experts in understanding the EWC and all test criteria, by providing detailed descriptions and additional fulfilled and not fulfilled examples for each test criterion.

Acknowledgments. This study was funded by Stiftung Innovation in der Hochschullehre.

Disclosure of Interests. The authors have no competing interests to declare that are relevant to the content of this article.

[1] The EWC and its web application is available in a first version in https://easy-web-check.bf-lernen.de/. At the moment it is only available in German.

References

1. Kollotzek, G.A.: Analyse von manuellen Prüfmethoden zur Bewertung der Barrierefreiheit von Websites nach BITV 2.0 (Master's thesis) (2021)
2. Kompetenzzentrum für digitale Barrierefreiheit. (2023) Easy Web Check. https://barrierefreiheit.hdm-stuttgart.de/easy-web-check/
3. Brauchle, R.: Analyse des Prüfverfahren auf Barrierefreiheit im Web „Easy Web Check" (Bachelor's thesis) (2024)
4. European Parliament and the Council of the European Union. (2016). Directive (EU) 2016/2102 of the European parliament and of the council of 26 October 2016 on the accessibility of the websites and mobile applications of public sector bodies. https://eur-lex.europa.eu/legal-content/EN/TXT/?uri=CELEX:32016L2102
5. European Parliament and the Council of the European Union. (2019). Directive (EU) 2019/882 of the European Parliament and of the Council of 17 April 2019 on the accessibility requirements for products and services. https://eur-lex.europa.eu/eli/dir/2019/882/oj/eng
6. European Telecommunications Standards Institute. (2021) EN 301 549 V3.2.1 (2021–03) Accessibility requirements for ICT products and services. https://www.etsi.org/deliver/etsi_en/301500_301599/301549/03.02.01_60/en_301549v030201p.pdf
7. WCAG: W3C. (2023). Web Content Accessibility Guidelines (WCAG) 2.2. https://www.w3.org/TR/WCAG22/
8. Abascal, J., Arrue, M., Valencia, X.: Tools for web accessibility evaluation. In: Web accessibility: a foundation for research, pp. 479–503. Springer, London (2019)
9. Brajnik, G.: Beyond conformance: the role of accessibility evaluation methods. In: Hartmann, S., Zhou, X., Kirchberg, M., (eds.), Lecture Notes in Computer Science. Web Information Systems Engineering – WISE 2008 Workshops, vol. 5176, pp. 63–80. Springer, Heidelberg (2008). https://doi.org/10.1007/978-3-540-85200-1_9
10. Henry, S., Abou-Zhara S.: WCAG-EM overview: website accessibility conformance evaluation methodology. W3C WAI (2020). https://www.w3.org/WAI/test-evaluate/conformance/wcag-em/
11. DIAS GmbH. (o.D.a) BIK BITV-/WCAG-Test. (Web). BIK BITV-Test. https://bitvtest.de/tests-und-beratung/bik-bitv-test-web
12. Graap, F., et al.: Freie Werkzeuge zur Beurteilung des Barrierefreiheitsgrades von Webseiten. In Mensch und Computer 2018-Workshopband (pp. 10–18420). Gesellschaft für Informatik eV (2018)
13. Acosta-Vargas, P., Luján-Mora, S., Acosta, T., Salvador-Ullauri, L.: Toward a combined method for evaluation of web accessibility. In: Proceedings of the International Conference on Information Technology & Systems (ICITS 2018), pp. 602–613. Springer, Cham (2018)
14. Mateen, S., Voß-Nakkour, S., Rustemeier, L.: Eine Studie zur Qualitätsbeurteilung von automatisierten Testwerkzeugen zur Prüfung auf Barrierefreiheit. In DELFI 2020–Die 18. Fachtagung Bildungstechnologien der Gesellschaft für Informatik eV (S. 157–168). Gesellschaft für Informatik eV (2020)
15. Bericht der Bundesrepublik Deutschland an die Europäische Kommission über die periodische Überwachung der Einhaltung der Barrierefreiheitsanforderungen von Websites und mobilen Anwendungen öffentlicher Stellen gemäß Artikel 8 der Richtlinie (EU) 2016/2102. (2025). https://www.bfit-bund.de/SharedDocs/Downloads/DE/Eu-Berichte/eu-bericht-pdf-24.pdf?__blob=publicationFile&v=3

A Personalized Digital Solution to Assist Task Organization and Time Management for People with Attention Deficit/Hyperactivity Disorder (ADHD)

Joana Oliveira[1], Tânia Rocha[1,2](✉), and João Barroso[1,2]

[1] University of Trás-os-Montes and Alto Douro, Vila Real, Portugal
{joanagilde,trocha,jbarroso}@utad.pt
[2] INESC TEC, Vila Real, Portugal

Abstract. This study details the development of an assistive technology (AT) designed to support individuals with ADHD in task organization and time management. Using a User-Centered Design (UCD) methodology, the process included persona-based and scenario-based evaluations to guide the development of the solution. The application features customizable functionalities tailored to individuals with ADHD, such as flexible task management, adaptive reminders, and focus strategies, with the aim of promoting user autonomy and well-being. Preliminary evaluations, based on usability and accessibility heuristics, suggest that the mobile application is intuitive and inclusive, addressing the specific needs of users with ADHD—particularly in terms of personalization and visual clarity. These results highlight the interface's potential to improve organizational and time management skills, emphasizing the importance of personalized digital solutions for this population. Future steps include real-user testing to validate and refine the solution.

Keywords: Assistive Technology (AT) · Attention Deficit/Hyperactivity Disorder (ADHD) · user-centered design (UCD) · organizations and time management · mobile application · UI/UX

1 Introduction/Background

Attention Deficit Hyperactivity Disorder (ADHD) has a significant impact, affecting multiple aspects of daily life for individuals with the condition [1–3]. A key characteristic of ADHD is its influence on task organization and time management [4]. While various time management and task organization applications are available on the market—such as Trello, Todoist, Focus@Will, and Forest—which offer features to enhance organization, productivity, and focus, these tools are not fully tailored to the specific needs of individuals with ADHD, highlighting the demand for more personalized solutions [5, 6]. Innovative approaches have also emerged, including the use of gamification

K. Mavrou and P. Encarnação (Eds.): AAATE 2025, *Technology for Inclusion and Participation for All: Recent Achievements and Future Directions*, pp. 372–377, 2025.
https://doi.org/10.1007/978-3-032-01628-7_44

and biofeedback techniques for ADHD users. These strategies emphasize the diversity of methods explored to foster concentration and autonomy, reinforcing the relevance of customized solutions [7, 8].

This study presents the development of a mobile application that integrates task organization and time management functionalities with user well-being support. It underscores the importance of customizable digital solutions tailored to the specific needs of users and documents the challenges encountered and the methodology applied, offering insights for future research and development in Assistive Technology (AT).

2 Methods

User-Centered Design (UCD) methodology was adopted to ensure that the developed solution effectively meets users' real needs [9–11]. The development process followed four main phases. In the planning and research phase, a literature review was conducted on existing applications with similar objectives, along with a user needs assessment based on the creation of personas and interaction scenarios. These were derived from secondary data and empirical evidence reported in the literature. The personas served as the foundation for building interaction scenarios that guided design decisions and the definition of core functionalities.

During the design and implementation phase, a functional Mind Map was developed to organize the application's core functionalities into five central areas: Task Organization, Time Management, Reminders and Alerts, Focus Strategies, and Emotional and Mental Well-being. This systematic categorization provided a holistic view of the needs of individuals with ADHD and served as the basis for the solution's design. The defined structure guided the creation of wireframes and mockups, ensuring that the functionalities were coherently organized and user-centered. In parallel, a User Journey Map was created, structured around four main stages of interaction: Discovery and Adoption, Exploration and Configuration, Regular Use, and Support and Sustainability. This structure enabled the identification, at each stage, of the most likely user actions, critical experience points, emotional responses, and improvement opportunities. This approach contributed to a deeper understanding of the target audience's needs and informed the development of functionalities focused on the ADHD user experience. The application was developed in C# within the Visual Studio environment.

The evaluation phase included two main components, one of which was an accessibility assessment. This was conducted using automated tools to verify compliance with WCAG 2.1 guidelines [12], focusing on aspects such as visible focus, readability, screen reader compatibility, alternative text for images, adaptation for users with color blindness, and multimedia content accessibility. This preliminary evaluation provided a solid foundation for future iterations focused on inclusive usability.

The second component consisted of a usability analysis based on Nielsen's heuristics [13], which are well recognized for their effectiveness in identifying usability issues throughout various stages of digital product development. This evaluation was conducted systematically by analyzing the solution's wireframes and mockups, without involving end-user testing. Each of the ten heuristic principles was considered, allowing for the identification of interface strengths and opportunities for improvement, with the aim of optimizing the user experience prior to implementation. As this is a preliminary phase of the project, real-user testing has not yet been conducted; however, the heuristic evaluation and interaction simulations based on personas enabled the identification of interface and functionality improvement opportunities.

3 Results

The analysis of existing solutions, including Trello, Todoist, Focus@Will, and Forest, highlights that while they effectively support general organization and productivity, they fall short in fully addressing the specific needs of individuals with ADHD. Many lack essential features such as task breakdown into smaller steps and personalized time management strategies, often prioritizing task completion over individualized progress [14–17].

Meanwhile, increasingly innovative approaches have emerged. Gamification and biofeedback are two such strategies that emphasize the reinforcement of positive habits through immediate feedback, enhanced concentration, and the development of emotional and cognitive self-regulation skills [7, 8].

To bridge this gap, three personas representing diverse user characteristics were developed, along with two interaction scenarios that guided the identification of key requirements, including task flexibility, interface simplification, adaptive reminders, and personalization options. The Mind Map outlined five core areas—Task Organization, Time Management, Reminders and Alerts, Focus Strategies, and Emotional and Mental Well-being—integrating features such as customizable checklists, personalized schedules, mindfulness techniques, and guided meditation.

Additionally, the User Journey Map identified four critical phases—Discovery and Adoption, Exploration and Configuration, Regular Use, and Support and Sustainability—detailing user actions, challenges, emotional responses, and improvement opportunities.

In the accessibility evaluation, improvement areas were identified in aspects such as color contrast, readability, keyboard navigation, and sensory feedback, in accordance with WCAG 2.1 guidelines.

In parallel, the usability evaluation based on Nielsen's Heuristics highlighted the need to improve system status visibility, user control, error prevention, and design consistency.

Finally, preliminary tests confirmed that the application offers an intuitive and responsive interface, with positive feedback regarding personalization and visual clarity. Future improvements will focus on layout design optimization to enhance usability and accessibility, ensuring an inclusive and effective user experience.

4 Discussion

The results obtained throughout this study reinforce the effectiveness of a user-centered approach in the development of AT targeted at populations with specific needs, such as individuals with ADHD. Personalization and inclusive design emerged as fundamental pillars in adapting the digital solution to the challenges commonly associated with ADHD, particularly in relation to time management, task organization, and emotional regulation.

The use of personas and interaction scenarios enabled a precise identification of both functional and usability requirements from the early stages of the project. This methodological strategy facilitated the definition of adaptive functionalities—such as customizable reminders, focus-enhancing strategies, and emotional well-being techniques—thus contributing to a more meaningful user experience, tailored to the actual needs of the target audience.

The conceptualization of the solution around five core areas—Task Organization, Time Management, Reminders and Alerts, Focus Strategies, and Emotional and Mental Well-being—reflects a holistic understanding of the daily challenges faced by users with ADHD. This functional categorization, guided by a mind map and a user journey map, made it possible to anticipate critical pain points in the user experience and propose design strategies that foster user autonomy.

Regarding accessibility, the evaluation conducted in accordance with WCAG 2.1 guidelines identified several critical areas for improvement, namely color contrast, keyboard navigation, and the adequacy of sensory elements. These aspects are particularly relevant for users with ADHD, who often present heightened sensitivity to visual and auditory stimuli. Eliminating barriers in this domain is essential to ensure a seamless, intuitive, and inclusive user experience.

In parallel, the usability evaluation, based on Nielsen's Heuristics, revealed opportunities for enhancement in areas such as system status visibility, user control and freedom, error prevention, and aesthetic and functional consistency. These dimensions directly affect the user's perceived efficacy and trust in the system, which are crucial for sustained engagement with the technology.

It is also important to note that, although real user testing was not conducted at this preliminary stage, the persona-based simulations combined with the heuristic evaluation provided robust insights into both the strengths of the interface and the aspects to be optimized in future iterations.

The findings suggest that this approach holds strong potential for providing personalized responses to the everyday challenges experienced by individuals with ADHD—particularly in promoting routine creation, reducing procrastination, and reinforcing self-regulation. However, empirical validation with real users will be a critical next step to assess the application's effectiveness in real-world contexts, particularly in terms of behavioral impact, long-term adherence, and overall user satisfaction.

In conclusion, the results underscore the importance of adopting user-centered, iterative, and inclusive design strategies in the development of digital Assistive Technologies. The future integration of complementary approaches—such as gamification or biofeedback—could further enhance the application's effectiveness in key domains such as motivation, focus, and emotional regulation, which are particularly critical in the context of ADHD.

5 Conclusion/Implications for the AT Field

Assistive Technologies play a fundamental role in fostering autonomy, organization, and emotional regulation in individuals with ADHD. Despite advances in the development of digital tools aimed at enhancing productivity and time management, existing solutions exhibit significant gaps, particularly in task flexibility, execution process structuring, and the personalization of temporal strategies.

Accordingly, this solution integrates the essential principles of User-Centered Design (UCD), Usability, Accessibility Standards Compliance, and Personalization. This multidimensional approach ensures that the system is functional, inclusive, and tailored to the cognitive and behavioral specificities associated with ADHD.

Preliminary results indicate that the application has the potential to improve user autonomy and well-being by facilitating structured routines, reducing procrastination, and reinforcing self-regulatory behaviors. This perspective underscores the importance of UCD in the development of effective and sustainable AT systems.

However, certain limitations must be acknowledged. Reliance on prototypes constrained the depth of usability testing, and manual accessibility assessment may not have identified all issues related to WCAG 2.1 guidelines. The absence of empirical testing with real users limits the validation and generalization of the proposed functionalities.

Future work should focus on iterative prototyping, testing with users with ADHD, and quantitative evaluation of usability, effectiveness, and satisfaction. Furthermore, the integration of innovative approaches such as gamification and biofeedback will be explored to enhance motivation and emotional regulation. These steps are essential to ensure the validity, inclusivity, and long-term impact of the solution within the field of Assistive Technologies.

Acknowlegments. This work is financed by National Funds through the Portuguese funding agency, FCT - Fundação para a Ciência e a Tecnologia, within project LA/P/0063/2020. DOI https://doi.org/10.54499/LA/P/0063/2020 | https://doi.org/https://doi.org/10.54499/LA/P/0063/2020.

References

1. Oliveira, L., Pereira, M., Medeiros, M., Serrano, A.: PHDA: O que há de novo no DSM-5. Revista Portuguesa de Pedagogia **49**(2), 75–94 (2016). https://doi.org/10.14195/1647-8614_49-2_4
2. Correia, J.F.R.: Atenção, memória e estruturação visuo-perceptiva em crianças e adolescentes com Perturbação de Hiperatividade com Défice de Atenção (PHDA). Mestrado, Universidade Católica Portuguesa (2015). http://hdl.handle.net/10400.14/22241
3. Gallagher, J., McDonnell, M.: JournalMate: an accessible academic reading tool for third-level students with ADHD. Assistive Technol. Shaping Sustainable Inclusive World **306**, 207 (2023)
4. American Psychiatric Association: Manual de Diagnóstico e Estatística das Perturbações Mentais – DSM-5 (5ª ed.). Dislex (2013). Disponível em: https://dislex.co.pt/images/pdfs/DSM_V.pdf
5. Lindstedt, H., Umb-Carlsson, Õ.: Cognitive assistive technology and professional support in everyday life for adults with ADHD. Disabil. Rehabil. Assist. Technol. **8**(5), 402–408 (2013). https://doi.org/10.3109/17483107.2013.769120
6. Sonne, T., Marshall, P., Obel, C., Thomsen, P.H., Grønbæk, K.: An assistive technology design framework for ADHD. In: OzCHI 2016: Proceedings of the 28th Australian Conference on Computer-Human Interaction, pp. 60–70. (2016). https://doi.org/10.1145/3010915.3010925

7. Alabdulakareem, E., Jamjoom, M.: Computer-assisted learning for improving ADHD individuals' executive functions through gamified interventions: a review. Entertainment Comput. **33**, 100341 (2020). https://doi.org/10.1016/j.entcom.2020.100341. ISSN 1875-9521
8. Boyd, W., Campbell, S.: EEG BIOFEEDBACK IN the schools the use of EEG biofeedback to treat ADHD in a school setting. J. Neurother. **2**(4), 65–71 (1998). https://doi.org/10.1300/J184v02n04_05
9. Kramer, J., Noronha, S., Vergo, J.: A user-centered design approach to personalization. Commun. ACM **43**(8), 44–48 (2000)
10. Lowdermilk, T.: User-Centered Design: A Developer's Guide to Building User-Friendly Applications, pp. 1–91. 1ª edição. O'Reilly Media Inc., Sebastopol, CA (2013)
11. Kujala, S., Kauppinen, M.: Identifying and selecting users for user-centered design. In: NordiCHI 2004: Proceedings of the Third Nordic Conference on Human-Computer Interaction, pp. 297–303 (2004). https://doi.org/10.1145/1028014.1028060
12. Mao, J.Y., Vredenburg, K., Smith, P.W., Carey, T.: The state of user-centered design practice. Commun. ACM **48**(3), 105–109 (2005). https://doi.org/10.1145/1047671.1047677
13. Nielsen, J.: 10 Usability Heuristics for User Interface Design. Nielsen Norman Group (1995). https://www.nngroup.com/articles/ten-usability-heuristics/
14. Trello. https://trello.com. Accessed 08 July 2024
15. Todoist. https://tost.com. Accessed 08 July 2024
16. Focus@Will. https://www.focusatwill.com. Accessed 10 July 2024
17. Forest. https://www.forestapp.cc. Accessed 10 July 2024

Transferring Linguistic Knowledge Representation to Automatically Adapt Easy-to-Read Text Guidelines from Spanish to German

Isam Diab[1(✉)], Klaus Miesenberger[2], and Mari Carmen Suárez-Figueroa[1]

[1] Ontology Engineering Group (OEG), Universidad Politécnica de Madrid, Madrid, Spain
isam.diab@upm.es, mcsuarez@fi.upm.es
[2] Institute Integriert Studieren, University of Linz, Linz, Austria
klaus.miesenberger@jku.at

Abstract. Ensuring cognitive accessibility in written texts is crucial for individuals with cognitive disabilities. The Easy-to-Read (E2R) Methodology provides guidelines for adapting texts, but its manual application is time-consuming and costly. This initial research explores the use of Large Language Models (LLMs) to infer and transfer rule-based linguistic knowledge representations (LKRs) from Spanish to German for two E2R guidelines: explanatory relative clauses and dialogue formatting. We evaluated three LLMs, finding that only Qwen2.5-7B-Instruct-1M generated a usable LKR, which still required extensive manual refinement. The results show that while the LLM was helpful in identifying structures, it struggled to produce grammatically correct adaptations. The refined rules performed better, but still needed human adjustment. A validation with two German E2R experts confirmed the potential of the proposal, while also pointing out common errors. Additionally, a user-based evaluation is designed and planned to assess whether the adapted texts improve readability for individuals with cognitive disabilities.

Keywords: Cognitive Accessibility · Easy-to-Read · Large Language Models · Natural Language Processing

1 Introduction and Motivation

Equal opportunities and universal access to information are fundamental rights[1] that every person should benefit from. However, certain groups of society, particularly those with cognitive or intellectual disabilities, present some difficulties related to reading comprehension processes. Therefore, prioritising cognitive accessibility becomes essential for promoting active participation in diverse

[1] Convention on the Rights of Persons with Disabilities (United Nations, 2006). Available at: https://www.ohchr.org/en/instruments-mechanisms/instruments/convention-rights-persons-disabilities.

K. Mavrou and P. Encarnação (Eds.): AAATE 2025, *Technology for Inclusion and Participation for All: Recent Achievements and Future Directions*, pp. 378–385, 2025.
https://doi.org/10.1007/978-3-032-01628-7_45

social domains, such as politics, education, employment, and culture. For this reason, the methodology called Easy-to-Read (E2R) was created and formalised in standards such as the Spanish UNE 153101:2018 [1], the German DIN SPEC 33429:2025-03 [6], and other European references [10,15]. The main goal of this methodology is to present clear and easy to understand text by providing a set of guidelines on content and design and layout of written materials. Nevertheless, the E2R methodology is currently applied in a manual fashion. Such a manual process is labour-intensive and costly, and it would benefit from having a technological support.

To overcome this limitation, research has explored the development of technological solutions to (semi)automatically analyse and simplify or adapt texts. Regarding technological support for addressing E2R guidelines and recommendations in Spanish texts, it is worth mentioning (a) Easy-to-Read Advisor [21], FACILE [22], Comp4Text [9], and E2R-Helper [7], for an E2R analysis of documents; and (b) Simplext [17], LexSIS [2], DysWebxia [16], EASIER [14], and FACILE [22], for creating adapted versions of original documents. In the German context, cognitive accessibility is addressed through two main variants: Leichte Sprache (LS), which aligns with E2R objectives, and Einfache Sprache (ES), a form of simplified language closer to plain language[2] According to Madina and colleagues [13], numerous corpora and tools have been created to support LS, such as Geasy [8], LeiSA [11], APA [18], and tools like EasyTalk [20] and SUMM[3]. However, these resources tend to focus primarily on lexical simplification or surface-level reformulation.

Our research line explores the application of Artificial Intelligence (AI) techniques to support both the analysis of guideline compliance in a given text and the (semi)automatic adaptation of documents into easy-to-read versions. In particular, we have developed declarative approaches that model adaptation rules as Linguistic Knowledge Representations (LKRs), focusing on two specific E2R guidelines: (a) the formatting of dialogues[4], which should follow a theatrical style by explicitly naming the speaker before each intervention [5]; and (b) the transformation of explanatory subordinate clauses into shorter sentences[5] [4].

Given that the creation of symbolic rules is a human task and that many languages share similar syntactic constructions, we hypothesise that LKRs developed for Spanish may be transferable to other languages with comparable features, such as German. However, LKRs are inherently language-specific and reflect morphological, syntactic, and orthographic conventions that do not generalise directly. Thus, this work investigates whether Large Language Models

[2] Definition and international standards for Plain Language are available at: https://www.iso.org/standard/78907.html.

[3] https://summ-ai.com/en/.

[4] Example of adaptation: *âĂŞBruno, I have told you to unpack right now! âĂŞthe mother ordered.* (Original dialogue) > *The mother ordered: Bruno, I have told you tu unpack right now!* (Theatrical style).

[5] Example of adaptation: *The nurse, who is 63 years old, is about to retire.* (Relative clause) > *The nurse is 63 years old. The nurse is about to retire.* (Adapted version).

(LLMs) can assist in the cross-linguistic transfer of LKRs by inferring German rule candidates based on the logic of existing Spanish rules. We focus on the two adaptation tasks previously addressed in Spanish (explanatory clauses and dialogues) and explore their implementation in German using a hybrid approach: LLMs generate initial versions of the rules, which are then manually corrected.

The rest of the paper is organised as follows: Sect. 2 presents our cross-language proposal for transferring LKRs from Spanish to German using LLMs. In Sect. 3 we describe the preliminary expert evaluation carried out to validate the generated LKRs. Finally, we present some conclusions and future work.

2 Proposal for a Cross-Language Transfer of LKRs from Spanish to German

As introduced in Sect. 1, this work investigates the transfer of rule-based E2R adaptations. In this section, thus, we present (a) the procedure followed to generate German LKRs from Spanish rules using LLMs, (b) the manual refinement process required to ensure linguistic validity and alignment with Leichte Sprache principles, and (c) an exploratory evaluation of the resulting systems.

2.1 LLMs-Based Inference of LKRs

To explore the feasibility of transferring LKRs from Spanish to German, we evaluated the performance of three LLMs available on the Hugging Face Hub[6]:

- `LeoLM/leo-hessianai-13b-chat` [12], a 13-billion-parameter model based on LLaMa-2, fine-tuned specifically for high-quality German-language dialogue and text generation.
- `google/flan-t5-base` [3], a multilingual instruction-tuned encoder-decoder model based on the T5 architecture, designed for general-purpose reasoning and cross-lingual tasks.
- `Qwen/Qwen2.5-7B-Instruct-1M` [23], a 7-billion-parameter multilingual model from the Qwen2.5 family, instruction-tuned for code generation, text manipulation, and multilingual understanding.

These models were selected based on four criteria: (a) open access to weights for reproducibility; (b) coverage of the German language; (c) suitability for generative tasks; and (d) ability to run on limited computational infrastructure. All models were executed locally using the `transformers` library, and a few-shot prompting [19] strategy was applied to guide the generation of German rules based on the Spanish originals. Each prompt[7] included three elements: (a) a brief explanation of the target linguistic phenomenon (explanatory relative clauses or dialogues) and its relevance for cognitive accessibility; (b) a description of the Spanish rule-based adaptation; and (c) the original rule expressed

[6] https://huggingface.co/.

[7] Prompts used are available at: https://zenodo.org/records/15411016.

in Python-like pseudocode. To better align with the training profiles of each model, the prompts were written in English for the two multilingual models, and in German for the German-trained model.

The three models produced different outputs. Despite being fine-tuned for German, `LeoLM/leo-hessianai-13b-chat` failed to perform any linguistic transfer and reproduced the Spanish rules verbatim. `google/flan-t5-base` returned incoherent or corrupted character sequences, likely due to tokenisation errors or instability in decoding. In contrast, `Qwen/Qwen2.5-7B-Instruct-1M` produced a syntactically structured response that retained the original rule logic and included German-language comments and variable names, so it was selected as the basis for subsequent manual refinement and implementation in German.

2.2 Manual Adjustment and Implementation

The LKR generated by `Qwen2.5-7B-Instruct-1M` was manually reviewed and corrected to ensure grammaticality and compliance with Leichte Sprache principles. The revision process focused on preserving the structural logic of the original rules while adapting them to German syntactic and morphological conventions. The manual refinement was carried out by an interdisciplinary team consisting of two native German speakers from Austria, both computer scientists, and a computational linguist from Spain. The readjusted rules were implemented in Python 3.11 using the spaCy[8] library, with the `de_core_news_lg` model for Natural Language Processing (NLP) tasks, such as part-of-speech tagging and dependency analysis.

For **explanatory relative clauses**, the most recurrent issue in the LLM-generated outputs was incorrect word order. The model frequently preserved the Spanish syntactic structure, leading to agrammatical constructions in German. For example, the sentence *Das Mädchen, dessen Cousinen Sie kennen, wird zur Party kommen* was adapted as *Das Mädchen wird zur Party kommen. Das Mädchen kennen Cousinen Sie*, which does not conform to standard German clause structure, particularly with regard to verb-final positioning and constituent order. Additional issues included inappropriate use of subordinating conjunctions (e.g. "weil dass" instead of the correct "weil"), missing punctuation marks at sentence boundaries, and inconsistencies in the treatment of relative pronouns.

In the case of **dialogues**, the primary modifications involved adapting the detection of quotation structures. The initial rules failed to account for typographic variation, particularly between dashes (–) and angular quotation marks (« »), both of which are used in German literary texts. These adjustments required refining the matching conditions in the rule set to handle diverse punctuation patterns while preserving consistency with Leichte Sprache stylistic conventions.

[8] https://spacy.io/.

2.3 Exploratory Results

We evaluated the effectiveness of the LKR-based systems, both the version directly inferred by the LLM and the manually refined implementation.

Regarding the **explanatory relative clauses**, the evaluation was conducted on a corpus of 95 sentences[9], manually annotated by two native German speakers from Austria. The corpus combined samples from the Leipzig Corpora Collection[10] with additional synthetically generated examples by ChatGPT. As shown in Table 1, the LLM-generated rules achieved high precision in clause identification but failed to produce grammatically valid adaptations. Most errors were caused by unadjusted Spanish word order and improper handling of German clause structure. In contrast, the manually revised rules demonstrated higher overall performance across all metrics and were able to produce correct adaptations in more than half of the cases. However, almost 46% of the outputs still presented grammatical issues, particularly related to verb placement.

Table 1. Evaluation results for explanatory relative clauses.

Metric	LLM-generated LKR	Manually Revised LKR
Accuracy	60.64%	69.15%
Precision	84.85%	91.89%
Recall	46.67%	56.67%
F1 Score	60.22%	70.10%

As for **dialogues**, we analysed passages from the German editions of *The Little Prince* and *The Boy in the Striped Pajamas*. The LLM-generated LKR successfully identified all dialogue turns in the first book, which uses dash-based quotation conventions. However, it failed to detect any dialogue in the second text due to the use of angular quotation marks (« »), a variation not covered by the original rule. After refining the detection criteria, the manually adjusted version achieved full coverage in both cases. Nevertheless, adapting the identified dialogues into theatrical format by explicitly introducing speaker names followed by colons proved to be challenging for both systems. The adaptation precision remained below 50%, due to inconsistent segmentation, omission of speaker cues, and misuse of punctuation.

3 Preliminary E2R Expert Evaluation

To complement the automatic evaluation, we conducted a qualitative validation with two experts in German Leichte Sprache. Each expert reviewed a set of 12 pairs of sentences (original and LKR-adapted versions) to evaluate the treatment

[9] The corpus is available at: https://zenodo.org/records/15411016.

[10] https://corpora.uni-leipzig.de/es?corpusId=deu_news_2024.

of explanatory relative clauses. The review was carried out individually using a spreadsheet[11] with the following elements for each sentence: (a) the original sentence in German; (b) the automatically adapted version; (c) a Yes/No judgment on whether the adaptation followed the E2R guideline; (d) a free text field for suggestions on how to improve the adaptation; and (e) an optional comment section for general observations.

Both experts provided detailed feedback on each adaptation. In general, only a small number of adaptations were considered fully compliant with the guideline. In addition, the comments revealed consistent patterns of error. The most frequent issues were ungrammatical structures, especially due to incorrect verb placement, along with repetitions, unnatural sentence segmentation, and referential inconsistencies following clause splitting. In addition, several adaptations did not align stylistically with the conventions of Leichte Sprache.

Despite these problems, both experts recognised the potential of the rule-based approach and offered concrete reformulations that were closer to the Leichte Sprache standards.

4 Conclusions and Future Work

In this study, we propose a cross-linguistic approach for transferring LKRs for E2R adaptation from Spanish to German. Specifically, we explored whether LLMs could assist in generating rule equivalents in German of two existing rule-based adaptations developed in Spanish: one for explanatory relative clauses and another for dialogue formatting. The results show that while LLMs were able to generate syntactically structured rule systems, their output lacked grammatical and stylistic adequacy for German. Most rules required substantial manual revision to conform to Leichte Sprache principles, and even after correction, many adaptations included structural or referential issues. These findings confirm that human intervention remains essential when adapting symbolic linguistic rules across languages. In addition to the automatic evaluation, two German E2R experts reviewed the adapted sentences and identified similar problems. Although only a few adaptations were found to be fully correct, both experts saw the value of the approach and suggested useful ways to improve the rewritten versions. Although the use of LLM accelerated the initial generation of rule logic, the process revealed a clear trade-off between automation and linguistic precision. This raises the question of whether transferring existing rules is more efficient than designing them natively for the target language. Nevertheless, the structured results generated by LLMs, when combined with expert feedback, can be a useful starting point for multilingual adaptation efforts. As future work, we plan to experiment with models specifically trained for German, extend the proposal to other E2R guidelines, and incorporate evaluations with end users, particularly individuals with cognitive disabilities. To this end, we have already designed the experimental setup for the upcoming evaluation phase, which will

[11] The spreadsheets with the experts' comments are available at: https://zenodo.org/records/15411016.

follow the same methodology used in our previous user validation studies [5]. Participants will be asked to complete a form[12] assessing the comprehensibility of adapted versus non-adapted sentences.

Acknowledgements. This research work has been also funded by the grant "Ayudas para la contratación de personal investigador predoctoral en formación para el año 2022" (Reference: PIPF-2022/COM-25762) funded by Comunidad Autónoma de Madrid (Spain). We also thank Negah Ibrahimkhel and Lukas Ruckenstuhl for their help in development.

References

1. AENOR: Lectura F cil. Pautas y recomendaciones para la elaboraci n de documentos (UNE 153101:2018 EX). Asociación Española de Normalización (UNE) (2018)
2. Bott, S., Rello, L., Drndarevic, B., Saggion, H.: Can Spanish be simpler? LexSiS: lexical simplification for Spanish. In: Proceedings of COLING 2012. The COLING 2012 Organizing Committee (2012)
3. Chung, H.W., et al.: Scaling instruction-finetuned language models. arXiv preprint arXiv:2210.11416 (2022)
4. Diab, I., Suárez-Figueroa, M.C.: First attempt to an automatic adaptation of explanatory structures in Spanish to easy-to-read. In: Poster Proceedings of the 40th Annual Conference of the Spanish Association for Natural Language Processing (SEPLN-P 2024). CEUR Workshop Proceedings, vol. 3846. CEUR-WS.org (2024). https://ceur-ws.org/Vol-3846/
5. Diab, I., Suárez-Figueroa, M.C., Peris, R.: Towards an automatic easy-to-read adaptation of dialogues in narrative texts in Spanish. In: Miesenberger, K., Peňáz, P., Kobayashi, M. (eds.) Computers Helping People with Special Needs, pp. 208–216. Springer, Cham (2024)
6. DIN: Empfehlungen für Deutsche Leichte Sprache (DIN SPEC 33429:2025-03). Deutsches Institut für Normung e.V. (DIN) (2025)
7. Díez, E., Rodríguez, N., Fern ndez, A., Alonso, M.A., Díez-Álamo, M.A., Sánchez, M.J., Wojcik, D.: E2R-helper (easy-to-read assistant) [Text in Spanish]. Instituto Universitario de Integraci n en la Comunidad - Universidad de Salamanca (2019). https://eapoyo-inico.usal.es/asistente-lectura-facil/
8. Hansen-Schirra, S., Nitzke, J., Gutermuth, S.: An intralingual parallel corpus of translations into German easy language (Geasy corpus): what sentence alignments can tell us about translation strategies in intralingual translation. In: New Perspectives on Corpus Translation Studies, pp. 281–298 (2021)
9. Iglesias, A., Cobián, I., Campillo, A., Morato, J., Sánchez-Cuadrado, S.: Comp4Text checker: an automatic and visual evaluation tool to check the readability of spanish web pages. In: Computers Helping People with Special Needs (ICCHP 2020). Lecture Notes in Computer Science, vol. 12376, pp. 258–265. Springer, Cham (2020). https://doi.org/10.1007/978-3-030-58796-3_31

[12] A sample of the form (in German) is available at: https://zenodo.org/records/15411016.

10. Inclusion Europe: Information for All. European standards for making information easy to read and understand. Inclusion Europe (2009)
11. Jekat, S.J., Massey, G.: Barrier-free communication: methods and products. In: Proceedings of the 1st Swiss Conference on Barrier–Communication, Winterthur, 15-16 September 2017. ZHAW Zürcher Hochschule für Angewandte Wissenschaften (2018)
12. LeoLM Team: LeoLM/leo-hessianai-13b-chat (2024). https://huggingface.co/LeoLM/leo-hessianai-13b-chat
13. Madina, M., Gonzalez-Dios, I., Siegel, M.: Easy-to-read language resources and tools for three European languages. In: Proceedings of the 16th International Conference on PErvasive Technologies Related to Assistive Environments, PETRA 2023, pp. 693–699. Association for Computing Machinery, New York (2023). https://doi.org/10.1145/3594806.3596530
14. Moreno, L., Alarcon, R., Martínez, R.: EASIER system. Language resources for cognitive accessibility. In: The 22nd International ACM SIGACCESS Conference on Computers and Accessibility, ASSETS 2020. Association for Computing Machinery (2020). https://doi.org/10.1145/3373625.3418006
15. Nomura, M., Nielsen, G.S.: International federation of library associations and institutions, library services to people with special needs section: guidelines for easy-to-read materials. IFLA Headquarters, The Hague (2010)
16. Rello, L., Baeza-Yates, R., Saggion, H.: DysWebxia: Textos más Accesibles para Personas con Dislexia. Procesamiento del Lenguaje Natural **51**, 205–208 (2013)
17. Saggion, H., Stajner, S., Bott, S., Mille, S., Rello, L., Drndarevic, B.: Making it simplext: implementation and evaluation of a text simplification system for Spanish. ACM Trans. Accessible Comput. **6**(4) (2015). https://doi.org/10.1145/2738046
18. Säuberli, A., Ebling, S., Volk, M.: Benchmarking data-driven automatic text simplification for German. In: Proceedings of the 1st Workshop on Tools and Resources to Empower People with Reading Difficulties (READI), pp. 41–48 (2020)
19. Song, Y., Wang, T., Cai, P., Mondal, S.K., Sahoo, J.P.: A comprehensive survey of few-shot learning: evolution, applications, challenges, and opportunities. ACM Comput. Surv. **55**(13s) (2023)
20. Steinmetz, I., Harbusch, K.: Enabling fast and correct typing in Leichte Sprache (easy language). In: Proceedings of The Fourth Widening Natural Language Processing Workshop, pp. 64–67 (2020)
21. Suárez-Figueroa, M.C., Ruckhaus, E., López-Guerrero, J., Cano, I., Cervera, Á.: Towards the assessment of easy-to-read guidelines using artificial intelligence techniques. In: Miesenberger, K., Manduchi, R., Rodriguez, M.C., Peňáz, P. (eds.) Computers Helping People with Special Needs (ICCHP 2020). Lecture Notes in Computer Science, vol. 12376, pp. 74–82. Springer (2020). https://doi.org/10.1007/978-3-030-58796-3_10
22. Su rez-Figueroa, M., Diab, I., Ruckhaus, E., Cano, I.: First steps in the development of a support application for easy-to-read adaptation. Univ. Access Inf. Soc. (2022). https://doi.org/10.1007/s10209-022-00946-z
23. Yang, A., et al.: Qwen2.5-1m technical report. arXiv preprint arXiv:2501.15383 (2025)

Assistive Technology and Accessibility Training

Integrating Digital Accessibility Into Software Engineering Education: A Case Study of Learning Outcomes

Baifan Zhou and Weiqin Chen(✉)

Oslo Metropolitan University, St. Olavs plass, P.O. Box 4, 0130 Oslo, Norway
{baifan.zhou,weiqin.chen}@oslomet.no

Abstract. There is a growing emphasis on integrating digital accessibility (DA) into higher education curricula. However, research on the learning outcomes of integrating DA into computing curricula remains limited. This study investigates the effectiveness of incorporating a DA-focused lecture and project requirements into a software engineering course. By analysing 62 student project reports, we evaluate students' understanding and application of DA principles across seven key questions, including general DA concepts, user-centric considerations such as disabilities and diversity, and developer-centric practices like adherence to standards and accessibility testing. Our findings reveal that while students generally grasped the importance of DA, their engagement with specific concepts and practical skills, such as accessibility testing and diversity considerations, was limited. These results highlight the need for targeted educational strategies and hands-on training to bridge gaps in knowledge and application.

Keywords: Digital Accessibility · Inclusive Design · Accessibility Education · Software Engineering

1 Introduction

Digital accessibility (DA) is increasingly recognised as a fundamental requirement in modern society, driven by legal and regulatory mandates across the globe. These laws aim to ensure equal access to digital technologies for all individuals, regardless of their abilities. However, while progress has been made in establishing standards and guidelines, a critical challenge persists: the lack of competence among IT students to understand, adopt, and apply digital accessibility principles in real-world scenarios [14].

Despite the growing importance of DA, there is a noticeable gap in research concerning the effectiveness of educational interventions in this area [2,8]. Specifically, there is limited knowledge about the learning outcomes associated with integrating DA into higher education curricula [2]. Questions remain unanswered about the extent to which students grasp digital accessibility principles and their ability to apply these principles in practice.

K. Mavrou and P. Encarnação (Eds.): AAATE 2025, *Technology for Inclusion and Participation for All: Recent Achievements and Future Directions*, pp. 389–396, 2025.
https://doi.org/10.1007/978-3-032-01628-7_46

To address this gap, we present a case study of integrating digital accessibility into the curriculum of an IT higher education program. The case study is based on a software engineering course taught in Norway. Our work combines qualitative and quantitative methods to assess the learning outcomes of including DA in the curriculum, providing insights into how such integration impacts students' understanding and application of accessibility principles. This study contributes to ongoing efforts to enhance accessibility education and align it with the broader goal of fostering inclusion and participation for all in digital spaces.

2 Related Work

To meet the growing demand from the IT industry for competency in digital accessibility [12], researchers have emphasised the importance of teaching digital accessibility to computing students at the undergraduate level [5,9]. In recent years, there has been an increased focus on integrating digital accessibility into higher education computing curricula [2,8,13]. Ko and Ladner [6] proposed three strategies for incorporating accessibility: modifying an existing lecture, adding a dedicated lecture, or introducing a new course. In a literature review, Baker et al. [2] identified three approaches to accessibility integration: designing courses entirely focused on accessibility, embedding accessibility as a focused theme in courses whose primary learning objectives are not accessibility, or including accessibility as an additional topic within existing courses.

Within computing programs, the majority of efforts to incorporate digital accessibility have been concentrated in Human-Computer Interaction (HCI) courses, as well as in courses focused on web design and web programming [2,10]. These courses typically address accessibility through topics such as accessibility or universal design principles, the understanding and use of assistive technologies, and compliance with established accessibility standards. However, efforts to integrate digital accessibility into software engineering courses remain limited, though accessibility considerations can play an important role in influencing architectural decisions, shaping development processes, and informing testing practices [4,7]. Among the 51 papers analysed in the systematic literature review by Baker et al. [2], only one reported on the integration of accessibility within a software engineering course.

Through their systematic literature review, Baker et al. [2] also found that digital accessibility is only covered in a small number of elective courses rather than in core computing courses, and is rarely evaluated for learning effectiveness. In the few instances where accessibility has been integrated into software engineering curricula, the focus has largely been on course design rather than on evaluating learning outcomes. An exception is Ludi [7], who assessed students' understanding of accessibility through their exam responses, providing an example of assessing learning effectiveness in the context of digital accessibility integration in software engineering education.

3 Method

To enhance the integration of digital accessibility (DA) into the curriculum, we introduced a dedicated lecture session on DA within a software engineering course. The lecture aimed to provide students with foundational knowledge of DA principles, focusing on both user-centric and developer-centric perspectives.

Following the lecture, students were required to document their understanding and application of DA principles in their project reports as part of their coursework. This requirement served as both a learning activity and a data collection method to assess the effectiveness of the intervention. We now elaborate on more details.

3.1 Course Description

The course is designed to provide students with the skills and knowledge required to develop modern software applications, integrating both back-end and front-end technologies. The course has about 250 students, forming more than 60 groups. Throughout the course, students participate in lectures, hands-on demos, and lab sessions. A key component of the course is the exam project, which requires students to work in groups to select and address one of three real-world problems. The exam project requires the students to collaboratively design and implement a minimal viable product of software application, and document their development process and results as a project report. The teaching activities foster active engagement of students with real-world challenges and align with enquiry-based learning principles.

3.2 Data Collection

We collected and analysed a total of 65 project reports submitted by students. The data and the analysis were completely anonymous. Out of these, three submissions were empty, resulting in 62 valid submissions for analysis. Each report was examined to evaluate the students' comprehension and practical application of DA principles.

3.3 Analytical Framework

The analysis was structured seven questions designed to verify whether students demonstrated an understanding of DA and addressed key concepts correctly. These questions targeted both general DA principles and specific user and developer perspectives:

- **General Understanding:**
  - Q1: Have the students demonstrated correct understanding of digital accessibility principles?
- **User Perspective:**

  - Q2: Have the students addressed digital accessibility in relation to disabilities, such as broken arms, cognitive impairment, dyslexia, vision impairment, hearing impairment, and motor impairment?
  - Q3: Have the students considered diversity in digital accessibility, including factors such as language barriers, cultural contexts, and economic constraints?
- **Developer Perspective:**
  - Q4: Have the students addressed the needs of diverse users in the user interface design, such as simplified navigation, customisability, and compatibility with assistive technologies?
  - Q5: Have the students referenced standards and guidelines for digital accessibility, such as WCAG (Web Content Accessibility Guidelines), ARIA (Accessible Rich Internet Applications), and the Norwegian Regulation for Universal Design of ICT Systems?
  - Q6: Have the students employed accessibility testing techniques, such as automated tools (Lighthouse, Axe, Wave), manual testing with assistive technologies (e.g., screen readers like NVDA, VoiceOver, or JAWS), and keyboard-only navigation testing?
  - Q7: Have the students implemented development best practices for digital accessibility, including the use of Semantic HTML (headings, lists, landmarks), sufficient colour contrast, responsive design, and clear error message?

3.4 Data Analysis

Each report was reviewed to determine whether the student addressed the aspects specified in the seven questions. A systematic coding approach was employed to mark whether each question was answered affirmatively based on evidence presented in the report. The results were aggregated to provide an overview of the extent to which students understood and applied DA principles, both in general and within the specific contexts defined by the user and developer perspectives. The affirmative answers to each question were also qualitatively analysed to identify the specific DA concepts, methods, techniques that were addressed and implemented in the reports.

4 Results

Results from the quantitative analysis are summarised in Table 1.

Regarding the general understanding of digital accessibility, out of the total 62 project reports, 48 groups (77.42%) demonstrated a correct understanding of digital accessibility (Q1). Their reports emphasised key aspects such as *"the importance of making the website inclusive and easy to use for people with special requirements"*, describing it as *"a key part of the UI design, ensuring that the application can be used by everyone, including those with disabilities"*, and

Table 1. Results summary of the verification questions from 62 valid reports

	Q1	Q2	Q3	Q4	Q5	Q6	Q7
Yes Count	48	24	1	34	11	13	40
Yes Percentage	77.42%	38.71%	1.61%	54.84%	17.74%	20.97%	64.52%
Yes Percentage given Q1 = Yes	100.00%	50.00%	2.08%	70.83%	22.92%	27.08%	83.33%

"ensuring applications can be used by a wide range of users". Some also discussed digital accessibility in terms of adaptability, diverse user needs and assistive technologies.

From the user perspective, among the 48 groups that addressed digital accessibility, 24 groups (38.71%) explicitly related to users with disabilities (Q2). Visual impairments, particularly colour blindness, were the most commonly discussed (20 reports), while motor and cognitive impairments received comparatively less attention (6 reports). Dyslexia and older users were each mentioned in one report. Only 1 group (1.61%) considered broader diversity factors such as language barriers, digital literacy and cultural contexts (Q3).

From the developer perspective, 34 groups (54.84%) addressed diverse user needs in interface design (Q4), highlighting strategies such as simplified navigation and compatibility with assistive technologies. Regarding accessibility standards and guidelines (Q5), 11 groups (17.74%) referenced frameworks, including WCAG and ARIA. For accessibility testing techniques (Q6), 13 groups (20.97%) utilised tools such as WAVE and Lighthouse, as well as methods like screen reader testing, keyboard-only navigation, and colour contrast checking. Moreover, 40 groups (64.52%) followed development best practices for accessibility (Q7). This included the use of semantic HTML, appropriate colour contrast, and alternative text for images. Several reports also implemented logical heading structure, consistent navigation, grouping related elements, clear instructions and error message (especially for forms), adequately sized text and interactive elements, and logical focus order for keyboard navigation.

When considering only the 48 groups that correctly mentioned digital accessibility (Q1), 50.00% of these groups addressed disabilities (Q2), 2.08% considered diversity (Q3), and 70.83% accounted for diverse needs in user interface design (Q4). Additionally, 22.92% referenced standards and guidelines (Q5), 27.08% employed accessibility testing techniques (Q6), and 83.33% applied development best practices (Q7).

5 Discussion

The results indicate varying levels of student engagement with digital accessibility (DA) principles, reflecting both successes and challenges in integrating DA into educational curricula. While 77.42% of project groups demonstrated some correct understanding of DA (Q1), the application of specific concepts, particularly from the user view, was inconsistent. For example, only 38.71% addressed disabilities (Q2), and a mere 1.61% considered diversity (Q3). These findings suggest that although many students demonstrate an awareness of disability, there is a considerable gap in understanding the full spectrum of user diversity. Enhancing this awareness, particularly by deepening students' understanding of varied user experiences, is essential to motivating the need for digital accessibility [5]. Integrating case studies and engaging diverse users in the learning process may help students better grasp both disability-specific and broader diversity-related challenges.

From the developer perspective, the relatively low percentages for referencing standards (17.74%, Q5) and employing accessibility testing techniques (20.97%, Q6) suggest that students may lack familiarity with practical tools and frameworks. Incorporating hands-on training with technologies such as automated testing tools (e.g., Lighthouse, Axe), assistive technologies and accessibility auditing practices could improve students' competence in applying and testing against accessibility standards. Promoting co-design processes with individuals with disabilities and fostering partnerships with IT-industry may also help bridge the gap between theoretical knowledge and real-world application, as demonstrated in [11].

The higher percentages of addressing special needs in user interfaces (54.84%, Q4) and implementing development best practices (64.52%, Q7) suggest that some aspects of DA are more accessible to students, potentially due to their overlap with usability principles commonly taught in web development. Leveraging this overlap, educators could emphasise the broader relevance of DA concepts across devices and platforms, wearables [15], IoT [1], and augmented and virtual reality (AR/VR) [3].

Lastly, the significant improvement in percentages when filtering for groups that considered DA generally (e.g., Q7 rises from 64.52% to 83.33%) underscores the importance of integrating DA principles early in the curriculum. Doing so not only enhances students' comprehension but also encourages the adoption of accessibility as a standard practice in their projects. Future research could explore the long-term impacts of accessibility education [16], such as students' ability to apply these principles in professional settings and the effects of integrating accessibility across multiple courses throughout the academic program.

In this study, the integration of digital accessibility (DA) into the software engineering course followed the strategy of adding a dedicated lecture, as proposed by Ko and Ladner [6], and adopted the approach identified by Baker et al. [2] of introducing accessibility as an additional topic within an existing course. Due to time constraints, the study did not involve direct engagement with individuals with disabilities. Instead, students were shown videos demonstrating how people with various disabilities interact with digital applications. Incorporating a participatory approach, such as involving users with disabilities in evaluating student work or co-designing lecture content, could enhance the real-world relevance and impact of such instructional efforts.

Another limitation of the study lies in the deductive nature of the coding process. The use of a pre-defined set of codes, based on the questions in the analytical framework, may have introduced bias by overlooking emergent themes. This limitation may have implications for the validity of the findings. In addition, the study relied solely on the analysis of students' written reports to assess their understanding and application of DA principles. The functional applications submitted by students were not evaluated. Combining the analysis of both the reports and the functional applications would likely have yielded more objective and comprehensive results.

6 Conclusion

This study highlights the importance of integrating digital accessibility (DA) education into the curriculum of software engineering courses to enhance students' understanding and application of accessibility principles. While students demonstrated a foundational grasp of DA concepts, the inconsistent application of specific principles highlights the need for targeted educational interventions, including practical training with accessibility standards, tools, and co-design methodologies.

The findings from the study also emphasise the critical role of higher education in preparing future developers and designers to create inclusive technologies. By fostering a deeper understanding of user needs and accessibility best practices, educational institutions can contribute to the development of innovative solutions that address diverse user requirements. Furthermore, integrating DA education into mainstream curricula not only prepares further IT-professionals with adequate digital accessibility skills to meet growing industry demand, but also promotes a broader cultural shift toward inclusivity in digital environments, paving the way for more equitable access to technology for all.

References

1. Abou-Zahra, S., Brewer, J., Cooper, M.: Web standards to enable an accessible and inclusive internet of things (IoT). In: Proceedings of the 14th International Web for All Conference, W4A 2017. Association for Computing Machinery, New York (2017). https://doi.org/10.1145/3058555.3058568
2. Baker, C.M., El-Glaly, Y.N., Shinohara, K.: A systematic analysis of accessibility in computing education research. In: Proceedings of the 51st ACM Technical Symposium on Computer Science Education, SIGCSE 2020, pp. 107–113. Association for Computing Machinery, New York (2020). https://doi.org/10.1145/3328778.3366843
3. Creed, C., Al-Kalbani, M., Theil, A., Sarcar, S., Williams, I.: Inclusive AR/VR: accessibility barriers for immersive technologies. Univ. Access Inf. Soc. **23**, 59–73 (2023). https://doi.org/10.1007/s10209-023-00969-0
4. El-Glaly, Y.N.: Teaching accessibility to software engineering students. In: Proceedings of the 51st ACM Technical Symposium on Computer Science Education, SIGCSE 2020, pp. 121–127. Association for Computing Machinery, New York (2020). https://doi.org/10.1145/3328778.3366914

5. Elglaly, Y.N., Baker, C.M., Ross, A.S., Shinohara, K.: Beyond HCI: the need for accessibility across the CS curriculum. In: Proceedings of the 55th ACM Technical Symposium on Computer Science Education, SIGCSE 2024, vol. 1, pp. 324–330. Association for Computing Machinery, New York (2024). https://doi.org/10.1145/3626252.3630788
6. Ko, A.J., Ladner, R.E.: Accesscomputing promotes teaching accessibility. ACM Inroads **7**(4), 65–68 (2016). https://doi.org/10.1145/2968453
7. Ludi, S.: Introducing accessibility requirements through external stakeholder utilization in an undergraduate requirements engineering course. In: 29th International Conference on Software Engineering (ICSE 2007), pp. 736–743 (2007). https://doi.org/10.1109/ICSE.2007.46
8. Nishchyk, A., Chen, W.: Integrating universal design and accessibility into computer science curricula–a review of literature and practices in Europe. Transforming our World Through Design, Diversity and Education, pp. 56–66 (2018). https://doi.org/10.3233/978-1-61499-923-2-56
9. Patel, R., Breton, P., Baker, C.M., El-Glaly, Y.N., Shinohara, K.: Why software is not accessible: technology professionals' perspectives and challenges. In: Extended Abstracts of the 2020 CHI Conference on Human Factors in Computing Systems, CHI EA 2020, pp. 1–9. Association for Computing Machinery, New York (2020). https://doi.org/10.1145/3334480.3383103
10. Shinohara, K., Kawas, S., Ko, A.J., Ladner, R.E.: Who teaches accessibility? A survey of U.S. computing faculty. In: Proceedings of the 49th ACM Technical Symposium on Computer Science Education, SIGCSE 2018, pp. 197–202. Association for Computing Machinery, New York (2018). https://doi.org/10.1145/3159450.3159484
11. Sitbon, L.: Engaging it students in co-design with people with intellectual disability. In: Extended Abstracts of the 2018 CHI Conference on Human Factors in Computing Systems, CHI EA 2018, pp. 1–6. Association for Computing Machinery, New York (2018). https://doi.org/10.1145/3170427.3188620
12. TeachAccess, AFB-TalentHub: Bridging the accessible technology skills gap. Technical report, Teach Access Survey & Analysis (2023)
13. Valūnaitė-Oleškevičienė, G., Çavuşoğlu, E., Duarte, C.: Accessibility in technology-enhanced curricula: a comparative study in Portugal, Turkey, and Lithuania, pp. 265–281. Springer, Cham (2024). https://doi.org/10.1007/978-3-031-60049-4_15
14. Vinadé, R., Marczak, S.: Educating for accessibility: insights into knowledge gaps and practical challenges in software engineering. In: Proceedings of the XXIII Brazilian Symposium on Software Quality, SBQS 2024, pp. 646–656. Association for Computing Machinery, New York (2024). https://doi.org/10.1145/3701625.3701689
15. Wentzel, J., Velleman, E., van der Geest, T.: Developing accessibility design guidelines for wearables: accessibility standards for multimodal wearable devices. In: Antona, M., Stephanidis, C. (eds.) Universal Access in Human-Computer Interaction. Methods, Techniques, and Best Practices, pp. 109–119. Springer, Cham (2016). https://doi.org/10.1007/978-3-319-40250-5_11
16. Zhao, Q., et al.: Comparison of methods for teaching accessibility in university computing courses. In: Proceedings of the 22nd International ACM SIGACCESS Conference on Computers and Accessibility, ASSETS 2020. Association for Computing Machinery, New York (2020). https://doi.org/10.1145/3373625.3417013

Modeling Transcription Task of Educational Documents for Visually Impaired Students

Lea Pacini[1(✉)], Jérôme Dupire[1], Isabelle Barbet[1], Olivier Pons[1], and Caroline Huron[2,3,4]

[1] Conservatoire National des Arts et Métiers, Cedric, Paris, France
lea.pacini@lecnam.net
[2] Inserm, SEED, U1138, Paris, France
[3] Learning Planet Institute, Paris, France
[4] Institut du Cerveau de l'Enfant, Paris, France

Abstract. Since February 11, 2005, a French law has guaranteed the right for all children with a disability to attend mainstream classes. As a result, teaching materials must be adapted to ensure they are accessible and usable by these students. Through shadowing observations and interviews with experts in the transcription of materials for visually impaired students, we have modeled the transcription process. This model aims to provide a comprehensive description of the different practices used by these experts. It enabled us to: 1) understand the challenges and 2) identify the main obstacles.

Keywords: Inclusion · Visual Impairment · Adaptation Process · Shadowing · Special Needs

1 Introduction

In France, the law of February 11, 2005 guarantees every child with a disability to attend the school closest to their home. This law also mandates classroom accommodations and the adaptation of teaching materials to meet disabled students' specific needs, allowing them to develop the same skills as their peers. As teaching materials are not always adapted to their different needs, they must be transformed so that students can access and use them independently in the classroom or at home. This transformation process, referred to as transcription-adaptation, is carried out by various organizations across France. Our study focuses on the adaptation of educational documents for visually impaired students, which can be achieved in two different ways: enhancing visual information (e.g., by producing large print documents) or presenting visual information through another sensory modality, such as auditory or tactile formats [1]. Although essential, this adaptation process entails significant costs in terms of both time and financial resources. Schools struggle to provide all students with disabilities with the

K. Mavrou and P. Encarnação (Eds.): AAATE 2025, *Technology for Inclusion and Participation for All: Recent Achievements and Future Directions*, pp. 397–405, 2025.
https://doi.org/10.1007/978-3-032-01628-7_47

transcriptions and adaptations they need to follow the curriculum. According to Castillan et al. [2], delays in delivering adapted materials can range from several days to several weeks. Moreover, transcription experts sometimes encounter difficulties in preserving the educational objectives of the content while ensuring it remains accessible. In an attempt to meet the high demand for adapted documents to a wide range of student profiles, we aim to develop tools that can partially or fully automate the adaptation process. To this end we sought to gain a deeper understanding of the practices and challenges involved, by conducting observation sessions and semi-structured interviews with experts in the field. Our study, funded by ANR, specifically addresses the adaptation of French language textbooks for visually impaired students, who represent just over 5,000 children for the 2023–2024 school year in France [3]. We seek to determine, for example, whether there is a standardised process shared among different experts in how to deal with requests and adapt content. Ultimately, our aim is to formulate recommendations for the development of a platform dedicated to the adaptation of educational materials, capable of meeting the various requirements. This work was supported by the French National Research Agency (ANR) MALIN project (ANR-21-CE38-0014).

2 Investigation Methods

Our research first involved designing a model of the transcription-adaptation process based on shadowing observations. We then sought to generalize and refine this model through semi-directed interviews with other professionals in the field.

Shadowing is a research technique in which a researcher closely follows a subject over a period of time to study what they actually do in their daily work, rather than what their role dictates [4,5]. It involves the real-time transcription of observations on all the activities of the person being observed. This method has the advantage of capturing information directly from the target audience and is flexible enough to allow the observer to ask questions for clarification and gain deeper insights into observed activities [4–6]. We conducted observations within two organizations. In organization A, the transcription department is composed of six experts, including two proofreaders who are visually impaired. Three experts agreed to be observed. One expert was observed over three sessions, and the other two over 2 sessions each. Each session lasted between 2 and 3 consecutive hours. Notes were taken either on a computer, or on a mobile phone when moving between different departments or buildings. In organization B, the transcription department is composed of five experts. Observations took place during a 1-hour session with a single expert. Notes were taken on a computer. Observations were conducted at the beginning of the school year, a period characterized by a high workload. We therefore adapted to the deadlines of the observed experts. Additional questions were asked during or after the observations to clarify certain observed practices.

The semi-structured interviews were conducted individually by videoconference, each lasting approximately one hour. This method was suitable for interviewing several experts located across the country and for gathering information about their practices to complete our data. Nine experts from various organizations across France participated. Recruitment was carried out through direct contact or via the French transcribers' mailing list. The interview grid was organized in three parts: (1) questions about the transcription department and the expert's work (working conditions, type of produced adaptations, ...) ; (2) a practical adaptation situation in which the expert was asked to explain how they would adapt a set of exercises presented to them; (3) questions regarding potential difficulties encountered, concluding with an open-ended question about possible solutions.

3 Results

Observed experts worked at a desk, with a stationary computer, a keyboard and a mouse. They often used two screens. Remote work is possible, in which case they may need to remotely access their worksite computer. The observations covered the entire adaptation process, from the initial transcription request to the final delivery, as well as related tasks such as specific internet searches or media printing. Despite the variety of practices, we identified the common stages of the process based on our investigations. These stages are described below.

3.1 Adaptation Process

Requests Reception. Requests are made by specialist or mainstream teachers, by e-mail or directly given to the transcriber or the transcription department. Documents to be adapted are mainly excerpts of textbooks or material created by teachers.

Process Preparation. Since the adaptation process takes time, transcribers usually request a 2-to-3 weeks period to guarantee the quality of the production. Preparation consists of two tasks, which may be carried out either simultaneously or iteratively. On one hand, requests must be managed and allocated among the different experts in the transcription department. Allocations are usually discussed among transcribers at the beginning of the academic year and can be impacted by various factors. Often, the identity of the students who will benefit from the adaptation is considered first, as each transcriber follows about ten students per year. The distribution of students among transcribers is based on the latter's skills and preferences. For example, some transcribers specialize in Braille adaptation; in such cases, they will primarily follow students who need Braille documents. Sometimes, the student's age or the type of material to be adapted influences the allocation. For instance, some transcribers prefer adapting textbooks, which are more often requested for elementary school children. Before requests are allocated, an evaluation of the estimated time and difficulty is systematically carried out to assess the workload associated with each task. Other requests, particularly those involving a heavy workload, are then divided on a case-by-case basis. For example, adapting drawings often involves recreating them entirely to ensure accessibility. These assessments are taken into account when distributing the workload among experts. On the other hand, some documents may be needed for the adaptation to be produced. Transcribers have access to scanned textbooks (pdf format) provided by publishers through a dedicated official platform. This platform enables document sharing, and contains some materials that have already been adapted. Similarly, some organizations use an internal database to share both source documents and adapted materials. Eventually, transcribers can also rely on their peers to obtain specific documents if they can't be found either on the official platform or in the internal database.

Text Retrieval. Once the relevant documents have been retrieved, experts need to extract the textual content. This can be done in three different ways.

(1) Optical Character Recognition (OCR) is applied either to the entire document or area by area. The latter approach better preserves the structure of the source document and is particularly useful when working with PDF files. The extracted text is then exported into a new document.
(2) Copy/paste from the source document to the adaptation document. This method requires frequent back-and-forth between the two documents.
(3) Direct text typing is used when textual content in the source document is actually an image (screenshot, handwritten scanned document, ...) or when the document layout complicates extraction (such as when text is divided into two or more columns, which is common in textbooks). Methods (2) and (3) require the creation of a new working document beforehand.

Regardless of the method used, the extracted text must be reviewed to correct any extraction errors (such as misrecognized special characters or mathematical equations). It is typically during this stage of the transcription process that some textual formatting— such as italic or forced line breaks—is removed.

Adaptation Choices. Most adaptation choices concern textual formatting or graphic content modifications. They are based on the medical assessments conducted by healthcare teams working with the students (such as orthoptists, occupational therapists, ...). These assessments are provided to the experts and detail the students' specific needs, including the required font, font size, line spacing, colors, and other elements.

Adaptation Choices for Textual Document can be divided into two categories : generic adaptation choices and Braille specific adaptation choices.

Generic adaptation choices relate to the formatting and structuring of the content. These modifications are based not only on the needs and preferences of the students but also on the type of source document, the type of adaptation requested, or even the transcriber's working habits. In this regard, it is usually easier for experts to first remove all formatting from the source document when possible (for example, when working with a word processor file), before reformatting it as needed. When a student's profile allows it, transcribers often save the required formatting as a template in order to speed-up the process. Below are some of the main adaptation choices made by transcribers.

(1) Document structuring: This is one of the most important aspects, as it allows students to navigate the document easily and to follow during classes. Since both large print and Braille format face similar space limitation issues, the printed adapted documents are significantly larger than the original document. Therefore, double pagination is required, indicating both the page number in the adapted document and the corresponding page number in the source document. Document structuring is also achieved through the use of heading styles, which similarly aim to facilitate navigation within the document. However, practices vary depending on several factors, such as document size, type of adaptation, and the transcriber's habits. Nevertheless, excessive use of heading style can make a printed Braille adaptation confusing, as the distinctions between styles rely primarily on spacing.
(2) Reorganization is applied in specific situations, mostly when the source document contains pictures. The title and caption of the picture are placed before it, to improve understanding of the content. Similarly, when questions relate to a specific piece of content, they are displayed before the content itself. This way, students know precisely what to look for, avoiding repeated back-and-forth between the content and the questions.
(3) Emphasizing text: text emphasis choices are made to align closely with the source document, as long as they are compatible with student's needs, while also taking into account experts' established practices and preferences. However, some types of emphasis, such as framing text or using italics, are often avoided by experts. In Braille adaptations, emphasis is generally kept to minimum, to "not slow down the reading, even more (when) there is emphasis elsewhere". For instance, during one of the observation sessions of a printed Braille adaptation, a transcriber initially decided to remove bold formatting from the text because it was "slowing the reading down". Later in the session, they decided to keep bold emphasis because "it is only one word". Indeed, the information load of the content needs to be taken into account when emphasizing, and more broadly when adapting. Regarding large print adaptations, text may be emphasized using bold or color as long as there are no contraindications in the student's medical records.

Braille adaptation, whether printed or digital, is carried out using Braille software. The most commonly used software by the experts we observed and interviewed is DBT. It allows the conversion of text from a word-processed document into Braille. Specific codes are used to structure the adapted document. Information such as double pagination, whether "uncontracted" or "contracted" Braille is used, transcribers' notes or Braille indicators are conveyed through these codes.

Adaptation Choices for Graphic Documents depend on their nature. Decorative elements, such as those commonly found in textbooks, are not adapted because they are not essential for understanding the content. For pictures with educational relevance, adaptation choices depend on the time required for adaptation, the complexity of the document, the skills of the transcribers, and the information that needs to be conveyed. Graphic documents for blind students can be adapted in two different ways: textual description or raised-line pictures. Textual description is used for documents that can be easily understood by reading, such as timelines: "(we need to) give the information in the easiest way". Transcribers also insert a "transcriber's note" in the adapted document to link the description to the corresponding source picture. Raised-line pictures are produced using a special ink that rises when heated, forming a picture in relief that can be recognized by touch... Geographical maps are often adapted in this way. Different areas are distinguished by adding tactile patterns (e.g., dots, lines, grids), which become raised after heating. However, patterns must be clearly distinguishable from one another, and the use of too many patterns can make the pictures confusing. For partially sighted students, adaptation focuses more on modifying the original picture: adjusting contrasts or colors, thickening outlines or even adding "comfort blank"—white spaces around outlines—to improve overall contrast, and thus enhance perception of the picture. Sometimes, pictures are too complex, and must be divided into simpler parts or enlarged, to pass on the information step by step. Regarding color adaptation, "It's important to determine whether the color conveys information or if it was added purely for aesthetic purposes".

Once edited, these images can be inserted directly into the main body of the text. Overall, adapting graphic content often involves partially or fully redrawing the source image to simplify it. In some cases, adaptation is not possible. This occurs particularly when textual content refers to an image that must be described, but a textual description would effectively give away the answer to the student, and creating a raised-line picture would be too costly or impractical. Furthermore, raised-line drawings require specific training and may. Not be suitable for younger students. In such cases, discussions with specialist teachers may take place to assess the relevance of the content. Transcribers may then propose not to adapt the content, to substitute it with another exercise training the same skills, or to adapt the content while explicitly indicating its limitations.

Verification. Verification involves reviewing and reassessing the entire document to ensure that it meets the specific needs of the student it is intended for, in terms of formatting and structure (e.g., font, font size, heading styles, etc.), Braille translation (correct recognition of Braille codes by the software), and the quality of raised-line graphics. This stage is carried out by transcribers and may occur several times during the adaptation process—for instance, when the document is particularly long. One of the experts we observed mentioned performing this check after completing each page, while noting that "it's up to the transcriber's discretion". Verification can also take place once the document has been fully adapted, especially if it is short or requires only minor adjustments, such as font and font size changes. Specialist teachers may also be consulted at this stage to finalize the adaptation to ensure that all educational objectives are met. For documents adapted into Braille, verification tends to be more time-consuming. In the DBT software, verification involves importing the word-processed document and converting it into Braille. The software then displays the Braille text and checks that Braille indicators have been correctly interpreted. In one of the observed organizations, an additional proofreading stage is included: the adapted documents are sent to proofreaders, who are both expert transcribers and expert users, as they themselves are visually impaired. These proofreaders review the Braille documents and provide feedback to their colleagues when sections are unclear due to overload or poor structure. This step also provides the original transcriber with an opportunity to discuss adaptation choices that may have been uncertain during the process. The validation stage may sometimes be skipped, particularly when the document is short, requires minor changes, or when time constraints apply.

Printing. Printed materials can include Braille documents, large print documents, or raised-line graphics. The latter are typically included as appendices to embossed Braille documents and are referenced within the Braille text. All components are often bound together to provide the student with a single, unified document. Large print documents may be intended for some partially-sighted students, but they can also be printed as a complement to Braille materials. In such cases, they are intended for the student's entourage—such as teachers or parents—who are generally not Braille readers.

Delivery. Delivery can occur in various ways, depending on the final format of the documents. Digital documents are either sent directly via email or uploaded onto a server or platform used for resource sharing. Printed materials (such as paper Braille documents or raised-line graphics) may be produced by the transcription organization and then either mailed or delivered to the specialist teachers, who are responsible for bringing them to the schools. In some cases, specialist teachers may also print the documents on-site.

Feedback. Feedback can be provided throughout the school year by students, mainstream teachers, and/or specialist teachers regarding the adaptations delivered. This can occur, for instance, when a student's condition evolves over time or when the adapted content is too specific for the student's exact needs to have been anticipated. Specialist teachers, who maintain contact with the transcribers, are responsible for conveying this feedback. Adaptations are then revised accordingly.

4 Discussion

Modeling the transcription activity has highlighted significant variations in practices between different organizations, but also within the same organization. These differences can be attributed to students' specific needs, as visual impairment can manifest in a wide range of ways. They are also linked to the diversity of software knowledge and skills among experts. Some are more accustomed to using word processing software (typically Microsoft Word), while others prefer the Adobe suite (particularly InDesign) to produce their documents. The high degree of content personalization combined with the diversity of practices among experts, leads to situations where, even for the same educational content and similar student needs, multiple different adaptations may be produced. Our model also highlights several obstacles to the efficiency of the adaptation process. In particular, the pooling of adapted documents is almost non-existent, often resulting in experts re-adapting documents that had already been processed by others. Moreover, the format of source documents, often provided as PDF, makes text retrieval difficult. The DBT verification workflow also entails several sequential steps—file saving, import into Braille software, validation of Braille code recognition, text-to-Braille conversion, output verification—which cumulatively impact production efficiency.

5 Conclusion

The objective of this study was twofold. First, we aimed to detail the adaptation process of educational content for visually impaired students, developing a model of transcription tasks based on observations and interviews. The stages we identified are consistent with the five stages previously described by [2]. Second, based on this model, our future work aims to identify areas where targeted interventions could help enhance and optimize the adaptation process. For instance, this could involve developing recommendations for presenting adapted educational content, in order to unify practices, or for structuring PDF files to ensure better content retrieval. Combined with better sharing of adapted documents, this approach could meet rising demand for accessible materials while preserving the expertise needed to personalize them.

References

1. Douglas, G., McLinden, M.: Access to print literacy for children and young people with visual impairment: findings from a review of literature. Eur. J. Spec. Needs Educ. **26**, 25–38 (2011). https://doi.org/10.1080/08856257.2011.543543
2. Castillan, L., Lemarié, J., Mojahid, M.: L'accessibilité des manuels scolaires numériques: l'exemple suédois, entre édition adaptée et édition inclusive. La nouvelle revue - Educ. Soc. Inclusives **87**, 149–163 (2020). https://doi.org/10.3917/nresi.087.0149
3. RERS - Repères et références statistiques: enseignements - formation - recherche: édition 2024. https://rers.depp.education.fr/data/2024/RERS.pdf

4. Quinlan, E.: Conspicuous invisibility: shadowing as a data collection strategy. Qual. Inq. **14**, 1480–1499 (2008). https://doi.org/10.1177/1077800408318318
5. McDonald, S.: Studying actions in context: a qualitative shadowing method for organizational research. Qual. Res. **5**(4), 455–473 (2005). https://doi.org/10.1177/1468794105056923
6. Lallemand, C., Gronier, G.: Méthodes de design UX: 30 mthodes fondamentales pour concevoir des expériences optimales. Eyrolles, Paris (2018)

Designing and Evaluating Disability Simulation Games for Positive Player Impact

Patricia Piskorek(✉) and Gottfried Zimmermann

Stuttgart Media University, 70569 Stuttgart, Germany
piskorek@hdm-stuttgart.de

Abstract. Research shows that traditional disability simulations can reinforce negative stereotypes by focusing solely on challenges. To address this issue, we developed three strategy-based games about students with visual, hearing, and mental health impairments, which follow previously designed guidelines. They were evaluated regarding Attractiveness, Knowledge Gain, and Empathy Gain. Nine participants, consisting of teachers, played two games each. Attractiveness was assessed using the User Experience Questionnaire (UEQ). Knowledge Gain was assessed by having participants identify information that could be taken from the games. Empathy Gain was assessed using a self-report questionnaire with a Likert Scale rating the capabilities of people with disabilities. Results show that participants found the games enjoyable and effectively acquired knowledge through playable mechanics. While participants' perceptions of students with disabilities did not show the anticipated quantitative improvements, qualitative feedback indicated an increased willingness to support these students. This suggests that disability simulations should prioritize showcasing strategies and successes.

Keywords: disability simulation · disability awareness · attitude change · serious games · game design

1 Introduction

In 2021, 15.9% of German students reported having a condition that made studying more difficult. These impairments include hearing, visual and mobility disabilities, learning disabilities, chronic illnesses and mental health conditions [1].

Adapting their teaching methods and materials to the needs of students with disabilities is not just a considerate approach of teaching staff, but a legal obligation: According to the "German Framework Act for Higher Education" ("Hochschulrahmengesetz"), § 2, Paragraph 4, universities "must ensure that students with disabilities are not disadvantaged in their studies and that they can make use of the university's offerings without external assistance as far as possible" (translated from German) [2].

To effectively address students with disabilities, however, teaching staff must first be sensitized to their needs. For this purpose, we developed "BlindDate", which is a German-language digital information platform on accessibility [3]. BlindDate was developed in a co-design process with students with disabilities to authentically reflect their

K. Mavrou and P. Encarnação (Eds.): AAATE 2025, *Technology for Inclusion and Participation for All: Recent Achievements and Future Directions*, pp. 406–413, 2025.
https://doi.org/10.1007/978-3-032-01628-7_48

experiences [4]. It enables users to meet virtual personas who represent students with various needs and highlight their common barriers and strategies. BlindDate is therefore primarily aimed at university lecturers, with the goal to raise awareness about the needs of their students and to point out accessibility barriers in their teaching. Interactive elements are used to give users a more varied way of engaging with the information and to foster empathy. We further considered integrating disability simulations into BlindDate, allowing teaching staff to adopt the viewpoint of their students.

Besides traditional resources such as books, websites, or presentations, simulations are a tool frequently employed in the context of disability education and have been since at least the 1970s [5]. Disability simulations place participants without disabilities in situations designed to help them experience what it is like to have a disability, albeit temporarily and under artificial conditions. Their primary objectives include enhancing knowledge about disabilities and fostering empathy among participants [6].

However, as long as disability simulations have existed, they have also faced criticism. The most common critique is the lack of evidence supporting the effectiveness of disability simulations [5, 7]. Some studies suggest that disability simulations can have negative rather than positive effects on participants. By emphasizing challenges associated with newly acquired disabilities [8], they fail to address the skills and strategies people with disabilities develop in living with impairment [5]. For example, participants in a 2015 study by Silverman et al. viewed blind individuals as less capable after completing blindfolded tasks [8].

To address critiques from the literature, guidelines to create more constructive disability simulations were developed [9]. In a "guided" simulation, the person with the disability acts as the expert, assisting participants by proposing strategies to solve tasks. In a "task-based" simulation, tasks are drawn from the daily lives of people with disabilities, avoiding exaggerated or artificial scenarios. In a "strategy-based" simulation, its focus shifts from encountering barriers due to disability, to overcoming them despite disability. The aim is not to create fully authentic simulations, but to teach participants how to create environments where people with disabilities can apply their strategies efficiently [9].

In a next step, three games that follow these guidelines were designed, implemented and evaluated for students with visual impairments ("Gabriel"), hearing impairments ("Hannah"), and mental health issues ("Maxi"). They were integrated into BlindDate both technically and conceptually. This work focuses on the evaluation methodology and results.

2 Methodology

The evaluation compared the typically negative effects of disability simulations reported in the literature against the positive effects of games designed according to the guidelines, thus implicitly validating the guidelines. The research question was whether the games achieve the intended positive effects, specifically regarding:

(1) Attractiveness: Are the games enjoyable to the players?
(2) Knowledge Gain: Are players able to teach in a strategy-supportive way?
(3) Empathy Gain: Are players willing to teach in a strategy-supportive way?

The study targeted teachers at higher education institutions, as they form the main target group of BlindDate. Teachers at secondary schools were included to broaden the participant pool. The study was conducted with nine test subjects, consisting of five female and four male teachers. They were aged between 28 and 62 years, with an average age of 35 years. The user study was conducted with one test subject at a time. It was done in an online format, using a video conferencing tool. The study was conducted in German, with all content subsequently translated for this work. Each participant played two of the three games in varying order. Empathy was assessed once at the beginning to establish a benchmark for comparison. Attractiveness, Knowledge Gain, and Empathy Gain were further measured immediately after playing each game.

2.1 Attractiveness

To evaluate the **Attractiveness** of the games, the User Experience Questionnaire (UEQ) [10] was chosen because of its established validity. The UEQ employs seven-point semantic differentials across six scales, of which three were selected for this evaluation: Attractiveness (6 items) measures the overall impression of the games, Stimulation (4 items) assesses how exciting and motivating the games are to use, and Novelty (4 items) assesses the perceived innovation and creativity of the games [10].

2.2 Knowledge

The aim of teaching knowledge is for the players to learn about the barriers and strategies of students with disabilities, in order to be able to teach in a way that supports the usage of these strategies.

To evaluate the **Knowledge Gain,** a list of key information extractable from the games was created, distinguishing whether each piece of information is conveyed through reading the textual content or playing the game mechanics. For example, the statement *"Gabriel works slower compared to students without visual impairments"* is of type "Playable", as participants directly experience this, i.e., being slow themselves, while playing the game. In contrast, the statement *"Gabriel often gets neck pain from constant head movement"* is of type "Readable", as the persona mentions this within the game dialogue.

Participants were then presented with seven statements per game and asked to identify which information could be taken from the respective game, regardless of whether the statements were true. For each game,

- two statements were of type "playable",
- two statements were of type "readable".
- two statements were true, but the information was not conveyed in the game,
- one statement was false.

2.3 Empathy

The aim of teaching empathy is for the players to gain a better perception of students with disabilities by understanding their barriers and strategies, and to enhance the willingness of teaching staff to accommodate these students.

Participants' **Empathy Gain** was measured using an adapted self-report questionnaire by Silverman et al. (2015). While they asked participants to rate *"how well [they] think the average blind person could do [an] activity, compared with the average non-disabled person"* [8], we adjusted the wording to match the impairment of the personas featured in the games, i.e., how well a *"student with visual impairment"*, *"student with hearing impairment"*, and *"student with mental health issues"* could do different activities.

The seven activities selected are commonly encountered in everyday university life: *"Can find their way around campus"*, *"Can follow an on-site lecture"*, *"Can follow an online lecture"*, *"Can upload a submission to the university's Learning Management System"*, *"Can write an academic paper with a bibliography"*, *"Can take a written exam"*, and *"Can take an oral exam"*.

Silverman et al. (2015) used a symmetrical 7-Point Likert Scale to measure perceptions, where 1 indicated that *"[the impaired person] could perform the activity much worse than a non-disabled person"*, 4 indicated that *"[the impaired person] could perform the activity just as well as a non-disabled person"*, and 7 indicated that *"[the impaired person] could perform the activity much better than a non-disabled person"* [8]. During the pilot phase of our study, test subjects did not perceive students with impairments to be superior to non-disabled students in any category. To capture more nuanced results, the scale was adjusted asymmetrically to provide greater detail on the negative side. The revised scale ranged from:

"[The impaired person] could perform the activity..."
1: "...much worse than a non-disabled person."
5: "...just as well as a non-disabled person."
7: "...much better than a non-disabled person."

Qualitative interview questions supported the quantitative methods.

3 Results

3.1 Attractiveness

The assessment of **Attractiveness** yielded good results, as indicated by the calculated mean values (see Table 1). With the exception of the Attractiveness of the game about Maxi, all measured values are in the positive range, with positive being defined by the UEQ standard interpretation as scores above 0.8.

The large confidence intervals observed (see Fig. 1) are mainly attributable to the small sample size. In the categories with high ratings, there was considerable agreement among participants, whereas there was less agreement in categories that were rated more poorly.

Table 1. The UEQ Mean Values for the Games about Hannah, Gabriel and Maxi

	Hannah	Gabriel	Maxi
Attractiveness	1,542 (best)	1,200 (2nd)	0,556 (3rd)
Stimulation	2,375 (best)	2,150 (2nd)	2,028 (3rd)
Novelty	2,313 (best)	1,600 (3rd)	2,278 (2nd)

The UEQ was effective in measuring Stimulation and Novelty, while the Attractiveness category allowed for much subjective interpretation. Some participants associated the given terms with specific in-game situations that they found negative. For example, participants reported that it was "annoying" and "unpleasant" to have an impairment. Others referred more broadly to the overall gaming experience which they found positive. For example, participants thought it was "enjoyable" and "pleasing" to experience the games and their topics.

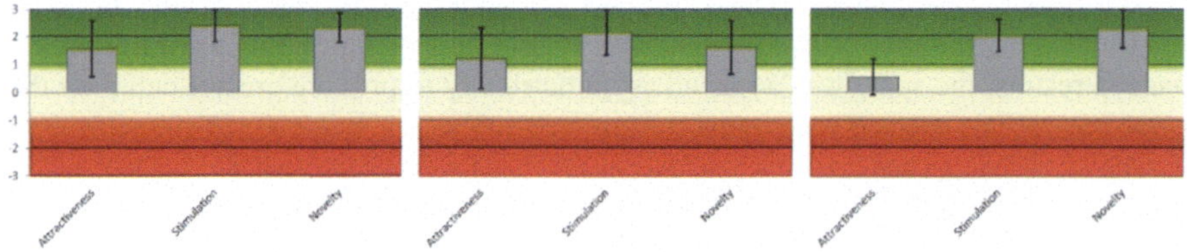

Fig. 1. The UEQ Graphs for the Games about Hannah, Gabriel and Maxi

3.2 Knowledge

The evaluation of the **Knowledge Gain** shows that the players internalize the learning content better when they experience it themselves as playable mechanics. All nine participants correctly ticked all statements of the "Playable" type for the games they played.

The acquisition of knowledge through reading texts in the games was slightly lower. Qualitative observations showed that subjects occasionally skipped reading the texts, either when they were preoccupied with the sudden experience of having a disability, or when they were focused on solving tasks quickly using newly discovered strategies.

3.3 Empathy

The analysis of the **Empathy Gain** did not reveal the anticipated outcome. Overall, participants' perception of students with disabilities worsened in 15 out of 18 cases.

For example, for Hannah, the student with hearing impairment, notable negative change occurred in the category *"Can follow an online lecture"* (see Table 2), as the game illustrated the difficulties of attending such a lecture with poor subtitle quality.

Conversely, categories which were not directly related to the game content, such as *"Can find their way around campus"* (see Table 3), saw little change in participants' ratings.

The games' focus on strategies aimed to demonstrate the general capability of students with impairments. However, participants often perceived the strategies as obstacles rather than as helpful. For example, one participant noted that online lectures were more challenging than expected for Hannah due to poor subtitle quality, without recognizing that subtitles were essential for her to access the lecture content at all.

Table 2. Participant's perception on Hannah's ability to follow an online lecture

[Hannah] can follow an online lecture	Before Play	After Play
Subject 2	3	3
Subject 4	4	4
Subject 6	**5**	**3**
Subject 8	**4**	**3**
Change		–3

Table 3. Participant's perception on Hannah's ability to find their way around campus

[Hannah] can find their way around campus	Before Play	After Play
Subject 2	5	5
Subject 4	4	4
Subject 6	5	5
Subject 8	4	4
Change		0

The game about Maxi, the student with mental health issues, represented the only instance in the user test where a participant indicated that a student with a disability could handle a situation not just as good as, but even better than a non-disabled student. This occurred in the category *"Can follow an online lecture"* (see Table 4). The participant reasoned that while isolation typically poses a challenge for non-disabled students, it provides meaningful support for Maxi.

Although the quantitative results do not suggest it, the simulations effectively raised awareness about supporting students with disabilities, as participants qualitatively expressed a greater willingness to offer such support. Participants additionally reported that supporting students with disabilities appeared less daunting than initially expected.

Table 4. Participant's perception on Maxi's ability to follow an online lecture

[Maxi] can follow an online lecture	Before Play	After Play
Subject 1	5	3
Subject 2	2	3
Subject 3	4	4
Subject 4	4	4
Subject 5	4	3
Subject 6	**3**	**6**
Subject 7	5	4
Subject 8	4	3
Subject 9	3	2
Change		−2

4 Discussion

This research aimed to address the criticisms of disability simulations presented in the literature by applying guidelines that ensure these simulations have a positive, rather than negative, impact. The evaluation of the developed games focused on three pillars: Attractiveness, Knowledge Gain, and Empathy Gain.

Participants found the games enjoyable and engaging, as reflected in both the quantitative results and qualitative feedback. There was a strong consensus within the target group on the positive aspects.

The game about Maxi excelled in the category of Novelty. This aligns with qualitative feedback in which participants noted that simulations of visual or hearing impairments are more commonly encountered, while simulations of mental health issues are more original. Some participants associated terms from the category of Attractiveness with specific in-game situations that they found unpleasant. While this does not reflect the intended measure as defined by the UEQ, the lower score for the game about Maxi can be seen as an implicit compliment to the game's immersive quality, as participants experienced genuine emotional reactions.

The conveyance of theoretical knowledge was effective. Gameplay mechanics proved to be a more successful medium for knowledge transfer than textual explanations. This finding suggests that disability simulations should prioritise showcasing not just challenges, but strategies and successes in overcoming those, by making them playable and experiential. This is crucial for ensuring that these strategies, and consequently a more positive perspective on students with impairments, are retained in memory.

Participants' perception of students with disabilities did not show the anticipated improvements, which seems to leave the main point of criticism from the literature unsolved. While the games demonstrated strategies and adaptations used by students with disabilities to highlight the students' capabilities, participants often focused on the challenges associated with these strategies rather than viewing them as supportive measures.

However, all participants qualitatively expressed a strong desire to support students with disabilities more.

5 Limitations and Outlook

Future research should consider alternative study designs to determine whether the guidelines consistently fail to achieve desired results or if the current study is insufficient in capturing them. The current study design sometimes poorly reflected changes in participants' ratings, particularly when initial ratings were low, leaving little room for further change.

A comparison between the BlindDate simulations and simulations that do not implement the guidelines might reveal that the latter result in even more negative outcomes. Similarly, multiple control groups could be established to play the games under different conditions, such as varying levels of prior knowledge about accessibility.

A larger number of participants is necessary for meaningful study results.

Disclosure of Intrest. The authors have no competing interests to declare that are relevant to the content of this article.

References

1. Steinkühler, J., et al.: Die Studierendenbefragung in Deutschland: best3. Deutsches Zentrum für Hochschul- und Wissenschaftsforschung (2023)
2. Federal Republic of Germany: Hochschulrahmengesetz (2019)
3. BlindDate Platform. https://barrierefreies-blinddate.de. Accessed 14 May 2025
4. Heitmeier, K.-A. et al.: Persona co-design for improving digital accessibility. In: Extended Abstracts of the 2023 CHI Conference on Human Factors in Computing Systems, vol. 227, pp. 1–7. Association for Computing Machinery, New York, NY, USA (2023)
5. French, S.: Simulation exercises in disability awareness training: a critique. Disabil. Handicap Soc. **7**(3), 257–266 (1992). https://doi.org/10.1080/02674649266780261
6. Hertel, J.-P., Millis, B.: Using Simulations to Promote Learning in Higher Education: An Introduction. 1st edn. Routledge, New York, NY, USA (2002). https://doi.org/10.4324/9781003448594
7. Flower, A., Burns, M., Bottsford-Miller, N.: Meta-analysis of disability simulation research. Remedial Spec. Educ. **28**, 72–79 (2007). https://doi.org/10.1177/07419325070280020601
8. Silverman, A., Gwinn, J.D., Van Boven, L.: Stumbling in their shoes: disability simulations reduce judged capabilities of disabled people. Soc. Psychol. Pers. Sci. **6**, 464–471 (2015). https://doi.org/10.1177/1948550614559650
9. Piskorek, P., Heitmeier, K.-A., Kersken, V., Zimmermann, G.: Re-conceptualizing disability simulations: a guided strategies-based approach. in: book of abstracts. In: 17th International Conference of the Association for the Advancement of Assistive Technology in Europe, AAATE 2023, pp. 265–267. Georgios Kouroupetroglou (Eds.), Paris, France (2023)
10. UEQ: User Experience Questionnaire. https://www.ueq-online.org/. Accessed 14 May 2025

MuVie: Innovative Methods and Digital Technologies for Blind Music Education – Needs Analysis

Anna Nicotra[1(✉)], Konstantina Geramani[2], Giovanni Bertoni[3], Helena Jakubowska[4], Chariklia Strouthou[5], Christiana Koutsoupidou[5], Ágota Szilágyi-Kispista[2], and George Ioannidis[2]

[1] Siena Jazz University, Fortezza Medicea, 10, 53100 Siena, Italy
info@sienajazz.it
[2] IN2 Digital Innovations GmbH, Auf Dem Hasenbank 23A, 88131 Lindau, Germany
info@in-two.com
[3] Associazione Giuseppe Paccini, Via da Campione 17, 37138 Verona, Italy
[4] Towarzystwo Muzyczne Im. Edwina Kowalika, Gubinowska 15, 02-956 Warsaw, Poland
[5] St Barnabas School for the Blind, 28Th October Street 30, 2012 Strovolo-Nicosia, Cyprus

Abstract. The European project MuVie - Music and Visual Impairment in Inclusive Education, funded by the Erasmus + programme, aims to make music education more accessible to blind and visually impaired students through the integration of advanced digital tools such as computers, Braille displays, screen readers and specialised music notation software. In its initial phase, the project carried out an in-depth analysis of the state of the art and a survey of the real needs of end-users through a questionnaire administered to students in four European countries - Italy, Germany, Poland and Cyprus - with the aim of collecting data on habits, barriers and strategies in music learning in the presence of visual impairment. The survey highlighted, on the one hand, the existence of good practices and, on the other hand, the numerous needs expressed by students, both in specialised institutions and in inclusive school settings. It also highlighted the central role of the family and inclusive school policies in the educational success of students. On the basis of these findings, the project launched a study on the applicability of artificial intelligence to improve the usability of Braille music and to increase the effectiveness of teaching, outlining the need for a dedicated digital platform to optimise access to resources, provide training for teachers and promote a more inclusive and sustainable music education.

Keywords: Braille Music · Artificial Intelligence in Music · Braille Music Notation

1 Introduction/Background

In recent years, technological innovation has opened up new prospects for improving accessibility and inclusion in music education. Digital tools such as accessible music notation software, interactive Braille displays, advanced screen readers and artificial

K. Mavrou and P. Encarnação (Eds.): AAATE 2025, *Technology for Inclusion and Participation for All: Recent Achievements and Future Directions*, pp. 414–420, 2025.
https://doi.org/10.1007/978-3-032-01628-7_49

intelligence-based systems are profoundly changing the way students with visual impairments learn and interact with music. In this context, the European MuVie project has been launched with the aim of developing innovative teaching methods and promoting the use of accessible technologies to enable blind and partially sighted students to study music more independently, effectively and continuously.

Inclusive education, particularly for students with visual impairments, has increasingly drawn the attention of scholars and practitioners seeking to leverage technology to create accessible learning environments. At the forefront of this movement is the European Agency for Special Needs and Inclusive Education, which serves as a central policy and knowledge hub, promoting best practices across member states and informing systemic reforms in special needs education [1].

In the domain of music education, researchers have underscored the critical role of technology in overcoming barriers for students with visual impairments. One key development is the use of music computer technologies (MCT) in both informatics and music instruction. As demonstrated by Gorbunova and Govorova [2], these tools—when complemented by speech synthesizers, Braille displays, and customized keyboard shortcuts—enable blind students to rapidly acquire technical skills and engage creatively with music curricula. This integration not only fosters academic achievement but also supports the development of digital literacy.

Complementing this technical perspective, a comparative qualitative study of music educators in Spain, Australia, Sweden, and the UK reveals the importance of inclusive pedagogical approaches that go beyond technological solutions. Teachers highlighted the need for emotional investment, reflective practice, and adaptive methods that empower marginalized youth to succeed musically [3]. The convergence of these pedagogical insights with assistive tools forms the backbone of effective inclusive music education.

Expanding the scope, Manirajee et al. conducted a systematic literature review identifying 20 key assistive technologies aimed at enhancing mobility, learning, and social inclusion for visually impaired individuals [4]. Their findings reveal both the promise and limitations of current technologies, particularly in developing regions such as Malaysia, where localized innovations remain sparse. This underlines a pressing need for context-specific design and participatory development models in assistive tech.

From a more specialized angle, the CONTRAPUNCTUS project addresses the complexity of Braille music notation, often perceived as counterintuitive by users. Their computer-based solution helps learners transition into Braille music fluency, making music reading more accessible through interactive, adaptive methods [5]. The project exemplifies how targeted technological interventions can resolve domain-specific accessibility issues.

Finally, initiatives like SMS AEC Music's online platform explore how digital innovation supports broader inclusive practices in music education. By curating digital tools and sharing success stories, the platform advocates for scalable, inclusive strategies that accommodate learners with diverse needs [6].

Fully in line with the European debate on accessibility in music education, MuVie systematically addresses existing challenges and proposes both technological and pedagogical solutions to improve the educational experience of visually impaired students.

The project is structured around three closely related areas: the creation of an accessible digital platform, the design of inclusive educational courses, and the creation of an online library of music scores in Braille. The aim is to create an integrated educational ecosystem in which technology, pedagogy and accessibility work in synergy to ensure equal opportunities for music learning at European level.

At the heart of the proposal is the recognition of the educational value of music in Braille, which is not simply an alternative form of reading, but a fundamental means of independent access to musical content.

2 Methodology

In order to develop innovative tools capable of improving the accessibility of music studies for blind and visually impaired students, the MuVie project has adopted an integrated methodological approach, divided into two closely linked phases.

The first phase involved a state-of-the-art analysis and comparison of the pedagogical methods and tools currently available for teaching music to visually impaired students. This part of the work included a reflection on the integration of the most advanced digital technologies, with a particular focus on the use of artificial intelligence in the transcription, reading and adaptation of musical scores. The aim was to assess the potential of intelligent digital solutions to make music learning more accessible, efficient and adaptable.

The second phase involved a systematic needs analysis, carried out through a structured survey of students in the four partner countries: Italy, Germany, Poland and Cyprus. The main survey tool was a questionnaire consisting of both open and closed questions aimed at collecting quantitative and qualitative data on key aspects of the educational experience. These included access to and use of Braille music notation, availability and use of assistive technologies, level of digital literacy, school and family support, and students' approach to music study.

To ensure a consistent and evidence-based survey, the research directly involved blind and, in some cases, partially sighted students in the different participating countries, with the aim of collecting at least twenty responses for each national context. However, the uneven distribution of this population and the lack of official lists made it impractical to use a classical statistical sampling method. A purposive sampling method was therefore adopted, based on professional networks, direct contacts, industry associations and voluntary availability.

The questionnaires were collected through a variety of channels. Information messages were disseminated in thematic groups (such as WhatsApp lists of blind teachers or musicians), individual telephone contacts were made with users and teachers already known to the partners, and focus groups or interviews were organised by telephone and videoconference, which made it possible to reach about two-thirds of the total number of responses. This approach not only clarified the context and objectives of the project, but also encouraged more informed participation, reducing the risk of approximate or misunderstood responses.

The data analysis focused primarily on the type of educational context attended by the students: inclusive schools, where students with visual impairments are integrated into mainstream classes, and specialised institutions, which offer a more structured educational programme. This criterion made it possible to identify more precisely the factors that influence effective accessibility to music education and to highlight elements that can be replicated or improved in the different school systems involved.

3 Results

The survey involved 91 blind or visually impaired students aged between 8 and 59, from both specialised institutions and inclusive schools, with a significant concentration in the 8–19 age group, but with a not insignificant number of young adults and adults who continue to study music through academic, professional or amateur courses. The decision to include participants beyond compulsory school age made it possible to extend the observation to musical experiences throughout the entire lifespan, offering a broader and more realistic view of how people with visual impairments access music.

The diversity of ages and educational contexts made it possible to collect comparable and, at the same time, differentiated data. The information obtained highlights both common structural issues and effective practices already in place, which are developing in environments capable of adequately supporting music study.

One of the most significant problems that emerged concerns access to teaching materials in Braille. Only 11% of students say they are very satisfied with the resources available; 23% describe them as unsatisfactory and a further 10% as very unsatisfactory. Furthermore, over 75% do not know where to find digital Braille sheet music online, and 65% say they have never used digital libraries dedicated to accessible music. This data highlights a shortcoming not only in the availability of resources, but also in their dissemination and in the ability of the education system to make them known and accessible to users.

The picture becomes even more complicated when looking at the actual study methods used by students. In many cases, music learning is based on informal strategies: listening by ear, voice recordings provided by teachers, personalised notes, and home transcriptions. While these methods demonstrate strong individual motivation, they are not always sustainable over time nor do they guarantee equal educational opportunities. Only 45% of respondents use digital tools to read or write music, while the remaining 55% continue to use traditional methods, often due to the lack of suitable tools or technical training. Despite this, there is evidence of resilience and autonomy: many students have developed personal techniques for studying music, demonstrating creativity and adaptability.

The analysis of the data also highlights a clear distinction between educational contexts. In specialised institutions, where the teaching of musical Braille is structured within the curriculum and there are trained teachers and specific resources, students report a higher level of satisfaction, as well as more regular and in-depth use of musical material. In these environments, access to music study is more continuous and systematic, fostering more solid and structured learning experiences.

However, positive cases have also been reported within inclusive schools, mainly linked to the personal commitment of teachers and the ability of the school context to adapt. Some teachers, even without specific training in musical Braille or the use of assistive technologies, manage to build personalised programmes, using the available resources creatively. In these situations, the spirit of collaboration between teachers, families and students can compensate, at least in part, for the lack of structured tools, allowing for a meaningful musical experience. There have also been reports of contexts in which students receive support from mediators, assistants or support teachers who are sensitive to the issue of music and able to facilitate inclusion even in the absence of dedicated resources.

One of the areas where the critical issues are most evident is in the training of music teachers. Many students report that their teachers say they do not have adequate knowledge of Braille music notation, do not have up-to-date materials and do not know how to integrate assistive technologies into their daily teaching. This hinders not only the use of digital tools, but also the possibility of building accessible and continuous music education. Despite this, in several cases, teachers are very willing to undergo training and refresher courses, showing a genuine interest in more inclusive practices.

Alongside systemic difficulties, the survey also revealed very clear and articulate expectations on the part of students. Many request a greater variety of accessible musical repertoire: scores for solo instruments, vocal pieces, classical music, jazz, pop, progressive exercises and materials designed for different levels of proficiency. There is a need not only for Braille transcriptions, but also for alternative formats – such as simplified notation, audio descriptions and scores with audio support – to enable those who are unfamiliar with musical Braille to access musical content. In particular, those with visual impairments also request hybrid solutions, with high-contrast or enlargeable visual scores accompanied by audio support or tactile feedback.

Another positive aspect is the presence, albeit minority, of schools and associations where accessibility to music is addressed in a collaborative manner. In some cases, families take the initiative to create tailor-made materials, while in others, students themselves experiment with tools and software, sharing their experiences with their peers. Some schools manage to integrate accessible collective musical activities, such as choirs, instrumental groups or guided listening workshops, which offer a concrete opportunity for participation. This type of experience shows that, even in the absence of a fully equipped system, there is room for manoeuvre to create inclusive pathways, provided that there is coordination between the various educational figures and a clear focus on accessibility.

4 Discussion

The analysis of the data collected in the four partner countries offers a complex and multifaceted picture, highlighting not only the technical difficulties related to the accessibility of music for blind and visually impaired students, but also cultural, organisational and educational issues. The critical issues that emerged – such as the limited availability of accessible digital resources, the fragmentation of available materials, the inconsistent use of Braille music notation and the often inadequate training of teachers – should not be seen as isolated obstacles, but as symptoms of a system that struggles to accommodate sensory diversity in music education in a structured way.

In this context, the approach adopted by the project is particularly significant. From its inception, MuVie has been designed as an initiative capable of connecting various levels of educational intervention: from the infrastructural dimension (the lack of usable digital resources) to the methodological dimension (the lack of familiarity with musical Braille) to the more relational dimension (the need to strengthen collaboration between teachers, students, families and institutions).

A particularly important element emerged from the comparison between different educational contexts. Although specialised institutions generally offer more structured courses and specific technologies, significant experiences have also been reported in many inclusive schools. In these cases, the personal commitment of teachers, organisational flexibility and the active involvement of families compensate, at least in part, for the absence of dedicated tools. This suggests that inclusion is not the exclusive prerogative of specialised structures, but can also be achieved in mainstream contexts, provided there is adequate awareness, training and support.

On the technological front, the introduction of advanced digital tools and the use of artificial intelligence represent a particularly promising avenue for development. Automation in Braille transcription, simplified access to interactive scores, and customisation of content based on user level are all actions that respond to concrete needs, as documented by the questionnaire responses. But beyond technological innovation, what emerges strongly is the need to make students active protagonists of their own learning path, thanks to tools that enhance autonomy and self-determination.

Finally, one of the most significant prospects outlined by the survey is the creation of a platform as a shared educational space, open not only for consulting materials, but also for their creation, adaptation and dissemination. In this sense, MuVie goes beyond the logic of a passive repository and presents itself as a dynamic laboratory, where users and operators co-design educational responses that are better suited to actual needs. The involvement of students and teachers in the production of content, the adoption of alternative formats and the possibility of customising study paths are the prerequisites for a truly accessible, flexible and inclusive music education.

5 Conclusions

The work carried out as part of the MuVie project has highlighted how accessibility to music for blind and visually impaired students cannot be addressed with partial or occasional solutions. The research has shown that, alongside well-known problems such as a lack of resources, the complexity of musical Braille and the limited availability of assistive technologies, there is a more pressing need for structural change in the way music education is conceived in educational contexts.

MuVie's proposal moves in this direction, articulating an integrated model that combines intelligent technologies, cooperative digital environments and flexible training courses. The platform under development will not only be a library of accessible sheet music, but a true educational ecosystem in which students, teachers and operators can interact, learn, produce and share musical content. The adoption of artificial intelligence-based tools, the ability to navigate interactive scores, generate automatic transcriptions or access simplified materials for different levels represent a quantum leap in the field of accessible education.

However, the value of the project lies not only in the tools it produces, but in the educational vision it promotes. MuVie demonstrates that inclusion is not an accessory feature to be applied at a later stage, but a design criterion to be adopted from the outset. Integrating accessibility into music education means recognising diversity as a generative element, capable of enriching the educational landscape with new practices, new languages and new forms of participation.

From this perspective, music is not just a discipline to be learned, but a space for expression and relationship that can become truly shared if supported by an inclusive educational structure. MuVie thus paves the way for a transformation that is not only technical but also cultural, proposing a paradigm in which technology, equity and innovation work together to build a European education system that is more just, open and accessible to all.

Acknowledgement. This work has been supported by the Grant Agreement KA220-BY-24-24-254847 of the Erasmus+ Programme in the European Union, managed by Nationale Agentur Erasmus+ Schulbildung, Bonn, Germany.

References

1. European Agency for Special Needs and Inclusive Education. Official Website. https://www.european-agency.org/
2. Gorbunova, I., Govorova, A.: Music computer technologies in informatics and music studies at schools for children with deep visual impairments: from the experience. In: Pozdniakov, S.N., Dagienė, V. (eds.) ISSEP 2018. LNCS, vol. 11169, pp. 381–389. Springer, Cham (2018). https://doi.org/10.1007/978-3-030-02750-6_29
3. Inclusive pedagogies in music education: A comparative study of music teachers' perspectives from four countries. Int. J. Music Educ. **26**(2), 113–124 (2008).https://doi.org/10.1177/0255761407088489
4. Manirajee, L., Shariff, S., Mohd Rashid, S.: Assistive technology for visually impaired individuals: a systematic literature review (SLR). Int. J. Acad. Res. Bus. Soc. Sci. **14**(2), 1–19 (2024). https://doi.org/10.6007/IJARBSS/v14-i2/20827
5. Nicotra, G., Quatraro, A.: CONTRAPUNCTUS Project: a new computer solution for braille music fruition. In: Proceedings of the International Conference on Computers Helping People with Special Needs (ICCHP 2008), pp. 303–309. Springer, Heidelberg (2008). https://doi.org/10.1007/978-3-540-70540-6_45
6. Technology and Inclusivity in Music Education, SMS AEC Music. https://sms.aec-music.eu/digitisation/technology-and-inclusivity-in-music-education/

MuVie: Braille Jazz Notation, Digital Library and Voice Interaction for Accessible Music

Anna Nicotra[1(✉)] and George Ioannidis[2]

[1] Siena Jazz University, Fortezza Medicea, 10, 53100 Siena, Italy
info@sienajazz.it
[2] IN2 Digital Innovations GmbH, Auf Dem Hasenbank 23A, 88131 Lindau, Germany
info@in-two.com

Abstract. The Erasmus + project MuVie promotes accessibility in music education for blind students through innovative digital tools. A survey of students, teachers, librarians and musicians in Italy, Cyprus, Germany and Poland focuses on issues such as the complexity of Braille notation, the limited availability of assistive technologies and the lack of accessible digital scores. The survey also explores the need for specific tools to support the teaching of music to visually impaired students. There is also a growing interest in jazz music among blind music students. To address these needs, among other initiatives, a new method for writing jazz chords in Braille has been introduced and integrated into the Braille Music Editor 2025 software, with support for importing and exporting in the MusicXML format, improving compatibility with major music software. In parallel, the project is developing a digital archive of Braille scores and exploring solutions based on Artificial Intelligence (AI) to optimise music reading, making music education more inclusive and autonomous, with a particular focus on jazz repertoire.

Keywords: Braille Music · Jazz braille · Inclusive Music Education · Assistive Music Software · Digital Music Resources

1 Introduction/Background

At the beginning of the 19th century, Louis Braille developed a tactile reading and writing system for the blind and also introduced a special code for musical notation. This important achievement finally made reading and writing music accessible to blind musicians, giving them the opportunity to pursue educational and artistic paths independently.

However, Braille music notation requires special transcriptions and the production of specific materials, which are often limited in availability and repertoire variety. With the development of information technology and the proliferation of digital devices, there has been a strong need for electronic sheet music structured in a way that makes it easier to read, explore and use. To this end, specific software modules and digital formats are being developed that allow flexible navigation of the score, bringing the experience of

K. Mavrou and P. Encarnação (Eds.): AAATE 2025, *Technology for Inclusion and Participation for All: Recent Achievements and Future Directions*, pp. 421–428, 2025.
https://doi.org/10.1007/978-3-032-01628-7_50

blind musicians closer to that of sighted musicians, especially in the context of study and live performance.

This scenario also includes the need to make accessible the jazz repertoire, a musical genre that has very different notational characteristics from classical music. Jazz is based on the use of lead sheets that contain chord symbols, essential rhythmic cues and plenty of space for personal improvisation. This concise and symbolic structure is difficult to represent in Braille, especially as there is no uniform symbolism for chords. While sighted musicians can quickly orient themselves within the structure of a piece and grasp its harmonic and rhythmic components, for blind musicians Braille reading remains sequential and requires greater effort to connect and interpret musical information.

The development of inclusive music resources has advanced considerably through both technological innovation and international standardization efforts, particularly in the domain of Braille music notation and digital sheet music exchange. These advancements aim to empower visually impaired musicians by enhancing access to written music through harmonized formats, comprehensive transcriptions, and digital tools.

One foundational contribution is the Standard Music Notation Code developed by the American Printing House for the Blind (APH), which provides an up-to-date framework for transcribing print music into braille [1]. Updated to align with Unified English Braille (UEB) standards, this publication introduces new notation systems and offers clearer, unified guidance for braille music transcribers.

Similarly, Bettye Krolick's New International Manual of Braille Music Notation represents a milestone in establishing globally accepted norms for braille music transcription [2]. Based on collaborative agreements reached between 1982 and 1992, this manual encapsulates a broad spectrum of standardized signs and their usage, making it a foundational reference for professionals working across linguistic and musical traditions.

The UK Association for Accessible Formats (UKAAF) has further supported the UEB transition with its 2019 guidelines, detailing how braille music should now be formatted and presented in the UK context [6]. These documents collectively highlight the international momentum toward standardized transcription practices, enabling consistency and interoperability for learners and professionals alike.

Complementing these transcription frameworks are repositories that provide free and accessible music resources, such as the RNIB Library [4]. Offering a vast selection of music scores in braille and other accessible formats—including classical, jazz, and popular genres—the RNIB Library ensures that blind and partially sighted musicians can engage with a full spectrum of musical works.

On the technological frontier, the CONTRAPUNCTUS project introduced a computer-assisted approach to tackling the often "anti-musical" complexity of braille music syntax [3]. Designed to gradually introduce learners to braille music through digital aids, the system represents a pedagogical and technological intervention that merges user-centered design with accessible content delivery.

Simultaneously, MusicXML, developed by the W3C Music Notation Community Group, serves as the de facto open standard for digital music notation exchange [7]. By enabling seamless interoperability among more than 250 applications, MusicXML bridges the gap between visual and accessible formats, facilitating conversions and storage across digital platforms. This standard plays a pivotal role in integrating accessible formats into mainstream music software ecosystems.

The practical application of accessible notation is also reflected in printed collections such as The New Real Book [5], which, while not explicitly designed for the visually impaired, exemplifies clear, musician-approved formatting. Such resources remain important benchmarks for quality and usability that inform accessible adaptation processes.With these needs in mind, the European MuVie project has begun to develop specific digital tools and encoding standards to improve access to music, even in the most complex genres such as jazz, and to promote a truly inclusive, dynamic and contemporary music education.

2 Methodology

The first phase of the MuVie project involved an in-depth survey to understand the main difficulties that blind students and musicians face when using sheet music. Among the most recurring needs was a strong demand for a significantly greater number of scores in accessible electronic format, not only for the classical repertoire but also for jazz and pop music. Equally clear is the need for more immediate and easier access to musical content.

In order to respond effectively to these needs, the MuVie project has launched a major project to transcribe music into Braille using the BMML (Braille Music Markup Language) electronic format, designed for digital archiving of music in Braille and interoperability between different systems. At the same time, two new add-on modules have been developed for the Braille Music Editor 2025 (BME) software, an editor specifically designed to manage and edit music in Braille. The software uses scripts that are fully compatible with major screen readers such as JAWS and NVDA, ensuring accessible and smooth interaction with digital music content.

The first module introduces the ability to enter jazz chords in Braille, using a standardised symbolism that includes 140 internationally recognised chord symbols. This feature allows blind users to write, read and listen to chords directly, facilitating harmonic study and musical performance. Compatibility with the MusicXML format further extends the possibilities for exchange and collaboration between blind and sighted musicians, facilitating the sharing of materials in educational, professional or amateur contexts.

The second module involves the development of an innovative hands-free control system for use with BME during musical performance. This feature allows users to navigate the score and interact with the editor without using their hands to read the Braille line. This is particularly useful for instrumentalists, such as pianists, who can consult the musical text without interrupting their performance to operate manual controls. The result is greater continuity, fluidity and naturalness in learning and practising music.

3 Results

3.1 Digital Library of Accessible Music

The literature review and the results of the MuVie survey highlight a persistent problem in inclusive music education: the significant disparity in the availability of accessible sheet music for blind musicians compared to the wide range available to their sighted colleagues. This structural deficiency is a fundamental barrier to equitable access to musical resources and, consequently, to learning and professional practice for blind students and musicians.

The problem is exacerbated in the field of jazz music, where notation has peculiarities that differ significantly from traditional music notation. Jazz lead sheets, characterised by chord symbols and basic rhythmic indications, favour the personal interpretation and improvisation that are characteristic of this musical genre. Translating this system of notation into Braille presents particular challenges because, unlike classical music notation, there is no universally accepted system for encoding jazz chords in Braille.

In response to this critical issue, the MuVie project will develop a digital library of 5,000 Braille pages of music scores in BMML format, covering a variety of genres including jazz. Although this is a limited collection compared to the extensive corpus available to sighted musicians, this archive represents a significant advance in many ways. The library has been designed using the BMML (Braille Music Markup Language) format, which ensures high interoperability with different reading systems and music software through its instant conversion to the widely used MusicXML format, thus promoting collaboration and inclusion in educational and performance contexts.

3.2 Jazz Chord Management

In order to ensure that jazz scores are transcribed correctly, a specific module for the representation of jazz chords was developed. The first phase of this activity involved a comparative analysis of the main braille notation guides and manuals (APH, NLS, Krolick) and the jazz standard represented by the New Real Book. Next, music notation software such as Finale, Sibelius and MuseScore were examined to understand how each of them handles the representation of chord symbols in their libraries. The information gathered from these sources enabled us to create a preliminary list of over 500 musical signs and symbols used in jazz notation, many of which are equivalent but represented in different ways.

The iterative process, carried out in collaboration with jazz composers and musicians, resulted in the standardisation of 140 jazz chord symbols, uniquely defined to ensure clarity and accessibility. For each of these 140 chords, the corresponding Braille notation has been created, encoding each symbol in a clear and consistent manner to ensure maximum accuracy and ease of use for the user.

Let's take a closer look at the functions implemented in the BME programme for entering chords into the score. First of all, all 140 chord symbols are available in a list accessible via an input window (Fig. 1).

After writing the notes, the user must first position himself at the end of each bar and use the Braille characters 6.36 to indicate the beginning of the chord section. The

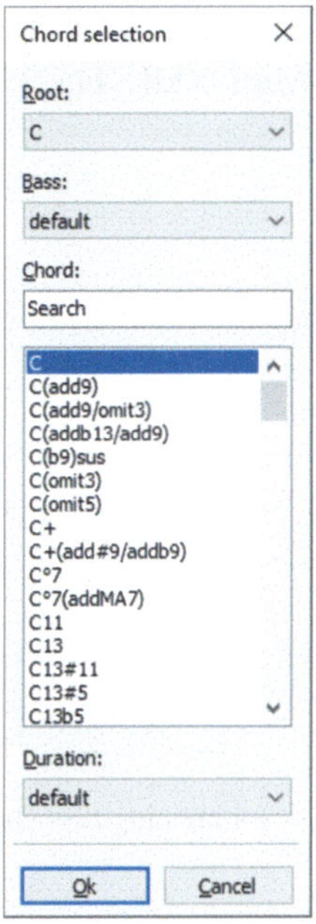

Fig. 1. Window for entering chords in the BME programme.

chord can then be entered using the menu, which opens a dialogue box. In this window, the user can set various parameters: the keynote (the fundamental of the chord), the bass (if different from the fundamental), the type of chord (selected from the list available) and the duration of the chord, which is automatically calculated by the programme if not specified. The user can proceed in the same way for all the chords in the bar and then move on to the next.

Thanks to the 'play notes' function, you can hear the corresponding sound each time the cursor moves over a note, interval or rest, improving interaction with the score. You can also hear the sound of the chord, making it easier to understand musical harmony (Fig. 2).

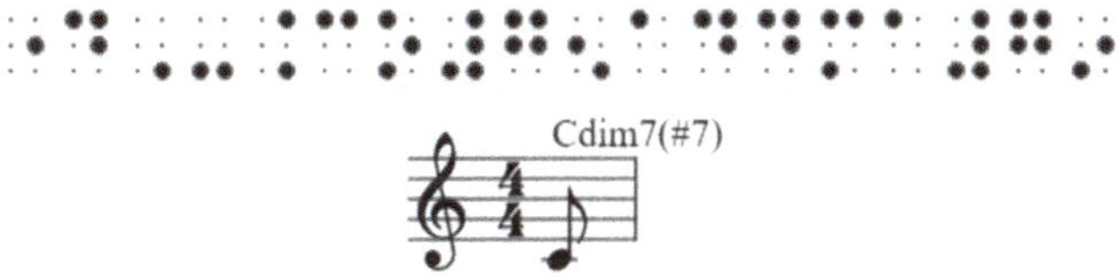

Fig. 2. Representation of a chord in Braille and traditional graphic notation

The BME also includes a chord library that allows you to create your own set of jazz chords. A window allows the user to add, delete or modify a chord, with the option of expanding the list of available chords. To create a new chord, you enter the name of the chord, the notes that make it up (to allow correct sound reproduction), the Braille representation and the chord text, which is useful for speech synthesis and for conversion to traditional graphical format via MusicXML (Fig. 3).

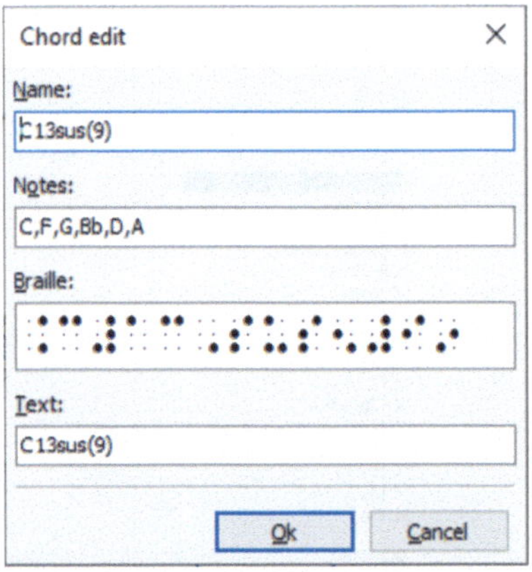

Fig. 3. Window for creating new chord symbols

3.3 Hands-Free Control in the BME Editor

Sighted musicians can quickly access and locate any point in the score by visually exploring the entire score. This mode of perception allows non-sequential and immediate access to musical information, which greatly facilitates orientation within the score and the contextualisation of individual elements within the overall structure.

Blind musicians, on the other hand, must read Braille with their hands, which severely limits their ability to play and read simultaneously. This constraint imposes a mode of score reading that is necessarily sequential and discontinuous with respect to performance, requiring alternation between moments of reading and moments of playing, with obvious slowdowns and greater effort during study.

The module allows the user to speed up the search for and reading of music through voice commands and to receive information through the auditory channel. This mode of interaction allows the user to explore the score with a freedom comparable to that of a sighted musician.

The module allows a wide range of interactions through intuitive voice commands, structured according to functional categories that reflect the typical needs of musical practice. The voice interface has been designed around four main areas of interaction: navigating within the score, controlling playback, changing interpretations and analysing the score.

When navigating the score, the system recognises commands such as 'Go to bar 48 of the right hand', 'Go back to the beginning', 'Go to section B' and 'Go to the chorus'. These commands allow non-linear access to the score, emulating the ability of sight-reading musicians to quickly move their gaze to any point in the score. This feature is particularly useful for studying specific sections or comparing musical elements that are far apart in the score.

In terms of playback control, the user can give instructions such as 'Play to end', 'Play to bar 52', 'Play left hand only' and 'Repeat last four bars'. This category of commands facilitates selective listening and targeted repetition of specific passages, effectively supporting the process of learning and memorising the score. The ability to isolate individual components of the composition (such as the left or right hand) is also an important tool for analysing and studying the relationships between different parts.

In the area of interpretative changes, the system recognises commands such as 'Set metronome to 120 BPM', 'Increase tempo by 10%', 'Decrease tempo by half' and 'Play only the chords of the current measure'. These functions allow you to experiment with different interpretations without interrupting the flow of the performance to change parameters manually. The real-time manipulation of elements such as tempo, dynamics and articulation is a significant aid to exploring the score.

Finally, for score analysis, users can request functions such as 'Isolate the main melody', 'List all dynamics present' and 'Identify the key of this section'. These advanced commands provide sophisticated analytical tools that facilitate structural and harmonic understanding of the composition, supporting both learning and interpretation through detailed analysis of the elements of the score.

In order to use this new way of navigating and interacting with the score, it is necessary to have Braille scores in electronic format, i.e. in BMML format. The online library described above is therefore not a simple static archive, but becomes functional by enabling the use of these new modes of reading and navigation, offering an immersive and multidimensional educational experience.

4 Conclusions/Implications for the AT Field

The MuVie project demonstrates how assistive technologies can evolve from tools designed solely to compensate for a disability to solutions capable of innovating educational and cultural methods, introducing new ways of accessing, learning and participating that enrich the entire educational system. The solutions developed in the project are not limited to replicating existing practices in an accessible way, but propose alternative interaction models that are natively designed to meet the needs of blind users in an effective and sustainable way.

One of the most important aspects is the standardisation of accessible digital formats, such as BMML, and their interoperability with widely used formats, such as MusicXML. This combination allows the creation of resources that can be used and shared by different users and platforms, promoting inclusion even in mixed environments where sighted and blind users coexist.

Another innovative contribution is the introduction of adaptive speech interfaces that allow navigation, analysis and musical interaction using speech. This mode, which also incorporates elements of artificial intelligence, is not limited to 'making the existing accessible', but opens up new forms of musical experience that can be extended to other educational and creative fields.

The project also highlights the importance of a systemic and integrated approach, where content, tools, skills and learning environments are designed as a coherent ecosystem. The effectiveness of assistive technologies, as demonstrated by MuVie, depends largely on their ability to adapt to real learning contexts and on the possibilities offered to users to personalise their usage strategies.

Another key aspect is the multidisciplinary collaboration fostered by the project: developers, educators, musicians, librarians and accessibility specialists have worked together to co-design tools based on real needs. This operating model can inspire future developments in assistive technologies and promote greater inclusion in other areas of education and culture.

Acknowledgement. This work has been supported by the Grant Agreement KA220-BY-24–24-254847 of the Erasmus+ Programme in the European Union, managed by Nationale Agentur Erasmus+ Schulbildung, Bonn, Germany.

References

1. American Printing House for the Blind (APH): Standard Music Notation Code. APH, Louisville, KY (2015)
2. Krolick, B.: New International Manual of Braille Music Notation. National Library Service for the Blind and Physically Handicapped, Library of Congress (1996)
3. Nicotra, G., Quatraro, A.: CONTRAPUNCTUS Project: a new computer solution for braille music fruition. In: Miesenberger, K., Klaus, J., Zagler, W., Karshmer, A. (eds.) Computers Helping People with Special Needs. ICCHP 2008. Lecture Notes in Computer Science, vol. 5105. Springer, Heidelberg (2008). https://doi.org/10.1007/978-3-540-70540-6_45
4. Royal National Institute of Blind People (RNIB): RNIB Library – Accessible Books and Music Scores (2025). https://www.rniblibrary.com/iguana/www.main.cls?p=b3ba52c6-5bac-4699-afb9-0dfb99409462&v=fff27ec4-b717-47c0-838f-bf5e2c5444ab
5. Sher, C. (ed.): The New Real Book, 1st edn. Sher Music Co., Petaluma (1988)
6. UK Association for Accessible Formats (UKAAF): Braille Music in Unified English Braille. April 2019 (2025). Available at: https://www.ukaaf.org/wp-content/uploads/2020/02/b017-ukaaf-braille-music-unified-english-braille-april-2019.pdf
7. W3C Music Notation Community Group: MusicXML: The Standard Open Format for Notation Exchange (2021). https://www.musicxml.com

Empowering Inclusive and Multimodal Practices in Preschool Music Education: Exploring the Assistive Technology Potential of Synth4kids Software

Yannis Mygdanis[1](✉) and Christina Charalambidou[2]

[1] National and Kapodistrian University of Athens, Athens, Greece
YMygdanis@music.uoa.gr
[2] European University Cyprus, Engomi, Cyprus
c.charalambidou@external.euc.ac.cy

Abstract. This study examines the design and implementation of educational scenarios created with Synth4kids, an innovative music-educational software designed for preschool education. Synth4kids offers an intuitive environment that fosters musical creativity, inclusion, and differentiated instruction through emerging and assistive capabilities. As an original and innovative tool designed for preschool education, it bridges accessibility gaps while promoting joyful, multisensory music-making. The intervention consisted of twelve weekly sessions, each lasting approximately 45 min, for four male children aged four to six years, at Baby Maestro Cyprus. Sessions were structured to integrate the AT and ET features of Synth4kids. In line with Universal Design for Learning principles, educational activities have been crafted to boost musical engagement in young children via personalized, play-based interactive experiences. The findings indicate that the early integration of assistive features, such as gaze control, motion tracking, tactile interfaces, and chromesthesia, significantly enhances accessibility, creativity, and social participation among learners with diverse profiles. Notably, incorporating Synth4kids into music teaching and learning activities demonstrates how AT tools promote agency, autonomy, and collaborative creativity, fostering new pedagogical approaches that enhance equity and engagement in music education. Overall, this study reveals that the thoughtful use of digital technologies can transform early childhood music education into an inclusive and dynamic learning environment.

Keywords: Music Education · Synth4kids Music Software · Assistive Technology · Emerging Technologies · Music Inclusion

1 Introduction

Emerging technologies (ETs) and digital media are continuously transforming the ways people interact with music, broadening the possibilities for musical expression, creativity, learning, and education [11, 19]. Today's children, as digital natives, grow up

K. Mavrou and P. Encarnação (Eds.): AAATE 2025, *Technology for Inclusion and Participation for All: Recent Achievements and Future Directions*, pp. 429–436, 2025.
https://doi.org/10.1007/978-3-032-01628-7_51

in multimodal, technology-rich environments that shape their musical experiences in fundamentally different ways compared to previous generations [13, 20]. Consequently, integrating educational technology into music education has been recognized as a critical factor in enhancing student participation, engagement, and the construction of inclusive learning environments [9, 15, 16].

Despite progress, inclusion and equity in music education continue to be challenging [8, 10, 16]. Traditional teaching methods, often centered on Western classical traditions, exclude students with diverse needs [12, 18]. Utilizing Assistive Technologies (ATs) and multimodal strategies can bridge these gaps, offering personalized pathways for musical engagement [6]. Educational technology provides innovative methods for engaging with musical content, facilitating instruction that accommodates learners' preferences, cognitive styles, and modes of expression [1, 2]. In early childhood settings, digital musical tools can promote equitable education from the outset [16]. Research shows that preschoolers thrive in multimodal environments that foster exploration, creativity, and social interaction, extending beyond conventional teaching methods [17]. Additionally, features in music software that enhance AT support participation for all learners, promoting inclusive education pathways [16].

This study focuses on creating and implementing educational scenarios using Synth4kids music-educational software in preschool environments to investigate how its AT features can promote inclusion, participation, and equity in early childhood music education. The subsequent sections examine how ATs and Universal Design for Learning (UDL) principles can transform early childhood music education.

2 Assistive Technology in Preschool Music Education

AT encompasses a range of tools, devices, and software designed to support individuals with diverse learning profiles by enhancing accessibility, participation, and independence [7]. AT can bridge the gaps created by traditional, monomodal teaching approaches in music education, providing new pathways for all students to engage meaningfully with musical content [15]. In preschool education, where foundational experiences shape children's later engagement with learning, the integration of AT is particularly critical. At this developmental stage, children exhibit various learning styles, sensory preferences, and cognitive profiles [20]. Educational music tools that incorporate AT features enable young learners to access musical activities through multiple modes of interaction—auditory, visual, kinesthetic, and tactile—thereby fostering differentiated and inclusive learning environments [6].

2.1 Universal Design for Learning in Early Childhood Music Education

Integrating AT features in preschool music education aligns with UDL principles, emphasizing flexibility and accessibility for all learners [16]. UDL recommends various methods of representation, engagement, and expression to accommodate diverse needs and learning styles. [4]. In music education, this means presenting concepts in a multisensory manner through notation, colors, gestures, sounds, and tactile feedback [16]. This multimodal approach enables children to engage with musical ideas using their unique sensory strengths. Varied engagement methods, including gaze control, hand gestures, motion tracking, and tactile interfaces, foster autonomy and motivation [16]. Lastly, diverse means of expression enable children to demonstrate their musical understanding by improvising with motion tracking or composing using color-coded systems inspired by chromesthesia. UDL empowers every child, regardless of cognitive, physical, or sensory characteristics, to engage meaningfully with music, developing agency and confidence through inclusive education experiences.

The early introduction of AT in music education supports the goals of inclusion, equity, and social participation. Research indicates that technology-enhanced musical environments promote participation and significantly enhance social, cognitive, and emotional skills in children with diverse abilities [17]. By creating accessible and engaging learning environments aligned with UDL principles, AT is essential in building inclusive early childhood music education settings [6].

2.2 Emerging Technologies and Inclusive Music Learning

ETs have transformed music creation, experience, and learning, creating new opportunities for inclusivity in music education [15, 19]. To that end, DAWs, interactive music apps, and web synthesizers offer flexible engagement with music, contrasting with traditional training and Western classical education [1, 2]. In classrooms, ET enables educators to cultivate various pathways for musical expression, addressing students' diverse needs [17, 20]. Incorporating multimodal interactions—auditory, visual, tactile, and kinesthetic—empowers students to engage in music without the limitations of traditional literacies, such as notation or instrument mastery [18]. This shift fosters inclusive teaching environments where every child can engage meaningfully with music.

ET also fosters differentiated instruction by enabling personalized learning experiences. Applications that facilitate sampling, sequencing, remixing, and composing music through intuitive, visually based interfaces make musical creativity accessible to learners from diverse backgrounds and abilities [6]. In this context, informal, DIY music practices can be adapted into educational activities that reflect students' real-world musical experiences and cultural identities [1]. When students' musical literacies and experiences are recognized and valued in the classroom, inclusive and equitable music learning environments are more likely to emerge [3, 9]. ETs serve as catalysts for such recognition, offering diverse entry points for musical engagement and promoting active participation, creativity, and social connection among learners. Therefore, harnessing novel technologies in music education transcends mere modernization, emerging as an essential pedagogical strategy aimed at fostering more inclusive, equitable, and profound learning experiences for all students.

3 The Case of Synth4kids as a Tool for Assistive Technology

Synth4kids is a music-educational software designed to support preschool children's musical engagement within an intuitive and multimodal environment [5, 15]. Emphasizing differentiated learning and inclusive practices, Synth4kids incorporates innovative tools that enhance accessibility and participation. Its multimodal interaction framework establishes it as an AT-oriented tool [16]. Users can interact with sound through a combination of visual, tactile, and auditory stimuli, accommodating diverse sensory and cognitive profiles. The user-friendly interface, characterized by simple icons, color coding, and immediate auditory feedback, promotes autonomy among young learners while minimizing barriers related to reading skills, motor control, or traditional musical notation [16, 17].

Synth4kids utilizes chromesthesia-inspired visual feedback, which links sounds to colors [14]. This fosters intuitive musical understanding and enhances expression for children with diverse learning styles, particularly those who require visual support. Additionally, it includes advanced interaction modes to improve accessibility for children with diverse abilities [14]. These include eye-tracking control, which enables users to trigger musical events through gaze direction; face-tracking and head movement-based controls, allowing for music interaction without manual input; and click-less interaction mechanisms that eliminate the need for fine motor precision [15]. Furthermore, Synth4kids can connect to tactile interfaces, such as Makey Makey, creating customized and differentiated tactile musical instruments [14, 15]. This functionality enables educators to design unique, accessible sound-making devices tailored to their students' physical and sensory needs, thereby enhancing inclusive participation and active musical engagement exploration. [16] (Fig. 1).

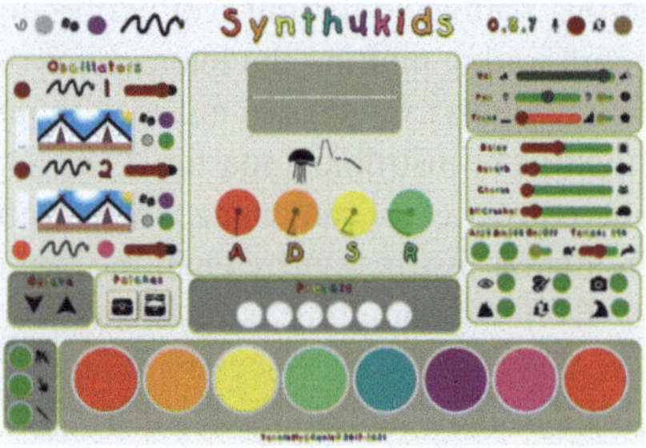

Fig. 1. Synth4kids main window

Through the integration of AT elements, Synth4kids demonstrates how ETs enrich music education for a variety of learners, offering enhanced inclusive opportunities. In addition to encouraging musical creativity, it advances wider educational objectives, fostering agency, social interaction, and cognitive development in preschool children within inclusive environments.

4 Aim, Research Question, and Sample

This study aimed to design and implement educational scenarios incorporating Synth4kids in preschool settings and explore how the integration of AT can foster inclusive participation and differentiated musical engagement in early childhood music education. The research investigated how digital multimodal environments could enhance access to musical experiences for children with varied learning profiles, promoting creativity, self-regulation, and social interaction within authentic classroom contexts. Research questions are as follows:

1. In what ways do Synth4kids' diverse interaction modalities (e.g., gaze control, motion tracking, tactile interfaces) enhance accessibility, autonomy, and embodied musical engagement among preschool children with various learning needs?
2. How do preschool children demonstrate active agency, creative expression, and multimodal interaction when engaging with the AT features of the Synth4kids environment in authentic educational contexts?

The research sample consisted of four male children aged four to six years, enrolled in the group music program for preschool and early childhood education at Baby Maestro Cyprus. Participation in the intervention was entirely voluntary and conducted with the informed consent of the children's parents or legal guardians, who were thoroughly informed about the research's aims and procedures. The children exhibited diverse developmental profiles, including cognitive, motor, and learning characteristics.

5 Methodology

This study employed a qualitative research design to investigate how AT features in Synth4kids promote inclusivity in early childhood music education. The research included four boys aged four to six who naturally engaged in intervention activities. The children exhibited different levels of prior musical experience, contributing to a varied developmental profile study.

The intervention encompassed twelve weekly sessions, each lasting approximately 45 min. Sessions were structured to integrate the AT and ET features of Synth4kids, including eye-tracking, hand-tracking, motion tracking, tangible interfaces such as Makey Makey, chromesthesia-inspired visual feedback, and non-conventional musical notation [5, 15]. Educational activities were designed in alignment with UDL principles, ensuring accessibility and differentiated engagement for all participants. Children participated in various musical activities, including sound creation, improvisation, and rhythm composition using gesture-based controls and touch-free involvement interactions.

Data collection utilized a multi-method approach to comprehensively understand the children's experiences. Diary-format structured observations were conducted throughout the twelve-week intervention, focusing on children's interactions with the AT features, engagement levels, and creative behaviors. Semi-structured interviews with educators and parents provided insights into the usability, accessibility, and pedagogical effectiveness of Synth4kids. Informal discussions with the children were also held to capture their perceptions and experiences with the digital tool in an authentic and developmentally appropriate manner. This triangulation of data sources enhanced the depth and credibility of the findings regarding the impact of Synth4kids on inclusive music learning.

6 Results and Discussion

The findings of this study confirm that Synth4kids significantly enhances accessibility, participation, and musical engagement in early childhood music education through its AT features. By integrating functionalities such as eye tracking, hand tracking, motion tracking, tangible interfaces, chromesthesia-based visual feedback, and non-conventional notation [5, 15], the software addresses common barriers to music-making and creates inclusive opportunities for children of diverse abilities to explore, develop, and perform music.

Eye-tracking emerged as a particularly impactful modality. Children could improvise and compose music without requiring fine motor skills, thereby expanding access for learners with motor impairments and promoting autonomy in music interaction. One participant expressed this vividly: "Look! My eyes are dancing with the sounds!" (FN1). Hands-free interaction enhanced cognitive engagement and independent musical decision-making, reinforcing the role of alternative input methods in inclusive music education. Motion and hand tracking further fostered kinesthetic involvement, linking bodily movement directly with sound creation. During one session, a participant joyfully swung his arms to modulate musical patterns and shouted, "I'm painting music in the air!" (FN2), illustrating the intuitive and embodied nature of musical engagement through movement.

Integrating tangible interfaces, such as Makey Makey sensors, expanded accessibility and encourages playful exploration. Children creatively transform everyday objects into musical triggers, such as bananas, aluminum foil, and sponges. After activating a sound by touching a piece of fruit, one participant exclaimed, "The banana sings when I touch its belly!" (FN3), reflecting the imaginative and multisensory experiences facilitated by tactile interaction. These approaches align with research highlighting the benefits of multisensory environments for enhancing musical understanding and participation.

Chromesthesia-based, color-coded feedback offered an intuitive way for young learners to understand pitch relationships and musical structures. By using color cues, children made independent musical choices and expressed emotional associations with sound, as seen when a child remarked, "Blue is a sleepy song, yellow is a jumping song!" (FN4). The alternative graphic-based notation system eliminated barriers to symbolic literacy and fostered composition and improvisation among the participants.

Synth4kids enhanced accessibility, offering opportunities for musical creativity and engagement. Children showed autonomy by creating original sound patterns through gaze, movement, and touch. Their interest and initiative to explore outside sessions and collaboration during tasks reflected high intrinsic motivation and musical involvement. Educators and parents noted that the playful environment empowered children to engage with music confidently. Observations indicated that even hesitant children became more active and expressive during the intervention. These findings suggest that integrating AT features into early childhood music education supports differentiated instruction aligned with UDL principles, fostering creativity, autonomy, and social connections among learners.

The observational data were organized into three core categories: accessibility and participation, creative autonomy, and social interaction. Accessibility features, such as eye-tracking and hands-free interaction, enabled children with limited motor abilities to participate in music-making, thereby enhancing their autonomy and engagement. Regarding creative autonomy, Synth4kids' multimodal design—particularly its use of motion and color—supported spontaneous improvisation and intuitive musical understanding. Socially, tactile interfaces encouraged peer collaboration and verbal expression through playful engagement. This study shows that playful digital tools with assistive modalities enhance participation and musical agency in early education. Rather than compensating for deficits, the AT features in Synth4kids foster creative autonomy, enabling children to intuitively express musical ideas [16]. The pedagogical implications are significant; UDL alignment supports diverse profiles without limiting access. Children's engagement indicates that AT-enhanced environments can promote inclusion and redefine musical literacy, urging educators to adopt strength-based pedagogies that empower all learners. These findings align with prior research on multimodal learning in inclusive music education [6, 20]. Overall, Synth4kids demonstrates that AT and ET can transform preschool music education into an inclusive, equitable, and dynamic environment for all children.

7 Conclusion

The findings of this study highlight the transformative potential of AT in early childhood music education, stating that Synth4kids significantly enhances accessibility, inclusion, and engagement through its multimodal and intuitive design [5, 15]. Synth4kids exemplifies how AT can bridge the gap between traditional music pedagogy and contemporary digital learning environments, fostering collaborative, playful, and creative musical engagement among young learners. By eliminating conventional barriers to participation, such as the need for fine motor skills, symbolic literacy, or specific sensory modalities, AT-enhanced digital tools like Synth4kids create inclusive spaces where children of all abilities can actively explore, develop, and express themselves through music. Multimodal interaction modes, including gaze-based control, gesture-based sound manipulation, tactile interfaces, and chromesthesia-inspired visual feedback, offer multiple entry points into musical experiences, supporting differentiated instruction and UDL principles.

The implications for the AT field are substantial, particularly within the contexts of early childhood education and inclusive pedagogy. This study suggests that AT should not be viewed merely as compensatory tools for overcoming deficits, but as dynamic educational resources that can actively enhance learning, creativity, and social participation. Synth4kids demonstrates that playful, technology-enhanced musical environments can revolutionize how inclusivity is realized in educational settings, expanding opportunities for meaningful engagement beyond traditional modes of limitations.

References

1. Bell, A.P.: Dawn of the DAW: The studio as a musical instrument. Oxford University Press, Oxford (2018)
2. Brown, A.R.: Music technology and education: Amplifying musicality, 2nd edn. Routledge, New York (2015)

3. Burnard, P., Dillon, S., Rusinek, G., Sæther, E.: Inclusive pedagogies in music education: a comparative study of music teachers' perspectives from four countries. Int. J. Music. Educ. **26**(2), 109–126 (2008)
4. CAST: Universal Design for Learning guidelines version 2.2. CAST, Wakefield, MA (2018)
5. Charalambidou, C., Mygdanis, Y.: Designing and applying educational scenarios using Synth4kids musical educational software in preschool education. Int. J. Integr. Technol. Educ. **13**(3), 65–76 (2024)
6. Chrysostomou, S.: Technology in the music classroom – Navigating through a dense forest. In: Ruthmann, A., Mantie, R. (eds.) The Oxford Handbook of Technology and Music Education, pp. 105–119. Oxford University Press, Oxford (2017)
7. Daniels, H., Garner, P.: Inclusive education. Routledge, London (2013)
8. Done, E.J., Andrews, M.J.: How inclusion became exclusion: policy, teachers and inclusive education. J. Educ. Policy **35**(4), 447–464 (2020)
9. Fraser, N.: Justice Interruptus: Critical reflections on the 'post-socialist' condition. Routledge, New York (1997)
10. Haustätter, R.S.: In support of unfinished inclusion. Scand. J. Educ. Res. **58**(4), 424–434 (2014)
11. Kokkidou, M.: Musical literacies in the aftermath of postmodernism: From ignorance to awareness. In: Kokkidou, M. & Dionysiou, Z. (eds.) Musical Literacy: Formal and Informal Forms of Music Teaching–Learning, pp. 3–36. GSME, Thessaloniki (2016)
12. Kokkidou, M., Mygdanis, Y.: New approaches in "Theory of Music" courses: preliminary findings of a pilot study in conservatoire settings. Musicologica Balcanica **1**(1), 319–337 (2020)
13. Mygdanis, Y.: Virtual instruments in music teaching and learning at kindergarten-age: an educational proposal using Synth4kids web-application. In: Panagopoulos, M., Papadopoulou, A. & Giannakoulopoulos, A. (eds.) Digital Culture & Audiovisual Challenges: Interdisciplinary Creativity in Arts and Technology, pp. 141–149. CEUR Workshop Proceedings, Aachen (2021)
14. Mygdanis, Y.: Designing an educational musical software combining traditional music-pedagogical methods and emerging technologies: The case of Synth4kids. Int. J. Integr. Technol. Educ. **11**(2), 103–120 (2022)
15. Mygdanis, Y.: Design and development of the original music educational software Synth4kids and integration in the "Music Theory" course of the conservatoire education in Greece. Unpublished doctoral dissertation, European University of Cyprus, Nicosia (2023)
16. Mygdanis, Y.: Educational technology as a pathway to foster inclusion and equity in music education. In: Al-Din, S. (ed.) Handbook of Research on Advancing Equity and Inclusion through Educational Technology, pp. 153–169. IGI Global, Hershey, PA (2023)
17. Mygdanis, Y., Kokkidou, M.: Collaborative D.I.Y. music production practices in conservatoire settings: findings from a pilot distance teaching–learning project. ICT in Muzical Field **12**(2), 7–22 (2021)
18. Mygdanis, Y., Papazachariou-Christoforou, M.: Exploring the integration of maker culture activities in the theory of music course at a Greek conservatoire. J. Music Technol. Educ. **16**(1–2), 99–118 (2023)
19. Tobias, E.S.: Learning with digital media and technology in hybrid music classrooms. In: Abril, C., Gault, B. (eds.) Teaching General Music: Approaches, Issues, and Viewpoints, pp. 112–140. Oxford University Press, New York (2016)
20. Young, S.: Critical new perspectives in early childhood music: Young children engaging and learning through music. Routledge, London (2018)

Building an Inclusive Future: Creating a Master's Program in Assistive Technology Through International Course Mapping

Mabel Giraldo(✉), Fabio Sacchi, and Serenella Besio

University of Bergamo, Bergamo, IT, Italy
mabel.giraldo@unibg.it

Abstract. The increasing relevance of assistive technologies (AT) in advancing inclusion and autonomy for persons with disabilities has highlighted the critical need for robust academic training in the field. Despite international efforts—such as WHO's GATE initiative and EU-funded projects like EUSTAT—AT education remains fragmented, with significant disparities in program structure, delivery, and epistemological foundations. This study presents the results of a qualitative document analysis of 20 academic programs across Europe and internationally, examining curricular content, pedagogical models, target audiences, and theoretical orientations. Findings reveal a diverse but disjointed landscape, where biomedical and technical paradigms often prevail over inclusive, user-centred approaches. Notably, emerging trends include the integration of AI, smart environments, and interdisciplinary frameworks. In response to these insights, the University of Bergamo has developed a new Master's Degree in *Designing Accessible and Inclusive Living Environments*, aiming to bridge current gaps through a person-centred, interdisciplinary curriculum rooted in technological innovation, inclusive pedagogy, and socio-cultural awareness.

Keywords: AT education · qualitative document analysis · master's program

1 Background

Interest in assistive technologies (AT), and in their potential to foster social inclusion and enhance the autonomy of persons with disabilities, has grown significantly in recent years [1]; consequently, organizations and associations within the AT sector increasingly emphasize the urgent need to educate and train competent professionals capable of addressing the complex challenges of this evolving field. The World Health Organization (WHO) launched initiatives like the Global Cooperation on Assistive Technology (GATE) in 2014, stressing the importance of adequate training for equitable AT access. UNESCO programs since 2015 have also highlighted the need for specialized educational pathways to enhance professional skills while promoting inclusion. In Europe,

The authors thank Kinzica Soldano for her contribution to the early stages of this study.

K. Mavrou and P. Encarnação (Eds.): AAATE 2025, *Technology for Inclusion and Participation for All: Recent Achievements and Future Directions*, pp. 437–445, 2025.
https://doi.org/10.1007/978-3-032-01628-7_52

the 1994 Horizontal European Activities in Rehabilitation Technology (HEART) study identified professional training as a key to AT development, advocating collaboration between educational institutions and the joint creation of devoted educational materials [1]. Aligned with these experiences, funded by the EU Commission some projects (e.g. Deloitte@Touche, 1994; Empowering Users Through Assistive Technology, 1998; and Telematic Multidisciplinary Assistive Technology Education TELEMATE, 1998) reinforced the need for stronger education and training, revealing disparities across EU states and underlining the necessity for increased skills in AT assessment and prescription. Among these, the EUSTAT project (Empowering Users through Assistive Technology, 1998) played a pivotal role in promoting not only professional training but also the empowerment of users themselves, persons with disabilities and their families—as active participants in AT adoption and education. This approach, well known within the AAATE community, anticipated a model of lifelong learning where users are central actors in the training process, enhancing autonomy and self-determination. Additionally, the Association for the Advancement of Assistive Technology in Europe (AAATE) reaffirmed this urgency at its 2023 congress, *Shaping a Sustainable and Inclusive World*, stressing the need for stronger academic training in AT. Despite this growing awareness, AT education remains fragmented, lacking structured, interdisciplinary and long-standing university programs [2–4]. Some studies [3, 4] stress that AT training should not be limited to specialists but extended to teachers, general practitioners, and career counsellors. These *gatekeepers* significantly influence access to AT, yet a lack of training limits their ability to grant full social participation for people with disabilities [3]. The EUSTAT model also recalls that training efforts should address not only current and future professionals but also end users, thereby fostering inclusive knowledge ecosystems that support co-design and participatory practices. Addressing this gap requires comprehensive, interdisciplinary programs that integrate the various AT dimensions and cater to diverse user needs. Properly trained professionals are essential for ensuring inclusion and autonomy in line with international policies. Recognizing this, University of Bergamo (Italy) decided to deliver a Master's Degree in *Designing Accessible and Inclusive Living Environments* starting in 2025/2026. This program aims at bridging the academic gap in AT education, by training experts to tackle accessibility and inclusion challenges through instructional educational methodology coming from humanities and technological areas. The design of this Master's Degree required a preliminary map of AT training courses in European and international universities. This study presents the ongoing results of this exploratory research guided by the following research questions (RQs). RQ1: What are the existing academic courses on AT at international and European levels? RQ2: What are the main features of these programs regarding academic level, delivery mode, curriculum, subjects, teaching methods, target audiences, and professional outcomes? RQ3: If stated, what is the disability and assistive technology epistemological background of these courses?

2 Methods

Given the exploratory nature of the RQs, a qualitative document analysis [5] was conducted to draw the state of the art, categorize primary sources, and provide a broad understanding of the topic. This method involves collecting, organizing, and interpreting various textual materials, including grey literature such as reports, articles, and training course announcements. To identify academic course announcements, promotional materials, and regulations, researchers conducted a web search using the following keywords and boolean operators: "international" OR "European" AND "Academic" OR "University Programs" AND "Assistive Technologies". The Retrieved Materials Were Analysed Based on Flick's [6] criteria – authenticity, credibility, representativeness, and meaning – using a structured reading form to extract key details on academic level, delivery mode, curriculum, and target audiences. Following Morgan [5], the study proceeded through the following steps: a) searching for relevant documents; b) gathering contextual information on the documents and the institutions; c) assessing document validity [6]; d) categorizing and analysing content; e) identifying key concepts across courses, such as disciplinary focus, didactic structure, and user engagement models; f) organizing findings to support both descriptive synthesis and generative reflection, aimed at informing the design of new educational proposals. In this perspective, the new Master's Degree was not only framed on data collected and analysed through document analysis but also conceived as a response to the observed gaps, aligning with best practices and inclusive frameworks highlighted by the programs reviewed.

3 Results

The qualitative document analysis identified 20 academic courses, whose references and features are detailed in Table 1. The review of AT training programs at international and European levels highlighted a diverse educational landscape in terms of structure, disciplinary background, professional orientation, and delivery modes. Together, these elements contributed to shaping a multifaceted, though fragmented, panorama of AT education across national and institutional boundaries.

3.1 Existing Initiatives (RQ1)

AT courses include Master's Degrees (12), graduate certificates (6), Bachelor's Degree (1) and professional certification (1), but no PhD programs were found. Most programs are in the U.S. (9), while others are in Europe (7), Australia (2), the Middle East (1), and Africa (1). In Italy, only one program is available for a.y. 2024–2025. Structured Master's programs are offered by "La Sapienza" University of Rome, the University of Dundee (UK), and Jönköping University (Sweden), while institutions such as Tufts University (USA), the University of New Hampshire (USA), and George Mason University (USA) provide specialized graduate certificates, which are considered preparatory for the Assistive Technology Professional (ATP) credentialing exam administered by RESNA. The set of programs also includes courses in inclusive design, rehabilitation engineering, and disability studies with dedicated AT modules.

3.2 Main Features (RQ2)

Table 1 below provides a comparative overview of the structural and pedagogical differences among the programs identified.

Table 1. Structural and pedagogical differences among identified programs.

University/Country	Program Title & Level	Delivery Mode	Duration	Focus Areas	Hands-On Components
University of Dundee (UK)	MSc Educational Assistive Technology	Online	1–2 years	AT in education, user needs, inclusive pedagogy	Virtual labs, case-based
Jönköping University (Sweden)	MSc Product Dev. – Assistive Technology	On campus	2 years	Product design, prosthetics, co-design, quality of life	Industrial projects
University College London (UK)	MSc Rehabilitation Engineering and AT	On campus	1 year	Clinical applications, prosthetics, interdisciplinary teamwork	Labs, design projects
Sapienza University of Rome (Italy)	Master in Assistive Technology	On campus	1 year	Inter-professional AT design and prescription	Labs, project-based
George Mason University (USA)	M.Ed. Curriculum & Instruction (AT)	Online	1.5–2 years	Educational AT, UDL, IEP integration	Online labs, field tasks
University of Pittsburgh (USA)	MS Rehabilitation Science & Technology	On campus	2 years	Rehab engineering, clinical AT, policy	Clinical practicum
Tufts University (USA)	Graduate Certificate in Assistive Technology	On campus	1 year	Daily living AT, device design	Makerspace projects
University of New Hampshire (USA)	Graduate Certificate in Assistive Technology	Online	1 year	AT tools and techniques across lifespan	Toolkits, home labs
CSUN (USA)	M.S. in Assistive Technology Engineering	Online	2 years	Design and testing of AT systems	Project portfolio
CSUN (USA)	M.S. in Assistive Tech Studies and Human Services	Online	2 years	AT service delivery, law, ethics	Capstone case projects
Jade University of Applied Sciences (DE)	B.Eng. in Assistive Technologies	On campus	3 years	Biomedical tech, HCI, inclusive design	Practicums, industry links
University of Linz (Austria)	MSc Inclusive Technologies (in development)	Hybrid	Planned 2 years	Inclusive design, eAccessibility	To be confirmed

(*continued*)

Table 1. *(continued)*

University/Country	Program Title & Level	Delivery Mode	Duration	Focus Areas	Hands-On Components
Curtin University (Australia)	MSc Health Sciences (Assistive Technology)	Online	2 years	Clinical AT, disability services, health IT	Online case studies
Dublin City University (Ireland)	MSc in Assistive Technology	Online/Blended	1 year	AT assessment, education, service delivery	Practice-based modules
University of Haifa (Israel)	MA in Assistive Technology and Inclusion	On campus	2 years	Inclusive education, cognitive AT	Practicum
Georgia Tech (USA)	MS in Human-Computer Interaction (Assistive Tech Track)	On campus	2 years	HCI, accessible design, usability	Lab work, design studio
University of Illinois at Chicago (USA)	Certificate in Assistive Technology and Accessible Design	Online	1 year	Universal design, accessibility laws, AT tools	Virtual design labs
NTU Athens (Greece)	MSc Biomedical Engineering (AT modules)	On campus	2 years	Engineering, sensor tech, medical devices	Labs, field work
University of Sydney (Australia)	Grad Certificate in Assistive Technology	Online	0.5–1 year	AT planning, user needs, digital tools	Online workshops
University of Pretoria (South Africa)	MSc Disability Studies (AT specialization)	Blended	2 years	Disability rights, AT in low-resource settings	Community-based practicum

The programs listed reveal a notable contrast: European curricula tend to emphasize policy frameworks and service delivery systems, whereas North American ones focus more on certification pathways and technological development, particularly device engineering. An increasing number of programs reflect the integration of artificial intelligence, smart environments, and tele-rehabilitation, signalling a broader transformation in the field of assistive technology (AT) research and application. Courses at institutions such as UCL, CSUN, and Jönköping University feature content on robotics, AI-driven interface design, and next-generation smart assistive systems. Emerging cross-cutting themes include the incorporation of AI and connected devices, interdisciplinary modules on inclusive design and usability, and a consistent emphasis on laboratory work and co-design practices. Some programs – such as those at the University of Dundee and Tufts University – embed user participation directly into their curricula, supporting participatory design methodologies. Pedagogical strategies vary widely. Engineering-oriented tracks often adopt project-based learning (PBL), while others prioritize clinical placements, case-based reflection, or virtual simulations. Distance learning programs tend to combine asynchronous instruction with interactive laboratories or home-based assignments using AT toolkits. Structural formats vary from traditional, sequential curricula to more flexible and modular clusters that support individualized learning trajectories. Experiential learning is a defining feature across most programs, commonly realized through capstone projects, internships, or supervised design challenges. Ten programs explicitly emphasize interdisciplinary collaboration – for example, between occupational therapy and engineering – or offer opportunities for international mobility. However, only four incorporate the structured involvement of individuals with disabilities in shaping the training pathway itself, which remains a significant and unresolved challenge. The mapping also enabled preliminary estimates of annual student cohorts (ranging from 15 to 80, where data were available), levels of professional accreditation (e.g., ATP certification in the U.S.), and program longevity. In addition, seven programs integrate remote learning environments and digital laboratories, reinforcing the trend toward accessible and scalable AT education, as documented in the literature [4, 8].

3.3 Epistemological Background (RQ3)

Most programs do not explicitly address the cultural or theoretical foundations underpinning their training approaches. However, an analysis of affiliated departments and disciplinary sectors reveals that many are grounded in a predominantly medical-rehabilitation model, which shapes prevailing conceptions of disability and technology. In some instances – particularly within education- or design-focused curricula – a social or user-centred perspective emerges, though it is not always formally articulated in program descriptions. A limited number of programs reference what literature recommends as ecosystemic perspective on disability, such as the biopsychosocial model promoted by the *International Classification of Functioning, Disability and Health*, fostered as a broader conceptualization and context-dependent phenomenon. These epistemological foundations – often implicit – shape how assistive technology is framed and taught, influencing whether it is seen as compensatory or as part of a broader system promoting autonomy. Similarly, growing evidence highlights the value of co-design approaches that actively involve persons with disabilities, ensuring relevance, usability, and contextual fit.

4 Discussion

The findings underscore several critical dimensions of assistive technology (AT) education that demand targeted attention and strategic advancement. Analysis of 20 academic programs reveals that AT is fundamentally interdisciplinary – encompassing healthcare, engineering, computer science, education, design, and the social sciences – yet often lacks structural coherence and shared epistemological foundations. Moreover, similar considerations pertain to the adoption of participatory approaches in AT development, as a growing body of evidence [7] highlights the imperative of co-design methodologies that meaningfully involve persons with disabilities. Such engagement is essential to ensure that AT are not only technically functional, but also contextually appropriate, socially relevant, and aligned with the lived experiences of users. Curricular models span from narrowly biomedical and monodisciplinary tracks to more integrated frameworks that include inclusive design and rehabilitation sciences. However, the persistence of departmental silos and single-discipline governance continues to pose a risk of fragmentation, constraining innovation and cross-sector collaboration [2]. The establishment of a shared international framework for AT education could help define essential competencies while preserving contextual and cultural adaptability. Pedagogical approaches across programs are equally heterogeneous: while many employ project-based learning, co-design practices, and clinical placements, others – particularly at the certificate level – still rely predominantly on traditional lecture-based instruction [4]. The increasing adoption of digital and hybrid formats enhances access, especially for learners in low-resource settings and working professionals yet simultaneously raises important concerns about the quality and equity of practical training, particularly where opportunities for hands-on experience and direct user engagement are limited. Systematic evaluation of how remote learning environments influence skill acquisition is essential, as is the deliberate inclusion of soft skills such as communication, empathy, and participatory design [8]. Technological innovation plays a growing and dynamic role in shaping AT curricula, with emerging domains such as artificial intelligence, smart devices, robotics, and tele-rehabilitation gaining prominence. However, critical engagement with the ethical and social dimensions of these technologies remains limited. Few programs explicitly address concerns such as algorithmic bias, responsible AI development, or the risks associated with excluding and including users from the design and evaluation of intelligent systems. Embedding structured modules focused on technological responsibility, inclusive innovation, and participatory design methodologies could help close this gap [5]. Challenges related to inclusiveness and global equity persist across the educational landscape. While online holds considerable promise for widening participation, enduring barriers – such as affordability, connectivity, language, and cultural relevance – continue to disproportionately affect learners in low- and middle-income countries. A more robust commitment is needed to ensure that persons with disabilities are not only considered as recipients or subjects of instruction, but are actively engaged as learners, experts, and co-educators [7]. This requires inclusive admissions strategies, targeted financial support, partnerships with Disabled People's Organizations (DPOs), and the creation of educational pathways that value diverse forms of knowledge and lived experience. Accessibility should be conceptualized not merely as a technical specification, but as a foundational cultural and pedagogical principle that informs both curricula and learning environments [6]. Given the wide-ranging target audience of AT programs – including engineers, therapists, designers, educators, healthcare providers, and service professionals – there is a clear need for modular, adaptable program structures capable of accommodating diverse professional goals, learning rhythms, and life trajectories. This calls for the expanded use of micro-credentials, stackable modules, and flexible certification frameworks that support lifelong learning and professional requalification. Ultimately, AT education must evolve not only in response to technological progress, but in alignment with broader imperatives of autonomy, equity, and social inclusion. It must equip future professionals not simply as technically skilled practitioners, but as critical, creative, and ethically grounded agents capable of co-designing inclusive futures with – and for – the full diversity of human experience.

5 Implications for Further AT Academic Courses Field

Three overarching priorities emerge from the qualitative document analysis for the advancement of AT education: (a) the definition of shared competence frameworks at both EU and international levels; (b) the meaningful involvement of users in the co-design of training pathways; and (c) the integration of technological, ethical, and contextual dimensions within cohesive and adaptable curricula. The document analysis also underscores the importance of collaborative action among academic institutions, policymakers, and industry stakeholders to promote knowledge exchange and strengthen AT education globally. The current educational landscape is characterized by a diverse array of academic programs designed to meet the needs of multiple professional profiles. Variations in structure, delivery mode, disciplinary focus, and intended audiences underscore the interdisciplinary nature of AT and its relevance across sectors. The inclusion of hands-on components plays a critical role in bridging theory and practice, while interdisciplinary collaboration fosters comprehensive approaches to emerging challenges – particularly in relation to artificial intelligence, human-computer interaction, and inclusive design. Despite these advances, many programs continue to lack a clearly defined cultural or theoretical orientation toward disability and technology. Effective AT education must go beyond technical competence to cultivate critical understanding of the social, political, and cultural dimensions that shape the development, deployment, and accessibility of assistive solutions. In response to these challenges and opportunities, a new Master's Degree in Assistive Technology has been conceived based on an interdisciplinary framework that aligns curricular design with technological innovation, evolving policy contexts, socio-economic impact, and participatory methodologies. Drawing on the *Matching Person and Technology model* and adopting a lifelong and life-wide perspective, the program recognizes the multiplicity of life contexts that influence individual experiences with technology. The Master's Degree in *Designing Accessible and Inclusive Living Environments* is grounded in the principles of accessibility and inclusion, aiming to break down disciplinary silos by integrating perspectives from engineering, rehabilitation, pedagogy, and design. It promotes a person-centred, context-aware vision of assistive technologies – acknowledging the need for solutions that evolve alongside users' biographies, environments, and aspirations. Moving forward, the development of AT education should be anchored in a dual commitment: to technological and pedagogical innovation on one hand, and to a deep, explicit engagement with the cultural and human dimensions of assistance, inclusion, and autonomy on the other.

References

1. Fagerberg, G.: From HEART to date. Technol. Disabil. **23**(3), 183–189 (2011)
2. Lenker, J.A.: The importance of professional education in assistive technology service delivery. Technol. Disabil. **9**(1–2), 57–68 (1998)
3. Jans, L., Scherer, M.J.: Assistive technology assessment: a handbook for professionals in disability rehabilitation. CRC Press (2006)
4. Slaughter, S., Waller, S., Griffiths, L.: Trends and challenges in assistive technology education: a global perspective. Assistive Technol. J. **35**(4), 201–219 (2023)
5. Morgan, H.: Conducting a qualitative document analysis. Qualitative Report **27**, 64–77 (2022)

6. Flick, U.: An Introduction to Qualitative Research 6th ed. (2018).
7. Scherer, M.J., Federici, S.: Matching person and technology (MPT): an assistive technology assessment and service provision model. In: The Palgrave Encyclopedia of Disability, pp. 1–13. Cham: Springer Nature Switzerland (2024)
8. Berke, W., Wiseman, T.: The e-learning answer. J. Nurs. Adm.Nurs. Adm. **33**(Suppl. 10), 26–29 (2003)

Assistive technology for sensory impairments

Developing and Pilot Testing a Hybrid Prototype for Audio-Tactile Graphics

Kirsten Rassmus-Gröhn[1], Taqwa Saeed[1], Charlotte Magnusson[1](✉), Ginger Claassen[2], Linda Mathé[2], and Tobias Schulze-Velmede[2]

[1] Department of Design Sciences, Lund University, 221 00 Lund, Sweden
charlotte.magnusson@design.lth.se
[2] Siemens AG., 80333 Munich, Germany

Abstract. Refreshable tactile displays offer a tactile alternative for users with visual impairments to access graphical information. However, their low resolution and uniform texture makes the integration of touch-triggered auditory feedback a potentially useful feature. In this work, we explore the design of such feature through co-design work on user requirements for a refreshable tactile display, using a hybrid prototype made of paper overlays in combination with an iPad app. This prototype has been used in a workshop at a school with blind youths in Germany, and we present preliminary results and recommendations.

Keywords: prototype · paper · interaction · tactile tablet · visually impaired

1 Introduction

Digital graphics are still difficult access for visually impaired computer users. The use of text descriptions is getting more pervasive when it comes to web pages, but all graphical material might not be readily explained in text only. Using refreshable tactile pin displays (RTDs) for imagery that is difficult to explain, like maps for example, is promising. However, the technology for producing tactile refreshable graphics is still costly, and this is an area where research is ongoing. One project aiming at producing an affordable RTD using novel technology is the EU project ABILITY. The ABILITY device is intended to have a touch screen with vibrotactile feedback as well as an RTD (pin display) with a conversational agent that responds to the user touch as depicted in Fig. 1.

An essential part of the work in this project has been to obtain the user requirements through different user studies. The user studies started with a series of initial interviews, which were followed by co-design workshops using paper prototyping with a human pretending to be the computer (Wizard of Oz) [1] for the interaction as is reported in [2]. Although these workshops generated useful discussions and results, it was difficult for the human to mimic the tight coupling of feedback needed in explorative gestures. Thus, we decided to create a more interactive prototype. We found that an iPad can sense touch events even when a swell paper is placed on top of the screen, making a combination of an iPad and swell paper a possibility to create a more interactive, hybrid prototype.

K. Mavrou and P. Encarnação (Eds.): AAATE 2025, *Technology for Inclusion and Participation for All: Recent Achievements and Future Directions*, pp. 449–456, 2025.
https://doi.org/10.1007/978-3-032-01628-7_53

Refreshing the tactile part of the prototype would still be limited to exchanging tactile overlays, but in-gesture feedback could be explored in more detail with such a prototype.

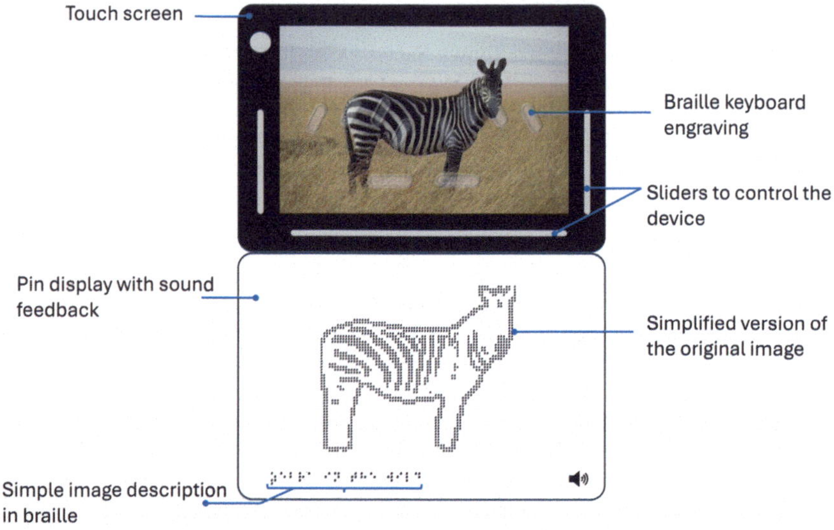

Fig. 1. The envisioned ABILITY device

2 Related Work

2.1 Audio-Tactile Graphics

The understanding and exploration of tactile graphics if facilitated when there is additional information such as text or speech [3]. Touch-sensitive tactile graphics with audio feedback have been established as a clutter-free method to accompany a tactile graphic with descriptive information. This technique has been used in a multitude of applications such as maps [4] and artworks [5]. The audio feedback is usually achieved through deterministic information that is played when a touch input is detected. One way to detect touch events is touch-sensitive layers on top or beneath the tactile graphics [6], another relies on cameras that detect the users' hand position [7]. Tactile overlays on top of touch screen-based devices such as tablets and smart phones to create tactile graphics with audio feedback has been used for navigation maps [6] and geometric figures [8]. In this work, we adopt this approach to gather user feedback regarding the development of the ABILITY system.

2.2 Refreshable Tactile Displays

Although interactive audio-tactile graphics present a significant improvement from tactile graphics that rely on braille descriptions, they are still limited by their static nature.

Image manipulation, eg. Zooming, panning and annotation, require a dynamic display. As such, RTDs [9] are considered a promising alternative that can revolutionise access to graphical material for people with VI. These displays allow users to display and create tactile graphics instantly, and unlike traditional/printed tactile graphics, there is no additional cost for each graphic. RTDs can also be equipped with voice feedback, speech recognition, dynamic interaction with graphics as well as gesture control. These features potentially transform how users interact with tactile materials [10, 11]. Moreover, RTDs' portability and ability to connect to computers and smart phones render them suitable for a multitude of applications. To date, they have been considered in several studies for education, entertainment, and mobility applications [12–15].

With RTDs the need for audio feedback is exacerbated due to their low resolution and uniform texture. Audio feedback is also important to support the RTD's ability to manipulate an image (e.g., zoom-in, panning). For instance, when a zoomed-in version of an image is loaded on the pin display, users may find it challenging to orient themselves in the new manipulated image without the audio feedback. This has been proven in the very recent work [11], where Wizard-of-Oz method is used to investigate the impact of equipping an RTD with a conversational agent. Their results are in line with our own results from using the same method with paper prototype [2, 16] and with findings from literature [17]. Some commercial RTDs have the ability to sense touch and provide feedback such as the Monarch RTD, which is equipped with infrared sensors that can sense the finger positions of the user [18].

3 Methods

This work is based on co-creative action research, involving a mix of expert users, engineers, researchers, UX-designers as well as teachers and pupils in a school for youths with visual impairment. Methods used are lo-fi prototyping, hi-fi prototyping and co-creative workshops.

Users from the target group have been involved in two stages so far. First, we had a 1-on-1 co-design session between a UX researcher and an expert user who reads braille and uses tactile graphics. The session was exploratory, using material that the researcher had prepared based on the preferences of the expert user, who had asked to experience a current map of Russia's occupation of Ukrainian territory. The map featured borders, neighbouring countries, the Black Sea, the capital and the occupied regions.

The app, which was implemented on iOS and used on an iPad, allowed the test team to draw shapes and assign text, which was then read by text-to-speech, to these shapes. The first version of the app used single touch and repeated the spoken information as long as the user was in touch with/moved over an object. Based on feedback from the 1-on-1 workshop, two additional ways of invoking the speech feedback were implemented: long press and voice command. That way, a user could explore freely with two hands and then point in the map (for example) and get feedback on what was under their finger. Since the built-in long press didn't allow the user to go from moving on the screen to long press without lifting the finger from the screen, a custom long press that allowed the user to stay in touch was implemented. The voice input used the built-in voice control option in iOS. Swell paper tactile overlays were created using a vector drawing program where images were drawn using a template of empty circles with the same spacing/dimensions as the envisioned RTD device. Circles mimicking raised pins were manually filled to create a dotted version of the desired image. This was printed, swelled and cut to suitable size. These tactile paper graphics were then placed on the screen with the app open, outlines were traced on paper to draw shapes on the screen as shown in Fig. 2. Text was added in the app directly with the overlay removed. This version of the prototype was used in the next sessions, the setup and implementation of a workshop at a school for blind youths. The workshop was exploratory, and the material used was the above described Ukranian map and a bar chart of energy consumption in Germany. Eight students, 15–24 years of age, 4 female and 4 male, participated in the workshop. The students were invited for one of two sessions. Five students participated in the first session and three students in the second. The group with five students was divided in two subgroups. Each group of 2–3 students participated in a session that took maximum 1 h. The students tested the prototype individually with the map or the bar chart or both, and then discussed the features and suggestions for improvements with each other and the test leaders.

Fig. 2. The prototype used with a swell paper and an iPad app.

4 Results

4.1 Session with Expert

In the 1-on-1 session, the immediate feedback was that the iPad "*talked too much*" and that it would be nice to explore only the tactile image first using more fingers and then ask for specific information in the map. The map information itself was considered enlightening, as the expert user commented that the occupied region was smaller than they had anticipated or understood from the news.

4.2 School Workshop

Regarding sound feedback, students had varied opinions. Some thought that direct continuous feedback is a faster way to explore images compared to tapping. Others, however, like the expert, thought it talked too much and preferred to trigger it on demand and suggested a button, double or triple tapping. Mostly, they preferred to start exploring images without auditory feedback and then toggle freely between receiving the feedback and stopping it. While one group of students preferred asking for the feedback through speech, this type of triggering was found too slow and not suitable for public places by others. Many of the students preferred to speed-up the speech feedback since they are typically used to fast voice-over and prefer to receive information quickly. This was particularly evident when they were exploring the bar chart, probably because the shape itself was less complicated. Overall, participants thought the speed should be adjustable. Another suggestion was that single, double and triple tapping should offer different levels/layers of details about the explored image. Finally, one participant wanted audio feedback to lead them to the desired spot on the map.

Participants thought that the presented solution was suitable for maps which are known to be challenging to study through text alone. When exploring maps, they were interested in seeing borders between the different countries, zooming in on cities, rivers and other map elements and to know the map coordinates. One student preferred to explore the border between two countries thoroughly and only receive audio information about the adjacent country when their hand crosses over to it. Some students proposed complementing the prototype with AI to be able to request further information such as "how many people live in the Ukraine?". One of the participants wanted a way to determine whether the tablet was appropriately oriented (i.e., not accidentally held upside down).

The students thought that exploring the tactile bar chart with sound was interesting. Some participants found the bars to be a bit too close to each other which made it hard to distinguish them from one another. Others, however, did not mind the closeness of the bars. Additionally, the participants preferred to have a description of the chart and to be able to ask an AI assistant about the chart characteristics as for example, the minimum point. Some suggested to have vibration to assist in finding the desired points in the graph. The students had other suggestions relevant to the interaction with the app. For instance, they were interested in having an alternative input method other than speech such as a keyboard. Some thought that having a dedicated button to activate and control audio feedback maybe easier and faster.

After the student workshop an informal demo for teachers at the school was held. In total, 6 teachers were shown the prototype and two expressed interest in using it for their own teaching. The prototype has since then been made available for the school teachers.

5 Discussion

Generally, the prototype worked as intended, discussion was lively, and useful information was obtained. We see that being able to get spoken feedback is very valuable, but also that the exact design of the interaction needs further work. Students generally want to explore the tactile patterns with both hands, something which potentially could cause the device to "talk too much". However, since it is possible to let users choose their settings it might make sense to keep the function of having speech on single touch events "all the time", and one where it talks on demand. Moreover, some students preferred the continuous feedback for the bar chart but not for the maps suggesting that different setting might be suitable for different applications. Some problems were encountered during the workshops. The long press was deemed too long (since it uses a custom function, this can be updated). Most of the time, touch events are registered through the overlay, but on occasion it appeared as if they did not register. The voice control function worked during pilot testing, but not in the school workshops. This may be due to several factors, sound environment, distance from the device, or potentially that touch events were not registered through the paper. Given there is paper over the device (i.e., you don't see what happens on the screen), troubleshooting is harder, but for future studies we will investigate this in more detail. Still, the presented hybrid solution has been valuable as is, and adds interactivity to the type of paper prototypes described in [2, 19] and [20]. We hope it can serve as inspiration for others developing this type of technology.

On a side note, this type of app could potentially be re-designed and developed to provide functionality similar to the Flexiboard [21] (no longer on the market) or the T3 Tactile Tablet [22]. It is an open question how valuable such an app would be, but the possibility exists. Additionally, some of the features that were proposed by the participants are already considered for the ABILITY device as for example the AI assistance. The envisioned ABILITY display will be equipped with an AI-based system that can generate descriptions and answer questions about the displayed images [23].

6 Conclusions

We present a hybrid prototype using tactile paper overlays and an iPad app, that can be used to investigate functionality in the development process of an RTD, and potentially also in the development of other interactive tangible or tactile systems. In addition, we provide some tentative results and observations on what users find important and useful in this kind of system.

References

1. W. L. in R.-B. U. Experience. 'The Wizard of Oz Method in UX'. Nielsen Norman Group. https://www.nngroup.com/articles/wizard-of-oz/ Accessed 01 Feb 2024
2. Magnusson, C., Saeed, T., Rassmus-Gröhn, K.: Lo-Fi Prototyping a refreshable pin display. In: Haptics: Understanding Touch; Technology and Systems; Applications and Interaction: 14th International Conference on Human Haptic Sensing and Touch Enabled Computer Applications, EuroHaptics 2024, Lille, France, June 30 – July 3, 2024, Proceedings, Part II, Berlin, Heidelberg: Springer-Verlag, pp. 426–432 (2024). https://doi.org/10.1007/978-3-031-70061-3_34
3. Stangl, A., Morris, M.R., Gurari, D.:"Person, shoes, tree. is the person naked?" what people with vision impairments want in image descriptions. In: Proceedings of the 2020 CHI Conference on Human Factors in Computing Systems, Honolulu HI USA: ACM, pp. 1–13 (2020). https://doi.org/10.1145/3313831.3376404

4. Wang, Z., Li, B., Hedgpeth, T., Haven, T.: Instant tactile-audio map: enabling access to digital maps for people with visual impairment. In: Proceedings of the 11th international ACM SIGACCESS conference on Computers and accessibility, in Assets '09. New York, NY, USA: Association for Computing Machinery, pp. 43–50 (2009). https://doi.org/10.1145/1639642.1639652
5. Cho, J.D.: A study of multi-sensory experience and color recognition in visual arts appreciation of people with visual impairment. Electronics **10**(4), Art. no. 4 (2021). https://doi.org/10.3390/electronics10040470
6. Götzelmann, T.: Visually augmented audio-tactile graphics for visually impaired people. ACM Trans. Access. Comput. **11**(2), 8:1–8:31 (2018). https://doi.org/10.1145/3186894
7. Fusco, G., Morash, V.S.: The tactile graphics helper: providing audio clarification for tactile graphics using machine vision. In: Proceedings of the 17th International ACM SIGACCESS Conference on Computers & Accessibility, in ASSETS '15. New York, NY, USA: Association for Computing Machinery, pp. 97–106 (2015). https://doi.org/10.1145/2700648.2809868
8. Maćkowski, M., Brzoza, P., Kawulok, M., Meisel, R., Spinczyk, D.: Multimodal presentation of interactive audio-tactile graphics supporting the perception of visual information by blind people. ACM Trans. Multimedia Comput. Commun. Appl. **19**(5s), 167:1–167:22 (2023). https://doi.org/10.1145/3586076
9. Vidal-Verdu, F., Hafez, M.: Graphical tactile displays for visually-impaired people. IEEE Trans. Neural Syst. Rehabil. Eng. **15**(1), 119–130 (2007). https://doi.org/10.1109/TNSRE.2007.891375
10. Prescher, D., Bornschein, J., Köhlmann, W., Weber, G.: Touching graphical applications: bimanual tactile interaction on the HyperBraille pin-matrix display. Univ. Access Inf. Soc. **17**(2), 391–409 (2018). https://doi.org/10.1007/s10209-017-0538-8
11. Reinders, S., Butler, M., Zukerman, I., Lee, B., Qu, L., Marriott, K.: When refreshable tactile displays meet conversational agents: investigating accessible data presentation and analysis with touch and speech. IEEE Trans. Visual Comput. Graphics **31**(1), 864–874 (2025). https://doi.org/10.1109/TVCG.2024.3456358
12. Leo, F., Cocchi, E., Brayda, L.: The effect of programmable tactile displays on spatial learning skills in children and adolescents of different visual disability. IEEE Trans. Neural Syst. Rehabil. Eng. **25**(7), 861–872 (2017). https://doi.org/10.1109/TNSRE.2016.2619742
13. Ohshima, H., Kobayashi, M., Shimada, S.: Development of blind football play-by-play system for visually impaired spectators: tangible sports. In: Extended Abstracts of the 2021 CHI Conference on Human Factors in Computing Systems, in CHI EA '21. New York, NY, USA: Association for Computing Machinery, pp. 1–6 (2021). https://doi.org/10.1145/3411763.3451737
14. Rao, H.V.,Modhrain, S. O.: 2Across: a comparison of audio-tactile and screen-reader based representations of a crossword puzzle. In: Proceedings of the 2020 CHI Conference on Human Factors in Computing Systems, in CHI '20. New York, NY, USA: Association for Computing Machinery, pp. 1–12 (2020). https://doi.org/10.1145/3313831.3376207
15. Zeng, L., Weber, G.: Exploration of location-aware you-are-here maps on a pin-matrix display. IEEE Trans. Hum. Mach. Syst. **46**(1), 88–100 (2016). https://doi.org/10.1109/THMS.2015.2477999
16. Saeed, T., et al.: Deliverable 2.3 Evaluation, ABILITY. EU Ability Project. https://www.ability-project.eu/public-deliverables
17. Zeng, L., Miao, M., Weber, G.: Interactive audio-haptic map explorer on a tactile display. Interact. Comput. **27**(4), 413–429 (2015). https://doi.org/10.1093/iwc/iwu006
18. Meet Monarch | American Printing House. https://www.aph.org/meet-monarch/ Accessed 07 May 2025

19. Miao, M., Köhlmann, W., Schiewe, M., Weber, G.: Tactile paper prototyping with blind subjects. In: Proceedings of the 4th International Conference on Haptic and Audio Interaction Design, in HAID '09. Berlin, Heidelberg: Springer-Verlag, pp. 81–90 (2009). https://doi.org/10.1007/978-3-642-04076-4_9
20. Prescher, D., Weber, G., Spindler, M.: A tactile windowing system for blind users. In: Proceedings of the 12th International ACM SIGACCESS Conference on Computers and Accessibility, in ASSETS '10. New York, NY, USA: Association for Computing Machinery, pp. 91–98 (2010). https://doi.org/10.1145/1878803.1878821
21. Flexiboard - Utgått | Abilia. https://www.abilia.com/sv/our-products/utgangna-modeller/utgangna-modeller-kognition/flexiboard-utgatt Accessed 07 Feb 2025
22. T3 Tactile Tablet. Touch Graphics. https://www.touchgraphics.com/store/p/t3-tactile-tablet Accessed 07 Feb 2025
23. Salous, M., Lange, D., Reeken, T.V., Heuten, W., Boll, S., Abdenebaoui, L.: Towards AI-based accessible digital media: image analysis pipelines and blind user studies. 2023 IEEE International Conference on Systems, Man, and Cybernetics (SMC), IEEE (2023)

MS-EXPLORER: Multimodal Spatial Access to Information for Visually Impaired People

Yacine Bellik(✉) and Celine Clavel

Université Paris-Saclay, CNRS, Laboratoire Interdisciplinaire Des Sciences du Numérique, LISN, 91405 Orsay, France
{Yacine.Bellik,Celine.Clavel}@lisn.fr

Abstract. The objective of the MS-EXPLORER project is to provide a new interaction paradigm for document access for visually impaired people. This paradigm aims to preserve spatial structure of information, in particular two-dimensional information such as mathematical formulas, maps, web pages, arrays, diagrams, etc. The approach consists in combining different interaction modalities to present the information and relies on the definition of three levels for information access. First, the *spatial level* allows the user to perceive the overall spatial structure of the information without being bothered by the details. Then, the *type level* allows having access to the type of information currently explored by the user's fingers (image, paragraph, list, array, menu, etc.). Finally, the *substantial level* allows getting the detailed content of the currently explored element. Different modalities and devices (dynamic tactile graphics, braille, speech synthesis) will be used to provide access to these different levels. A first prototype has been developed and a preliminary study to determine how to render the spatial structure has been conducted.

Keywords: Multimodal · Spatial · Access · Visually · Impaired

1 Introduction/Background

Blind people can have access to digital documents using specific software: screen readers. Screen readers allow to present in a linear way, either through speech synthesis or braille, the content of a document or elements of a graphical interface to a blind user. However, it is not so easy for a blind user to explore 2D structures in this way [10]. The user faces many problems such as disorientation and difficulties to memorize and to build a correct mental model of information. This represents a major obstacle to access to advanced-level studies for blind people. A major problem with the screen reader approach is that it leads to losing the spatial structure of information, which is very important for different kinds of information such as maps [2, 9], mathematical formulas [1], graphs, arrays, electronic schemas, diagrams [3], etc.

K. Mavrou and P. Encarnação (Eds.): AAATE 2025, *Technology for Inclusion and Participation for All: Recent Achievements and Future Directions*, pp. 457–464, 2025.
https://doi.org/10.1007/978-3-032-01628-7_54

2 Methods

To preserve the spatial structure of information, we propose to combine different modalities and devices with three access levels to information (see Fig. 1). This paradigm allows visually impaired people to explore a document in a 2D space (the refreshable dot-matrix display) in the same way they do with a static embossed document [7], except that the document here is dynamic and that they can interact with it.

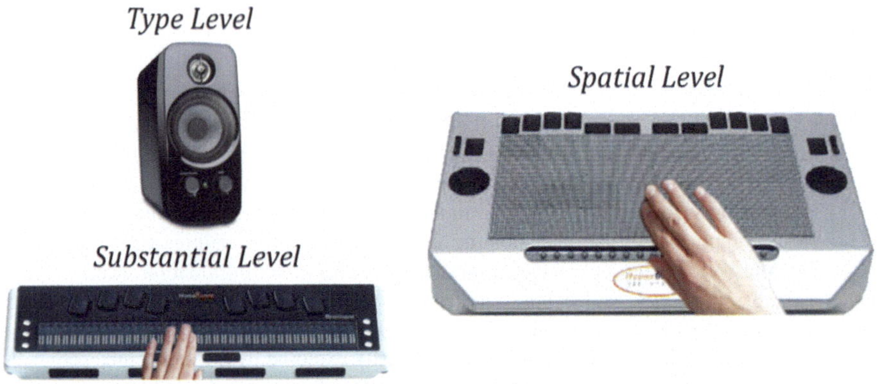

Fig. 1. The 3 access levels to information

2.1 Spatial Level

Using a refreshable dot-matrix display, this level allows the user to perceive the overall spatial structure of the information (size, position…) without being bothered by the details.

2.2 Type Level

This level allows the user to have access to the type of information currently explored by the user's fingers (image, paragraph, list, array, menu, etc.). This level can be rendered either by using speech synthesis or braille.

2.3 Substantial Level

This level allows the user to get the detailed content of the currently explored element such as the description of an image or the text of a paragraph. This level can also be rendered either by using speech synthesis or braille.

3 Challenges

This paradigm raises different scientific and technical challenges.

3.1 Spatial Structure Rendering

A first challenge is to determine the best way to render the spatial structure: which geometrical shapes should be used and how to render them (empty, solid, vibrant...) (see Fig. 2).

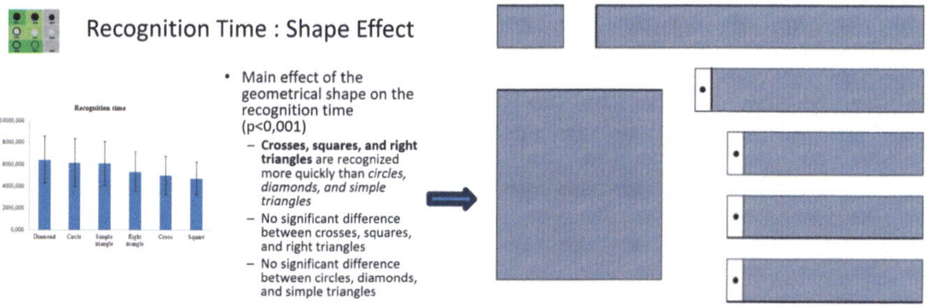

Fig. 2. Spatial structure rendering.

3.2 Exploration Strategy

It will be interesting to analyze the exploration strategy that users will use to explore the document. This may have an impact on the rendering technic that should be used. Also, it will be interesting to compare such a strategy with the one involved when exploring an embossed document [8].

3.3 Modalities and Devices Assignation

Different modalities and devices can be used for each access level. For instance, speech synthesis may be used for type level and braille for substantial level, but the opposite is also possible. It is also possible to use speech synthesis or braille for both levels. For the spatial level different devices [4, 5] can also be used (refreshable dot-matrix display, braille mouse, Laterotactile Display...). It will be interesting to study different scenarios to analyze the user's preferences regarding the modalities and devices assignation for each level.

3.4 Fingers Position and Identification

It is necessary to be able to detect accurately the fingers position and to identify them to define a strategy for choosing which information under which finger (right index, left index, both...) should be communicated to the user.

3.5 Navigation Interaction Techniques

Due to the small size of dot-matrix displays and similar devices, there is an important need for efficient interaction techniques that allow fast and easy navigation in the spatial structure (zooming / panning / scrolling...). Such interaction techniques for visually impaired people still have to be developed and evaluated.

4 Preliminary User Studies

A first prototype allowing to create documents using the described paradigm has been developed (see Fig. 3). This software called MS-EXPLORER EDITOR allows to easily create documents following the described paradigm using a 3600-dot-matrix display (60×60 dots) from metec AG (see Fig. 4) and Microsoft Speech Synthesis. Concerning braille rendering, the software allows displaying braille characters in any area of the dot-matrix display. An important benefit of this metec device is that it is equipped with sensors allowing to determine the position of the fingers on the dot-matrix display.

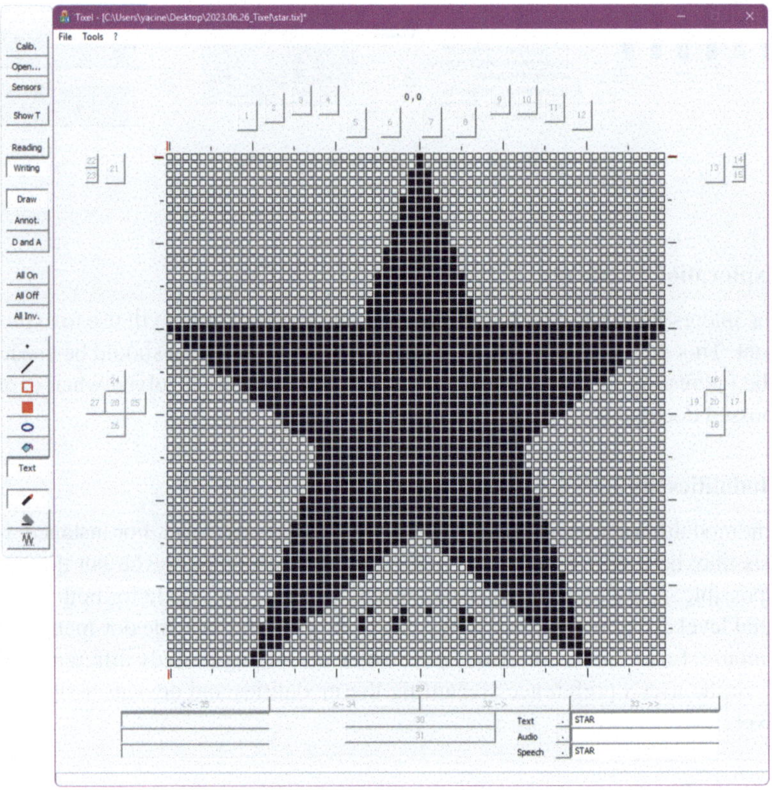

Fig. 3. MS-EXPLORER EDITOR.

Of course, the final objective is to be able to generate automatically the spatial structure from a source document, but as a first step toward this objective, we have used this editor to conduct two preliminary studies.

The first study has been conducted with 40 sighted subjects (31 men and 9 women), aged from 18 to 40 (M age = 23.7; SD age = 5.2). During the experiment, subjects were able to touch the dot-matrix display but cannot see what is displayed on it thanks to a raised cover that hides it.

Fig. 4. Metec AG 3600-dot-matrix display.

A second study has been conducted with 52 visually impaired subjects (31 men and 21 women), aged from 17 to 85 (M age = 46.8; SD age = 19.7).

In this paper, we present the results of the first study. The results of the second study are still under analyze process and will be presented in future publication.

The objective of the first study was to address the first challenge (see Sect. 3.1) to determine the rendering features that allow the fastest and most reliable recognition of geometrical shapes. Six different shapes were used in the experiment. These shapes are the same as the ones used in [6]: square, circle, simple triangle, right triangle, diamond and cross (see Fig. 5).

Fig. 5. The six shapes used in the study.

Six different rendering methods were used depending on the way the outline and the inside of the shapes are displayed (static, vibrating or empty). Figure 6 illustrates the 6 rendering methods.

Fig. 6. The six Rendering Methods (RM) used in the study.

Finally, we used three different sizes: small, medium, and large which correspond respectively to 2, 3 and 4-cm bounding square.

During the experiment, the geometrical forms varied according to the shape, the size, and the rendering method. The order of the shapes, sizes and rendering methods was randomly generated across participants. In all, each participant had to recognize 324 shapes (6 forms x 3 sizes x 6 rendering methods x 3 sessions).

5 Results

For each shape, we recorded time to recognize it (in milliseconds) and participant's answer. results show that recognition rate do not vary according to geometrical form, size, and rendering method. The global mean of the recognition rate is 95%.

However, we observed a main effect of the rendering method on the recognition time ($F(5, 195) = 73{,}237, p < 0{,}001$ see Fig. 7). Post hoc comparisons suggested that the best configuration is when the rendering method combines static outline with empty inside (RM3). Participants recognize faster the shapes with this configuration compared to other configurations. In addition, post hoc comparisons suggested that the worst configuration is when the rendering method combines vibrating outline with vibrating inside.

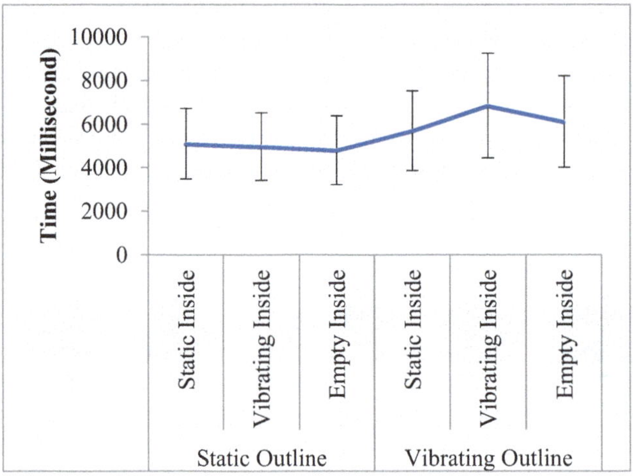

Fig. 7. Recognition time according to the combination of inside and outline rendering.

We observed also a main effect of the geometrical shape on the recognition time ($F(5, 195) = 39{,}295, p < 0{,}001$ see Fig. 8). Post hoc comparisons suggested that participants tended to recognize more quickly crosses, squares, and right triangles than circles, diamonds, and simple triangles. There was no significant difference between crosses, squares, and right triangles. In addition, there was no significant difference between circles, diamonds, and simple triangles. This can be explained by the fact that most users had a strategy that consists in looking for right angles to identify the form.

Finally we observed a main effect of the size on the recognition time ($F(2, 78) = 86{,}157; p < 0{,}001$). Post hoc comparisons suggested that participants recognized more slowly, small shapes than medium or large shapes. There was no significant difference between the medium and large shapes.

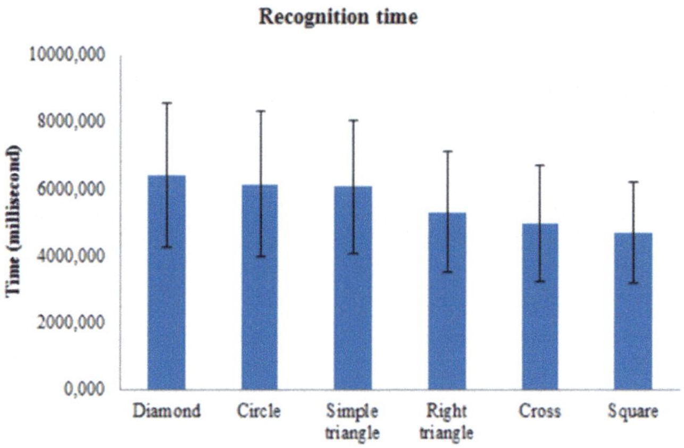

Fig. 8. Recognition time according to the geometrical shapes.

6 Discussion

Objective results showed the forms are well recognized regardless of the geometrical shape, the size, or the rendering method. Furthermore, the best rendering method is to use shapes with right angles and with static outline and empty inside. Indeed, the recognition times appears to be significantly better with crosses, squares, and right triangles than with circles, diamonds, and simple triangles. This result provides an interesting cue about the exploration strategy that participants followed. Some of them said that they start by looking for right angles in the shape, which help them to identify rapidly the form (only 1 right angle for right triangles, 4 for squares and a lot (12) for crosses). According to some subjective commentaries, solid shapes can be used when learning a new shape. However, when trying to identify an already known shape, empty shapes allow faster identification. The rendering methods that include vibrations seem to disrupt the participants even if there is no impact on the recognition rate. Participants spend more time recognizing the shapes rendered with vibrations than those rendered with static outline and empty or static insides.

7 Conclusion

The MS-EXPLORER project is a long-term project as several challenges are still to be addressed. However, some results have already been achieved. A first prototype allowing to experiment the proposed paradigm has been developed and the results of a first preliminary study have been presented. A very interesting feature of the metec AG dot-matrix display is that it contains sensors allowing to detect touch coordinates on the dot-matrix surface. Future work will focus on the exploitation of this feature which will allow us to start addressing remaining challenges.

References

1. Bitter, M.: Braille in Mathematics Education. Master Thesis, Radboud University Nijmegen, Netherlands (2013)
2. Brock, M.: Interactive Maps for Visually Impaired People: Design, Usability and Spatial Cognition. PhD Thesis, Université de Toulouse 3 Paul Sabatier (2013)
3. Torres, M.J.R., Barwaldt, R.: Approaches for diagrams accessibility for blind people: a systematic review. In: 2019 IEEE Frontiers in Education Conference (FIE), Covington, KY, USA, pp. 1-7 (2019), https://doi.org/10.1109/FIE43999.2019.9028522
4. Yang, W., Huang, J., Wang, R., Zhang, W., Liu, H., Xiao, J.: A survey on tactile displays for visually impaired people. IEEE Trans. Haptics **14**(4), 712–721 (2021). https://doi.org/10.1109/TOH.2021.3085915
5. Feitl, S., Kreimeier, J., Götzelmann, T.: Accessible Electrostatic Surface Haptics: Towards an Interactive Audiotactile Map Interface for People with Visual Impairments. In: Proceedings of the 15th International Conference on PErvasive Technologies Related to Assistive Environments, Corfu Greece: ACM, pp. 522–531 (2022). https://doi.org/10.1145/3529190.3534781
6. Levesque, V., Hayward, V.: Tactile Graphics Rendering Using Three Laterotactile Drawing Primitives. In: Proc. of the 16th Symposium on Haptic Interfaces for Virtual Environment and Teleoperator Systems. Reno, NV, USA (2008)
7. McCallum, D., Ungar, S.: An introduction to the use of inkjet for tactile diagram production. Br. J. Vis. Impair. **21**(2), 73–77 (2003)
8. Jordan, C.K., et al.: Data for all: Tactile graphics that light up with picture-perfect resolution. Sci. Adv. **8** (2022). https://doi.org/10.1126/sciadv.abq2640
9. Zeng, L., Weber, G.: Exploration of location-aware you-are-here maps on a pin-matrix display. IEEE Transa. Hum. Mach. Syst. **46**(1), 88–100 (2015)
10. Petit, G.: Conception, prototypage et évaluation d'un système pour l'exploration audio-tactile et spatiale de pages web par des utilisateurs non-voyants. PhD Thesis, Université de Montréal (2013)

DEEP: An Automatic Bidirectional Translator for Italian Sign Language

Nicolas Tagliabue, Elisa Colletti, Roberto Tedesco, Francesco R. Dani, Sonia Cenceschi, and Alessandro Trivilini(✉)

University of Applied Sciences and Arts of Southern Switzerland, Via La Santa 1, Viganello, Lugano, Switzerland
{nicolas.tagliabue,elisa.colletti,roberto.tedesco,francesco.dani, sonia.cenceschi,alessandro.trivilini}@supsi.ch

Abstract. DEEP is a novel bidirectional Italian sign language (LIS)-to-Italian automatic translation system. Sign languages are notoriously hard to recognize by a computer vision system, and it is also quite difficult to generate signs in a smooth and comprehensible way. DEEP focuses on two case study scenarios: pharmacy and passport/identity office, providing a vertical, high-quality dataset, and implementing two pipelines that permit to translate LIS to Italian (with a BLEURT of 0.81), and Italian to LIS (generating a very good avatar, as reported by several deaf subjects and LIS interpreters). Whereas the LIS-to-Italian pipeline uses a generic translator, the Italian-to-LIS pipeline is more constrained, relying on a predefined list of known sentences. Despite this limitation, we argue the whole system could be a pragmatic and useful choice for several scenarios.

Keywords: Italian sign language · LIS · NLP · Automatic translation · Deep learning

1 Introduction

Many deaf people currently live in the Italian part of Switzerland or in Italy, and many of them use the Italian sign language (LIS) as their only mean of communication. Some of them can be considered LIS mother tongue, with a very limited knowledge of written or spoken Italian. The consequence is them being inevitably isolated, experiencing daily life difficulties in interacting with municipalities, pharmacies, and public administrations.

The support of human interpreters is not only expensive but also difficult to implement, as they may not always be available, or the specific situation could prevent from finding one in adequate timing. DEEP aims to develop a communication system to facilitate the interaction and communication between deaf people, speaking LIS, and local and national institutions, speaking Italian. In particular, DEEP leverages digital video forensic techniques for recognizing

K. Mavrou and P. Encarnação (Eds.): AAATE 2025, *Technology for Inclusion and Participation for All: Recent Achievements and Future Directions*, pp. 465–472, 2025.
https://doi.org/10.1007/978-3-032-01628-7_55

LIS signs by means of a video camera. A new, high quality, annotated dataset has been created as a prerequisite to train the LIS-to-Italian translator. A video demonstrating the system is available at this link.

2 Background on Sign Language Translation

To enable two-way communication, a Sign Language Translation (SLT) system needs to effectively perform two key functions: Sign Language Recognition (SLR) and Sign Language Production (SLP).

SLR systems can be divided into three categories: Isolated Sign Language Recognition (ISLR), Continuous Sign Language Recognition (CSLR) based on glosses, and gloss-free CSLR.

The gestures of the signer can be captured using various methods: gloves, video cameras, and 3D sensors. However, the literature strongly indicates that video cameras are the most practical solution.

Early research mainly focused on ISLR, but these systems were too limited and we ignore them in this brief review. Gloss-based CSLR rely on *glosses*: symbols used for depicting discrete gestures in textual format; although useful, gloss notations are also regarded as a partial and imprecise portrayal of manual communication [6]; moreover, the process of creating gloss-annotated datasets is a time-consuming and labor-intensive endeavor.

Thus, recent advancements in SLR are moving towards gloss-free methods. These new methods utilize techniques such as sentence embeddings and self-supervised learning to improve translation accuracy and efficiency. Notable contributions include the sign2(sem+text) model [1], CLIP-inspired approaches [2], the GloFE framework [3], and a privacy-aware model [4]. These innovations are pushing the boundaries of gloss-free SLR, achieving state-of-the-art results on various datasets. DEEP follows this approach for its LIS-to-Italian pipeline, whose model and dataset are thus gloss-free.

The field of SLP remains a complex challenge. One of the most notable breakthroughs in this domain has been the development of Progressive Transformers [7]. The same authors also introduced a two-stage deep learning method [8]. Finally, the work in [9] describes an end-to-end approach for translating text into expressive sign language videos.

Focusing on LIS, most of the works concerns SLP [10–12]. About SLR, LIS2Speech [13] translates isolated LIS signs into Italian text and speech. Additionally, we identified Algho[1], a virtual assistant claimed to provide bidirectional translation between spoken language and LIS. However, the commercial availability of this product remains uncertain. The interactive demo they published on their web site seems to be limited to SLP. To the best of our knowledge, there is no other recent literature on SLT and SLP, for LIS.

The works on SLR we presented above cannot be deployed to final users, due to the limited dataset they adopt. Instead, we aim at releasing a SLR that can

[1] https://www.alghoncloud.com/funzionalita/artificial-human-lis/.

be used in the field. About SLP, our approach is pragmatic, as we depend on pre-signed LIS sentences for the two use cases we focus on; in other words, we trade off some flexibility in exchange for highly effective LIS generation.

Being designed to be deployed to final users, the UI *usability* is crucial. In particular, we designed the UI for the deaf user to be as simple as possible. In fact, the UI requires no action from the users to begin or end a conversation; they simply stand in front of the camera and start signing.

Moreover, our prototype stands out from existing systems due to its ability to operate in realtime and bidirectionally, without requiring the users to respect turns. Finally, to the best of our knowledge, it is the only system capable of translating complete sentences between LIS and Italian, whereas other systems are limited to individual signs (or are not testable, like the commercial system mentioned above). These features make it a significant step forward, enhancing accessibility and social inclusion for the LIS community.

3 Methods: The DEEP Pipelines and Dataset

The pipelines for bidirectional translation are shown in Fig. 1 and Fig. 2.

3.1 The LIS-to-Italian Pipeline

The LIS-to-Italian pipeline starts with a video camera (1920 × 1080, 60 FPS); the frames are continuously monitored by the Subject detection module that waits for a human body to stand in front of the camera. Then, a motion detection trigger starts video recording, until a resting pose is detected. The recorded video is analyzed by Google MediaPipe (Holistic model), which generates a 3D skeleton for each frame. The next module executes time interpolation for any missing nodes (when MediaPipe loses tracking) and calculates a set of 110 *measures*.

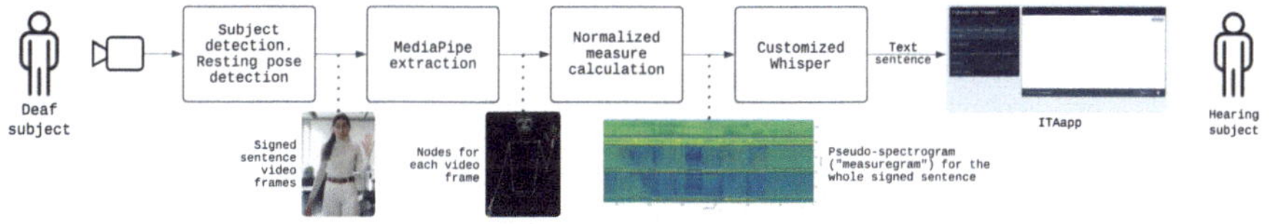

Fig. 1. LIS to Italian Translation Pipeline

We have three measure typologies: 3D points (e.g., wrist position), polar coordinates (e.g., for hand orientation) and distances (e.g., between upper and lower lips); each measure is normalized on a body's invariant characteristic (e.g., the torso height) to make them non-dimensional and more independent of signer's body traits, and then further normalized so that each measure falls in a subset of the $(-1, 1)$ range.

The result is what we call a "measuregram". We then adopted a customized version of OpenAI Whisper from HuggingFace, for the generation of the Italian transcription. Whisper is an ASR that starts from a spectrogram; we repurposed it for accepting our "measuregram" and trained it from scratch using our DEEP parallel dataset of LIS-Italian sentences.

Finally, the resulting Italian sentence is sent to ITAapp, the chat-like web application used by the hearing subject.

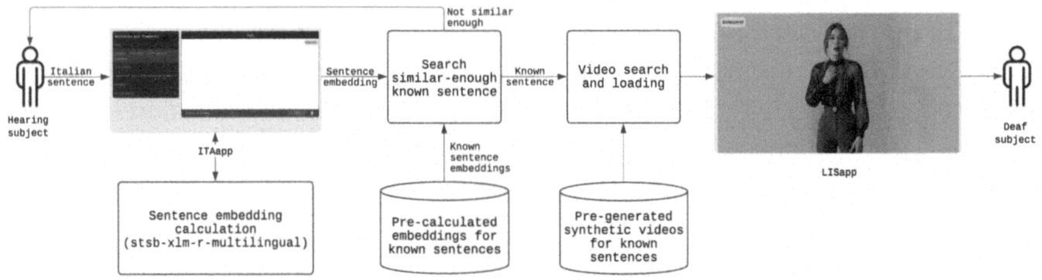

Fig. 2. Italian to LIS Translation Pipeline

3.2 The Italian-to-LIS Pipeline

The Italian-to-LIS pipeline permits the hearing user to insert an Italian sentence for which an embedding is then calculated (leveraging the stsb-xlm-r-multilingual model from HuggingFace[2]). A set of precalculated embeddings is stored, coming from the Italian sentences in the DEEP dataset, and a module searches among them the most similar to the user's sentence. If a suitable sentence is found (i.e., the distance is below a predefined threshold), the selected sentence is used as a key to search the archive of pre-generated synthetic videos.

Such synthetic videos are generated starting from the DEEP collection of recorded LIS sentences and leveraging the on-line service provided by Viggle.ia[3]. This approach allows for the preservation of privacy for individuals in the DEEP dataset, while still producing a high-quality synthetic avatar signer.

Finally, the selected video is sent to the deaf users' LISapp web application.

4 The LISapp and ITAapp

We built a preliminary test platform, which will then evolve to a production-ready kiosk. We focused on simplifying the interaction between the deaf user and the system, as an intuitive UI is essential for the system's effectiveness.

The LISapp web application is designed with a simple layout that features an avatar prominently in a single section. Once communication begins, a chat

[2] huggingface.co/sentence-transformers/stsb-xlm-r-multilingual.
[3] viggle.ai.

Fig. 3. LISapp UI for the deaf subject.

Fig. 4. ITAapp UI for the hearing subject.

overlay appears in the top-left corner of UI. The design, as illustrated in Fig. 3, ensures a seamless and intuitive user experience for the deaf subject.

It is important to note that, through the use of *triggers*, the deaf subject does not need to interact with the UI manually: everything operates automatically, offering a simple and intuitive user experience. Because the two translation pipelines are fully independent, the deaf subject can begin signing at any time, even while the avatar is signing, ensuring a seamless interaction.

The ITAapp interface consists of four sections (see Fig. 4): (1) a collection of commonly used messages; (2) the conversation; (3) a text entry field; (4) a button for speech recognition functionality (leveraging the Whisper ASR).

Whenever the hearing user types or says a sentence, the text is shown in (2) and sent to the Italian-to-LIS translation pipeline, which generates the video of the synthetic avatar signing the Italian sentence, in LIS; such a video is then sent to the LISapp, together with the original Italian sentence.

The experimental kiosk setup (Fig. 5) features a straightforward installation with just a monitor and a camera. The structure of the kiosk was developed based

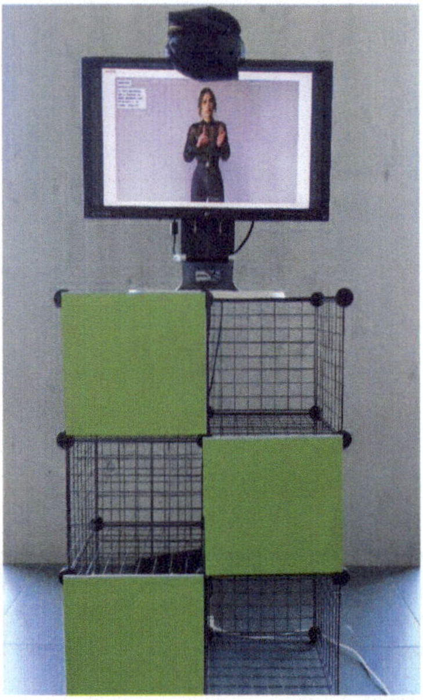

Fig. 5. The kiosk for the deaf subject.

on the average heights of participants; however, the monitor can be adjusted vertically to accommodate each individual's needs.

5 The DEEP Dataset

We created a parallel dataset, composed of 36 818 samples of LIS video recordings (~62 h, 1920 × 1080, 60 FPS, with shutter speed carefully calibrated to minimize motion blur) annotated with the corresponding Italian sentences.

Since our goal was to make the DEEP system available in two well-defined scenarios (pharmacy and passport/identity office), we started from 3075 sentences commonly used in such scenarios and asked 17 subjects (13 deaf subjects and 4 LIS interpreters) to sign as many sentences as possible. Then, we augmented the dataset with 55 458 synthetic samples by combining recorded sentences and single recorded words signed in dactylology, adding ~21 h.

6 Results and Discussion

We measured the effectiveness of the LIS-to-Italian pipeline using some standard metrics on the test set, obtaining: BLEU = 61.8 (max is 100), BLEURT = 0.81 (max is 1), and chrF = 71.8 (max is 100). Such values are quite good if compared to metrics usually obtained on datasets for sign languages; for example, the state of the art for the popular How2Sign dataset is BLEU = 12.0, BLEURT = 0.49, chrF = 31.8 [1]. While figures calculated on different datasets are not comparable, we present them to highlight the overall difficulty of the task and to demonstrate the strong results we achieved with our specific, vertical dataset.

We also made a study engaging 7 LIS signers (6 deaf subjects and 1 LIS interpreter) who signed 29 LIS sentences from our DEEP dataset (a total of 203 signed sentences); we found that 61.58% of these LIS sentences were translated into Italian exactly as they appeared in the dataset, while 11.81% differed in wording but still accurately conveyed the intended meaning (resulting in 73.39% of correct translations).

For assessing the Italian-to-LIS pipeline effectiveness, we carried out a research involving the same 7 LIS signers, who were invited to evaluate 31 LIS utterances from the DEEP dataset (217 sentences in total), performed by our avatar. Among these sentences, 16 incorporated dactylology, whereas 15 did not. The meaning communicated by the virtual avatar matched or was sufficiently clear compared to the Italian sentences in 97.23% of cases. Regarding dactylology signs, the meaning was identical or adequately clear in 93.75% of instances. Summing up, the quality of the virtual avatar was considered highly satisfactory.

7 Conclusion/Implications for the AT Field

The implications of our work for the AT field are twofold. First, the laboratory results of DEEP bidirectional translation system were very promising and will be validated on the field, aiming to measure the effectiveness of both pipelines in a more realistic, comprehensive, and systematic way. Secondly, given the scarcity of datasets for LIS [5], the DEEP dataset, with 62+21 h of multimodal LIS content (video, audio, text transcriptions, dactylology words, and skeleton key points) represents a valuable contribution to the field.

Finally, although the current focus is on LIS, the techniques developed in this research are language-agnostic and could be adapted to other sign languages, broadening the scope and impact of this work. Note that there is nothing preventing the implementation of intra-language translation pipelines (e.g., between LIS and English): the only limiting factor is the availability of a suitable dataset.

Acknowledgements. Special thanks to Daniele Raffa of Handy Systems for making this project possible. This research was funded by Innosuisse, the Swiss Innovation Agency (grant (www.aramis.admin.ch/Grunddaten/?ProjectID=50430) 100.656 IP-ICT). DEEP would not have been possible without the experience, feedback, and active contributions to data collection from members of the deaf community.

References

1. Hamidullah, Y., Genabith, J., España-Bonet, C.: Sign language translation with sentence embedding supervision. In: Proceedings of the 62nd Annual Meeting of the Association for Computational Linguistics (Volume 2: Short Papers), pp. 425–434 (2024). https://aclanthology.org/2024.acl-short.40/

2. Zhou, B., et al.: Gloss-free sign language translation: improving from visual-language pretraining (2023). https://arxiv.org/abs/2307.14768
3. Lin, K., Wang, X., Zhu, L., Sun, K., Zhang, B., Yang, Y.: Gloss-free end-to-end sign language translation. In: Proceedings of the 61st Annual Meeting of the Association for Computational Linguistics (Volume 1: Long Papers), pp. 12904–12916 (2023). https://aclanthology.org/2023.acl-long.722/
4. Rust, P., Shi, B., Wang, S., Camgoz, N., Maillard, J.: Towards privacy-aware sign language translation at scale. In: Proceedings of the 62nd Annual Meeting of the Association for Computational Linguistics (Volume 1: Long Papers), pp. 8624–8641 (2024). https://aclanthology.org/2024.acl-long.467/
5. Marchisio, M., Mazzei, A., Sammaruga, D.: Introducing deep learning with data augmentation and corpus construction for LIS. CLiC-it (2023). https://ceur-ws.org/Vol-3596/paper31.pdf
6. Müller, M., Jiang, Z., Moryossef, A., Rios, A., Ebling, S.: Considerations for meaningful sign language machine translation based on glosses. In: Proceedings of the 61st Annual Meeting of the Association for Computational Linguistics (Volume 2: Short Papers), pp. 682–693. Association for Computational Linguistics, Toronto (2023). https://doi.org/10.18653/v1/2023.acl-short.60
7. Saunders, B., Camgoz, N.C., Bowden, R.: Progressive transformers for end-to-end sign language production. In: Proceedings of Computer Vision – ECCV 2020, pp. 687–705. Springer, Cham (2020)
8. Saunders, B., Camgoz, N.C., Bowden, R.: Mixed SIGNals: Sign language production via a mixture of motion primitives. arXiv:2107.11317 [cs.CV] (2021)
9. Qi, F., Duan, Y., Zhang, W., Xu, C.: SignGen: end-to-end sign language video generation with latent diffusion. In: Proceedings of Computer Vision – ECCV 2024, Milan, Italy, 29 September–4 October 2024, pp. 252–270. Springer, Heidelberg (2024). https://www.ecva.net/papers/eccv_2024/papers_ECCV/papers/06988.pdf
10. Lombardo, V., Battaglino, C., Damiano, R., Nunnari, F.: An avatar-based interface for the italian sign language. In: Proceedings of 2011 International Conference on Complex, Intelligent, and Software Intensive Systems, pp. 589–594 (2011). https://doi.org/10.1109/CISIS.2011.97
11. Mazzei, A.: Translating Italian to LIS in the rail stations. In: Proceedings of the 15th European Workshop on Natural Language Generation (ENLG), pp. 76–80. Association for Computational Linguistics, Brighton (2015). https://doi.org/10.18653/v1/W15-4712
12. Colonna, E., et al.: Towards Italian sign language generation for digital humans. In: Proceedings of the Eighth Workshop on Natural Language for Artificial Intelligence (NL4AI 2024) Co-located with 23th International Conference of the Italian Association for Artificial Intelligence (AI* IA 2024). CEUR-WS.org (2024)
13. Mercurio, G.: LIS2SPEECH LIS translation in written text and spoken language. Master's thesis. Politecnico di Torino, Torino, Italy (2021). http://webthesis.biblio.polito.it/id/eprint/18096

Guiding Methods for On and Off-Screen Subtitles in XR: Enhancing Accessibility for Deaf and Hard-of-Hearing Users

Ahmed Katar, Sebastian Koch(✉), and Gottfried Zimmermann

Hochschule der Medien, Nobelstr. 10, 70569 Stuttgart, Germany
sebastian@koch.cf

Abstract. Developing effective subtitle guidelines for Extended Reality (XR) is essential to ensure accessibility for Deaf and Hard-of-Hearing (DHH) users. This study synthesizes existing 2D guidelines and addresses 3D-specific challenges, focusing on subtitle placement in virtual space and guiding methods for off- and on-screen sound localization.

We developed a Virtual Reality (VR) prototype in Unity simulating a business meeting with ten use cases, each combining different subtitle placements (world-spaced vs. screen-spaced) with a range of guiding methods. A mixed-method evaluation with ten able-bodied participants assessed factors such as distraction, speaker identification, comprehensibility, and cognitive load.

Quantitative results revealed a statistically significant preference for world-spaced subtitles, which were rated as less distracting ($p < 0.05$) and more effective for speaker identification ($p < 0.05$). Guiding methods such as icon bubbles were consistently preferred for their unobtrusiveness and spatial clarity, while others like speech bubbles were often perceived as disruptive. Preferences varied by context, with screen-spaced subtitles favored in exploratory scenarios.

These findings highlight the need for scenario-specific, adaptable, and user-centered subtitle systems to improve accessibility in immersive XR environments.

Keywords: HCI and Non-Classical Interfaces · Accessibility · Deaf and Hard-of-Hearing users

1 Introduction

XR environments provide immersive experiences that can improve communication, collaboration, and entertainment. However, accessibility remains a critical challenge, particularly for DHH individuals. While 2D subtitle guidelines are well-established [4–6], their direct application to XR contexts is insufficient due to the spatial complexity of virtual environments. Traditional subtitles often fail to address challenges such as off-screen sound localization, making it difficult for users to identify speakers and their positions within the scene.

K. Mavrou and P. Encarnação (Eds.): AAATE 2025, *Technology for Inclusion and Participation for All: Recent Achievements and Future Directions*, pp. 473–479, 2025.
https://doi.org/10.1007/978-3-032-01628-7_56

Prior research has explored various enhancements, including visual indicators and spatial positioning techniques [2,3,7]. These approaches show potential, but no clear consensus has emerged on which combination of subtitle positioning and guiding methods best supports comprehension without increasing cognitive load or reducing immersion.

A central design choice in XR subtitling is the spatial referencing of subtitles. Screen-spaced subtitles remain fixed in the user's field of view (FOV), typically at the bottom of the screen, regardless of head movement. In contrast, world-spaced subtitles are anchored within the 3D environment and dynamically positioned near the virtual speaker. These differing approaches may influence users' ability to follow dialog, identify speakers, and remain immersed. This study investigates which of these methods better supports accessible communication in XR, with a particular focus on the needs of DHH users.

We investigate how subtitling can be effectively adapted for XR applications, focusing on the challenge of off-screen sound sources. By synthesizing existing 2D and 3D guidelines and empirically evaluating multiple off-screen guiding methods, our aim is to provide practical insights for inclusive XR subtitle design, especially for DHH users.

2 Methods

This study employed a mixed-method approach, combining quantitative user testing with qualitative feedback to evaluate the presentation of subtitles in XR. To inform the prototype design and evaluation framework, we consulted authoritative sources, including FCC guidelines, IEEE standards, Microsoft accessibility documentation, and peer-reviewed publications accessed through academic databases. A targeted keyword search strategy was used to identify relevant literature, such as *subtitle guidelines 3D*, *accessibility in XR* and *off-screen sound localization in VR*.

Based on the findings, we developed a VR prototype in Unity, simulating a business meeting with ten use cases incorporating world-spaced and screen-spaced subtitles. As illustrated in Fig. 1, the user is positioned at one end of a table, facing a virtual screen where a presenter delivers a formal presentation with multiple colleagues in the room. Subtitle positioning was combined with various guiding methods, including radar visualizations, directional arrows, and icon bubbles—please see the supplemental document for details on each use case. Ten able-bodied participants (aged 20–58, $M = 31.9$) evaluated the prototype using a Meta Quest 3 headset. The group included 7 male and 3 female participants; 80% reported no prior VR experience, while 20% had limited experience. All were proficient in reading English.

Participants observed the scenario and assessed key factors, including comprehensibility, cognitive load, temporal coordination, distraction, speaker identification, and aesthetic satisfaction. Each category was assessed using Likert scale ratings (1 = low, 5 = high). Example questions included: "Could you understand the content of the off-screen speaker?" (comprehensibility), "How much mental effort was needed to assign cues?" (cognitive load), and "How easily could you identify the speaker?" (speaker identification). Qualitative reflections on subtitle adaptation across content types were also collected. Statistical analyses were conducted to compare preferences and performance metrics across subtitle positioning and guiding method conditions.

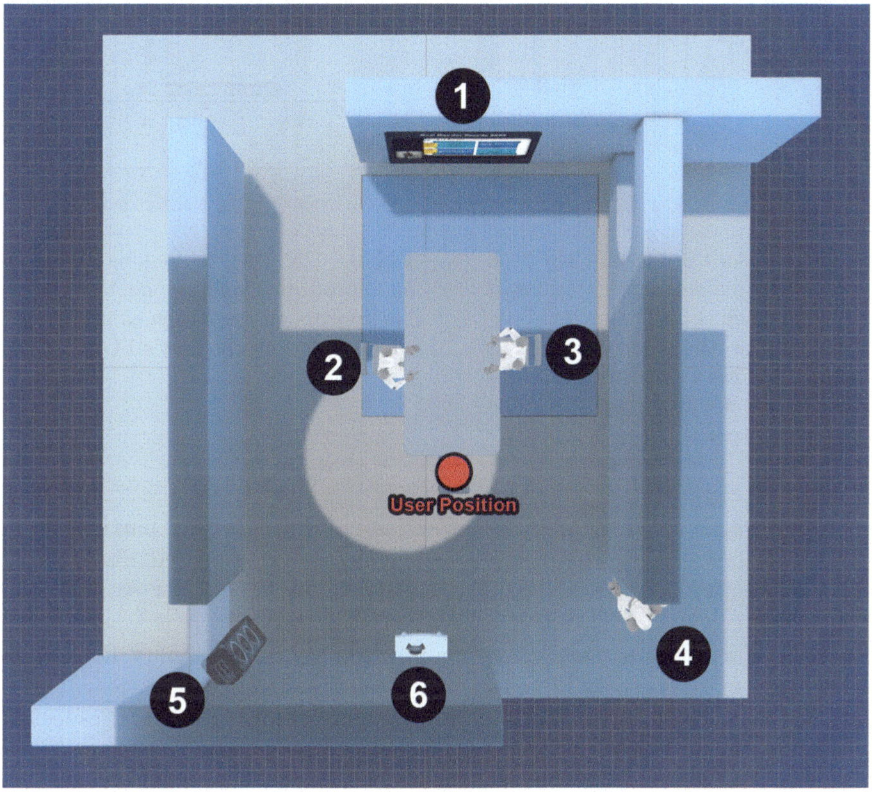

Fig. 1. Meeting room from a bird's eye view. The red dot indicates the start position of user when entering the use case and each number indicates a sound source that will emit a sound at one or multiple times during the scenario. These include the moderator (1) on the screen, three colleagues positioned to the left (2), right (3), and diagonally behind to the right (4), as well as a loudspeaker (5) diagonally behind to the left and an alarm clock (6) directly behind the user.

3 Results

The user evaluation ($N = 10$) revealed a statistically significant preference for world-spaced subtitles, particularly when paired with guiding methods such as icon bubbles (see Figs. 2 and 3). World-spaced subtitles were rated higher for their ability to reduce distraction ($p < 0.05$) and improve speaker identification ($p < 0.05$). Although not all comparisons reached statistical significance, world-spaced subtitles also showed a positive trend toward being more comprehensible and less cognitively demanding across multiple scenarios.

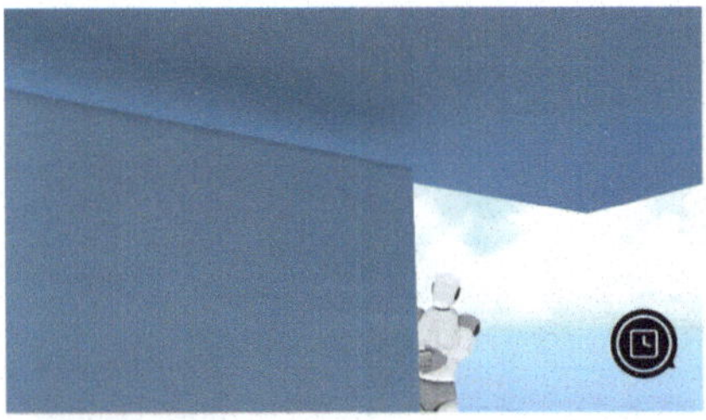

Fig. 2. The user has turned their head away from the alarm clock, placing it outside of their FOV. An icon bubble appears at the screen edge. The inner circle shows a clock symbol, while the outer ring points toward the sound source in 3D space.

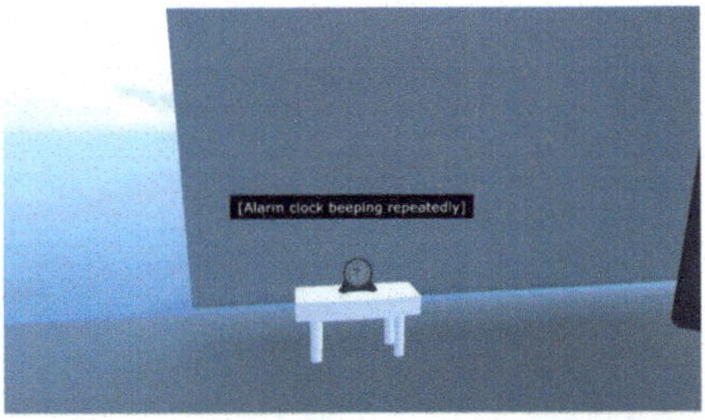

Fig. 3. The user is looking directly at the alarm clock, so no icon bubble appears on the screen as the sound source is within the user's FOV.

Screen-spaced subtitles were perceived as more distracting, particularly in multi-speaker scenarios where subtitles overlapped or appeared disconnected from the speaker's location. Despite the overall preference for world-spaced subtitles, screen-spaced subtitles were favored in exploratory scenarios, such as documentary-style scenes, where maintaining subtitle visibility helps users explore the environment while reading.

Qualitative feedback further emphasized that guiding methods, especially icon bubbles, were helpful for locating off-screen sound sources and contributed to a greater sense of control and situational awareness. Some participants, however, reported occasional difficulty reading world-spaced subtitles when they were occluded or positioned too far from the center of view.

These findings suggest that while world-spaced subtitles enhance immersion and speaker association, screen-spaced subtitles retain value in content types where persistent subtitle visibility is critical. The combined use of spatial positioning with subtle guiding methods may offer a balanced solution for improving accessibility in XR.

Table 1 summarizes the mean user ratings across subtitle types and guiding methods for all measured dimensions.

A series of statistical comparisons between world-spaced and screen-spaced subtitle conditions revealed significant effects across multiple user experience metrics. Specifically, participants reported lower cognitive load ($p < 0.05$), reduced distraction ($p < 0.05$), and improved speaker identification ($p < 0.05$) when using world-spaced subtitles. Additionally, aesthetic satisfaction was significantly higher in the world-spaced condition ($p < .001$). These results indicate

Table 1. Mean (M) and standard deviation (SD) of user ratings by guiding method and subtitle type (WS = World-spaced, SS = Screen-spaced). Ratings are on a 5-point Likert scale. *Column abbreviations:* Comp. = Comprehensibility, Cog. Load = Cognitive Load, Distr. = Distraction, Spkr ID = Speaker Identification, Aesthetic = Aesthetic Satisfaction.

Guiding Method	Type	Comp.	Cog. Load	Distr.	Spkr ID	Aesthetic
Radar (Dyn. FOV)	WS	3.10 ± 1.10	2.60 ± 0.97	3.90 ± 1.45	3.10 ± 1.37	3.60 ± 1.17
Radar (Stat. FOV)	WS	3.30 ± 1.25	3.30 ± 1.49	3.00 ± 1.49	3.50 ± 1.65	3.60 ± 0.97
Radar (Dyn. FOV)	SS	3.30 ± 1.57	2.80 ± 1.55	2.80 ± 1.75	3.20 ± 1.62	2.60 ± 1.35
Radar (Stat. FOV)	SS	3.20 ± 1.69	2.80 ± 1.23	2.50 ± 1.51	3.70 ± 1.16	2.90 ± 1.29
Extended Arrow	SS	3.10 ± 1.29	2.70 ± 1.34	2.60 ± 1.43	3.10 ± 1.37	2.50 ± 1.08
Speech Bubble	SS	2.50 ± 1.27	1.50 ± 0.85	1.80 ± 1.14	2.20 ± 1.23	2.00 ± 1.15
Screen Arrow	WS	4.00 ± 0.94	3.80 ± 1.32	3.50 ± 1.08	3.50 ± 1.27	3.20 ± 1.14
Screen Arrow	SS	2.70 ± 1.57	2.10 ± 0.88	2.00 ± 1.25	2.50 ± 1.35	2.60 ± 1.43
Icon Bubble	WS	4.50 ± 0.53	4.10 ± 1.20	3.70 ± 1.49	4.40 ± 0.97	4.10 ± 1.20
Icon Bubble	SS	3.70 ± 1.25	3.80 ± 1.14	3.55 ± 1.74	3.70 ± 1.06	3.10 ± 1.10

a consistent preference for world-spaced subtitles across both cognitive and perceptual dimensions.

4 Discussion

Balancing guidance and immersion is essential for effective subtitle design in XR environments. Our findings show that guiding methods such as speech bubbles and extending arrows were often perceived as visually intrusive, potentially disrupting the immersive experience. Subtitle preferences varied depending on the scenario, suggesting that a one-size-fits-all approach may not be suitable for XR. This reinforces the strong need for adaptive subtitle systems that adjust dynamically to content type, user behavior, and contextual factors.

Guiding methods also played a crucial role in how users interpreted and followed subtitle cues. Participants responded most positively to icon bubbles, which provided spatial information without overwhelming the visual field. For example, five participants selected the icon bubble with world-spaced subtitles as their top-rated use case. Conversely, the speech bubble with screen-spaced subtitles received the lowest rating ($M = 2.05$) and was described by several participants as confusing or disruptive. One participant noted, *"Radar + Bubble + WS would be the best combination"*, while another mentioned, *"Text disappears too quickly"* in less favored scenarios.

In line with this, nine participants emphasized the importance of context-adaptive subtitle behavior. Screen-spaced subtitles were favored in documentary-style scenarios, where persistent visibility supports free exploration. Conversely, world-spaced subtitles were preferred in speaker-focused scenes or narrative-driven contexts. Participants also proposed future directions such as cancelable music subtitles or emotion-based visual cues (e.g., highlighting anger with red colors). These insights reinforce the design implication that subtitling systems in XR should be scenario-specific, minimally intrusive, and customizable.

5 Limitations

This study involved only able-bodied participants, limiting the generalizability of the findings to users with disabilities. This research simulated the experience of DHH users, yet prior work by Agulló & Matamala[1] suggest that actual DHH users may have different preferences, such as radar mechanisms, despite their increased cognitive demand, underscoring the varied accessibility needs that future research should address.

Another limitation is the absence of a pre-study phase. A preliminary evaluation could have identified usability issues earlier and allowed for refinement of guiding methods. For instance, the speech bubble method exhibited excessive jitter, which negatively impacted its rating; this could potentially be improved with smoother animations. Similarly, the extending textbox method expanded in both directions, which created confusion about where the user should look. Limiting the extension to only the relevant direction could make the cue clearer and reduce cognitive load.

The small sample size ($N = 10$) limits statistical generalizability. While this number was sufficient to identify trends and gather qualitative feedback, a larger participant pool would be required to achieve statistically more robust conclusions and better represent user diversity.

Finally, the evaluation involved only static sound sources. While this design simplified the study setup, it does not reflect the dynamic nature of many real-world XR scenarios, where speakers or sound sources may move. Incorporating dynamic spatial audio and motion in future prototypes would allow for a more comprehensive assessment of guiding methods, including potential issues such as tracking accuracy, increased complexity, or user disorientation.

6 Conclusion

This study contributes to the field of assistive technology by providing empirical evidence on how subtitle presentation and guiding methods affect accessibility and user experience in XR. Our findings indicate that user preferences for subtitle placement vary depending on context, with world-spaced subtitles generally preferred in speaker-focused scenarios and screen-spaced subtitles favored in exploratory settings. These results underscore the importance of adaptable subtitle systems that respond both to environmental conditions and to individual user needs.

Implementing customizable settings for guiding mechanisms and subtitle positioning could enhance accessibility across a wide range of XR applications, including education, remote collaboration, and entertainment. In addition, the development of standardized XR subtitle guidelines grounded in user studies would support industry practitioners and researchers in creating more inclusive immersive experiences.

Based on our findings, we recommend the use of world-spaced subtitles combined with unobtrusive guiding methods such as icon bubbles, which participants consistently rated as effective and minimally distracting. In contrast, guiding techniques that relied on large, animated elements, such as jittery speech bubbles or bidirectionally extending text boxes, were often perceived as disruptive to the immersive experience and should be avoided or refined. These insights provide actionable guidance for XR developers aiming to implement accessible and user-friendly subtitle systems.

Future research should explore machine learning-driven approaches for real-time subtitle adaptation, as well as the integration of multimodal feedback, such as haptic cues, to support diverse user groups. By prioritizing user-centered design and inclusive research practices, the field can ensure that emerging immersive technologies are both innovative and accessible for all.

Disclosure of Interests. The authors have no competing interests to declare that are relevant to the content of this article.

References

1. Agulló, B., Matamala, A.: Subtitles in virtual reality: guidelines for the integration of subtitles in 360° content". Íkala Rev. Lenguaje Cultura **25**(3), 643–661 (2020). https://doi.org/10.17533/udea.ikala.v25n03a03 https://doi.org/10.17533/udea.ikala.v25n03a03
2. Agulló, B., Montagud, M., Fraile, I.: Making interaction with virtual reality accessible: rendering and guiding methods for subtitles. Artif. Intell. Eng. Design Anal. Manuf. **33**(4), 416–428 (2019). https://doi.org/10.1017/S0890060419000362
3. Lin, Y.-C., et al.: Tell me where to look: investigating ways for assisting focus in 360° video. In: Proceedings of the 2017 CHI Conference on Human Factors in Computing Systems, CHI 2017, Denver, Colorado, USA, pp. 2535–2545. Association for Computing Machinery, New York (2017). https://doi.org/10.1145/3025453.3025757. ISBN 978-1-4503-4655-9
4. Matamala, A., Orero, P.: Listening to Subtitles: Subtitles for the Deaf and Hard of Hearing. Peter Lang Verlag, Lausanne (2010). https://doi.org/10.3726/978-3-0351-0147-8, https://www.peterlang.com/document/1052116
5. Microsoft Corporation. Xbox Accessibility Guidelines. Version 3.2 (2023). https://learn.microsoft.com/en-us/gaming/accessibility/guidelines. Accessed 13 Apr 2025
6. Netflix, Inc. Timed Text Style Guide: Supplemental Service & Marketing Assets (2022). https://partnerhelp.netflixstudios.com/hc/enus/articles/115000239632-Timed-Text-Style-Guide-Supplemental-Service-Marketing-Assets. Accessed 13 Apr 2025
7. Rothe, S., Tran, K., Hussmann, H.: Positioning of subtitles in cinematic virtual reality. In: Bruder, G., Yoshimoto, S., Cobb, S. (eds.) ICAT-EGVE 2018 - International Conference on Artificial Reality and Telexistence and Eurographics Symposium on Virtual Environments. The Eurographics Association (2018). https://doi.org/10.2312/egve.20181307. ISBN 978-3-03868-058-1

Author Index

K. Mavrou and P. Encarnação (Eds.): AAATE 2025, *Technology for Inclusion and Participation for All: Recent Achievements and Future Directions*, pp. 481–483, 2025.
https://doi.org/10.1007/978-3-032-01628-7

The manufacturer's authorised representative in the EU is Springer
Nature Customer Service Centre GmbH, Europaplatz 3, 69115 Heidelberg,
Germany. If you have any concerns regarding our products, please
contact ProductSafety@springernature.com

Printed and bound by CPI Group (UK) Ltd, Croydon, CR0 4YY
01/10/2026
02225974-0018